Enzymes of the Cholinesterase Family

Enzymes of the Cholinesterase Family

Edited by

Daniel M. Quinn
University of Iowa
Iowa City, Iowa

A. S. Balasubramanian
Christian Medical College and Hospital
Vellore, Tamil Nadu, India

Bhupendra P. Doctor
Walter Reed Army Institute of Research
Washington, D.C.

and

Palmer Taylor
University of California at San Diego
La Jolla, California

Plenum Press • New York and London

Library of Congress Cataloging-in-Publication Data

Enzymes of the cholinesterase family / edited by Daniel M. Quinn ...
[et al.].
        p.   cm.
        "Proceedings of the Fifth International Meeting on
Cholinesterases, held September 24-28, 1994, in Madras, India"--T.p.
verso.
    Includes bibliographical references and index.
    ISBN 0-306-45135-2
    1. Cholinesterases--Congresses.   I. Quinn, Daniel M.
II. International Meeting on Cholinesterases (5th : 1994 : Madras,
India)
    [DNLM: 1. Acetylcholinesterase--metabolism--congresses.
2. Butyrylcholinesterase--metabolism--congresses.   QU 136 E611
1995]
QP609.C4E56   1995
574.19'253--dc20
DNLM/DLC
for Library of Congress   95-39179
                                                            CIP

Front cover: A model of the complex between the snake venom toxin fasciculin (light color) and acetylcholinesterase (dark color), constructed by H.K.L. van den Born *et al.* [(1995) *Protein Science 4*, 703–715]. Photography by Z. Radić of the University of California at San Diego.

Proceedings of the Fifth International Meeting on Cholinesterases,
held September 24–28, 1994, in Madras, India

ISBN 0-306-45135-2

© 1995 Plenum Press, New York
A Division of Plenum Publishing Corporation
233 Spring Street, New York, N. Y. 10013

PREFACE

The Fifth International Meeting on Cholinesterases convened in Madras, India, in September of 1994. The long and rich history and culture of India provided an excellent setting for the meeting. More than 120 delegates from Asia, Australia, Europe and North America heard 54 oral presentations and viewed 54 posters on current research on enzymes of the cholinesterase family. The aim of this book is to compile the presentations of the Fifth International Meeting on Cholinesterases into a volume that describes recent investigations on the structure and catalytic function of acetylcholinesterase (AChE), butyrylcholinesterase (BuChE) and related enzymes, as well as studies on the molecular and cellular biology of these enzymes and the genes which encode them.

Cholinesterases enjoy a long and storied history in diverse areas. In basic biochemical research, AChE is one of the best studied, though yet enigmatic, of enzymes. The efficient catalytic function of this enzyme presents the biochemist with a fundamental challenge in understanding the relationship between structure and function. AChE and BuChE belong to a family of proteins, the α/ß hydrolase fold family, whose constituents evolutionarily diverged from a common ancestor. Proteins in this family have a wide range of physiological functions. In commerce, AChE is a prime target for agricultural insect control, and for the development of therapeutic agents for Alzheimer's disease. On the national security front, AChE is the target of chemical warfare agents, "nerve gases," which pose a threat on the battlefield of nations at war and in the city centers of nations at peace. Recent events, chronicled in the international news media, provide a stark reminder of this fact.

The broad reach of cholinesterase function and research affects the biological scientist and nonscientist alike. From biotechnology to the military, one is hard pressed to find a family of enzymes whose effect on the human condition is more pervasive. Various presentations at the meeting described structure-activity relationships for the interaction of cholinesterases with toxic organophosphorus agents. Novel efforts at antidotal therapy were described that aim to use cholinesterases in the presence of nucleophilic oximes to detoxify organophosphorus inhibitors. Strategies for effective inhibition of acetylcholinesterase in the central nervous system, as an approach to the treatment of the cognitive disfunctions associated with Alzheimer's disease, were described, as were the structure-activity relationships of new anti-Alzheimer's agents.

The cell and molecular biology of cholinesterases is advancing rapidly. The relationships between AChE gene structure and the various molecular forms of the enzyme were described at the meeting, as were the mechanisms of gene expression, cellular biosynthesis, and assembly of the cholinesterases. Various noncatalytic roles for cholinesterases were described, which include involvement in nervous system development and in neurotransmission involving dopamine-containing nerve terminals.

Reports of structure-function relationships not only described the fine detail that is emerging on the structural basis of the rapid catalytic mechanisms effected by cholinesterases, but also provided the bases for several controversies. X-ray crystallography of catalytically relevant ligand complexes with AChE provided visual verification of the roles, suggested by site-directed mutagenesis and other studies, for loci in the active site that are responsible for acyl group specificity, quaternary ammonium recognition, oxyanion recognition and acid-base catalysis. The crystal structure of AChE provided the template for theoretical evaluation of the role of the electrical field of the enzyme in ligand binding. Good spirited controversies arose on whether the electrical field of the enzyme plays a significant role in the rapidity of AChE catalysis, and whether an anthropomorphic "back door" mechanism is required for release of cationic products.

This Preface outlines just a fraction of the wide range of topics discussed at the Fifth International Meeting on Cholinesterases in Madras, India. The chapters that follow provide a clear view of the breadth and vigor of research on enzymes of the cholinesterase family. The meeting in Madras was a once in a lifetime opportunity for an international delegation to discuss and debate such research in a congenial atmosphere. Our Indian hosts are to be commended for the graciousness and organizational effort that were key to the success of the meeting.

Daniel M. Quinn
A. S. Balasubramanian
Bhupendra P. Doctor
Palmer Taylor

CONTENTS

Part I. GENE STRUCTURE AND EXPRESSION OF CHOLINESTERASES

A. Presentations

B. Posters

Part II. POLYMORPHISM AND STRUCTURE OF CHOLINESTERASES

A. Presentations

B. Posters

Part III. MECHANISM OF CATALYSIS OF CHOLINESTERASES

A. Presentations

B. Posters

Part IV. CELLULAR BIOLOGY OF CHOLINESTERASES

A. Presentations

B. Posters

Part V. STRUCTURE-FUNCTION RELATIONSHIPS OF ANTICHOLINESTERASE AGENTS

A. Presentations

B. Posters

Part VI. NONCHOLINERGIC FUNCTIONS OF CHOLINESTERASES

A. Presentations

B. Posters

Part VII. PHARMACOLOGICAL UTILIZATION OF ANTICHOLINESTERASES

A. Presentations

B. Posters

Part VIII. APPENDICES

Appendix I

Appendix II

Appendix III

Appendix IV

ANTISENSE OLIGONUCLEOTIDES SUPPRESSING EXPRESSION OF CHOLINESTERASE GENES MODULATE HEMATOPOIESIS IN VIVO AND EX VIVO

Hermona Soreq,[1] Efrat Lev-Lehman,[1] Deborah Patinkin,[1] Mirta Grifman,[1] Gal Ehrlich,[1] Dalia Ginzberg,[1] Fritz Eckstein,[2] and Haim Zakut[3]

[1] Department of Biological Chemistry, The Life Sciences Institute
The Hebrew University, Jerusalem 91904, Israel
[2] Max Planck Institute for Experimental Medicine
Gottingen D-37075, Germany
[3] Department of Obstetrics and Gynecology, The Edith Wolfson Medical
   Center, Holon, The Sackler Faculty of Medicine
Tel Aviv University, Israel

INTRODUCTION

Antisense oligonucleotides are short, synthetic DNA chains designed to match the sequence of their target RNA in the opposite orientation [hence "antisense" (Stein & Cheng, 1993); Fig. 1]. They are often protected against degradation by introducing sulfur atoms in their internucleotidic bonds to form phosphorothioates (Eckstein, 1985). They are actively taken up by cells, where they find their target mRNA chains and associate with them to form RNA:DNA double-strands. Phosphorothioated oligonucleotides further induce, within these cells, a specific ribonuclease - RNase H, selectively destroying double stranded RNA chains. Therefore, antisense drugs can prevent mRNA from being translated into protein in at least two ways - physical interference with the translation machinery and the initiation of mRNA destruction. Experimental antisense drugs are currently being tested in HIV patients, where oligonucleotides destroy viral RNA, in leukemias, where they are targetted towards onco-genes, and in many other diseases of the bone marrow (Stein & Cheng, 1993). Bone marrow cells are particularly convenient targets for these drugs, as they are reached *in vivo* within an hour from injection time. Moreover, these cells proliferate rapidly and differentiate to more mature forms, making them susceptible to rapid changes in gene expression.

Bone marrow cells from all known vertebrates (red blood cells, lymphocytes, platelet progenitors) express cholinesterases. When injected with acetylcholine analogues or with carbamate cholinesterase inhibitors, proliferation of megakaryocytes (platelet progenitors) was altered in rodents (reviewed in Soreq & Zakut, 1993). Moreover, farmers using

Enzymes of the Cholinesterase Family, Edited by Daniel M. Quinn et al.
Plenum Press, New York, 1995

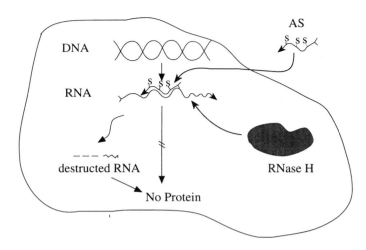

Figure 1. Phosphorothioated antisense oligonucleotides induce destruction of their target mRNA. Phosphorothioate groups are denoted by S; 3'-ends of sequences are marked by arrows. See text for details.

organophosphorous anticholinesterases as insecticides are at an increased risk of developing leukemia, and 25% of leukemic patients were found to carry cholinesterase genes with abnormal copy numbers and structures. Altogether, this implied that interference with the expression of blood cell cholinesterases is associated with enhanced bone marrow proliferation. To test this hypothesis, we employed the antisense approach.

There are two distinct cholinesterase genes in all vertebrates. The two human genes, ACHE and BCHE, were cloned, mapped to specific chromosomal sites and expressed in transgenic organisms (Soreq & Zakut, 1993). Interestingly, these two genes are very different in their base composition, ACHE being rich in G,C residues and BCHE in A,T ones (Soreq & Zakut, 1993). This ensures that antisense agents targetted to the RNA product of one of these genes will never interact with the RNA product of the other - a selectivity which, in this gene family, favors the antisense approach. In contrast, most inhibitors targetted to one of the cholinesterase proteins, like carbamates and organophosphates, will also interact with the other, since these proteins are 50% identical and >85% homologous.

THE EXPERIMENTAL APPROACH

We have studied the bone marrow of patients with somatically mutated and amplified ACHE and BCHE genes associated with very low platelet counts and abnormal development of megakaryocytes, the platelet progenitors from which platelets are formed (Zakut et al., 1992). When an antisense agent blocking BCHE gene expression was added to murine bone marrow cells grown in culture in the presence of the cytokine interleukin 3 (IL-3) megakaryocyte progenitor development was severely inhibited (Patinkin et al., 1990).

When synthesized in their natural phosphate backboned structure, antisense agents are rapidly hydrolyzed by exonucleases. To increase their intracellular stability, sulfur atoms may be substituted for oxygen in the internucleotidic bonds. However, such phosphorothioated oligonucleotides and/or their subsequent degradation products may be toxic to the host cell in a manner unrelated to their sequence. Therefore, each antisense agent to be developed needs to be tested for its cytotoxicity and sequence specificity for the relevant cell types.

When incubated with bone marrow cells, radioactively labeled phosphorothioated oligonucleotides bound to small, dividing cells. Incubation of bone marrow cellular proteins

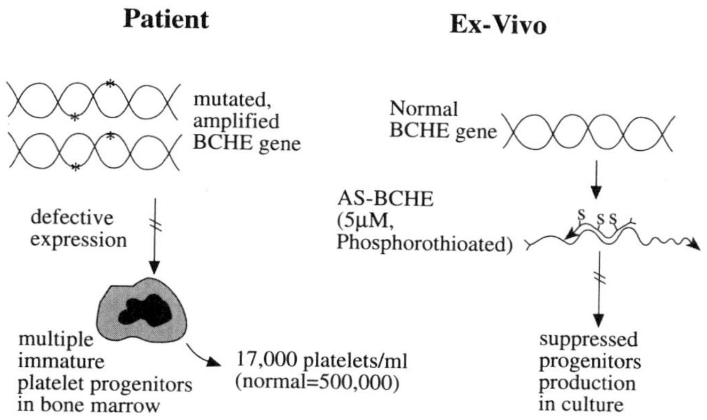

Patient   **Ex-Vivo**

mutated,
amplified
BCHE gene

Normal
BCHE gene

AS-BCHE
(5μM,
Phosphorothioated)

Figure 2. Ex-vivo antisense inhibition of BCHE gene expression suppresses platelet progenitors in a manner resembling the in vivo platelet deficiency in patients with abnormal BCHE genes. See text for details.

defective
expression

multiple
immature
platelet progenitors
in bone marrow

17,000 platelets/ml
(normal=500,000)

suppressed
progenitors
production
in culture

with these agents also resulted in protein labeling (Ehrlich et al., 1994). This proved to be unrelated to the nucleotide sequence of the employed oligonucleotides - a non-specific interaction which could explain part of the interference with *ex-vivo* bone marrow proliferation that was observed for AS-BCHE in its totally phosphorothioated form.

To prevent this cytotoxicity, the following arguments were made: (1) Phosphorothioate protection is important for nuclease induction and oligonucleotide stabilization. (2) This protection may also be cytotoxic and induce part or all of the observed interference with cell development. (3) Since RNA destruction is primarily initiated at the 3'-end, it can be blocked effectively at that end. (4) When three 3' bonds are blocked, 80% of the phosphorothioate groups can be saved. This should create less toxic, yet effective agents.

When administered in similar doses, thus partially protected agents reduce production of platelet progenitors as efficiently as their fully protected counterparts. When incubated with bone marrow proteins, they display no binding, their addition to cultures apparently does not interfere with the functioning of other genes than those targetted and no non-specific reduction in cell proliferation could be observed when irrelevant oligomers were tried (Figure 3). Altogether, this technological improvement provides us with better tools to work with for antisense targetting of the cholinesterase mRNAs (Ehrlich et al., 1994).

Antisense Agent Protection

Partially Phosphorothioated   Fully Phosphorothioated

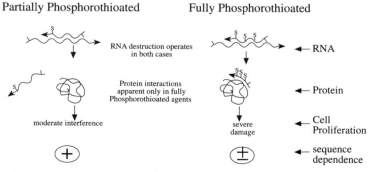

RNA destruction operates
in both cases

RNA

Protein interactions
apparent only in fully
Phosphorothioated agents

Protein

moderate interference

severe
damage

Cell
Proliferation

sequence
dependence

+   ±

Figure 3. Partial phosphorothioate protection of antisense oligonucleotides prevents non-sequence dependent interaction with proteins and reduces non-specific cytotoxicity. See text for details.

RESULTS

When mice were injected in vivo with the oligomer blocking BCHE gene expression (AS-BCHE), their platelet progenitors revealed reduced levels of BCHEmRNA (as tested by in situ hybridization). This ensured that specific RNA destruction occurred; at the same time, unrelated genes (i.e. actin) remained fully expressed, reflecting low toxicity of these antisense agents. When bone marrow cells from the injected mice were cultured megakaryo-cyte colony development was severely inhibited (40% reduction). *In vitro* cultures with AS-BCHE further displayed a sharp shift in differentiation from predominantly megakaryo-cyte to myeloid lineages (Patinkin et al., 1994).

Intraperitoneal injection of AS-ACHE, blocking expression of the BCHE-related acetylcholinesterase protein, resulted in much more dramatic changes. A sole injection of $5\mu g/g$ weight caused drastic reductions in the fractions of bone marrow erythrocytes and lymphocytes at 12 days post-treatment, as well as reciprocal increases in myeloid cells, changes which were almost totally reversed by day 208. However, in the *in vivo* situation it is virtually impossible to determine absolute numbers of bone marrow cells. Therefore, we could not conclude whether erythroid development was inhibited, if myeloid cell prolifera-tion was enhanced, or both. To answer these questions, we administered such oligonu-cleotides *ex-vivo*, in primary cultures of murine bone marrow cells.

Bone marrow cell cultures have several advantages: (1) They reveal the absolute number of proliferating stem cells present at plating time, each of which develops into a colony, whereas the other terminally differentiated cells die in culture (essentially, as they would *in vivo* as well). (2) They are subject to easy modulation by cytokines. In the presence of IL-3, both platelet progenitors and myeloid cells will develop, but with added erythro-poietin and transferrin erythroid cells are produced in large numbers. This enables examina-tion of the dependence of antisense effects on cell composition and growth factors. (3) The *ex-vivo* cultures are a clinically important model system for progenitor cells produced for transplantation, an increasingly popular procedure for treatment of cancer patients following drastic chemotherapy or irradiation.

When added to erythropoietic cell cultures, AS-ACHE caused a dose-dependent increase in colony and cell counts and increased the fraction of myeloid cells at the expense of erythroid cells and megakaryocytes (Soreq et al., 1994), as shown in Figure 4. With IL-3 alone, AS-ACHE decreased colony counts but not cell numbers, and diverted up to 50% of the cells into erythroid blasts (which could not develop further for the lack of erythropoietin). Additional tests revealed transient reduction in ACHEmRNA, followed by a 10-fold increase by day 4; a general change in the pattern of mRNA transcripts; a 10-fold increase in DNA yield and prevention of the DNA fragmentation which appeared in non-treated cultures; and generally healthier appearance of cells under AS-ACHE treatment (less vacuoles, larger nuclei, smoother cell surface). AS-BCHE did not cause such changes, it only reduced megakaryocytes at the expense of myeloid cells (Soreq et al., 1994). Figure 4 compares the effects of AS-ACHE to those of AS-BCHE in erythropoietic cultures, revealing the wider scope and dominant nature of AS-ACHE over erythroid and megakaryocyte development.

DISCUSSION

In essence, our findings confirm the earlier indications of a regulatory role of the cholinesterases in bone marrow development (Lifson-Lapidot et al., 1989). The accumulated evidence suggests that AS-ACHE blocks the differentiation pathway in a different site than AS-BCHE; whereas AS-BCHE interferes primarily with megakayocyte development (and

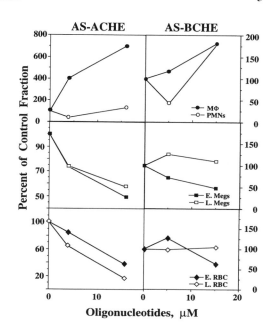

Figure 4. AS-ACHE, but not AS-BCHE decreases erythroid cell production *ex vivo*. Normal fractions of cell lineages are presented as 100%. Cultures were grown with erythropoietin. M: macrophages; PMNs: polymorphonuclears; E.L.Megs: early, late erythroid cells.

hence platelet production), AS-ACHE apparently blocks the production of the multi-potential progenitors leading toward both erythroid and megakaryocyte cells. Under such conditions, the cultured bone marrow cells can only proliferate or develop into myeloid cells, which indeed they do.

The hypothesis emerging from the above experiments is that the cholinesterases possibly participate in directing proliferating stem cells toward a differentiated state that eventually will terminate in programmed death. Important implications arise for bone marrow transplantations: this *ex vivo* procedure includes a phase of cell culture very much like the *in vitro* culture technique detailed above. It would be extremely helpful to amplify the number of proliferating stem cells prior to their re-introduction into the patient - thus shortening considerably the hospitalization time and improving the patient's condition while reducing the volume of cells to be injected.

If reproduced in humans, our procedure further offers the opportunity of improving the proliferative state of bone marrow in patients following chemotherapy or irradiation for any pathology, not only hematopoietic in nature.

Cholinesterases are known to be expressed in the embryonic stage of development of many other cell types undergoing terminal differentiation (i.e., muscle and nerve cells). It would be intriguing to examine if, in those tissues as well, these interesting enzymes are involved in controlling the shift from proliferation to differentiation.

ACKNOWLEDGMENTS

This research was supported, in part, by the United States Army Medical Research and Development Command (contract DAMD 17-94-C-4031, to H.S. and H.Z.), the German Israeli Fund (to H.S. and F.E.) and the Israel Ministry of Science (to H.S., F.E. and H.Z.).

REFERENCES

Eckstein, F. (1985) Ann. Rev. Biochem. 54:367.

Ehrlich, G., Patinkin, D., Ginzberg, D., Zakut, H., Eckstein, F., & Soreq, H. (1994) Antisense Research and Development 4:173.

Lev-Lehman, E., Hornreich, G., Ginzberg, D., Gnatt, A., Meshorer, A., Eckstein, F., Soreq, H., & Zakut, H. (1994) Gene Therapy 1:127.

Lifson-Lapidot, Y., Prody, C.A., Ginzberg, D., Meytes, D., Zakut, H., & Soreq, H. (1989) Proc. Natl. Acad. Sci. USA 86:4715.

Patinkin, D., Lev-Lehman, E., Zakut, H., Eckstein, F., & Soreq, H. (1994) Cellular and Molecular Neurobiology, in press.

Patinkin, D., Seidman, S., Eckstein, F., Benseler, F., Zakut, H., & Soreq, H. (1990) Molec. Cell. Biol. 10:6046.

Soreq, H., Patinkin, D., Lev-Lehman, E., Ginzberg, D., Eckstein, F., & Zakut, H. (1994) Proc. Natl. Acad. Sci. USA 91:7907.

Soreq, H., & Zakut, H. (1993) Human Cholinesterases and Anticholinesterases, Academic Press, New York.

Stein, C.A., & Cheng, Y.C. (1993) Science 261:1004. 300 p. (1993).

Zakut, H., Lapidot-Lifson, Y., Beeri, R., Ballin, A., & Soreq, H. (1992) Mutation Research 276:275.

PROPERTIES OF CLASS A ACETYLCHOLINESTERASE, THE ENZYME ENCODED BY *ACE-1* IN *CAENORHABDITIS ELEGANS*

Martine Arpagaus,[1,2] Nathalie Schirru,[1] Emmanuel Culetto,[1,2]
Vincenzo Talesa,[1,3] Xavier Cousin,[1,4] Arnaud Chatonnet,[1] Yann Fedon,[1]
Jean-Baptiste Berge,[2] Didier Fournier,[2] and Jean-Pierre Toutant[1*]

[1] Equipe "Cholinestérases", INRA
2, place Viala, 34060 Montpellier, France
[2] Biologie des Invertébrés, INRA
BP 2078, 06606 Antibes, France
[3] Department of Expl. Medicine, University of Perugia
Via del Giochetto, Perugia, Italy
[4] Neurobiologie, ENS
46 rue d'Ulm, 75230 Paris 05, France

INTRODUCTION

Three genes (*ace-1*, *ace-2*, and *ace-3*) encode acetylcholinesterases (AChEs) in the nematode *Caenorhabditis elegans* (review in Johnson, 1991). The three enzymes differ in their catalytic properties and are called AChE of class A, B and C respectively. Classes A and B AChEs are major components accounting for at least 95% of the total AChE activity. They are both expressed in the motor system and in particular at the cholinergic excitatory neuromuscular junctions since homozygous double null mutants ace-1-/ace-2- are uncoordinated and hypercontracted.

The existence of several AChE genes in nematodes contrasts with the presence of a single AChE gene expressed in the central nervous system of *Drosophila* (Hall and Spierer, 1986). In vertebrate AChEs, the use of alternative exons in a single gene results in the production of transcripts encoding either a hydrophobic (H) or a 'tailed' (T) catalytic subunit (Massoulié et al., 1993).

---

[*] Address correspondence to Dr. Toutant: Départment de Physiologie animale, 2, place Viala, Centre INRA de Montpellier, 34060 Montpellier Cedex 1, France. Tel: (33)67 61 26 87; fax: (33) 67 54 56 94.

Enzymes of the Cholinesterase Family, Edited by Daniel M. Quinn et al.
Plenum Press, New York, 1995

C. elegans is a good model for studying developmental and molecular aspects of gene expression. We are interested in the structure and function of the different classes of AChE as well as in the regulation of their expression in the adult worm and during development. We report here some properties of class A AChE deduced from ace-1 sequence and its in vivo and in vitro expression (Arpagaus et al., 1994).

CLONING OF ACE-1

A fragment of genomic ace-1 was amplified by polymerase chain reaction (PCR) using degenerate oligonucleotides deduced from conserved sequences in vertebrate cholinesterases (EDCLYLN and FGESAG). Screening of a cDNA library with this PCR fragment led to the isolation of a full-length cDNA. The amino acid sequence of ACE-1 (the product of ace-1) is shown figure 1.

PROPERTIES OF CLASS A AChE

ACE-1 possessed 620 amino acids (71.4 kDa) with a putative peptide signal of 31 residues (vertical arrow in fig. 1, see also hydropathy profile below). This would lead to a mature protein of 589 amino acids with three possible sites of N-glycosylation. The sequence shows that important residues necessary to define a cholinesterase are present (legend of fig. 1).

That ACE-1 is an AChE is indicated by the conservation of aromatic residues lining the active site gorge that determine in part substrate and inhibitor specificity (see Massoulié et al., 1993, for review). Twelve out of the fourteen residues conserved in vertebrate AChEs were found. S85(Y70) and G307(F288) that are not aromatic in ACE-1 are not aromatic in Drosophila AChE and mammalian BChEs. This likely explains why ACE-1 hydrolyzes ACh best but also hydrolyzes butyrylthiocholine (Arpagaus et al., 1994). ACE-1 was inhibited by excess substrate although at higher concentrations than vertebrate AChEs. All residues that have been implicated in this type of inhibition are found in conserved position in ACE-1. In particular W300 (279) which could be critical for the binding of excess substrate is present. The components: W300(279)-D87(72)-Y353(334)-Y349(330) that are parts of a putative allosteric relay from the peripheral site to the active serine in human AChE (Shafferman et al., 1992) are found in ACE-1.

Two free cysteines are found (618 and 392). C618 is in a favorable position to form an interchain S-S bond as in other cholinesterases. Location of C392 in the tertiary structure of ACE-1 was tentatively reconstituted by analogy to the structure of Torpedo AChE (Sussman et al., 1991). It was found at the beginning of the α helix αF'3 and is thus situated at the periphery of the molecule facing the other subunit in a potential dimer. Participation of C392 and C618 in oligomer formation through interchain S-S bonds is currently being investigated by directed mutagenesis.

PHYLOGENETIC RELATIONSHIPS

Comparisons of ACE-1 sequence to the EMBL SwissProt data bank showed 41-42% identity with vertebrate AChEs and BChEs and only 35% with Drosophila AChE (Arpagaus et al., 1994). Note that the hydrophilic insertion of Drosophila AChE (Hall and Spierer, 1986) was not taken into account in this comparison. The lower percentage of identity thus results from a higher degree of variation in the sequence as exemplified by the conservation of

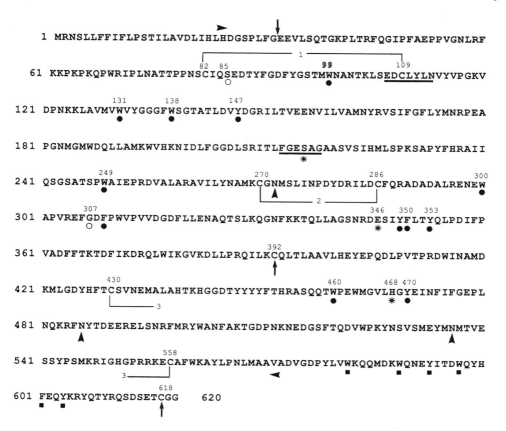

Figure 1. Amino acid sequence of ACE-1 deduced from the open reading frame of *ace-1*. Numbering of ACE-1 starts at the initiator M. Important conserved features include: a) S220(*200*), H468(*440*) and E346(*327*) of the catalytic triad (asterisks) and W99(*84*) as the choline binding site, b) three intrachain disulfide bonds present in all cholinesterases. One free cysteine (618, vertical arrow) is found in the C-terminus. The only other free cysteine is indicated by the arrow at position 392. Three putative N-glycosylation sites are shown by vertical arrowheads. Filled circles show 12 conserved aromatic residues among the 14 lining the active gorge in vertebrate AChEs. Open circles show the two positions that are not conserved. Black squares in the C-terminus indicate a series of aromatic residues conserved in hydrophilic T subunits of cholinesterases. Sequences EDCLYLN and FGESAG that were used to design oligonucleotides for initial PCR amplification are underlined. The portion of sequence used for the construction of phylogenetic tree in figure 2 is shown between horizontal arrowheads.

aromatic residues of the gorge (twelve conservations in ACE-1, ten in *Drosophila*). The phylogenetic tree in fig. 2 was constructed according to the Neighbor Joining Method by using the programs NJTREE and NJDRAW provided by Drs. Ferguson and Lin (University of Texas, Houston). Here also, areas of questionable homologies in all sequences such as N-terminal peptide signals or regions of alternative splicing in C-termini were not used. The comparison was thus performed from the first to the last consensus sequences found between all proteins (such region in ACE-1, covering 550 amino acids is shown between horizontal arrowheads in fig. 1).

The tree confirms the intermediate position of ACE-1 between vertebrate ChEs and *Drosophila* AChE. However, the codon used in ACE-1 for the active serine is TCA (TCG in

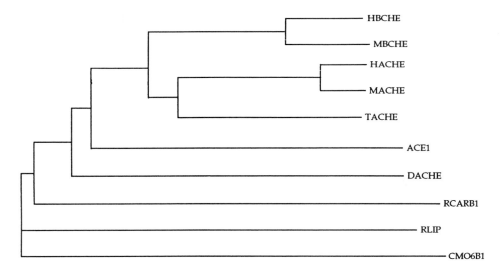

Figure 2. Phylogenetic relationships of ACE-1 with other esterases. The following peptide sequences (with their EMBL SwissProt accession numbers) were used. *Torpedo marmorata* (P07692, TACHE), mouse (P21836, MACHE) and human (P22303, HACHE) AChEs. Human (P06276, HBCHE) and mouse (MBCHE, Rachinsky et al.,1990) BChEs. *Drosophila* AChE (P07140, DACHE); rat lysophospholipase (EC 3.1.1.5, P07882, RLIP) and carboxylesterase (EC 3.1.1.1, P10959, RCARB1); ACE-1 (X75331) and a carboxylesterase from *C. elegans* (EC 3.1.1.1; Fedon et al., 1993; CM06B1).

Drosophila) whereas AGY is used in vertebrate ChEs (Brenner, 1988). Thus it appears that ACE-1 and *Drosophila* AChE belong to the same subfamily, in spite of their lower percentage of identity. This could be the result of different selection pressures exerted on AChE between animal groups. In this respect, it is important to note that ACh is the excitatory neurotransmitter in the motor systems of vertebrates and nematodes but not in arthropods, where it is used mainly in the sensory systems. Functional requirements for ACh hydrolysis in motor and sensory systems may be very different and might result in different selection pressures on important functional amino acids of AChE.

MOLECULAR FORMS OF CLASS A AChE

The hydropathy profile of ACE-1 (Figure 3) shows that the N-terminus is hydrophobic (signal peptide) and the C-terminus is hydrophilic (arrow).

Thus ACE-1 polypeptide is referred to as a 'tailed' subunit (T, Massoulié et al., 1993). Regular disposition of aromatic residues in C-termini of T subunits in vertebrate ChEs may explain that these 'hydrophilic' subunits form amphiphilic monomers or dimers of type II (Bon et al., 1988) provided that the C-terminal sequence adopts an α helix conformation (Massoulié et al., 1993; Bon, Cornut, Dufourcq, Grassi and Massoulié, in preparation). So far a motif WxxxFxxWxxYxxxWxxxFxxY of *Torpedo* AChE has been found in conserved position in T subunits of quail AChE (Anselmet et al., 1994) and both rabbit AChE and BChE (Jbilo et al., 1994). Sequence alignment of C-terminus between *Torpedo* and ACE-1 T peptides shows conservation of several aromatic residues (*) with two additional Y (°) in ACE-1 that may be included in the helix (c.f. Figure 4).

So far, we have no evidence that such amphiphilic forms of type II occur for ACE-1 by this mechanism. In particular, the only molecular forms identified when *ace-1* was

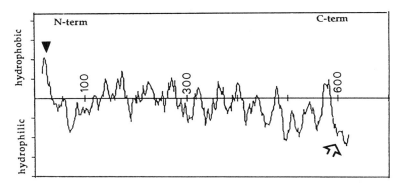

Figure 3. Hydropathy profile of ACE-1.

expressed *in vitro* in the system recombinant baculovirus/*Spodoptera* cells (Arpagaus et al., 1994) were hydrophilic G1 and G4 recovered either in the supernatant or in pelleted cells. That *ace-1* produces only hydrophilic forms should be confirmed by transfection in mammalian cells that are known to produce amphiphilic forms of type II when transfected with *Torpedo* T subunit cDNA (Duval et al., 1992).

We compared further molecular forms of class A AChE produced *in vitro* (Sf9 cells infected with baculovirus containing *ace-1*) and *in vivo* (in null mutants *ace-2⁻*, a way for suppressing potential contamination by AChE of class B). Figure 5A shows that G1 and G4 forms were produced *in vitro* (5 and 11.5S). Both forms did not interact with detergents and were thus hydrophilic. *In vivo*, G1 form was hydrophilic (4.5S). A prominent 13.5S peak was observed with a shoulder at 11.5S (fig. 5B). As in the related nematode *Steinernema* (Arpagaus et al., 1992), the fast-migrating peak (13.5S) was shown to be composed in majority of complex tetrameric association of T subunits disulfide-linked to hydrophobic non catalytic element(s). The major component of the 13.5S peak was thus amphiphilic. Figure 5C shows a non denaturing gel electrophoresis where mild treatments with increasing concentrations of proteinase K induced a progressive conversion of original amphiphilic tetramers (a) to hydrophilic ones (b). Migration of both forms found *in vivo* can be directly compared to simple G4 (c) and G1 forms produced *in vitro*.

Figure 6 summarizes the structure of ACE-1 molecular forms synthesized *in vivo* or in Sf9 cells. The latter appear unable to produce non catalytic components. Number and position of S-S bonds remain hypothetical.

PERSPECTIVES

Two sites of transcription start were determined by 5'RACE at -168 and -196 nt upstream of the initiator ATG of *ace-1*. We then cloned and sequenced a 1.2kb fragment of the 5' flanking region. No TATA box and no CAAT box were found at consensus distances

```
WKQQMDKWQNEYITDWQYHFEQYKRYQTYRQSDSETCGG(620)   ACE-1

WKPEFHRW-SSYMMHWKNQFDQYSRHENCAEL(575)          Torpedo AChE
 *      *  *  *  *  *  ○  ○
```

Figure 4. Alignment of C-terminal sequences of T subunits in *Torpedo* and *C. elegans* AChEs.

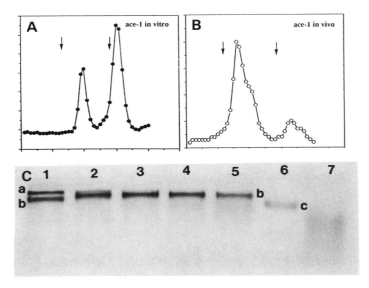

Figure 5. Molecular forms of ACE-1 expressed *in vitro* and *in vivo*. A and B. Gradient sedimentation. Markers are β-galactosidase (16S, arrow on left) and alkaline phosphatase (6.1S, right). C. Nondenaturing gel electrophoresis of class A AChE extracted from null mutants ace-2⁻, purified on edrophonium-sepharose (lane 1) and digested by increasing concentrations of proteinase K (0.5, 5, 10 and 25 μg/ml, 30 min, 20°C, lanes 2-5). G4 and G1 forms produced *in vitro* (lanes 6 and 7) were isolated on a preparative gradient of purified supernatants. a, b and c indicate complex amphiphilic G4, complex hydrophilic G4 and simple hydrophilic G4.

of transcription starts. These two sites are located within sequences resembling those of 'initiator elements' found in other TATA-less promoters. We are studying the promoter activity of the 5' region a) by heterologous expression of constructs in mammalian cells (10T1/2) to delineate important functional sequences and b) by injection in germinal cells of *C. elegans* for tissue (cell)-specific expression. ACE-1 has been reported to be expressed prominently in muscle cells (Herman and Kari, 1985) and we are studying the mechanism of this specific expression.

We have also identified the origin of the null mutation in *ace-1* initially isolated by Johnson et al. (1981, allele *p1000*). A transition TGG—>TGA introduces a stop codon (opale) in place of TGG encoding W99. A truncated protein is thus produced with no enzymatic activity. The mutation also leads to a significant reduction of *ace-1* mRNA. We

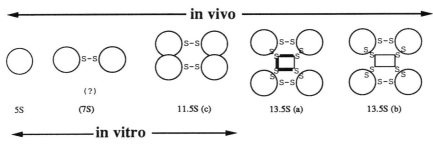

Figure 6. Molecular forms of ACE-1 produced *in vivo* and *in vitro*.

are testing the role of mRNA surveillance system encoded by the *smg* genes in this destabilization.

All trials to isolate *ace-2* or *ace-3* by sequence homology have been unsuccessful so far.

ACKNOWLEDGMENTS

This work was supported by grants from the *Institut National de la Recherche Agronomique* (AIP *Xénobiotiques*). V.T. was supported by the Human Capital and Mobility Program of the European Community (network Cholinesterases).

REFERENCES

Anselmet A., Fauquet M., Chatel J.M., Maulet Y., Massoulié J., Vallette F.M. (1994) J Neurochem 62, 2158-2165

Arpagaus M., Richier P., Bergé J.B., Toutant J.P. (1992) Eur J Biochem 207, 1101-1108

Arpagaus M., Fedon, Cousin X., Chatonnet A., Bergé J.B., Fournier D., Toutant J.P. (1994) J Biol Chem 269, 9957-9965

Bon S., Toutant J.P., Méflah K., Massoulié J. (1988) J Neurochem 51,786-794

Brenner S. (1988) Nature 334, 528-530

Duval N., Massoulié J., Bon S. (1992) J Cell Biol 118, 641-653

Fedon Y, Cousin X., Toutant J.P., Thierry-Mieg D., Arpagaus M. (1993) DNA Sequence 3, 347-355

Hall L.M.C., Spierer P. (1986) EMBO J 5, 2949-2954

Jbilo O., L'Hermite Y., Talesa V., Toutant J.P., Chatonnet A. (1994) Eur J Biochem, in press.

Johnson C.D., Duckett J.G., Culotti J.G., Herman R.K., Mennely P.M., Russell R.L. (1981) Genetics 97, 261-279

Johnson C.D. (1991) in *Cholinesterases* (Massoulié J et al., eds). Pp 136-140. American Chemical Society, Washington, D.C.

Herman R.K., Kari C.K. (1985) Cell 40, 509-514

Massoulié J., Pezzementi L., Bon S., Krejci E., Vallette F.M. (1993) Prog Neurobiol 41, 31-91

Rachinsky T.L., Camp S., Li Y., Ekström T.J., Newton M., Taylor P. (1990) Neuron 5, 317-327

Shafferman A., Velan B., Ordentlich A., Kronman C., Grossfeld H., Leitner M., Flashner Y., Cohen S., Barak D., Ariel N. (1992) EMBO J. 11, 3561-3568.

Sussman J.L., Harel M., Frolow F., Oefner C., Goldman A., Toker L., Silman I. (1991) Science 253, 872-879

LIGAND SPECIFICITY AND GENE EXPRESSION IN THE CHOLINERGIC SYNAPSE

A Comparison between Acetylcholinesterase and the Nicotinic Acetylcholine Receptor

Palmer Taylor, Zoran Radić, Hans-Jürgen Kreienkamp, Zhigang Luo, Natilie A. Pickering, and Shelley Camp

Department of Pharmacology
School of Medicine
University of California, San Diego
La Jolla, California 92093-0636

INTRODUCTION

Upon first inspection, acetylcholinesterase (AChE) and the nicotinic acetylcholine receptor (nAchR) share several functional and structural characteristics. Both recognize the neurotransmitter, acetylcholine, and elicit the respective actions of catalysis of the neurotransmitter and neurotransmitter-induced channel opening in a millisecond time frame. Their relative stoichiometries at various synapses are carefully regulated within precise limits. Homologous toxins, the fasciculins and α-neurotoxins (i.e., α-cobratoxin and α-bungarotoxin), which have nearly identical tertiary structures, inhibit the two respective proteins at concentrations in the picomolar to subpicomolar range. Finally, the two proteins show coordinated expression during development.

The parallels end here, for these proteins show a diversity in structure, ligand specificity and gene expression sufficient to suggest that they are neither products of divergent evolution from a common primordial protein nor have they shown a convergence in overall structural properties. Rather, they seem to have evolved by independent pathways; this, in turn, requires complex and auxiliary mechanisms to achieve coordinated function in an intact animal. For example, AChE has developed a mechanism to tether itself to both basement and plasma membranes through disulfide linkages to structural subunits. Presumably, this enables AChE to associate in particular synapses in a concentration and structural disposition important for optimal synaptic function. In this article we examine some of the recently reported features of these two proteins that point to their diversity.

Enzymes of the Cholinesterase Family, Edited by Daniel M. Quinn et al.
Plenum Press, New York, 1995

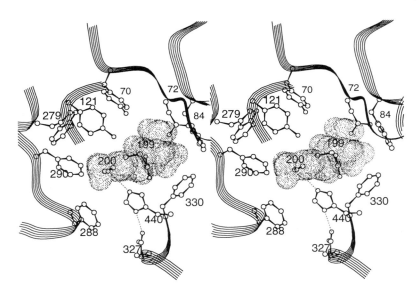

Figure 1. Stereodiagram of *Torpedo* acetylcholinesterase showing the active center gorge and residues critical to the catalytic triad (E_{327},H_{440},S_{200}), Acyl Pocket (F_{288},F_{290}), Choline Subsite (W_{84},E_{199},F_{330}), Peripheral Anionic Site (W_{284}, Y_{70}, Y_{121}, D_{72}) .

Structure and Ligand Specificity of AChE

The recent elucidation of the three dimensional structure of AChE (Sussman et al., 1991) has provided the cornerstone for understanding the overall structure of the enzyme and its complexes. Physicochemical studies, chemical modification and site-specific mutagenesis have further defined domains in the protein critical for function (Taylor and Radić, 1994). The active center of the enzyme is close to being centrosymmetric in the subunit and lies within a gorge of 18-20Å in depth. The active center may be subdivided into the catalytic triad, the acyl pocket and the choline subsite (fig. 1 and table 1). Two of the three residues of the catalytic triad were defined prior to the crystal structure determination, $H_{447(440)}$ and $S_{203(200)}$ (Gibney et al., 1990), and their alignment with $E_{334(327)}$ reveals a spatial orientation similar to other serine hydrolases (Sussman et al., 1991).[*]

The acyl pocket is outlined by two phenylalanines, $F_{295(288)}$ and $F_{297(290)}$, which exist as leucine and isoleucine in butyrylcholinesterase (BuChE). As expected, many of the features of BuChE catalysis can be achieved by these two respective substitutions. What is perhaps more surprising is that the substitution of $F_{297}I$ is sufficient to convert the enzyme from control by substrate inhibition typical of AChE to substrate activation typical of BuChE. Also, stereospecificity of the R and S enantiomers of cycloheptylmethyl phosphonothiocholine can be reversed by replacing the phenyl group at the 297, but not the 295 position (Pickering et al., 1994).

The choline subsite in the active center contains a cluster of aromatic residues dominated by $W_{86(84)}$. Although some controversy has arisen as to the molecular forces prevailing in the stabilization of substrates and inhibitors by W_{86} (Shafferman et al., 1993;

---

[*] Throughout the manuscript the mammalian AChE numbering system will be used followed by *Torpedo* numbering, where applicable, in parentheses.

Nair et al., 1994), it is clear that both dispersion forces of W_{86} and other aromatic residues in this domain and coulombic forces of $E_{202(199)}$ are important to the stabilization energy conferred by this region (Radić et al., 1992). Another aromatic residue in this region is a tyrosine or phenylalanine at the 337 position in AChE, whereas the corresponding residue is alanine in BuChE. This residue is critical to the specificity of huperzine as a selective AChE inhibitor and ethopropazine as a selective BuChE inhibitor (Saxena et al., 1994; Radić et al., 1993).

Another region forms the peripheral anionic site which is located at the lip of the gorge (cf. table 1). This site has long been known to control catalysis allosterically, presumably by inducing a conformational change in the gorge (Radić et al., 1993; Ordentlich et al., 1993). This site or some portion of this site is also presumably involved in substrate inhibition seen in AChE (Radić et al., 1991). The fluorescent ligand, propidium, has been the frame of reference for examining the peripheral site (Taylor and Lappi, 1975). Recently, 61 amino acid peptides isolated from the venom of mambas, termed fasciculins I, II and III, have been found to be peripheral site inhibitors (Karlsson et al., 1984; Marchot et al., 1993). Fasciculin binds to the region found near the lip of the gorge and defined by an aromatic cluster of W_{286}, Y_{124} and Y_{72}. Upon binding to this region, which likely constitutes part of the peripheral site, fasciculin, like propidium, behaves as an allosteric, non-competitive inhibitor. Substantial evidence has accumulated from fluorescence, circular dichroism and site-specific mutagenesis that the peripheral site and active center are conformationally linked (Epstein et al., 1979; Berman et al., 1981; Ordentlich et al., 1993; Radić et al., 1994).

STRUCTURE AND LIGAND SPECIFICITY OF THE nAChR

In the case of the nAChR, structural definition at atomic level resolution is lacking, yet important information on the binding sites has emerged from the analysis of sequences, subunit composition, electron microscopic image reconstruction and mutagenesis (cf: Unwin 1993). The ligand binding sites have been shown by chemical modification (Changeux et al., 1992; Karlin, 1993; Cohen et al., 1991), selective subunit expression (Blout and Merlie, 1989; Sine and Claudio, 1992) and site-specific mutagenesis (Czajkowski et al., 1993; Sine, 1993) to exist at the subunit interface with binding contacts arising from the α,γ and α,δ interfaces. Since γ and δ differ in sequence and they interact with the same face of α (fig. 2), it is possible to ascertain the domains in γ and δ responsible for ligand specificity. α-Conotoxin M shows a 10^4 preference for the $\alpha\delta$ binding site over the $\alpha\gamma$ site (Kreienkamp et al., 1994), and, in conjuction with Steven Sine's laboratory at the Mayo Clinic, we have been able to assign three regions in the extracellular domain of the γ and δ subunits that face α (Kreienkamp et al., 1995a). The approach, coupled with analogous studies for d-tubocurarine (Sine, 1993) not only enable one to assign residues involved in differences in ligand specificity, but also permit the assignment of residues to the negative face of γ or δ (fig. 2). Studies of the residues of the α subunit responsible for ligand specificity allow assignment of residues on the positive face of α.

A second approach to this question involves the assignment of residues responsible for subunit assembly. Here we can analyze assembly of the dimers, $\alpha\gamma$ or $\alpha\delta$, for they exhibit high affinity, agonist displaceable α-neurotoxin binding not observed with the α-subunit alone. Since γ is the only subunit to reside between α subunits (fig. 1), $\alpha\gamma$ association but not $\alpha\delta$ association, will lead to formation of tetramers (i.e. $\alpha\gamma\alpha\gamma$). Accordingly, residues responsible for assembly on both the (+) and (-) faces of the homologous subunits can be defined initially by constructing chimera of γ and δ and subsequently by specific residue replacements (Kreienkamp et al., 1995b). All of the regions responsible for dictating subunit assembly appear in the first 200 residues; this region is believed to be extracellular. Candidate

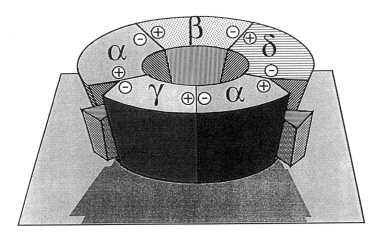

Figure 2. Subunit assembly of the acetylcholine receptor. Shown is a representation of the extracellular region of the four homologous subunits, α,β,γ and δ, in the pentameric arrangement of $\alpha_2\beta\gamma\delta$. Two ligand binding sites exist at the $\alpha\gamma$ and $\alpha\delta$ interfaces. Since the subunits are homologous, each of the clockwise or (+) faces should have corresponding residues in homologous positions and interact with corresonding loci on the (-) or counter clockwise face of the neighboring subunit.

membrane-spanning sequences begin after residue 200. Several residues on the (+) face of α and the (-) face of γ and δ dictate both ligand specificity and assembly (Kreienkamp et al., 1995a and b). The importance of intersubunit contacts for ligand specificity and for maintenance of cooperativity of the nAchR becomes immediately apparent from these studies.

The binding of acetylcholine and the α-neurotoxins is competitive on the nicotinic receptor, and agonist and α-neurotoxin are presumed to occupy overlapping sites in the respective $\alpha\gamma$ and $\alpha\delta$ interfaces of the nicotinic receptor. The fasciculins are very similar in structure to the α-neurotoxins, both in terms of sequence identity and three dimensional structure (Le Du et al., 1992); yet, in contrast to the competitive relationship between α-neurotoxin and agonist on the receptor, fasciculin and acetylcholine bind to distinct sites on AChE. The two sites on AChE are allosterically linked.

Proteins closely related to mammalian AChE and the mammalian muscle nAChR are known to be resistant to fasciculin (Karlsson et al., 1994) and to α-neurotoxin (Barchan et al., 1992; Neumann et al, 1989), respectively. In the case of BuChE resistance to fasciculin, three aromatic residues important to fasciculin binding in AChE are absent in BuChE (Radić et al., 1994). By contrast, snake and mongoose muscle nAChR's achieve their resistance primarily by glycosylation of residues in the vicinity of the ligand recognition site (positions 189 and 111 in snake and 187 in the mongoose) (Kreienkamp et al., 1993). This allows the smaller agonist molecules to enter the active center gorge while precluding the large α-neurotoxins from forming a stable complex. Hence, several features distinguish the configuration of nAChR and AChE binding sites for acetylcholine, and these differences are very much evident in evolution.

Gene Expression of Acetylcholinesterase and Nicotinic Receptors

Expression of both AChE and the nAChR is intimately controlled by differentiation in both nerve and muscle (Massoulié et al., 1993). In muscle, for example, both AChE and the nAChR increase upon differentiation from myoblasts to myotubes and localize within

Table I. Domains in acetylcholinesterase responsible for ligand specificity

	Ligands	Role	AChE residues	Changes in BuChE
Catalytic triad	Acyl substrates	Attack of carbonyl carbon or phosphyl phosphorus on acyl substrates	$E_{334(327)}$, $H_{447(440)}$, $S_{203(200)}$	None
Acyl pocket of the active center	Substrates phosphorylating inhibitors	Dictates acyl substrate length and phosphorylation stereospecificity	$F_{295(288)}$, $F_{297(290)}$	$F_{288}L$, $F_{290}I$ or V
Choline subsite of the active center	Substrates Edrophonium Ethopropazine Huperzine	Site of organic cation association determines selectivity of ethopropazine for BuChE and huperzine for AChE	$Y_{337(330)}$, $W_{86(84)}$, $E_{202(199)}$	Y or $F_{337}A$
Peripheral anionic site	Propidium Fasciculin	Allosterically regulates catalysis and specificity at the active center	$W_{286(279)}$, $Y_{72(70)}$, $Y_{124(121)}$, $D_{74(72)}$	$W_{286}R$ or A, $Y_{72}N$, $Y_{124}Q$

the synapse after formation of nerve-muscle contacts (cf Buonanno and Merlie, 1986; Rotundo et al., 1995; Cartaud and Changeux, 1993). In C2-C12 muscle cells, parallel expression of AChE and the nAChR occurs during fusion as myoblasts differentiate to myotubes. Both proteins and their encoding mRNA's are present in low quantities in myoblasts (Fuentes and Taylor, 1993); upon differentiation both the encoding mRNA's and subsequently AChE and nAChR proteins increase considerably with the formation of myotubes. However, we observe a departure in mechanism of expression for the increase in nAChE mRNA is transcriptionally controlled, whereas the increase in AChE mRNA arises from stabilization of a rapidly turning over mRNA (Fuentes and Taylor, 1993). These differences have been demonstrated by: (a) comparative rates of transcription using nuclear run-on analysis (Fuentes and Taylor, 1993), (b) rates of transcription following transfection of reporter genes to which the promoter regions of the *AChE* and *nAchR* genes are coupled to a luciferase reporter (Li et al., 1993) and (c) the capacity of only the AChE mRNA to show superinduction upon treatment with cycloheximide (Fuentes and Taylor, 1993).

Compared to transcriptional regulation, little is known about the signalling processes which stabilize or destabilize specific mRNA's. To this end we have examined factors which may influence AChE mRNA stability. Our initial findings show that AChE mRNA turns over with a very short half-time (<30 min) in the myoblasts. Upon fusion to myotubes, the half time increases to ~4 hrs. Message stabilization results in a corresponding increase in mRNA. Superinduction is most evident during the initial fusion process, immediately after exposing the cells to differentiation conditions (days 1 and 2). This, in turn, suggests that protein synthesis inhibitors may block the formation of a rapidly turning over protein that destabilizes the AChE mRNA. Experiments are now under way to identify the cis elements in the mRNA responsible for AChE mRNA instability.

Upon examining factors responsible in the signalling pathway for the increase in AChE mRNA associated with differentiation, we observed that L-type Ca^{2+} channel inhibitors such as nifedipine, but not N-type such as the ω-conotoxins, block the differentiation induced increase in AChE mRNA (Luo et al., 1994). Moreover, ryanodine and inhibitors of the ryanodine channel block the mRNA increase at concentrations in the nanomolar range (Table II). This clearly implicates the L-type Ca^{2+} channel-ryanodine receptor complex in providing the release of intracellular Ca^{2+} which serves in some capacity to stabilize the

Table II. mRNA and functional levels of AChE and nAChR in C2-C12 myoblasts and myotubes after treatment with Ca^{2+} channel ligands

	mRNA Levels and GeneProduct (% of Control)[#]							
	Day 0 Myoblasts Untreated	Day 3 Myotubes Untreated	Day 3 Ryanodine (100nM)	Day 3 FLA (10 μM)	Day 3 Nifedipine (5 μM)	Day 3 Nifedipine (10 μM)	Day 3 Diltiazem (10 μM)	Day 3 Verapamil (10μM)
AChE								
mRNA	6±2	100	24±5	39±5	42±14	13±6	28±9	31±12
activity	6±2	100	26±2	34±3	32±2	13±4	43±12	29±9
nAChR								
mRNA	21±5	100	101±13	128±15	97	ND	ND	ND
α-toxin	8±0.3	100	89±1	ND	ND	ND	ND	ND

The labeled antisense mRNA probes protected by the mRNA were analyzed by densitometric analysis normalized with respect to the density of U1 protected bands. Data are presented as percentage of maximum (100%) mRNA levels taken from control cells differentiated for three days. Unless otherwise specified, values reported are the means ± SEM averaged for at least four independent determinations.
AChE was measured by the Ellman assay using acetylthiocholine as a substrate and was expressed as nmoles/min/mg cell protein. nAChR was measured as $[^{125}I]$-α-bungarotoxin sites/mg cell protein.
ND - Not determined.
[#]Data from Luo et al., 1994.

AChE mRNA. We suspect that Ca^{2+} is inhibitory to the production of specific mRNA destabilizing factors, but we have yet to obtain definitive evidence on this question. Under the same conditions, no inhibition of nAChR receptor production is observed with either nifedipine or ryanodine (Luo et al., 1994).

Signalling through calcium suggests that Ca^{2+}-dependent phosphorylation or dephosphorylation steps might serve to modify certain *trans*-acting elements which affect mRNA stability. Our recent studies show that rapamycin and cyclosporine specifically altered AChE during development (Z. Luo, unpublished observations). These agents suggest that calcineurin and certain low molecular weight, ryanodine-sensitive channel binding proteins may be involved in the signal transduction pathway leading to mRNA stabilization.

The influence of Ca^{2+} channel inhibitors and inhibitors which affect "downstream" Ca^{2+} modulated proteins appear specific for the AChE mRNA and hence nAChR and AChE gene expression differ both in the signalling cascade and the functional level (transcriptional versus post-transcriptional) at which gene expression is influenced. Hence, although nAChR and AChE expression may appear to be coordinated, regulation of expression involves distinct signalling processes and control points in the biosynthesis of the two proteins.

ACKNOWLEDGMENTS

Supported by USPHS Grants GM18360 and 24437 and DAMD 17-91C-1058.

REFERENCES

Barchan, D., Kochalsky, S., Neumann, D., Vogl, Z., Ovadia, M., Kochva, E., and Fuchs, S., 1992, *Proc. Natl. Acad. Sci. USA* 89:7717-7721.
Berman, H.A., Becktel, W., and Taylor, P., 1981, *Biochemistry* 20:4803-4810.

Blout, P., and Merlie, J.P., 1989, *Neuron* 3:349-357.

Buonanno, A., and Merlie, J.P., 1986, *J. Biol. Chem.* 261:11482-11455.

Changeux, J.P., 1993, *Eur. J. Neuroscience* 5:191-202.

Cartand, J., and Changeux, J.P., Galzi, J.L., Devillers-Thiery, A., and Bertran, D., 1992, *Quart. Rev. of Biophys.* 25:395-432.

Cohen, J.B., Sharp, S.D., Liu, W.S., 1991, *J. Biol. Chem.* 266:23354-23364.

Czajkowski, C., Kaufmann, C., and Karlin, A., 1993, *Proc. Natl. Acad. Sci. USA* 90:6285-6289.

Epstein, D.A., Berman, H.A., and Taylor, P., 1979, *Biochemistry* 18:4749-4754.

Fuentes, M.E., and Taylor, P., 1993, *Neuron* 10:679-687.

Gibney, G., Camp, S., Dionne, M., MacPhee-Quigley, K., and Taylor, P. 1990, *Proc. Natl. Acad. Sci. USA* 87:7546-7550.

Harel, M., Schalk, I., Ehret-Sabatier, L., Bouet, F., Goldner, M., Silman, I., and Sussman, J.L., 1993, *Proc. Natl. Acad. Sci. USA* 90:9031-9035.

Karlin, A., 1993, *Curr. Opinion in Neurobiol.* 3:299-309.

Karlsson, E., Mougma, P.M., and Rodriquez-Ithurralde, D., 1984, *J. Physiol.* (Paris) 79:232-240.

Kreienkamp, H.-J., Sine, S.M., Maeda, R.K., and Taylor, P., 1994, *J. Biol. Chem.* 269:8108-8114.

Kreienkamp, H.-J., Sine, S.M., and Taylor, P., 1995a, submitted.

Kreienkamp, H.-J., Maeda, R., Sine, S.M., and Taylor, P., 1995b, submitted.

Le Du, M.H., Marchot, P., Bougis, P.E., and Fontecilla-Camps, J.C., 1992, *J. Biol. Chem.* 267:22172-22130.

Li, Y., Camp, S., Rachinsky, T., Bongiorno, C., and Taylor, P., 1993, *J. Biol. Chem.* 268:3563-3572.

Luo, Z., Fuentes, M.E., and Taylor, P., 1994, *J. Biol. Chem.* 269:27216-27223.

Marchot, P., Khelif, A., Ji, Y.M., Mansuelle, P., and Bougis, P.E., 1993, *J. Biol. Chem.* 268:12458-12467.

Massoulié, J., Pessementi, L., Bon, S., Krejei, E., and Vallette, F.M., 1993, *Prog. Neurobiol.* 41:31-91.

Nair, H.K., Seravalli, J., Arbuckle, T. and Quinn, D.M., 1994, *Biochemistry* 33:8566-8576.

Neumann, D., Barchan, D., Horowitz, M., Kochva, E., and Fuchs, S., 1989, *Proc. Natl. Acad. Sci. USA* 86:7255-7259.

Ordentlich, A., Barak, D., Kronman, C., Flashner, Y., Leitner, M. and Shafferman, A., 1993, *J. Biol. Chem.* 267:14270-14274.

Pickering, N.A., Berman, H.A., and Taylor, P., 1994, this volume.

Radić, Z., Duran, R. Dajas, F., Vellom, D.C., Li, Y., and Taylor, P., 1994, *J. Biol. Chem.* 269:11233-11239.

Radić, Z., Gibney, G., Kawamoto, S., MacPhee-Quigley, K., Bongiorno, C., and Taylor, P., 1992, *Biochemistry* 31:9760-9767.

Radić, Z., Pickering, N., Vellom, D.C., Camp, S. and Taylor, P., 1993, *Biochemistry* 32:12074-12084.

Radić, Z., Quinn, D.M., Vellom, D.C., Camp, S., and Taylor, P. this volume.

Rosenberry, T.L., Eastman, J., and Haas, P. this Volume.

Rotundo, R.R., Rossi, S.G., Gudinho, R.O., Vazquez, A.E., and Trivedi, B. this volume.

Saxena, A., Qian, N., Kovach, I.M., Kozikowski, A.P., Pong, Y.P., Radić, Z., Vellom, D.C., Doctor, B.P. and Taylor, P., 1994, *Protein Science*, in press.

Sine, S.M., 1993, *Proc. Natl. Acad. Sci. USA* 90:9436-9440.

Sussman, J.L., Harel, M., Frolow, F., Oefner, C., Goldman, A., and Silman, I., 1991, *Science* 253:872-879.

Taylor, P., and Radić, Z., 1994, *Ann. Rev. Pharmacol.* 34:281-320.

Taylor, P., and Lappi, S., 1975, *Biochemistry* 14:1989-1997.

Unwin, M., 1993, *Cell* 72:31-41.

Vellom, D.C., Radić, Z., Li, Y., Pickering, N.A., Camp, S, and Taylor, P., 1993, *Biochemistry 32:12-17.*

BUTYRYLCHOLINESTERASE TRANSCRIPTION START SITE AND PROMOTER

Omar Jbilo, Jean-Pierre Toutant, Arnaud Chatonnet, and Oksana Lockridge

INRA
Montpellier 34060, France
Eppley Institute
University of Nebraska Medical Center
Omaha, Nebraska 68198-6805

INTRODUCTION

Butyrylcholinesterase (BCHE) and acetylcholinesterase (ACHE) genes are expressed independently in adult tissues. This can be inferred from the fact that human BCHE mRNA levels are high in liver, lung, and brain while human ACHE mRNA levels are high in brain and muscle (Jbilo et al, 1994). Other evidence is that AChE and BChE activities do not correlate with each other (Edwards and Brimijoin, 1982). However, it has been suggested that BChE activity regulates the expression of AChE during development of the chicken neural tube (Layer, 1991). The explanation for the difference in expression of BChE and AChE must lie in their promoters. In the present work we have identified the minimal promoter for human and rabbit BCHE (Jbilo et al, 1994).

The gene for human BCHE was isolated and mapped by Arpagaus et al (1990). Human BCHE is the largest cholinesterase gene with 73 to 80 kb. The BCHE gene is large because its introns are large; intron one contains 6.5 kb, while introns two and three each contain approximately 32 kb. The BCHE gene has a minimum of 4 exons. Exons 2, 3, and 4 encode a 28 amino acid signal peptide and a protein of 574 amino acids. Exon 1 was only partially characterized by Arpagaus et al (1990). It was known that exon 1 contained at least 120 bp. However, it was not known how far exon 1 extended in the 5' direction and it was not known whether there was an additional upstream exon. The present work has determined that exon 1 extends an additional 29 bases upstream from the known cDNA boundary to give a total of 149 bp in exon 1 of human BCHE. Rabbit BCHE was found to contain a total of 131 bp in exon 1. No additional upstream exon has been found. The minimal promoter activity was mapped to a region less than 200 bp upstream of the transcription start site in human BCHE and 500 bp in rabbit BCHE.

Enzymes of the Cholinesterase Family, Edited by Daniel M. Quinn et al.
Plenum Press, New York, 1995

METHODS

Primer Extension Analysis

Oligonucleotide primers complementary to exon 1 regions were labeled with P32 and annealed to RNA from human liver or rabbit heart. Primers were extended with reverse transcriptase. Elongated products were fractionated through a 6% polyacrylamide gel containing 8 M urea.

Amplification of the 5'-Ends of Human and Rabbit cDNA

The 5'-AmpliFINDER RACE kit from Clontech was used to amplify unknown sequences at the 5' ends of cDNA. Total RNA from human liver and rabbit heart was the starting material for synthesis of BCHE cDNA. First strand cDNA was synthesized with a BCHE primer complementary to exon 2. The anchored cDNA was amplified by PCR, cloned into a plasmid, and sequenced.

RNase Protection

Genomic clones including 10 kb of 5' flanking sequence were isolated, mapped, and partially sequenced. The 2 kb region immediately upstream of the known cDNA boundary was used for RNase protection experiments. Four fragments from this 2 kb region were cloned into a plasmid for the purpose of producing radiolabeled cRNA. The four P32-cRNA probes were hybridized with human liver RNA; a separate set of four cRNA probes was hybridized with rabbit heart RNA. The hybridized RNA was digested with RNase A and RNase T1, thus degrading single stranded RNA but leaving double stranded RNA intact. The size of protected fragments was determined by electrophoresis through 5% polyacrylamide gels in 6 M urea.

Northern Blot Analysis

Northern blots of human tissue mRNA were purchased from Clontech. Blots were hybridized with exon specific probes of human BCHE. Northern blots of rabbit tissues were prepared from mRNA isolated with the Fast Track mRNA isolation kit (Invitrogen).

Determination of Minimal Promoter Length

Regions of DNA to be tested for promoter activity were cloned into a plasmid on the 5' side of the chloramphenicol acetate transferase (CAT) gene. Six human BCHE fragments varying in length from 58 to 1076 bp were tested. A 570 bp fragment of rabbit DNA was tested for promoter activity. Promoter activity was measured as CAT activity in transfected HeLa cells.

RESULTS AND DISCUSSION

A diagram of the gene for human BCHE with our latest information about the promoter and transcription start site is in Figure 1.

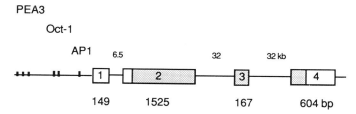

Figure 1. Gene for human BCHE. The minimal promoter is the 200 bp region immediately upstream of exon 1. The AP1 site at -71 is within the minimal promoter. Two Oct-1 sites at -403 and -494 as well as three PEA3 sites at -2169, -2241, and -2260 are upstream of the minimal promoter. The four exons code for hydrophilic, tetrameric G4 BCHE. Exon 1 contains 149 bp, exon 2 has 1525 bp, exon 3 has 167 bp, and exon 4 has 604 bp. The stippled regions encode 574 amino acids. The 28 amino acid signal peptide is within exon 2 as is the catalytic triad, Ser 198, His 438, and Glu 325. Two polyA addition sites are in exon 4. Intron lengths are approximately 6.5, 32, and 32 kb for introns 1, 2, and 3 respectively.

Transcription Start Site

Figure 1 shows that exon 1 is located exactly in the position where it was placed by Arpagaus et al (1990). New information from primer extension, 5' amplification, and RNase protection has shown that exon 1 has a total of 149 bp. This is 29 bp longer than was known from sequencing of our cDNA clones (McTiernan et al, 1987). There is a single transcription start site in a homologous position in the human and rabbit BCHE genes. No upstream exons were found.

Northern Blots

The most abundant BCHE transcript had a size of 2.5 kb in human tissues and a size of 3.5 kb in rabbit tissues. The size of 2.5 kb for human BCHE mRNA is consistent with the size of 2446 bases calculated from the present data and the cDNA sequence. There are 157 bases from the cap site to the initiator ATG, 84 bases in the signal peptide, 1722 bases coding for amino acids 1 through 574, and 483 bases in the 3'-untranslated region from the end of the coding sequence to the beginning of the poly(A)tail following the second poly(A) addition site. The first poly(A) addition site is 180 bases upstream and would yield an mRNA size of 2266 bases. Jbilo et al (1994) did not find a human mRNA with 2.3 kb, suggesting that this site, if used, has several hundred adenylic acids in its poly(A) tail.

The larger size of rabbit BCHE mRNA, compared to human BCHE mRNA, must be due to a longer 3'-untranslated region in rabbit BCHE, since their 5'-untranslated regions have the same length.

Initiator Element

The DNA sequence around the transcription start site in human and rabbit BCHE is shown in Figure 2.

There is no TATA box 25 to 30 nucleotides upstream of the start site. Thus, the BCHE promoter is a TATA-less promoter. In place of a TATA box, BCHE has an initiator element that directs start site placement and initiates transcription. This initiator element overlaps the transcription start site (Smale and Baltimore, 1989).

The sequence of the initiator element in BCHE, in mouse ACHE and in other genes is shown in Figure 3.

```
   .         .         .   AP1 -71.         .         .
TGAATCCTGT-AATACTTCTGTAAATGAATCACTATTGCATTTATACCATTAGCCTCTGAA   H
:::::: ::  :::: :: ::::::::::::::   :: :::    ::::::::::::::::
TGAATC-TGGCAATATTTTTGTAAATGAATCA---TT-CAT----ACCATTAGCCTCTGAA   R

   .                                     .         .         .
CAGATT-----------------------------TCAAGTTGCTGCTGCCAACTCTCGCG   H
:::::                              ::::::::::::::::::::::::::
CAGATGTCAGAAAGTAAAGATTGGCTGCCTCTCGATCAAGTTGCTGCTGCCAACTCTCGCG   R

      -1+1
       ⇓ Transcription start                   M  S  V  Q  S  N
AGCTTTGTCAGTAACAGTTGATTGTTACATTCAGTAACACTGAATGTCAGTGCAGTCCAAT   H
:::::::::::::::::::::::::                 :::::::::::::::::::
AGCTTTGTCAGTAACAGTTGATTGT-----------------AATGTCAGTGCAGTCCAAT   R
       ⇑ Transcription start                   M  S  V  Q  S  N
      -1+1

 L  Q  A  G  A  A  A  A  S  C  I  S  P  K  Y  Y  M  I  F  T
TTACAGGCTGGAGCAGCAGCTGCATCCTGCATTTCCCCGAAGTATT-ACATGATTTTCACT   H
::::::::::::::::::::::::::::::::::::: ::::: :: ::::::::: :: :
TTACAGGCTGGAGCAGCAGCTGCATCCTGCATTTCCCTGAAGTGTTTACATGATTTCCAGT   R
 L  Q  A  G  A  A  A  A  S  C  I  S  L  K  C  L  H  D  F  Q

 P   C  K  L  C  H  L  C  C  R  E  S
CCTT-GCAAACTTTGCCATCTTTGTTGCAGAGAATCGG INTRON 1                 H
:::: : :::::::: :::::::: :::::::  ::::
CCTTAG-AAACTTTATCATCTTTGCTGCAGAG--TCGG INTRON 1                 R
S  L   E  T  L  S  S  L  S  Q  S   R
```

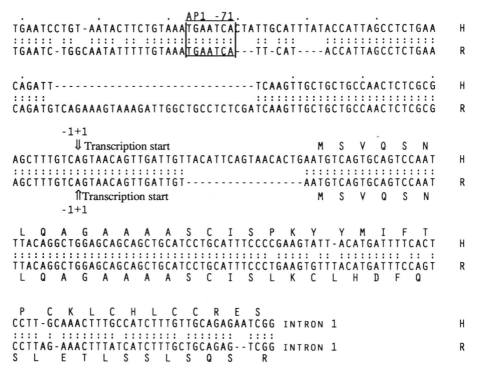

Figure 2. Transcription start site in human and rabbit BCHE. The top line shows the human (H) sequence and the bottom line the rabbit (R) BCHE sequence. The cap site in both sequences is numbered +1 (A+1). Amino acids are shown starting from the methionine located 69 amino acids before the N terminus of mature BChE, because these 69 codons are in-frame with the coding sequence. However, the functional translation start site is most likely the methionine located 28 amino acids before the N terminus, in exon 2, which is not included in this figure. The AP1 binding site at nucleotide -71 is boxed.

The motif TCA(+1)NT is the minimal sequence conserved around the transcription start sites in genes with well characterized initiator elements. Mouse ACHE also lacks a TATA box, although *Torpedo californica* ACHE and *Drosophila* ACHE have potential TATA boxes (Li et al, 1993).

							+1								
TdT	G	C	C	C	T	C	A	T	T	C	T	G	G	A	G	A	G
AdML	G	T	C	C	T	C	A	C	T	C	T	C	T	T	C	C	G
LFA	A	T	G	A	T	C	A	T	T	T	T	C	C	T	C	T	T
MOUSE ACHE	G	C	T	G	T	C	A	C	T	G	T	C	G	G	C	T	C
HUMAN BCHE	T	T	T	G	T	C	A	G	T	A	A	C	A	G	T	T	G
RABBIT BCHE	T	T	T	G	T	C	A	G	T	A	A	C	A	G	T	T	G

Figure 3. Initiator element sequences. Boxed nucleotides are conserved. A minimal conserved motif TCA(+1)NT is found around transcription start sites (A+1). TdT, terminal deoxynucleotidyltransferase. AdML, adenovirus major late promoter. LFA, human lymphocyte function-associated antigen-1. Mouse AChE, acetylcholinesterase.

Minimal Promoter

The minimal promoter was defined as the smallest DNA fragment that promoted CAT activity. A 200 bp region immediately upstream of the transcription start site of human BCHE promoted CAT activity. In rabbit BCHE the minimal promoter was a 500 bp region immediately upstream of the transcription start site.

Transcription Factor Binding Sites

Both the human and rabbit BCHE genes had the same transcription factor binding sites in the same positions. The fact that the binding sites were conserved in both species strengthens the likelihood that these binding sites are real. An AP1 site (TGAATCA) was located at -71, two Oct-1 sites (ATTTGTAT and ATTTGCAT) were at -403 and -494, and three PEA3 sites (AGGAAG) were at -2169, -2241, and -2260. Many developmentally regulated mammalian genes have PEA3 binding motifs in their regulatory regions (Martin et al., 1992). PEA3 and AP1 function synergistically and cooperatively for efficient activation of transcription (Wasylyk et al, 1990). The Oct-1 transcription factor regulates ubiquitously expressed genes (Schöler et al, 1989), a fact which is consistent with our finding that BCHE mRNA was expressed in all tissues tested with the exception of human placenta (Jbilo et al, 1994).

Conclusion

The binding sites for potential transcription factors are different in BCHE and ACHE, except for AP1 which is present in *Torpedo* ACHE (Ekström et al, 1993) as well as in human and rabbit BCHE. The differences in binding sites for transcription factors could explain the differences in the expression pattern of the two enzymes. In summary, it appears that the BCHE 5' flanking region possesses several binding sites for transcription factors that could account for its very early expression during development and cell proliferation.

ACKNOWLEDGMENTS

This work was supported in France by grants from the Ministére de la Recherche et de la Technologie (91-T-0439), the Institut National de la Recherche Agronomique and Association Française contre les Myopathies (MNM93-1250), and in the United States by U.S. Army Medical Research and Development Command Grants DAMD17-91-Z-1003 and DAMD17-94-J-4005 (to O.L.).

REFERENCES

Arpagaus, M. , Kott, M., Vatsis, K.P., Bartels, C.F., La Du, B.N., and Lockridge, O., 1990, Structure of the gene for human butyrylcholinesterase. Evidence for a single copy, *Biochemistry* 29:124-131.

Edwards, J.A., and Brimijoin, S., 1982, Divergent regulation of acetylcholinesterase and butyrylcholinesterase in tissues of the rat. *J. Neurochem.* 38: 1393-1403.

Ekström, T.J., Klump, W.M., Getman, D., Karin, M., and Taylor, P., 1993, Promoter elements and transcriptional regulation of the acetylcholinesterase gene. *DNA Cell Biol.* 12: 63-72.

Jbilo, O., Bartels, C.F., Chatonnet, A., Toutant, J-P, and Lockridge, O., 1994, Tissue distribution of human acetylcholinesterase and butyrylcholinesterase messenger RNA. *Toxicon* 32: In Press.

Jbilo, O., Toutant, J-P, Vatsis, K.P., Chatonnet, A. and Lockridge, O., 1994, Promoter and transcription start site of human and rabbit butyrylcholinesterase genes. *J. Biol. Chem.* 269: 20829-20837.

Layer, P.G., 1991, Cholinesterases during development of the avian nervous system. *Cell. Mol. Neurobiol.* 11: 7-33.

Li, Y., Camp, S., Rachinsky, T.L., Bongiorno, C., and Taylor, P., 1993, Promoter elements and transcriptional control of the mouse acetylcholinesterase gene. *J. Biol. Chem.* 268: 3563-3572.

Martin, M.E., Yang, X.Y., and Folk, W.R., 1992, Expression of a 91-kilodalton PEA3-binding protein is down-regulated during differentiation of F9 embryonal carcinoma cells. *Mol. Cell. Biol.* 12: 2213-2221.

McTiernan, C., Adkins, S., Chatonnet, A., Vaughan, T.A., Bartels, C.F., Kott, M., Rosenberry, T.L., La Du, B.N., and Lockridge, O., 1987, Brain cDNA clone for human cholinesterase. *Proc. Natl. Acad. Sci.* 84: 6682-6686.

Schöler, H.R., Hatzopoulos, A.K., Balling, R., Suzuki, N., and Gruss, P., 1989, A family of octamer-specific proteins present during mouse embryogenesis: evidence for germline-specific expression of an Oct factor. *EMBO J.* 8: 2543-2550.

Smale, S.T. and Baltimore, D., 1989, The "Initiator" as a transcription control element. *Cell* 57: 103-113.

Wasylyk, B., Wasylyk, C., Flores, P., Begue, A., Leprince, D. and Stehelin, D., 1990, The c-ets proto-oncogenes encode transcription factors that cooperate with c-Fos and c-Jun for transcriptional activation. *Nature* 346: 191-193.

THE C-TERMINAL ALTERNATIVE REGIONS OF ACETYLCHOLINESTERASE

Jean Massoulié, Alain Anselmet, Suzanne Bon, Françoise Coussen, Eric Krejci, and Claire Legay

Laboratoire de Neurobiologie Moléculaire et Cellulaire
CNRS URA 1857
Ecole Normale Supérieure
46 rue d'Ulm
75005 Paris, France

ALTERNATIVE SPLICING OF THE 3' EXONS AND AChE POLYMORPHISM

Vertebrates possess a single gene for acetylcholinesterase (AChE, E.C. 3.1.1.7), but this gene generates multiple types of transcripts through a combination of different processes: the choice of several promoters; alternative splicing in the 5' untranslated region; alternative splicing at the 3' extremity of the coding region; the choice of several polyadenylation sites, including variable lengths of 3' untranslated sequence (Ekström et al., 1993). The choice of promoter reflects the regulation of transcription in different cell types, and the choice of polyadenylation site may determine in part the stability of the mRNAs, but the resulting protein is not affected by these processes.

In this chapter, we will focus on the significance of the distinct short C-terminal regions, which are encoded by alternative exons and associated with the common catalytic domain. Depending on their distinct C-termini, the catalytic subunits then undergo different post translational modifications, and generate molecular forms with specific quaternary structures and anchoring properties (Massoulié et al., 1993). The resulting variety of molecular forms allows several modes of attachment of the enzyme, and thus probably ensures an optimal positioning of the enzyme in synaptic structures. These splicing choices are developmental controlled, tissue-specific, and may differ between species, as discussed later.

THE 3' SPLICING PATTERN OF THE AChE GENE

The structure of the mammalian AChE genes has been analyzed in the case of man, mouse (Li et al., 1991), and rat (Legay et al., 1993b). Three different types of coding

Enzymes of the Cholinesterase Family, Edited by Daniel M. Quinn et al.
Plenum Press, New York, 1995

Figure 1. Splicing pattern of the 3' region of the murine AChE gene (after Li et al., 1991 and Legay et al., 1993b, not to scale). Non coding regions are represented by narrow rectangles and coding regions by wide rectangles: dark rectangles correspond to the common catalytic domain and differently hatched rectangles to the three possible C-terminal regions. The thin lines represent constitutive introns; constitutive splicing is schematized by continuous lines and alternative splicing by dotted lines. Polyadenylation sites are indicated by black triangles (pA).

sequences may be generated as schematically illustrated in Figure 1, and called R, H and T (Massoulié et al., 1992). All three transcripts seem to terminate with the same polyadenylation sites, and contain varying lengths of 3' untranslated sequences.

In the "read through", or R transcript, the sequence following the last common exon (exon 4) is maintained, apparently without splicing. In the "hydrophobic", or H transcript, exon 4 is spliced to exon H, which encodes a signal sequence for attachment of a glycophosphatidylinositol (GPI) anchor. In the "tailed", or T transcript, exon 4 is spliced to exon T, downstream of exon H. Exon T encodes a C-terminal region which allows the quaternary association of catalytic AChE subunits with collagenic or hydrophobic subunits, generating the collagen-tailed and hydrophobic-tailed forms.

AChE transcripts may be characterized by RT-PCR, using specific primers located in the R, H and T regions. The fetal rat liver was thus shown to possess all three types of mRNAs (Legay et al., 1993b). The organization of the AChE mammalian gene implies that R transcripts also contain the downstream H and T exons, and that H transcripts contain the T exon. R transcripts resemble unspliced pre-messenger RNA, except that constitutive introns have been removed.

In *Torpedo*, three transcripts R, H and T, constructed in the same manner, have also been observed (Sikorav et al., 1988; Maulet et al., 1990). The gene is, however, much longer, and additional polyadenylation sites may be used, so that the observed R transcript, for example, did not contain the H or T sequences.

AChE TRANSCRIPTS AND DIFFERENCIATION OF MAMMALIAN MUSCLE

We studied the splicing pattern of AChE mRNAs in the mouse diaphragm, during development (C. Legay, M. Huchet, J. Massoulié and J.P. Changeux). RT-PCR showed that the diaphragm of 14-day mouse embryos contains mostly T transcripts, but also detectable levels of R and H transcripts. Their proportions were determined by competitive PCR, using as competitor a deleted gene fragment, producing smaller, distinguishable amplification products. We thus found that the R transcript represented about 5% of AChE mRNAs, and the H transcript only less than 1%. This is unusual for two reasons. Firstly, the R transcripts had only been observed previously in hematopoietic organs, e.g. bone marrow and liver and erythroleukemia cells, and their proportion was lower than that of H transcripts (Li et al., 1991; Legay et al., 1993b). This suggests that R transcripts might result from a splicing defect, when the synthesis was oriented towards H mRNAs. In the embryonic diaphragm, however, R transcripts seem to be produced by a specific biosynthetic pathway: during development, the minor R and H transcripts progressively disappeared, so that only T mRNAs were expressed at the time of birth, as previously found in mature rat muscle (Legay et al., 1993a). In situ hybridization showed that the T transcript became progressively focalized in 18-day embryos, the AChE mRNAs are clearly concentrated in the zone of neuromuscular junctions of the diaphragm and other skeletal muscles.

The presence of R transcripts was also observed in cultures of C_2 cells. Surprisingly, the level of R transcript was even higher in differentiated myotubes than in myoblasts.

ARE R TRANSCRIPTS PHYSIOLOGICALLY SIGNIFICANT?

There is no homology between the peptides encoded by *Torpedo* and mammalian R sequences. Whereas the *Torpedo* sequence predicts a rather long C-terminal peptide (60 amino acids) containing a hydrophobic region, the rat sequence predicts a shorter, non hydrophobic peptide (29 amino acids). A comparison of the mammalian sequences indicates a significant divergence, even between rat and mouse.

While the H and T peptides both contain cysteine residues which are involved in inter subunit disulfide bridges, the R sequences contains no cysteine, suggesting that, if expressed, the corresponding catalytic subunit would remain monomeric. Because its existence has not been demonstrated *in vivo*, we are developing antibodies specifically directed against the R peptide, which should be useful to detect and characterize this enzyme form. Our analysis of AChE transcripts suggests that it should be present in embryonic muscle and C_2 cell cultures.

THE H PEPTIDE CONTAINS A GPI ADDITION SIGNAL

The H peptides of *Torpedo* and mammalian AChEs do not show any significant homology, in contrast with the catalytic domain of the enzyme. They possess, however, important common features. Their C-terminal extremity constitutes a GPI addition signal and they contain a cysteine residue that forms an interchain disulfide bridge, so that H subunits produce GPI-anchored dimers in both species (Silman and Futerman, 1987).

The AChE H_C sequences constitute extremely favourable models for the analysis of GPI addition signals. The structural features of GPI signals have previously been analyzed by mutagenesis, principally in two model systems, based on the DAF protein and on placental alkaline phosphatase, by the groups of I. Caras and S. Udenfriend (Moran and Caras, 1991a, b, 1992; Gerber et al., 1992; Kodukula et al., 1992, 1993). The GPI signal is composed of two elements, a C-terminal hydrophobic sequence, and an upstream cleavage/attachment site. Shortly after completion of the polypeptide chain, a C-terminal fragment is cleaved, and a pre-assembled glycolipid is linked through an amide bond to the new C-terminal residue, called ω. The criteria proposed by the two groups for a potential cleavage site are partially divergent. Both groups agree that ω must be a small residue (G, A, S, V, D, N) and that proline is not allowed at ω or $\omega+1$ positions. In addition, Caras and coll. have essentially emphasized the fact that $\omega+1$ should also be a small residue, and that ω should be located within a window, from -10 to -12 residues upstream of the hydrophobic sequence. On the other hand, Udenfriend and coll. have emphasized the fact that $\omega+2$ must be, in the order of decreasing efficiency, A, G, S, or less favourably T, V, C, other residues being excluded. The importance of the $\omega+2$ position was also recently recognized by Moran and Caras (1994). The $\omega/\omega+1/\omega+2$ residues define a triplet cleavage/attachment site.

An inspection of the rat H peptide shows the presence of a pair of small residues at -12 from the hydrophobic sequence, but this is followed by a proline at the $\omega+2$ position and therefore does not satisfy the consensus proposed by Udenfriend and coll. It also contains two pairs of overlapping triplets, which constitute potential cleavage/attachment sites according to Udenfriend and coll. but are located respectively upstream and downstream of the -10/-12 window. Each pair of overlapping sites could be suppressed, in principle, by the single replacement by a proline of their common residue (at position ω in one site, ω+2 in the other). The results showed that the enzyme is fully glypiated with the upstream sites, and only partially with the downstream sites; in the latter case, the total activity is also decreased. Surprisingly, a small fraction of GPI-anchored can still be detected when both sets of sites are suppressed. In the case of this double mutant, we could restore complete glypiation by introducing at -10/-12 the triplet Ser-Gly-Thr (SGT), but not the triplet Ser-Glu-Gly (SEG). The first one (SGT) corresponds to the cleavage/attachment site of the wild type DAF protein (Moran et al., 1991) and of various functional constructions analyzed by the group of Caras, but the presence of Thr at ω+2 was found to be unfavourable by the group of Udenfriend. Conversely, the second triplet (SEG) apparently satisfies the criteria defined by Udenfriend and coll., but not those of Caras and coll. who insist on a pair of adjacent small residues. In agreement with previous studies on other proteins, we found that the amount of AChE protein exposed at the cell surface that could be labelled by immunofluorescence was directly related to the production of GPI-anchored enzyme.

We also analyzed the *Torpedo* H sequence, by studying the association of the Q_N/H_C chimeric protein with AChE T subunits (see below), and reached similar conclusions, particularly the existence of multiple potential cleavage sites. Thus, we showed that Cys547 constitutes a good possible ω residue, but that a number of other sites also exist downstream of this position. This may explain that Gibney et al. (1988) identified this residue as the C-terminus of the mature GPI G_2 form of *Torpedo* AChE, while other studies suggested that the cleavage site was further downstream: pronase can remove the GPI anchor but not the Cys547-Cys547 disulfide bond (Sikorav et al., 1988) and one serine residue per chain was found in a GPI-peptide (Mehlert et al., 1993). Our results in transfected COS cells suggest that the coexistence of several cleavage sites may be used in *Torpedo* AChE, resulting in a mixture of molecules with different C-termini.

Taken together, our results show that several potentially functional sites may coexist in a GPI signal peptide, that they are not necessarily located at a -10/-12 distance from the hydrophobic region, that different sites may possess different efficiency, and that neither of the sets of criteria previously proposed are fully satisfactory.

It appears likely that the discrepancies between the two sets of criteria largely result from differences in the context of the model proteins. It is therefore essential to concurrently analyze several protein systems. The AChE H subunits offer considerable advantages in this respect, because the characterization of GPI-anchored and unprocessed molecules benefits from the sensitivity of AChE assays, using sedimentation in sucrose gradients and digestion by the specific phospholipase, PI-PLC. This makes it possible to detect and evaluate GPI-anchored AChE, even if it represents less than 5% of the total activity and cannot be visualized by immunofluorescence. In addition, AChEs from different species offer the possibility to compare widely different sequences, derived from isologous genes.

THE T_C PEPTIDE: AMPHIPHILIC PROPERTIES AND QUATERNARY ASSOCIATIONS

The mature T subunits retain their C-terminal T_C peptide of 40 residues. At position -4 from the C-terminus, this T_C peptide contains a cysteine, involved in interchain disulfide bonds between pairs of catalytic subunits in tetramers, or between a catalytic subunit and a structural subunit in collagen-tailed and hydrophobic-tailed forms (for references, see Massoulié et al., 1993). Unlike that of H_C peptides, the sequence of T_C peptides is very well conserved among AChEs of vertebrates. In addition, the C-terminal region of butyrylcholinesterase (BChE) which is not subject to alternative splicing, is closely homologous to the AChE T_C peptide.

The T subunits of AChE, when expressed in COS cells, generate several types of molecular forms: in the absence of associated structural subunits, they produce mostly monomers G_1 and dimers G_2 together with a smaller proportion of tetramers G_4. The G_1 and G_2 forms are amphiphilic, as indicated by the influence of detergents on their sedimentation coefficient; they correspond to amphiphilic forms of type II, which are naturally abundant in the muscles and nervous tissues of mammals and birds (Bon et al., 1991). When co-expressed with the collagenic subunit from *Torpedo*, Q, the T subunits of *Torpedo* or mammalian AChE are assembled into collagen-tailed forms. The amphiphilic properties of the G_1 and G_2 forms, as well as the capacity of T subunits to form homo- or hetero-oligomers, depend on the presence of the T_C peptide. This is clearly shown by the fact that a truncated molecule, obtained by introducing a stop codon at the beginning of the T exon, produced only non-amphiphilic monomers (Duval et al., 1992a).

Because the T_C peptide does not contain any classical hydrophobic sequence, we examined whether the amphiphilic properties might result from the post-translational addition of hydrophobic groups, as in the case of the H_C peptide, which induces the addition of a GPI anchor. We therefore chemically coupled synthetic T_C peptides to non amphiphilic tetramers of *Electrophorus* AChE. The resulting modified molecules clearly acquired amphiphilic properties, demonstrating that the structure of the peptide is responsible for these interactions (S. Bon, I. Cornut, J. Dufourcq, J. Grassi and J. Massoulié, in preparation). The primary sequence of the T_C peptide suggest that it may form an amphiphilic α-helix. Indeed, fluorescence spectroscopy and circular dichroism showed that the peptide was partially α-helical in solution, and that this structure was increased in the presence of detergent or phospholipid micelles.

It is thus likely that an amphiphilic α-helix is responsible for the interactions of G_1 and G_2 forms of type II with detergent micelles. It is not certain, however, that this amphiphilic α-helix is sufficient to anchor these forms in phospholipid bilayers and attach them to cell membranes. In fact, immunofluorescence labelling showed that this enzyme remains mostly intracellular in cells transfected with the rat T subunit, both in transient expression in COS cells and in stable expression in rat basoleukemia cells, which produced essentially G_1 and G_2 forms (F. Coussen, J. Massoulié and C Bonnerot, in preparation). It is therefore possible that the monomers and dimers of T subunits essentially represent precursors of higher oligomers: soluble tetramers, hydrophobic-tailed tetramers and collagen-tailed forms.

In the amphiphilic α-helical structure, three conserved tryptophan residues, evenly spaced by seven residues, are exposed on the hydrophobic sector of the helix. These residues contribute importantly to this hydrophobic surface, but may also play a role in quaternary interactions. For example, the tryptophan side chains of two helices might intercalate, forming a "tryptophan zipper". We plan to explore the association between catalytic T subunits and their interaction with structural anchoring subunits by site-directed mutagenesis.

It will be possible to analyze the formation of collagen-tailed forms by co-expression of the AChE T subunit with the collagenic tail subunit, Q, since cDNA clones encoding Q subunits have been obtained from *Torpedo* electric organ (Krejci et al., 1991) and more recently from rat muscle (E. Krejci, J. Sketelj, F. Coussen, S. Bon and J. Massoulié, unpublished results). It has been shown that the N-terminal domain preceeding the collagenic region, Q_N, is involved in the interaction with catalytic T subunits as demonstrated with a chimeric protein, Q_N/H_C, in which it was linked to the H_C peptide. The co-expression of this Q_N/H_C protein with T subunits of *Torpedo* or rat AChE produces GPI-anchored tetramers

(Duval et al.,1992b). We are presently trying to define which structural features of the Q_N domain are necessary and sufficient to attach an AChE tetramer (S. Bon, F. Coussen and J. Massoulié, in preparation). We also explored the GPI addition signal in this system. We thus found that the association between Q_N/H_C and T subunits is strictly dependent on glypiation, indicating that it occurs in a subcellular compartment that is not reached by the unprocessed precursor, which is retained in the reticulum (F. Coussen, S. Bon and J. Massoulié, in preparation). This result confirms, by an entirely different method, the conclusion of Rotundo (1984), that collagen-tailed forms, which are built in the same way, are assembled in the Golgi apparatus.

EVOLUTION AND DISTRIBUTION OF THE DIFFERENT C-TERMINI OF CHOLINESTERASES

In contrast with the *Torpedo* and mammalian AChE genes, invertebrate cholinesterase genes were only found to encode one type of catalytic subunit. For example, *Drosophila* AChE consists exclusively of H subunits. In *Caenorhabditis elegans* , distinct genes, *ace-1* and *ace-2* encode T and H subunits, respectively (Arpagaus et al., 1992, 1994). Thus, the duality of H and T anchoring subunits exists in invertebrates, and the originality of the vertebrate AChE is to combine the two in a single gene. In some vertebrates, however, the AChE gene does not seems to contain the H exon: examination of the genomic sequence located between the last common exon and the T exon, in the quail AChE gene, does not reveal the presence of a possible H exon where it would be expected to occur (A. Anselmet and J. Massoulié, unpublished result). In addition, the BChE genes, which probably arose from a duplication at the origin of vertebrates, have only been reported to produce T subunits (Arpagaus et al., 1990).

Thus, some organisms function with only one type of AChE subunit, either H or T. Furthermore, in species which possess both types, their distribution is quite variable: for example, skeletal muscles of mammals and birds only express T subunits, and their neuromuscular junctions essentially contain collagen-tailed and hydrophobic-tailed forms. On the contrary, *Torpedo* muscles contain essentially GPI-G_2 dimers of H subunits. The multiplicity of anchoring modes is clearly not an absolute requirement of cholinergic transmission, but probably allows a fine adaptation of the synaptic structure, particularly for a precise temporal control of excitation. These physiological aspects are now open for experimental exploration, particularly in transgenic mice.

ACKNOWLEDGMENTS

This work was supported by Centre National de la Recherche Scientifique (CNRS), the Ecole Normale Supérieure (ENS), the Direction des Recherches et Etudes Techniques (DRET), the Association Française contre les Myopathies (AFM). We particularly thank Jacqueline Leroy and Anne Le Goff for their technical assistance and Jacqueline Pons for the preparation of this manuscript.

REFERENCES

Arpagaus, M., Kott, M., Vatsis, K.P., Bartels, C.F., La Du, B.N. and Lockridge, O., 1990, Structure of the gene for human butyrylcholinesterase. Evidence for a single copy. *Biochemistry* 29:124-131.

Arpagaus, M., Richier, P., Bergé, J.B. and Toutant, J.P., 1992, Acetylcholinesterases of the nematode Stein-ernema carpocapsae. *Eur. J. Biochem.* 207:1101-1108.

Arpagaus, M., Fedon, Y., Cousin, X., Chatonnet, A., Bergé, J.B., Fournier, D. and Toutant, J.P., 1994, cDNA sequence, gene structure, and *in vitro* expression of *ace-1*, the gene encoding acetylcholinesterase of class A in the nematode *Caenorhabditis elegans . J. Biol. Chem.* 269:9957-9965.

Bon, S., Rosenberry, T.L. and Massoulié, J., 1991, Amphiphilic, glycophosphatidyl-inositol-specific phos-pholipase C (PI-PLC)-insensitive monomers and dimers of acetylcholinesterase. *Cell. Molec. Neuro-biol.* **11**, 157-172.

Duval, N., Massoulié, J. and Bon, S., 1992a, H and T subunits of acetylcholinesterase from *Torpedo*, expressed in COS cells, generate all types of globular forms. *J. Cell Biol.* 118:641-653.

Duval, N. Krejci, E., Grassi, J., Coussen, F., Massoulié, J. and Bon, S., 1992b, Molecular architecture of acetylcholinesterase collagen-tailed forms; construction of a glycolipid-tailed tetramer. *EMBO J.* 11:3255-3261.

Ekström, T.J., Klump, W.M., Getman, D., Karin, M. and Taylor, P., 1993, Promoter elements and transcriptional regulation of the acetylcholinesterase gene. *DNA and Cell Biol.* 12:63-72.

Gerber, L.D., Kodukula, K. and Udenfriend, S., 1992, Phosphatidylinositol glycan (PI-G) anchored membrane proteins. *J. Biol. Chem.* 267:12168-12173.

Gibney, G., MacPhee-Quigley, K., Thomson, B., Vedvick, T., Low, M.G., Taylor, S.S. and Taylor, P., 1988, Divergence in primary structure between the molecular forms of acetylcholinesterase. *J. Biol. Chem.* 263:1140-1145.

Kodukula, K., Cines, D., Amthauer, R., Gerber, L.D. and Udenfriend, S., 1992, Biosynthesis of phosphatidyli-nositol-glycan (PI-G)-anchored membrane proteins in cell-free systems: cleavage of the nascent protein and addition of the PI-G moiety depend on the size of the COOH-terminal signal peptide. *Proc. Natl. Acad. Sci. USA* 89:1350-1353.

Kodukula, K., Gerber, L.D., Amthauer, R., Brink, L. and Udenfriend, S., 1993, Biosynthesis of glycosylphos-phatidylinositol (GPI)-anchored membrane proteins in intact cells: specific amino acid requirements adjacent to the site of cleavage and GPI attachment. *J. Cell Biol.* 120:657-664.

Krejci, E., Coussen, F., Duval, N., Chatel, J.M., Legay, C., Puype, M., Vandekerckhove, J., Cartaud, J., Bon, S. and Massoulié, J., 1991, Primary structure of a collagenic tail subunit of *Torpedo* acetylcholi-nesterase: co-expression with catalytic subunit induces the production of collagen-tailed forms in transfected cells. *EMBO J.* 10:1285-1293.

Legay C., Bon, S., Vernier, P., Coussen, F. and Massoulié, J., 1993a, Cloning and expression of a rat acetylcholinesterase subunit; generation of multiple molecular forms, complementarity with a *Tor-pedo* collagenic subunit. *J. Neurochem.* 60:337-346.

Legay, C., Bon, S. and Massoulié, J., 1993b, Expression of a cDNA encoding the glycolipid-anchored form of rat acetylcholinesterase. *FEBS Lett.* 315:163-166.

Li, Y., Camp, S., Rachinsky, T.L., Getman, D. and Taylor, P., 1991, Gene structure of mammalian acetylcholi-nesterase. *J. Biol. Chem.* 266:23083-23090.

Massoulié, J., Sussman, J., Doctor, B.P., Soreq, H., Velan, B., Cygler, M., Rotundo, R., Shafferman, A., Silman, I. and Taylor, P., 1992, Multidisciplinary Approaches to Cholinesterase Function. 285-288.

Massoulié, J., Pezzementi, L., Bon, S., Krejci, E. and Vallette, F.M., 1993, Molecular and cellular biology of cholinesterases. *Prog. Neurosci.* 41:31-91.

Maulet, Y., Camp, S., Gibney, G., Rachinsky, T.L., Ekstrom, T.J. and Taylor, P., 1990, Single gene encodes glycophospholipid-anchored and asymmetric acetylcholinesterase forms: alternative coding exons contain inverted repeat sequences. *Neuron* 4:289-301.

Mehlert, A., Varon, L., Silman, I., Homans, S.W. and Ferguson, M.A.J., 1993, Structure of the glycosyl-phos-phatidylinositol membrane anchor of acetylcholinesterase from the electric organ of the electric fish, *Torpedo californica. Biochem. J.* 296:473-479.

Moran, P. and Caras, I.W., 1991a, A nonfunctional sequence converted to a signal for glycophosphatidylinositol membrane anchor attachment. *J. Cell Biol.* 115:329-336.

Moran, P. and Caras, I.W., 1991b, Fusion of sequence elements from non-anchored proteins to generate a fully functional signal for glycophosphatidylinositol membrane anchor attachment. *J. Cell Biol.* 115:1595-1600.

Moran, P. and Caras, I.W., 1992, Proteins containing an uncleaved signal for glycophosphatidylinositol membrane anchor attachment are retained in a post-ER compartment. *J. Cell Biol.* 119:763-772.

Moran, P. and Caras, I.W., 1994, Requirements for glycophosphatidylinositol attachment are similar but not identical in mammalian cells and parasitic protozoa. *J. Cell Biol.* 125:333-343.

Rotundo, R.L., 1984, Asymmetric acetylcholinesterase is assembled in the Golgi apparatus. *Proc. Natl. Acad. Sci. USA* 81:479-483.

Sikorav, J.L., Duval, N., Anselmet, A., Bon, S., Krejci, E., Legay, C., Osterlund, M., Reimund, B. and
    Massoulié, J., 1988, Complex alternative splicing of acetylcholinesterase transcripts in *Torpedo*
    electric organ; primary structure of the precursor of the glycolipid-anchored dimeric form. *EMBO J.*
    7:2983-2993.
Silman, I. and Futerman, A.H., 1987, Modes of attachment of acetylcholinesterase to the surface membrane.
    Eur. J. Biochem. 170:11-22.

DEVELOPMENTAL EXPRESSION OF ACETYLCHOLINESTERASE IN SKELETAL MUSCLE

Zoran Grubič[1] and Armand F. Miranda[2]

[1] Institute of Pathophysiology, Faculty of Medicine
University of Ljubljana, 61105 Ljubljana, Slovenia
[2] Department of Pathology, Columbia University
College of Physicians and Surgeons, New York, New York 10032

During skeletal muscle differentiation *in vivo* and *in vitro* AChE activity fluctuates and there are alterations in the spatial localization of the enzyme within the muscle fiber. Histochemical studies demonstrated AChE activity in the cytoplasm of mononucleated myogenic cells of the chick embryo (Gerebtzoff, 1959). Electron microscopy of the rabbit myodermatome revealed localization of AChE reaction product in mononuclear myoblasts at the nuclear envelope, the endoplasmic reticulum and some tubules of the Golgi complex (Tennyson et al., 1971). Acetylcholine receptor (AChR) was also found in myoblasts at this stage and choline acetyltransferase activity was detectable in neural tube-spinal ganglia (Grubič et al., 1984). Therefore, components of the cholinergic system are expressed in neural tissues and muscle, long before the establishment of functional neuromuscular contacts. It has been hypothesized that at this early stage, AChE and butyrylcholinesterase act as tropic or trophic factors for ingrowing motor axons (Layer et al., 1988). When myotubes fuse to form multinucleate myotubes, there is a significant increase in AChE activity, as observed in studies of avian and mammalian muscle in vivo (Mumenthaler and Engel, 1961; Tennyson et al., 1973) and in culture (Rieger et al., 1980; Inestrosa et al., 1983). The localization of AChE in noninnervated myotubes also varies at different stages of development, and may be species-specific. Uniform diffuse distribution, with no preferential localization was observed in aneurally grown chick (Engel, 1961) and rat (Rieger et al., 1980) cultures, while in quail (Rotundo et al., 1992), mouse muscle C2 cell line (Inestrosa et al., 1982) and *Xenopus* muscle cultures (Moody-Corbett and Cohen, 1981) patchy accumulations of AChE were noted at the sarcolemma. When developing myofibers are innervated, AChE becomes concentrated at the neuromuscular junction, although some enzyme also remains detectable extrajunctionally. The relative amounts and patterns of the different molecular forms of extrajunctional AChE activity in adult muscle differ between species and also vary between different types of skeletal muscle within the same species (reviewed by Toutant and Massoulie, 1987, and Massoulie et al., 1993).

These observations raise many questions concerning the functional significance and molecular mechanisms controlling developmentally-regulated changes of AChE activity and

Enzymes of the Cholinesterase Family, Edited by Daniel M. Quinn et al.
Plenum Press, New York, 1995

its distribution. Alterations may result from regulatory factors that act at distinct developmental stages and control AChE activity selectively in restricted domains within the muscle syncytium. Such explanation may imply that at different stages of myogenesis the expression of AChE is selectively activated or inactivated in different pools of nuclei, so that the patterns of AChE localization reflect the distribution of nuclei that are active at the time of observation. Alternatively, it is conceivable that all myonuclei are synthesizing AChE and that temporal variations in enzyme activity and its localization reflect developmental variations of targeting mechanisms that determine the distribution of the enzyme which is uniformly synthesized by all myonuclei. A combination of both principles might also explain the above phenomena. Recent studies in skeletal muscle contributed significantly toward the elucidation of mechanisms modulating AChE expression.

EXPRESSION OF ACETYLCHOLINESTERASE IN THE SKELETAL MUSCLE PRIOR TO INNERVATION

There is increasing evidence that gene products remain localized in the vicinity of the nuclei responsible for their synthesis. For example, *in vitro* studies of interspecific muscle heterokaryons demonstrated that a component of the surface membrane, the Golgi apparatus mediating its transport, and a sarcomeric myosin heavy chain remain localized in the vicinity of the myonuclei of their origin (Pavlath et al., 1989). Moreover, genetic analysis of segregation of chimeric AChE subunits in allelic heterokaryons produced in avian muscle culture demonstrated that transcription, translation and assembly of AChE molecules occurred close to the nuclei of origin, and that the assembled enzyme becomes selectively inserted into a segment of the surface membrane overlying these nuclei (Rotundo, 1990; Rossi and Rotundo, 1992; Rotundo et al., 1992). In situ hybridization performed in chick muscle culture by Tsim et al. (1992) confirmed the territoriality of AChE transcripts to the nuclei of their origin. The same investigators also found by quantitative autoradiography that *Ache* mRNA is more widely distributed in the perinuclear cytoplasm than transcripts encoding the α-subunit of AChR. The number of AChE transcripts per nuclear domain was estimated to be 176, about six times greater than the α-subunit mRNA of the acetylcholine receptor (AChR). In human muscle-rat spinal cord cocultures, an *in vitro* model introduced by Kobayashi and Askanas (1985), we demonstrated by in situ hybridization and enzyme cytochemistry that both *Ache* mRNA and the active enzyme are present in most mononuclear myoblasts. After myoblast fusion, AChE reaction product was more abundant and distributed all along the length of the myotubes. We also noted, at the time of fusion, that there was intense perinuclear mRNA staining around virtually all myonuclei, indicating upregulation of *Ache* mRNA at this stage of development (Grubič et al., 1994 and in preparation). It could be concluded, therefore, that increased AChE activity observed after myoblast fusion reflects increased mRNA content. Our results are in accord with quantitative studies of Fuentes and Taylor (1993), who demonstrated by Northern blot and the Ellman assay that increased *Ache* mRNA and AChE activity are increased following fusion of C2-C12 mouse myoblasts.

EXPRESSION OF ACETYLCHOLINESTERASE IN THE SKELETAL MUSCLE AFTER INNERVATION

A major change in AChE localization occurs during synaptogenesis, both *in vivo* and in culture. At this stage, focal patches of AChE appear at the developing neuromuscular contacts. Studies of the AChE patches formed on the surface of myotubes in the mouse C2 cell line demonstrated that they consist of tailed asymmetric A_{12} forms that are associated with the basal lamina, bound primarily to the polysaccharide chains of heparan sulfate proteoglycans (Inestrosa et al., 1982; Inestrosa et al., 1992). In studies of adult quail muscle, junctional A_{12} AChE molecules could only be extracted after treatment with purified collagenase and not by extraction buffers that are conventionally used for solubilizing the A_{12} molecular forms of AChE, indicating that the attachment of AChE to the extracellular matrix of the neuromuscular junction (NMJ) is particularly firm (Rossi and Rotundo, 1993). In apparent contradiction with this report, more recent studies in frog muscle showed that more than 80% of junctional AChE, located at the sarcolemma could be solubilized in high ionic strength, detergent-free buffer (Anglister et al., 1994). However, because junctional regions isolated by free-hand dissection always contain a high proportion of extrajunctional muscle, most or all of the extractable enzyme may be extrajunctional, which brings the two studies into mutual accord.

Several studies performed in muscle-nerve cocultures of mammals (Koenig and Vigny, 1978; De la Porte et al., 1993; Kobayashi et al., 1987) and birds (Rubin et al., 1980) suggest that nerve contact is necessary for the formation of AChE patches. However, AChE patches were also found on myotubes in aneural cultures of xenopus and quail (Rotundo et al., 1992) and on older myotubes of the C2 mouse cell line (Silberstein et al., 1982). These studies in aneural muscle cultures indicate that, at least in some species, the muscle cell can independently produce the molecular components needed for the formation of patch-like AChE-basal lamina complexes. Rotundo et al. (1992) observed that AChE patches are formed following the onset of spontaneous muscle contractions, suggesting that contraction-mediated events are triggering their formation. However, the precise molecular mechanisms that could explain both contraction-induced formation of patches and elimination of extrajunctional sarcolemmal patches during maturation of the NMJ are still unclear. Conceivably, spontaneous contractions (i.e. not mediated by the nerve) are sufficient for the formation of unstable AChE-basal lamina complexes, which are transient, unless they are stabilized by nerve-derived factors secreted at the synaptic region. For example, Agrin, a nerve-derived protein that is bound to the basal lamina is capable of inducing the accumulation of AChRs as well as AChE molecules at the region of the NMJ (reviewed by Hall and Sanes, 1993). Burden (1993) recognized the synaptic basal lamina as an ideal location for signals involved in differentiation of the NMJ, since the molecules located there have limited diffusion and their local deposition by either nerve or muscle can provide precise spatial information to the myofiber. His conclusion was based on studies of synapse-specific gene expression, compared with gene expression in syncytial cells of the blastoderm of *D. melanogaster* and the syncytial germ line of *C. Elegans*. The detailed molecular mechanisms underlying the accumulation of AChE in the synaptic AChE-basal lamina complex and how it evolved during evolution are still unknown. In studies performed in innervated human muscle cultures, we observed that nerve-mediated muscle contractions coincide with the downregulation of *Ache* mRNA, virtually disappearing all along the length of the contracting fibers, except for at few nuclei located in the innervation region. This downregulation was accompanied by the appearance of focal patches of AChE activity at the neuromuscular contacts (Grubič et al., 1994 and in preparation). On the basis of these observations we hypothesize that the construction of a stable long-lived AChE-synaptic basal lamina complex is a part of an adaptation of the muscle fiber to the life-long postinnervation period, enabling the supply of sufficient functional AChE at the NMJ, at low rates of energy consuming *Ache* mRNA synthesis (Figure 1).

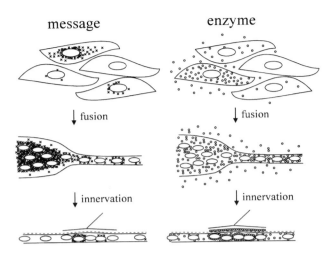

Figure 1. Expression of *Ache* mRNA (message) and AChE activity (enzyme) in innervated culture.

CONTROL OF AChE EXPRESSION IN THE DEVELOPING SKELETAL MUSCLE

Fuentes et al. (1993) studied the transcription of AChE in mouse muscle. They observed that the γ subunit *Achr* mRNA and *Ache* mRNA both increase at the time of fusion, but that their transcriptional regulation appears to be different. They noted that the increase in *Achr* mRNA is due to an increase in the rate of synthesis, whereas the increase in *Ache* mRNA results from the stabilization of the preexisting message. Whether, and if so to what extent a similar mechanism, acting in reverse, leads to *Ache* mRNA downregulation of extrajunctional *Ache* mRNA at onset of contractions remains to be established. In any case, as determined by RT PCR in quail, transcripts are either undetectable or sparse in the noninnervated regions (Jasmin et al., 1993). The half life of *Ache* mRNA estimated in C2-C12 mouse muscle cell line is about 4-5 hours (Fuentes and Taylor, 1993), which is approximately one tenth the half life of the mature enzyme in cultured chick muscle (Rotundo and Fambrough, 1980). Restriction of *Ache* mRNA to junctional region after functional innervation was also demonstrated by in situ hybridization in innervated human muscle cultures (Grubič et al. 1994, and in preparation). We also observed high nonuniformity of nuclear *Ache* mRNA staining at all stages of myogenesis, except at fusion when apparently all nuclei exhibit intensive perinuclear staining. This finding is in accord with the hypothesis that AChE transcription proceeds intermittently, as previously proposed by Jasmin et al. (1993), based on RT PCR studies.

The trafficking of AChE has been studied in developing skeletal muscle of the frog, using recombinant DNA techniques (Ben Aziz et al., 1993; Seidman et al., 1994). It was found that microinjection of constructs containing human AChE DNA, driven by a cytomegalovirus promoter, in oocytes and developing embryos of *Xenopus laevis* resulted in overexpression of human AChE in the myotomes. The heterologous human AChE accumulated within the same subcellular compartments as native frog AChE. There was a four-fold overexpression of AChE at NMJs in 3-day old injected embryos and a 30% increase in the length of the postsynaptic membrane. Besides demonstrating an important role of AChE in vertebrate synaptogenesis, these experiments also show that trafficking and targeting of muscle AChE is an evolutionary conserved process. At first glance this conclusion appears to contradict the earlier observations of Rotundo et al. (1992), who used species-specific antibodies for localization of surface AChE in mosaic fibers, prepared by coculturing quail and mouse myoblasts. They claimed that the mechanism(s) controlling the targeting of AChE to the cell surface overlying the nucleus of origin is very selective and specific. This apparent contradiction is best explained on one hand, by the existence of specific nuclear domains, preventing the expression of AChE derived from one nuclear domain to be processed through a synthetic pathway by nuclei in another domain and, on the other hand, by non-specificity of the synthetic pathway, allowing native and heterologous AChE to be expressed by the same synthetic pathway from nuclei within a particular nuclear domain.

Localization patterns of *Ache* mRNA and catalytically active AChE at different stages of myogenesis *in vitro* discussed in the text are summarized in the scheme in Figure 1.

Most of the mononuclear myogenic cells express variable amounts of *Ache* mRNA, forming a gradient from the nuclear envelope towards the cell periphery. Catalytically active AChE is also present in most of the mononuclear cells, but its distribution is somewhat more uniform than *Ache* mRNA. At sites where myoblasts appear to be actively fusing the cell nuclei, which at this stage have become smaller and more compact, are generally surrounded by large amounts of *Ache* mRNA. In more mature parts of the the differentiating muscle syncytia, most myonuclei continue to express variable amounts of *Ache* mRNA, accompanied by high AChE activity, which is diffusely distributed along the length of the myotube. At this stage of development, most or all of the synthetized AChE is secreted into the surrounding medium. After functional innervation, muscle *Ache* mRNA is downregulated, perhaps as a result of contraction-mediated neural factor(s). In the junctional region, downregulation is presumably prevented by some nerve derived factors so that a small fraction of junctional nuclei appear active, expressing *Ache* mRNA in a nonuniform fashion, as also observed at earlier developmental stages. At the time of functional innervation, AChE is retained in the myofiber and no longer released into the medium. Myonuclei accumulate beneath the nerve contact. AChE, bound to specialized synaptic basal lamina components, accumulates in the synaptic cleft. Some AChE activity remains diffusely distributed along the junctional region of muscle fiber, more concentrated in perinuclear areas and the nuclear envelope. Small amounts of diffusely distributed AChE also persist extrajunctionally. There seem to be species-specific differences in localization and amount of extrajunctional AChE. Multiple neuromuscular contacts, sometimes seen at early stages of innervation, disappear following maturation of a single NMJ. Nonuniform *Ache* mRNA distribution may reflect intermittent, asynchronous transcription at the level of individual nuclei, so that at any given time some nuclei are active, while others are silent.

REFERENCES

Anglister, L., Haesaert, B., and McMahan, U. J., 1994, Globular and asymmetric acetylcholinesterase in the synaptic basal lamina of skeletal muscle. *J. Cell Biol.* 125:183-196.

Askanas, V., Kwan, H., Alvarez, R. B., King Engel, W., Kobayashi, T., Martinuzzi, A., and Hawkins, E. F., 1987, De novo neuromuscular junction formation on human muscle fibres cultured in monolayer and innervated by foetal rat spinal cord: ultrastructural and ultrastructural - cytochemical studies. *J. Neurocytol.* 16:523-537.

Ben-Aziz-Aloya, R., Seidman, S., Timberg, R., Sternfeld, M., Zakut, H., and Soreq, H., 1993, Expression of a human acetylcholinesterase promoter-reporter construct in developing neuromuscular junctions of Xenopus embryos. *Proc. Natl. Acad. Sci.* U.S.A. 90: 2471-2475.

Burden, S. J., 1993, Synapse-specific gene expression, *Trends in Genetics* 9:12-16.

De la Porte, S., Ragueh, F., Eymard, B., Courbin, P., Chapron, J., and Koenig, J., 1993, Effect of sera from myasthenia gravis patients and of alpha- bungarotoxin on acetylcholinesterase during in vitro neuromuscular synaptogenesis. *J. Neurol. Sci.* 117:92-102.

Engel, W. K., 1961, Cytological localization of cholinesterase in cultured skeletal muscle cells. *J. Histochem. Cytochem.* 9:66-72.

Fuentes, M. E., and Taylor, P., 1993, Control of acetylcholinesterase gene expression during myogenesis. *Neuron* 10:679-687.

Gerebtzoff, M. A., 1959, Morphogenetic study of acetylcholinesterase in nervous system and skeletal muscle. In: Cholinesterases, pp. 36-56, Pergamon Press, Inc., Elmsford, New York, N. Y.

Grubič, Z., Tennyson, V. M., Chang, H. W., Kremzner, L. T., and Penn, A. S., 1984, α-bungarotoxin binding to the myotome and choline acetyltransferase activity in the rabbit embryo, *J. Comp. Neurol.* 222: 452-460.

Grubič, Z., Komel, R., Walker, W. F., and Miranda, A. F., 1994, Localization of acetylcholinesterase and its mRNA during early development and innervation of human muscle cocultured with embryonic rat spinal cord. *J. Neurochem.* 63 (Suppl. 1): S32.

Hall, Z. W., and Sanes, J. R., 1993, Synaptic structure and development: the neuromuscular junction. *Cell* 72/*Neuron* 10 (Suppl.): 99-121.

Inestrosa, N. C., Silberstein, L., and Hall, Z. W., 1982, Association of the synaptic form of acetylcholinesterase with extracellular matrix in cultured mouse muscle cells. *Cell* 29:71-79.

Inestrosa, N. C., Miller, J. B., Silberstein, L., Ziskind-Conhaim, L., and Hall, Z. W., 1983, Development and regulation of 16S acetylcholinesterase and acetylcholine receptors in a mouse muscle cell line. *Exp. Cell Res.* 147:393-405.

Inestrosa, N. C., Gordon, H., Esko, J. D., and Hall, Z. W., 1992, Binding of Asymmetric (A12) Acetylcholinesterase to C2 Muscle cells and to CHO mutants defective in glycosaminoglycan synthesis, In: Multidisciplinary approaches to cholinesterase function, pp. 25-32. Eds. A., Shafferman, and B., Velan. Plenum Press: New York and London.

Jasmin, B. J., Lee, R. K., and Rotundo, R. L., 1993, Compartmentalization of acetylcholinesterase mRNA and enzyme at the vertebrate neuromuscular junction. *Neuron* 11:467-477.

Kobayashi, T., and Askanas, V., 1985, Acetylcholine receptors and acetylcholinesterase accumulate at the nerve-muscle contacts of de novo grown human monolayer muscle co-cultured with fetal rat spinal cord. *Exp. Neurol.* 88:327-335.

Kobayashi, T., Askanas, V., and King Engel, W., 1987, Human muscle cultured in monolayer and cocultured with fetal rat spinal cord: importance of dorsal root ganglia for achieving successful functional innervation. *J. Neurosci.* 7:3131-3141.

Koenig, J., and Vigny, M., 1978, Neural induction of the 16S acetylcholinesterase in muscle cell cultures. *Nature* 271:75-77.

Layer, P. G., Alber, R., and Rathjen, F. G., 1988, Sequential activation of butyrylcholinesterase in rostral half somites and acetylcholinesterase in motoneurones and myotomes preceding growth of motor axons. *Development* 102:387-396.

Massoulie, J., Pezzementi, L., Bon, S., Krejci, E., and Vallette, F. -M., 1993, Molecular and cellular biology of cholinesterases. *Prog. Neurobiol.* 41:31-91.

Moody-Corbett, F., and Cohen, M. W., 1981, Localization of cholinesterase at sites of high acetylcholine receptor density on embryonic amphibian muscle cells cultured without nerve. *J. Neurosci.* 1:596-605.

Mumenthaler, M., and Engel, W. K., 1961, Cytological localization of cholinesterase in developing chick embryo skeletal muscle. *Acta Anat. (Basel)* 47:274-284.

Pavlath, G. K., Rich, K., Webster, S. G., and Blau, H. M., 1989, Localization of muscle gene products in nuclear domains. *Nature* 337:570-572.

Rieger, F., Koenig, J., and Vigny M., 1980, Spontaneous contractile activity and the presence of the 16S form of acetylcholinesterase in rat muscle cells in culture: reversible suppressive action of tetrodotoxin. *Dev. Biol.* 76:358-365.

Rossi, S. G., and Rotundo, R. L., 1992, Cell surface acetylcholinesterase molecules on multinucleated myotubes are clustered over the nucleus of origin. *J. Cell Biol.* 119: 1657-1667.

Rossi, S. G., and Rotundo, R. L., 1993, Localization of "Non-extractable" Acetylcholinesterase to the Vertebrate Neuromuscular Junction, *J. Biol. Chem.* 268: 19152-19159.

Rotundo, R. L., and Fambrough, D. M., 1980, Synthesis, transport and fate of acetylcholinesterase in cultured chick embryo cells. *Cell* 22:583-594.

Rotundo, R. L., with the assistance of Gomez, A. M., 1990, Nucleus-specific translation and assembly of acetylcholinesterase in multinucleated muscle cells. *J. Cell Biol.* 110:715-719.

Rotundo, R. L., Jasmin, B. J., Lee, R. K., and Rossi, S. G., 1992, Compartmentalization of acetylcholinesterase mRNA and protein expression in skeletal muscle *in vitro* and *in vivo*: implications for regulation at the neuromuscular junction. In: Multidisciplinary approaches to cholinesterase function, pp. 217-222. Eds. A., Shafferman and B., Velan. Plenum Press: New York and London.

Rubin, L. L., Schuetze, S. M., Weill, C. L., and Fischbach, G. D., 1980, Regulation of acetylcholinesterase appearance at neuromuscular junctions in vitro. *Nature* 283:264-267.

Seidman, S., Ben Aziz-Aloya, Timberg, R., Loewenstein, Y., Velan, B., Shafferman, A., Liao, J., Norgaard-Pedersen, B., Brodbeck, U., and Soreq, H., 1994, Overexpressed Monomeric Human Acetylcholinesterase Induces Subtle Ultrastructural Modifications in Developing Neuromuscular Junctions of Xenopus laevis Embryo, *J. Neurochem.* 62:1670-1681.

Silberstein, L., Inestrosa, N. C., and Hall, Z. W., 1982, Aneural muscle cells make synaptic basal lamina components, *Nature* 295:143-145.

Toutant, J. -P., and Massoulie, J., 1987, Acetylcholinesterase. In: Mammalian ectoenzymes, pp. 289-328. Eds. A. J. Kenny and A. J. Turner. Elsevier, Amsterdam.

Tsim, K. W. K., Greenberg, I., Rimer, M., Randall, W. R., and Salpeter, M. M., 1992, Transcripts for the acetylcholine receptor and acetylcholine esterase show distribution differences in cultured chick muscle cells. *J. Cell Biol.* 118:1201-1212.

Tennyson, V. M., Brzin, M., and Slotwiner, P., 1971, The appearance of acetylcholinesterase in the myotome of the embryonic rabbit. An electron microscope cytochemical and biochemical study. *J. Cell Biol.* 51:703-721.

Tennyson, V. M., Brzin, M., and Kremzner, L. T., 1973, Acetylcholinesterase activity in the myotube and muscle satellite cell of the fetal rabbit. J. Histochem Cytochem. 21:634-652.

Vallette, F-M., De la Porte, S., and Massoulie, J., 1991, Regulation and distribution of acetylcholinesterase molecular forms *in vivo* and *in vitro,* In: Cholinesterases: Structure, Function, Mechanism, Genetics and Cell Biology, pp. 146-151. Eds. J. Massoulie, F. Bacou, E. A., Barnard, A., Chatonnet, B. P., Doctor, and D. M. Quinn. American Chemical Society: Washington D. C.

ALTERNATIVE EXON 6 DIRECTS SYNAPTIC LOCALIZATION OF RECOMBINANT HUMAN ACETYLCHOLINESTERASE IN NEUROMUSCULAR JUNCTIONS OF XENOPUS LAEVIS EMBRYOS

Meira Sternfeld, Shlomo Seidman, Revital Ben Aziz-Aloya,
Michael Shapira, Rina Timberg, Daniela Kaufer, and Hermona Soreq

Department of Biological Chemistry
The Hebrew University of Jerusalem
Israel

ABSTRACT

We have expressed and characterized catalytically active recombinant human acetylcholinesterase (rHAChE) produced in microinjected oocytes of Xenopus laevis. However, the highly specialized, single-cell nature of the oocyte limits its usefulness in addressing questions regarding tissue-specific processing and the biological roles of cloned nervous system proteins. Therefore, to study the role of 3' alternative splicing in regulating tissue-specific expression of the human ACHE gene encoding AChE, we established an *in vivo* model in transiently transgenic Xenopus embryos. Following injection into *in vitro* fertilized Xenopus eggs, whole-mount cytochemistry and electron microscopy revealed that ACHE DNA bearing the alternative 3' exon E6 (ACHE-E6) induced prominent overexpression of catalytically active AChE in myotomes of 2- and 3-day-old embryos, including neuromuscular junctions (NMJs). NMJs from ACHE-E6-injected embryos displayed, on average, 4-fold greater AChE-stained areas (SA) and 80% increased post-synaptic lengths (PSL) compared to age-matched uninjected controls. Perhaps more significantly, ACHE-E6 overexpression stimulated the appearance of a class of large NMJs (PSL>4 mm) rarely observed in control embryos, apparently at the expense of small (PSL<3 mm) NMJs. Homogenates prepared from these embryos demonstrated increased binding of biotinylated a-bungarotoxin, indicating enhanced expression of the endogenous Xenopus acetylcholine receptor and suggesting coordinated regulation of cholinergic proteins in the developing NMJ. When exon E6 was replaced by ACHE DNA encoding the pseudo-intron I4 and 3' alternative exon E5, overexpressed rHAChE accumulated in epidermal cells, but not in muscle or NMJs. These findings, therefore, attribute an evolutionarily conserved NMJ-accumulating role for exon E6 and provide *in vivo* evidence for tissue-specific management of alternative AChEs.

REFERENCES

Ben Aziz-Aloya, R., Seidman, S., Timberg, R., Sternfeld, M., Zakut, H., & Soreq, H. (1993) Proc. Natl. Acad. Sci. USA 90:2471-2475.

Seidman, S., Ben Aziz-Aloya, R., Timberg, R., Loewenstein, Y., Velan, B., Shafferman, A., Liao, J., Norgaard-Pederson, B., Brodbeck, U., & Soreq, H. (1994) J. Neurochem. 62: 1670-1681.

Shapira, M., Seidman, S., Sternfeld, M., Timberg, R., Kaufer, D., Patrick, J., Soreq, H. (1994) Proc. Natl. Acad. Sci. USA, In press.

DEVELOPMENTAL REGULATION OF ACETYLCHOLINESTERASE MRNA IN THE MOUSE DIAPHRAGM

Alternative Splicing and Focalization

Claire Legay,[1] Monique Huchet,[2] Jean Massoulié,[1] and
Jean-Pierre Changeux[2]

[1] Neurobiologie Cellulaire et Moleculaire, CNRS URA 1857
Ecole Normale Supérieure
46 rue d'Ulm
75005 Paris, France
[2] Neurobiologie Moleculaire, CNRS URA 1284
Institut Pasteur
25 rue du Dr Roux
75015 Paris, France

ABSTRACT

A single gene generates distinct catalytic subunits of acetylcholinesterase (AChE, EC 3.1.1.7), by alternative splicing of exons encoding C-terminal peptides (Massoulié et al., 1993). The T ("tailed") subunits produce a variety of molecular forms, including collagen-tailed molecules and hydrophobic-tailed tetramers, the H ("hydrophobic") subunits produce membrane-bound dimers, and the R ("readthrough") subunits have only been inferred from the existence of cDNAs which retain the intervening sequence following the common exon in the genome (Li et al., 1993). In adult mammals, only mRNAs encoding the T subunit have been detected in muscles and brain (Legay et al., 1993). We have used in situ hybridization and the polymerase chain reaction (PCR) to analyze the nature and distribution of AChE mRNAs along myofibers in the mouse diaphragm, during development. While adult muscle only contains the T transcript, expressed by junctional nuclei (Jasmin et al., 1993), we were surprized to find that, from embryonic day 13 (E13) until birth, the diaphragm contains the three types of AChE mRNAs: T is already predominant at this stage, R represents a few percent of mRNA and H is detectable in still lower proportion. This is the first report of the existence of the R transcript in muscle, and we obtained similar results with the C2 myogenic cell line, both as myoblasts and differentiated myotubes. As early as E13, T mRNAs preferentially accumulate in the midline, where the first neuromuscular contacts are forming.

REFERENCES

Jasmin, B.J., Lee, R.K. & Rotundo, R.L. (1993) Neuron **11**, 467-477.
Legay, C., Bon, S. & Massoulié, J. (1993) FEBS Lett. **315**, 163-166.
Li, Y., Camp, S., & Taylor, P. (1993) J. Biol. Chem. **268**, 5790-5797.
Massoulié, J., Pezzementi, L., Bon, S., Krejci, E. & Vallette, F.-M. (1993) Progr. in Neurobiol. **41**, 31-93.

ACETYLCHOLINESTERASE AND BUTYRYLCHOLINESTERASE EXPRESSION IN ADULT RABBIT TISSUES AND DURING DEVELOPMENT

Omar Jbilo,[1] Yann L'Hermite,[1] Vincenzo Talesa,[1,2] Jean-Pierre Toutant,[1] and Arnaud Chatonnet[1]

[1] Différenciation cellulaire et Croissance
Centre INRA de Montpellier
2 place Viala, 34060 Montpellier Cedex 1, France
[2] Department of Experimental Medicine
University of Perugia, Italy

A large cDNA fragment covering the complete sequence of the mature catalytic subunit of rabbit acetylcholinesterase (AChE, EC 3.1.1.7.) has been cloned and sequenced. This sequence was compared to that of rabbit butyrylcholinesterase (BChE, EC 3.1.1.8.; Jbilo, O. & Chatonnet, A., 1990, *Nucleic Acids Res. 18*, 3990). Amino acids sequences of AChE and BChE had 51% identity. They both possessed a choline binding site W(*84*), a catalytic triad S(*200*)-H(*440*)-E(*327*) and six cysteine residues (*67-94; 254-265; 402-521*) in conserved sequence positions to those that form three intrachain disulfide bonds in all cholinesterases (by convention, numbering of amino acids is that of *Torpedo* AChE). Rabbit AChE had a larger number of aromatic residues lining the active site gorge than rabbit BChE (14 versus 8) and a smaller number of potential N-glycosylation sites (3 versus 8). Both catalytic subunits had a hydrophilic C-terminus (catalytic subunits of type T) characterized by a regular disposition of 7 aromatic residues in conserved position to 7 residues found in T subunits of *Torpedo* AChE. Expression of acetycholinesterase and butyrylcholinesterase genes (ACHE and BCHE) was studied in rabbit tissues and during development by a correlation of Northern blot analysis and enzymatic activities. The correlation was rendered difficult by the presence of an eserine-resistant esterase active on butyrylthiocholine in serum, liver and lung. This enzyme was characterized. When the contribution of this carboxylesterase was taken into account, brain was found as the richest source of BChE followed by lung and heart. Rabbit liver had a very low content of BChE that correlated with the low BChE activity in plasma. During development BCHE transcripts were detected as early as day 10 *postcoitum* whereas ACHE transcripts appeared only on day 12.

ACKNOWLEDGMENT

Sequence of rabbit ACHE cDNA was submitted to Genbank with accession number: U 05036. Present results will appear in a future issue of Eur. J. Biochem (1994). This work was supported by the *Institut National de la Recherche Agronomique.* O. J. was funded by the *Association pour la Recherche sur le Cancer.*

MUTATIONS IN THE CATALYTIC SUBUNIT OF ACETYLCHOLINESTERASE DO NOT APPEAR RESPONSIBLE FOR CONGENITAL MYASTHENIC SYNDROME ASSOCIATED WITH END-PLATE ACETYLCHOLINESTERASE DEFICIENCY

S. Camp,[1] A. G. Engel,[2] D. K. Getman,[1] S. Bon,[3] J. Massoulié,[3] and P. Taylor[1]

[1] Department of Pharmacology, University of California, San Diego
La Jolla, California 92093-0636
[2] Department of Neurology, Mayo Clinic
Rochester, Minnesota 55905
[3] Ecole Normale Superieure
Paris, France

A congenital myasthenic condition was described several years ago that is characterized by a deficiency in end-plate acetylcholinesterase (AChE); the absence of the inflammatory immune responses of myasthenia categorizes the condition as congenital (Engle, et al., 1977). Since that time several additional patients with a similar disorder have been identified. Patient blood samples were drawn for DNA extraction and isolation of lymphocytes which have been immortalized by Epstein-Barr virus transformation. To ascertain the genetic basis of the disease, we have examined the structure of the gene encoding the catalytic subunits of AChE by comparing patients and control individuals. Southern analysis revealed no differences between patient and control DNA, suggesting no major chromosomal rearrangements had occurred. PCR amplification of genomic DNA yielded clones covering exon 4 and the alternatively spliced exons 5 and 6; this region was analyzed by nuclease protection and sequencing. While allelic differences were detected, we found no differences in exonic or intronic areas which might give rise to distinctive splicing pattern differences in patients and control individuals. The *ACHE* gene was cloned from genomic libraries from one of the patients and a control. Identical patterns of mRNA production and species of expressed AChE were revealed when the cloned genes were transfected into COS and HEK cells. Cotransfection of the genes expressing the catalytic subunits from either patient or control with a gene expressing the tail unit from Torpedo AChE (Krejci et.al., 1991) yielded asymmetric species (18 + 14 S) which require assembly of catalytic subunits and tail unit. Thus, the catalytic subunits of AChE expressed in the congenital myasthenia syndrome appear identical in sequence, arise from similar splicing patterns and will assemble with a tail unit to form a heterologous species. These findings point to the genetic abnormality likely existing in the tail unit or in a protein responsible for assembly of the catalytic subunit with the tail subunit.

REFERENCES

Engle, A.G., Lambert, E.H., and Gomez, M.R.(1977) *Ann. Neurol.* 1: 315-330.
Krejci, E., Coussen,F., Duval, N., Chatel, J.M., Legay, C., Puype, M., Vandekerckhove, J., Cartaud, J., Bon, S. and Massoulié, J. (1991) *EMBO J.* 10: 1285-1293.

REGULATION OF ACETYLCHOLINESTERASE (AChE) mRNA BY RYANODINE-SENSITIVE AND L-TYPE CALCIUM CHANNELS DURING MYOGENESIS *IN VITRO* AND MUSCLE DEVELOPMENT *IN VIVO*

Zhigang Luo,[1] Martine Pincon-Raymond,[2] and Palmer Taylor[1]

[1] Department of Pharmacology-0636
University of California at San Diego
La Jolla, California 92093
[2] Institut National de la Sante et de la Recherche Medicale, U. 153,
17 Rue du Fer-a-Moulin, 75005 Paris, France

The increase in AChE mRNA associated with muscle differentiation is due to stabilization of a labile mRNA rather than enhanced transcription (Fuentes and Taylor, Neuron, 10, 679-687, 1993). In a search for the signaling pathways responsible for message stabilization, we examined the roles of ryanodine-sensitive and L-type Ca^{2+} channels in regulation of AChE mRNA during myogenesis. Treatment of C2-C12 cells in cultures during myogenesis with nM ryanodine, μM ryanodine receptor antagonists, FLA 365, and μM L-type, but not N-type, Ca^{2+} channel antagonists blocked the differentiation-induced increase in AChE mRNA and protein in a dose dependent manner. Ryanodine and the L-type Ca^{2+} channel blockers did not influence the enhanced expression of the nicotinic acetylcholine receptors (nAChR) associated with fusion. The ryanodine block was fully reversible within 24 hours upon removal of the drug indicating the functional integrity of the cells. Measurements of transcription rates using run-on transcription showed that ryanodine and L-type Ca^{2+} channel antagonist, nifedipine, did not change the rate of *AChE* gene transcription indicating that the block in the increased mRNA associated with differentiation was due to reduced stabilization of the labile mRNA. These findings indicated that ryanodine-sensitive Ca^{2+} channels in sarcoplasmic reticulum and L-type Ca^{2+} channels in T-tubules of skeletal muscle link to play important roles in regulation of AChE mRNA during myogenesis (Luo *et al.*, J. Biol. Chem., 1994, in press). To confirm the importance of this signaling pathway in AChE regulation in intact skeletal muscle, we examined AChE mRNA levels in skeletal and cardiac muscles from muscular dysgenic mice lacking the skeletal, but not the cardiac, L-type Ca^{2+} channel receptors. RNA protection experiments indicated 50-80%

reductions in AChE mRNA levels in leg muscles from new born and day 18 embryonic mutant mice as compared to control mice. Similar reductions in AChE activity were also observed. However, no reduction in mRNA levels of nAChR γ-subunit was observed in leg muscles and mRNA levels and AChE activity were not altered in cardiac tissues from mutant mice. These findings provide evidence to indicate that L-type Ca^{2+} channels also play an important role in regulation of AChE expression in intact skeletal muscle. The differential regulation of mRNA levels of AChE and nAChR suggests distinct mechanisms of regulation controlled by L-type Ca^{2+} channels in intact skeletal muscles.

ACKNOWLEDGMENT

Supported by GM 18360 and GM 24437.

REGULATION OF AChE GENE EXPRESSION IN NEURONALLY INDUCED MOUSE P19 CELLS

B. A. Coleman and P. Taylor

Department of Pharmacology
University of California at San Diego
La Jolla, California 92093-0636

Mouse P19 embryonic carcinoma cells are uncommitted, multipotent cells which can be induced to terminally differentiate along the neuroectodermal lineage. Neurons, glia and astrocytes develop after free-floating aggregates, cultured 4-6 days in the presence of retinoic acid (RA), are plated onto tissue culture dishes. Neuronal cultures up to 90% purity can be obtained by treatment with mitotic inhibitors. We have used this cell line to examine the developmentally regulated expression of acetylcholinesterase (AChE). AChE is not expressed in untreated cells, but can be detected within 24 hr of plating neuronally-induced cultures. The level of AChE activity increases during neuronal differentiation. This correlates with the increase in cells exhibiting neuronal morphology. Histochemical and immunohistochemical results indicate that AChE activity is associated with cells exhibiting neuronal morphology, but not with glia-like cells. AChE mRNA and enzyme activity increase in parallel. Splicing of AChE mRNA is almost entirely to exon 6, consistent with that observed in mouse brain. Run-on transcription assays demonstrate that the AChE gene is transcribed at a similar rate in undifferentiated monolayer cells, RA-treated aggregates, and differentiated P19 neurons. Transcription inhibition experiments suggest that changes in AChE mRNA stability play a role in regulating expression of the enzyme. RA-treated aggregates maintained in culture for more than six days begin to secrete large amounts of tetrameric AChE. However, if the cells are plated in tissue culture dishes to allow for morphological differentiation, approximately half of the G4 enzyme becomes membrane bound. These results demonstrate that AChE is synthesized early in neuronal development but is not localized to the neuronal cell membrane until later in the differentiation process. These results further suggest that the 20 kD hydrophobic subunit, which most likely tethers AChE to the P19 neuronal membrane, is generated concomitant with morphological differentiation. Thus, P19 cells appear to provide a useful model for examining expression of AChE during neuronal development. (Supported by USPHS GM18360, 24437 and American Heart Association Postdoctoral Fellowship No. 93-58).

REGULATION OF HUMAN ACETYLCHOLINESTERASE GENE EXPRESSION

D. Getman, K. Inoue, and P. Taylor

Department of Pharmacology
University of California, San Diego
La Jolla, California 92093-0636

Acetylcholinesterase (AChE) in human cells is encoded by a single gene, *ACHE*, located on chromosome 7q22. We have investigated the transcriptional regulation of *ACHE* by defining the mRNA cap sites and characterizing essential promoter and enhancer elements. We find putative promoter elements reside immediately upstream of the first untranslated exon, and are characterized by GC rich sequences containing consensus binding sites for transcription factors SP1, EGR-1 and AP-2. *In vitro* transcription studies and RNase protection of mRNA from human NT2/D1 teratocarcinoma cells reveal two closely spaced transcription cap sites. Both cap sites are located at a consensus initiator element similar to the terminal transferase gene initiator element. Transient transfection experiments show removal of three bases of this initiator sequence reduce promoter activity 98% in NT2/D1 cells. *In vitro* transcription and transient transfection of a series of *ACHE* promoter 5' deletion mutants linked to a luciferase reporter show one SP1 site at -70 is essential for promoter activity. Purified recombinant SP1 protein protects this site in DNase footprinting experiments. Footprinting also shows purified recombinant transcription factor AP-2 binds 17 base pairs upstream of the initiator element. *In vitro* transcription studies with AP-2 protein and transient co-transfection of wild type AP-2 cDNA with *ACHE* promoter-luciferase reporter constructs show AP-2 functions as a repressor of *ACHE* transcription. Retinoic acid treatment of NT2/D1 cells induces AChE enzyme activity greater than 30 fold and mRNA levels 50 fold, while the rate of gene transcription remains unchanged. These data indicate *ACHE* transcription is activated by SP1 or similar factors, AP-2 represses *ACHE* transcription *in vitro*, and mRNA levels are controlled by changes in mRNA stability. (Supported by USPHS GM18360).

REFERENCE

Getman, D., Eubanks, J., Camp, S., Evans, G., Taylor, P. 1992. The human gene encoding acetylcholinesterase is located on the long arm of chromosome 7, Am. J. Human Genetics 51: 170-177.

PROMOTER ELEMENTS OF THE MOUSE ACETYLCHOLINESTERASE GENE

Regulation during Muscle Differentiation

Annick Mutero and Palmer Taylor

Department of Pharmacology
University of California, San Diego
La Jolla, California 92093-0636

The increase in acetylcholinesterase expression during muscle differentiation of myoblasts to myotubes was shown previously to reflect a greater stability of the messenger RNA. We have employed myogenic transcription factors to transform non-muscle cells into myoblasts in order to investigate the regulation of the acetylcholinesterase gene during early determination of the muscle phenotype. The AChE promoter region was analysed by deletion analysis, point mutagenesis and gel mobility-shift assays. Accordingly, stimulation of E-boxes present at -335 bp from the start of transcription and in the first intron were found not to accelerate transcription. We find that a GC-rich region (at -105 bp to -59 bp from the start of transcription) containing binding sites for the transcription factors Sp1 and Egr-1 is essential for the promoter activity. Mutation of the Sp1 sites dramatically reduces the promoter activity while mutation of the Egr-1 sites has little effect. We show that Sp-1 and Egr-1 compete for binding to overlapping sites and that an increase in Egr-1 reduces the expression of the AChE gene. Also, an AP2 binding site present immediately 5' of the major transcription start site is found to repress promoter activity. We find that AP2 is associated with this site in cells not commited to the muscle phenotype and a distinct AP-2 -like protein may be associated to myoblasts and myotubes. (Supported by IBRO, Conseil regional P.A.C.A., and USPHS GM18360).

REFERENCES

Fuentes, M.E., and Taylor, P. 1993. Control of Acetylcholinesterase Gene Expression during Myogenesis. Neuron, 10: 679-687.

Li, Y., Camp, S., Rachinsky, T.L., Bongiorno, C., and Taylor, P. 1993. Promoter Elements and Transcriptional Control of the Mouse Acetylcholinesterase Gene. J. Biol. Chem. 268: 3563-3572.

STRUCTURES OF COMPLEXES OF ACETYLCHOLINESTERASE WITH COVALENTLY AND NON-COVALENTLY BOUND INHIBITORS

J. L. Sussman,[1] M. Harel, M. Raves,[1] D. M. Quinn,[2] H. K. Nair,[2] and I. Silman[3]

[1] Department of Structural Biology, The Weizmann Institute of Science
Rehovot 76100 Israel
[2] Department of Chemistry, The University of Iowa
Iowa City, Iowa 52242
[3] Neurobiology, The Weizmann Institute of Science
Rehovot 76100 Israel

INTRODUCTION

The principal biological role of acetylcholinesterase (AChE, acetylcholine hydrolase, EC 3.1.1.7) is to terminate signal transmission at cholinergic synapses by rapid hydrolysis of the neurotransmitter, acetylcholine (ACh) (Barnard, 1974). In keeping with this requirement, AChE possesses a remarkably high specific activity, especially for a serine hydrolase (Quinn, 1987), functioning at a rate approaching that of a diffusion-controlled reaction (Bazelyansky et al., 1986). Early kinetic studies indicated that the active site of AChE consists of two subsites, the 'esteratic' and 'anionic' subsites, corresponding to the catalytic machinery and the choline-binding pocket, respectively (Nachmansohn and Wilson, 1951). A second, 'peripheral', anionic site exists, so named because it appears to be distant from the active site (Taylor and Lappi, 1975). The elucidation of the three-dimensional structure of *Torpedo* AChE (Sussman et al., 1991) served to confirm these earlier studies, and has showed that AChE contains a catalytic triad similar to that present in other serine hydrolases (Steitz and Shulman, 1982). Unexpectedly, it also revealed that this triad is located near the bottom of a deep and narrow cavity, ~20 Å deep, which has been named the 'active-site gorge'. The cavity is lined by the rings of fourteen aromatic residues which are conserved in the AChE sequences published so far (Gentry and Doctor, 1991). Much of the subsequent research on structure-function relationships in AChE has been concerned with the functional significance of the gorge and with the role of the aromatic rings which account for more than 50% of its surface area (Axelsen et al., 1994). Thus, structural evidence (Axelsen et al., 1994; Sussman et al., 1991), as well as evidence obtained by modeling (Harel et al., 1992), by

Enzymes of the Cholinesterase Family, Edited by Daniel M. Quinn et al.
Plenum Press, New York, 1995

chemical modification (Harel et al., 1993; Schalk et al., 1992; Weise et al., 1990), and by site-directed mutagenesis (Harel et al., 1992; Ordentlich et al., 1993; Radic et al., 1993; Shafferman et al., 1992; Vellom et al., 1993), all point to important roles for certain of these conserved aromatic residues in both the 'esteratic' and 'anionic' subsites of the active site, and in the 'peripheral' anionic site.

In the following, we will present structural evidence, derived from X-ray data collected for complexes of AChE with a competitive inhibitor and with a transition-state analog, which substantiates the involvement of aromatic residues in the binding of the quaternary group of ACh in the so-called 'anionic' subsite of the active site, and provides experimental corroboration for a plausible orientation of ACh within the active site which we had suggested on the basis of computer-docking (Sussman et al., 1991).

MATERIALS AND METHODS

Crystalline AChE complexed with the transition state analog (N,N,N-trimethylammonio)trifluoroacetophenone (TMTFA) (Nair et al., 1993) was obtained by soaking native trigonal crystals of AChE, grown from ammonium sulfate solutions (Sussman et al., 1991), for 12 days in a solution containing 2 mM TMTFA. Diffraction data were collected to 2.8Å resolution on a Siemens-Xentronics area detector mounted on a Rigaku rotating anode. The structure was determined by the difference Fourier technique and refined using simulated annealing and restrained least-squares (Brünger, 1992). This structure was compared with those of the native crystals of AChE obtained by crystallization from ammonium sulfate of purified enzyme eluted from the affinity column with decamethonium (DECA) (Sussman et al., 1988), with complexes of the same crystals of AChE soaked with edrophonium (EDR) and DECA, and also with native crystals obtained by crystallization from PEG of purified enzyme eluted from the affinity column with the monoquaternary ligand, tetramethylammonium (TMA). These structures were compared with a model of ACh docked into the active site of AChE (Sussman et al., 1991). For the formulae of ACh and of the ligands employed, see Figure 1.

The overall conformation of the AChE-TMTFA structure is very similar to that of the two native structures and of the two non-covalent complexes, all of which are very similar to each other. The highest positive peak (7σ) of electron density in the difference Fourier map of the AChE-TMTFA complex was at a bonding distance from Ser200Oγ, and its shape accommodates a TMTFA molecule. A model of TMTFA was generated by using the EDR coordinates to build the trimethylammoniophenyl fragment, and the ACh coordinates to

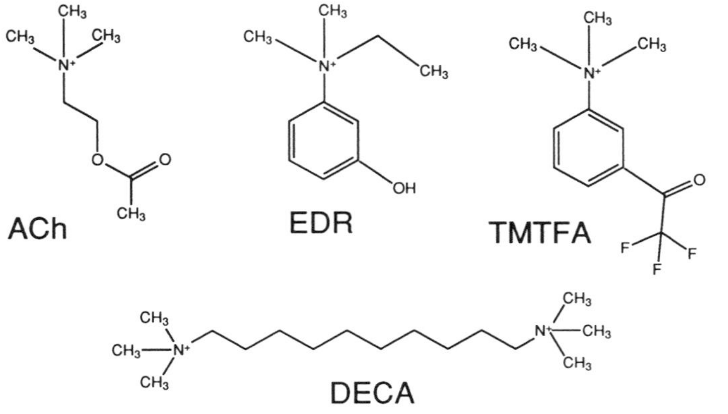

Figure 1. Chemical structures of the neurotransmitter acetylcholine (ACh); a monoquaternary competitive inhibitor of acetylcholinesterase, edrophonium (EDR); a transition-state analog (N,N,N-trimethylammonio)trifluoroacetophenone (TMTFA); a bisquaternary competitive inhibitor, decamethonium (DECA).

build the trifluoroaceto fragment, albeit with the C-F distance adjusted to 1.5Å. The final refinement parameters for the AChE-TMTFA complex are: R factor = 0.188, R_{free} = 0.235, number of water molecules = 99, number of sulfate molecules = 1, rms bond = 0.007 A, rms angle = 1.016°.

RESULTS AND DISCUSSION

In our initial report of the three-dimensional structure of *Torpedo* AChE (Sussman et al., 1991), it was suggested, on the basis of manual docking, that the primary interaction of the quaternary group of the substrate, ACh (Figure 1), is with the indole ring of the conserved tryptophan residue, Trp84. Indeed, Figure 2 shows that if ACh is docked in an all-*trans* configuration, with the Cα carbon of its acyl group positioned to make a tetrahedral bond with Oγ of Ser200, and the acyl oxygen pointing towards a putative oxyanion hole, composed of the main-chain nitrogens of Gly118, Gly119 and Ala201, the quaternary group is within van der Waals distance (~3.5 Å) of Trp84. This assignment was at odds with the accepted notion that the 'anionic' binding site for the positive quaternary group contains a cluster of negative charges (Nolte et al., 1980). However, support for such an assignment was provided by both affinity labeling of Trp84 in the *Torpedo* enzyme (Weise et al., 1990) and by a study which used both model host-guest systems and theoretical considerations to argue that aromatic rings might play an important role in receptors for quaternary ligands, via an interaction of their π electrons with the positive charge (Dougherty and Stauffer, 1990). Nevertheless, we felt that it was important to confirm this assignment at the X-ray level by studying appropriate AChE-ligand complexes.

We earlier determined the structures of complexes with *Torpedo* AChE of the monoquaternary ligand, EDR (Figure 1) and the bisquaternary ligand, DECA (Figure 1),

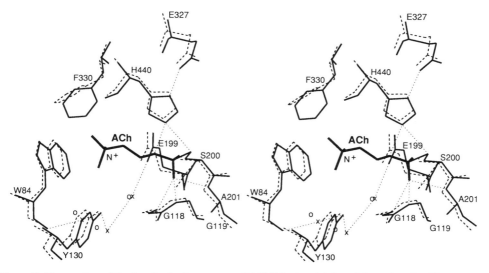

Figure 2. Stereo view of the two 'native' structures of AChE in the vicinity of the active site. The structure obtained for enzyme purified by elution from the affinity column with TMA is shown in solid lines, and that for enzyme for which DECA was the eluant (1ACE) is shown in dashed lines. A model of ACh docked in the active site (Sussman et al., 1991), with its quaternary group (N⁺) pointing towards the indole ring of W84, is shown for reference. Key waters found in the vicinity of the active site are shown as "X" for the new native structure, and as "O" for 1ACE. Dotted lines represent hydrogen bonds.

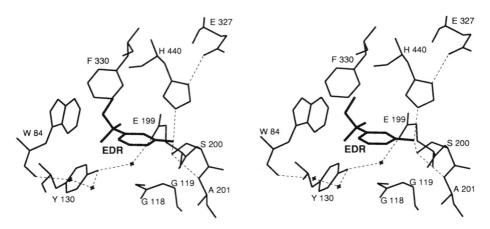

Figure 3. Stereo view of the crystal structure of the AChE:EDR complex in the vicinity of the active site (Harel et al., 1993).

both of which are strong reversible inhibitors of AChE (Harel et al., 1993). In the EDR-AChE complex (Figure 3), the quaternary group of the ligand nestles adjacent to the indole of Trp84, in a position equivalent to that assigned to the quaternary group of ACh in the docking procedure (Sussman et al., 1991). In the affinity labeling study mentioned above (Weise et al., 1990), this tryptophan was covalently labeled by the aziridinium ion, which is similar in structure to EDR, and EDR protected against the labeling by aziridinium. Our data thus demonstrate good correspondence between the crystal structure and that in solution. The *m*-hydroxyl group is positioned between His440N$^{\epsilon 2}$ and Ser200Oγ, thus making hydrogen bonds of 3.0 and 3.5 Å, respectively, to two of the three members of the catalytic triad. There is also a 3.5-Å hydrogen bond to Gly119N, which is part of the oxyanion hole. This provides a structural basis for the observation that such *meta*-substituted anilinium ions are much more potent inhibitors of AChE than the homologous ligand which lacks the hydroxyl group (Wilson and Quan, 1958).

DECA, which is a bisquaternary compound, lies along the aromatic gorge, and it can be seen from Figure 4 that it is in close contact with aromatic groups at both the top and bottom of the gorge.

We already reported that, although the overall conformation of AChE does not change upon binding EDR or other ligands (Harel et al., 1993) nor do most of the side-chains of the active site move appreciably, the phenyl ring of F330 does assume a number of significantly different conformations. Our interpretation of these movements was complicated by the realization that our original structure, Brookhaven Protein Data Bank IDCODE - 1ACE (Sussman et al., 1991), actually represented a native crystal containing a residual of DECA in the active site gorge (Axelsen et al., 1994; Harel et al., 1993). We decided, therefore to collect an additional data set from crystals obtained using AChE eluted from the affinity column employed for purification with the weak **mono**quaternary inhibitor, TMA. The overall structure of native AChE thus prepared is essentially identical to that of 1ACE. In particular, the shape of the aromatic gorge was preserved, showing that it was not influenced by the presence of endogenous DECA.

If we compare the position of the phenyl ring of F330 in the various structures, we see that in the EDR-AChE complex that it makes a quaternary-aromatic and/or aromatic-aromatic (Perutz, 1993; Verdonk et al., 1993) contact with the ligand, and a very similar interaction is seen in the TMTFA-AChE conjugate. In 1ACE, and in the DECA-AChE

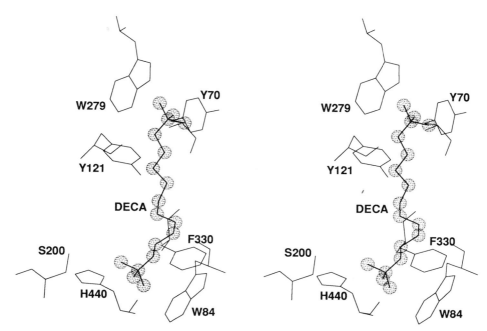

Figure 4. Stereo representation of DECA in the active site gorge of *T. californica* AChE.

complex, the F330 phenyl appears to be aligned parallel to the wall of the gorge, presumably being restrained in that position by the hydrocarbon position of the DECA (Harel et al., 1993). In the native data set, obtained with crystals of AChE which had been eluted with TMA, the phenyl ring assumes a position intermediate between these two conformations which may be regarded as representing its ligand-free conformation.

Transition-state analogs are designed to provide a stable adduct which closely resembles the transition state formed during substrate hydrolysis (Wolfenden, 1976). A number of studies have described the design, synthesis and functional characterization of transition-state analogs for AChE [see, for example (Dafforn et al., 1977; Gelb et al., 1985)]. One such compound, TMTFA (Figure 1), is a potent time-dependent inhibitor of both *Electrophorus* and *Torpedo* AChE (Brodbeck et al., 1979; Nair et al., 1993). In aqueous solution, TMTFA is an equilibrium mixture of the free ketone and of the corresponding ketone hydrate, with only the free ketone serving as an inhibitor of the enzyme. The second-order rate constant for inhibitor binding k_{on}, is 6-7 x 10^9 $M^{-1}s^{-1}$, and probably monitors a diffusion-controlled process. In contrast, dissociation is very slow, with $k_{off} =$ $10^{-5}s^{-1}$ for *Electrophorus* AChE and $10^{-4}s^{-1}$ for *T. californica* AChE. The corresponding inhibition constants, K_i, based on the concentration of the free ketone, are 1.3 and 15 fM, respectively (Nair et al., 1993).

The TMTFA molecule occupies the position of 6 water molecules that are found in the 2.25 Å native structure. Upon binding to TMTFA, the Ser200Oγ swings from its position in the native structure to become a partner in the tetrahedral conformation at a distance of 1.4Å from the aceto C.

In the TMTFA-AChE complex (Figure 5), the quaternary group lies virtually in the same position as in the EDR-AChE complex, interacting with the aromatic groups of Trp84 and Phe330. The latter group, which assumes different orientations in various AChE-ligand complexes studied (Harel et al., 1993), so as to interact favourably with the bound ligand,

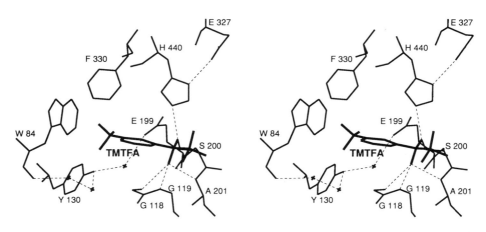

Figure 5. Stereo view of the crystal structure of AChE:TMTFA complex in the vicinity of the active site.

adopts virtually the same orientation in the TMTFA-AChE complex as in the EDR-AChE complex. The keto function, which is in the *meta* position, relative to the quaternary group, appears well positioned to form a hemiketal with Ser200Oγ, with its carbon atom also being in hydrogen-bonding distance of His440Nε2. The overall orientation thus seems to be very similar to that of ACh modeled in its all-*trans* conformation (Sussman et al., 1991). Moreover, the oxygen atom of TMTFA is within hydrogen-bonding distance of the main-chain nitrogens of Gly118, Gly119 and Ala201, which we had suggested might serve as the 'oxyanion' hole for the carbonyl of ACh (Sussman et al., 1991). Although the three fluorine atoms on the proximal carbon presumably serve to enhance the stability of the transition-state analog due to their high electronegativity, close inspection of the structure of the TMTFA-AChE complex reveals that all three also appear to make specific interactions with the protein, near the aromatic residues Phe288 and Phe290.

The structure presented in Figure 5 shows that the TMTFA-AChE complex indeed displays the structural characteristics to be expected of a transition-state complex. Detailed inspection of the structure should provide valuable insights into the structural features of the active site which are largely responsible for the catalytic power of this potent hydrolase.

The AChE active site provides an interaction surface that is complementary to the extended conformation of ACh, of which TMTFA is a constrained mimic. The CF_3 group of TMTFA occupies an acyl binding pocket that is comprised of five convergent amino acid residues that give AChE its acetyl ester specificity. These structural elements of AChE catalytic power and substrate specificity accord nicely with the voluminous literature on this important enzyme. Four binding loci, the quaternary ammonium and acyl binding sites, the oxyanion hole and the catalytic triad, converge on, envelop, and hence sequester TMTFA from the solvent, and, by analogy, ACh in the transition state. The geometrical constraints thus invoked may be the reason why the active site of such a rapid enzyme is located deep within the enzyme molecule at the bottom of the aromatic gorge.

ACKNOWLEDGMENTS

We thank Lilly Toker for preparation of the AChE, Alexander Faibusovitch and Clifford Felder for help in computation. This project was supported by the U.S. Army Medical Research and Development Command under Contract DAMD17-93-C-3070, the

Association Franco-Israélienne pour la Recherche Scientifique et Technologique (AFIRST), the Minerva Foundation, Munich, Germany and the Kimmelman Center for Biomolecular Structure and Assembly, Rehovot and NIH grant NS21334 (to DMQ).

REFERENCES

Axelsen, P.H., Harel, M., Silman, I. and Sussman, J.L. (1994) Prot Sci 3:188-197.

Barnard, E.A., 1974, in *The Peripheral Nervous System* (Hubbard, J.I., Ed.), pp. 201-224, Plenum, New York.

Bazelyansky, M., Robey, C. and Kirsch, J.F. (1986) Biochemistry 25:125-130.

Brodbeck, U., Schweikert, K., Gentinetta, R. and Rottenberg, M. (1979) Biochim Biophys Acta 567:357-369.

Brünger, A.T. (1992) X-PLOR Version 3.1 A System for Crystallography and NMR. Yale University Press, New Haven and London.

Dafforn, A., Anderson, M., Ash, D., Campagna, J., Daniel, E., Horwood, R., Kerr, P., Rych, G. and Zappitelli, F. (1977) Biochim Biophys Acta 484:375-385.

Dougherty, D.A. and Stauffer, D.A. (1990) Science 250:1558-1560.

Gelb, M.H., Svaren, J.P. and Abeles, R.H. (1985) Biochemistry 24:1813-1817.

Gentry, M.K. and Doctor, B.P., 1991, in *Cholinesterases: Structure, Function, Mechanism, Genetics and Cell Biology* (Massoulié, J., Bacou, F., Barnard, E., Chatonnet, A., Doctor, B.P. and Quinn, D.M., Ed.), pp. 394-398, American Chemical Society, Washington, DC.

Harel, M., Schalk, I., Ehret-Sabatier, L., Bouet, F., Goeldner, M., Hirth, C., Axelsen, P., Silman, I. and Sussman, J.L. (1993) Proc Natl Acad Sci USA 90:9031-9035.

Harel, M., Sussman, J.L., Krejci, E., Bon, S., Chanal, P., Massoulié, J. and Silman, I. (1992) Proc Natl Acad Sci USA 89:10827-10831.

Nachmansohn, D. and Wilson, I.B. (1951) Adv Enzymol 12:259-339.

Nair, H.K., Lee, K. and Quinn, D.M. (1993) J Am Chem Soc 115:9939-9941.

Nolte, H.-J., Rosenberry, T.L. and Neumann, E. (1980) Biochemistry 19:3705-3711.

Ordentlich, A., Barak, D., Kronman, C., Flashner, Y., Leitner, M., Segall, Y., Ariel, N., Cohen, S., Velan, B. and Shafferman, A. (1993) J Biol Chem 268:17083-17095.

Perutz, M.F. (1993) Phil Trans R Soc A 345:105-112.

Quinn, D.M. (1987) Chem Rev 87:955-979.

Radic, Z., Pickering, N.A., Vellom, D.C., Camp, S. and Taylor, P. (1993) Biochemistry 32:12074-12084.

Schalk, I., Ehret-Sabatier, L., Bouet, F., Goeldner, M. and Hirth, C., 1992, in *Multidisciplinary Approaches to Cholinesterase Functions* (Shafferman, A. and Velan, B., Ed.), pp. 117-120, Plenum Press, New York.

Shafferman, A., Velan, B., Ordentlich, A., Kronman, C., Grosfeld, H., Leitner, M., Flashner, Y., Cohen, S., Barak, D. and Ariel, N., 1992, in *Multidisciplinary Approaches to Cholinesterase Functions* (Shafferman, A. and Velan, B., Ed.), pp. 165-175, Plenum Press, New York.

Steitz, T.A. and Shulman, R.G. (1982) Ann Rev Biophys Bioeng 11:419-444.

Sussman, J.L., Harel, M., Frolow, F., Oefner, C., Goldman, A., Toker, L. and Silman, I. (1991) Science 253:872-879.

Sussman, J.L., Harel, M., Frolow, F., Varon, L., Toker, L., Futerman, A.H. and Silman, I. (1988) J Mol Biol 203:821-823.

Taylor, P. and Lappi, S. (1975) Biochemistry 14:1989-1997.

Vellom, D.C., Radic, Z., Li, Y., Pickering, N.A., Camp, S. and Taylor, P. (1993) Biochemistry 32:12-17.

Verdonk, M.L., Boks, G.J., Kooijman, H., Kanters, J.A. and Kroon, J. (1993) J Computer-Aided Mol Design 7:173-182.

Weise, C., Kreienkamp, H.-J., Raba, R., Pedak, A., Aaviksaar, A. and Hucho, F. (1990) EMBO J 9:3885-3888.

Wilson, I.B. and Quan, C. (1958) Arch Biochem Biophys 73:131-143.

Wolfenden, R. (1976) Ann Rev Biophys Bioeng 5:271-306.

ELECTROSTATIC PROPERTIES OF HUMAN ACETYLCHOLINESTERASE

Daniel R. Ripoll,[1] Carlos H. Faerman,[2] Richard Gillilan,[1] Israel Silman,[3] and Joel L. Sussman[4]

[1] Cornell Theory Center
[2] Biochemistry, Molecular and Cell Biology
Cornell University, Ithaca, NY 14853
[3] Department of Neurobiology
[4] Department of Structural Biology
Weizmann Institute of Science, Rehovot 76100 Israel

INTRODUCTION

We recently proposed (Ripoll et al., 1993) an electrostatic mechanism of substrate attraction for the enzyme acetylcholinesterase (AChE). The proposed mechanism was based on the calculated electrostatic properties of *Torpedo californica* AChE (TAChE), which show that the active site is entirely embedded in a region that has the lowest values of the electrostatic potential. Recent experimental results on human recombinant AChE (HrAChE), designed to test this hypothesis (Shafferman et al., 1994), show that HrAChE mutants, in which as many as seven negative groups near the entrance to the active-site gorge have been neutralized, have catalytic activities comparable to those of the wild type enzyme. Moreover, these authors presented computational data showing that the isopotential surface, -1kT/e, for their seven-residue mutant (M7), was much smaller than the corresponding surface for the wild type enzyme. They concluded, therefore, that electrostatic effects do not contribute to the catalytic rate of this enzyme. In our earlier study on TAChE (Ripoll et al., 1993) we pointed out that it is important to take into consideration not only the isopotential surfaces, but also the electric field, before drawing any conclusion about the influence of electrostatic forces on AChE activity. In our present study, we calculate the electrostatic field vectors for wild type HrAChE (WT) and for the mutant in which seven negative acidic residues have been changed to neutral residues (M7).

MATERIALS AND METHODS

Since HrAChE and TAChE display over 56% sequence identity and 74% sequence homology (Barak et al., 1992; Cygler et al., 1993), a model of HrAChE was constructed, based upon the crystallographic coordinates of TAChE (Sussman et al., 1991), in a manner

Enzymes of the Cholinesterase Family, Edited by Daniel M. Quinn et al.
Plenum Press, New York, 1995

similar to that used to derive a model for human butyrylcholinesterase (Harel et al., 1992). We used the program DELPHI (Gilson and Honig, 1988; Gilson et al., 1988; Klapper et al., 1986) to calculate the electrostatic potential and field of HrAChE on a grid of points lying both inside and outside the gorge. Evaluation of the electrostatic potential and field was carried out using the finite-difference algorithm in DELPHI, designed to solve the Poisson-Boltzman equation numerically. A cubic grid of 65 x 65 x 65 points was defined that includes the protein molecule and the adjacent solvent. Both the solvent and the protein were treated as homogeneous dielectric media. The former was represented, using the Debye-Huckel model, by a high-dielectric medium, $\varepsilon_{solv} = 78.3$, containing counterions (ionic strength of 0.01) while the latter was approximated by a cavity of low dielectric constant $\varepsilon_{solv} = 4$. The shape of this cavity was computed from the HrAChE model structure. An initial calculation, consisting of 200 iterations using the linear approximation of the Poisson-Boltzmann equation and 50 iterations of the non-linear approximation, was carried out with the protein cavity occupying 33% of the grid volume. This calculation was followed by a second one where 66% of the grid volume was filled using the focusing feature available within DELPHI.

RESULTS AND DISCUSSION

We note that the 'hot' area at the active site entrance and gorge on M7 (the effective attractive part), still lies in the region of lowest electrostatic potential values, as in the wild-type enzyme. The actual potential values are, however, smaller than the ones in the wild type HrAChE.

Figure 1 shows the electric field lines for both M7 and for the wild-type enzyme. It can be seen that there is still a sizable electrostatic field for M7 and a large number of the field lines are directed towards a region close to W286 (corresponding to W279 in *Torpedo* AChE), which is part of the peripheral anionic site (Eichler et al., 1994; Harel et al., 1992). In order to quantify the effect of the mutations on the electrostatic properties of HrAChE, we calculated the mean value of the electrostatic potential and the electrostatic flux on a square grid (20A length). The values for the former are 0.83 kT/e (WT) and 0.09 kT/e (M7) and the values for the latter are 0.083 kT/e (WT) and 0.019 kT/e (M7), respectively (see Figure 2). Thus, while the mean electrostatic potential is reduced 10-fold for the M7 enzyme, the mean flux is reduced only 4-fold. The square grid was positioned 16A away from the C_α of W286, and the calculated flux did not vary much as a function of the grid position. These calculations show that the mutated enzyme may still be capable of attracting the substrate, acetylcholine, using an electrostatic mechanism of guidance towards the active site. These

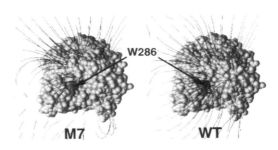

Figure 1. Electric field lines for acetylcholinesterase for both M7 and for the wild-type enzyme.

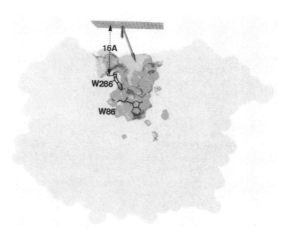

Figure 2. Mean electrostatic flux. The figure shows a detailed part of the molecular surface of the model of HrAChE, associated with the active site gorge, and an outline of the remainder of the enzyme. The arrows represent the mean value of the electrostatic flux on a square grid ($20 \times 20 Å^2$) positioned at the top of the gorge mouth and $16 Å$ away from the C_α of W286. The large arrow (0.083 kT/e) corresponds to the wild type HrAChE while the small one (0.019 kT/e) corresponds to the M7 mutant.

electrostatic effects may also be important in understanding how AChE clears the products of catalysis (Gilson et al., 1994) from the active site.

CONCLUSIONS

Our calculations of the electrostatic properties of some of these mutants and comparison with the wild type enzyme indicate that, in spite of the changes of the relatively low-value isopotentials, the most negative values of the electrostatic potential are still close to the catalytic residues (the most attractive region for the substrate). More important, the electrostatic field generated by the native enzyme is still substantial in the mutated enzyme and is directed toward the active site gorge. Independent calculations of the diffusional encounter rate constants for dumbbell models of the ligand also support the notion that the electrostatic field indeed guides the cationic substrate to the mouth of the active site gorge (Antosiewicz et al., 1995).

ACKNOWLEDGMENTS

This project was supported by the Cornell Theory Center, which receives major funding from the National Science Foundation and International Business Machines Corporation with additional support from New York State Science and Technology Foundation, from members of the Corporate Research Institute and from the NIH Grant #P41RRO-4293, the U.S. Army Medical Research and Development Command under Contract DAMD17-93-C-3070, the Association Franco-Israélienne pour la Recherche Scientifique et Technologique (AFIRST), the Minerva Foundation, Munich, Germany and the Kimmelman Center for Biomolecular Structure and Assembly, Rehovot.

REFERENCES

Antosiewicz, J., Gilson, M.K., Lee, I.H. and McCammon, J.A. (1995) Biophys J 68:62-68.
Barak, D., Ariel, N., Velan, B. and Shafferman, A., 1992, *in Multidisciplinary Approaches to Cholinesterase Functions* (Shafferman, A. and Velan, B., Ed.), pp. 195-199, Plenum Press, New York.

Cygler, M., Schrag, J.D., Sussman, J.L., Harel, M., Silman, I., Gentry, M.K. and Doctor, B.P. (1993) Prot Sci 2:366-382.

Eichler, J., Anselmet, A., Sussman, J.L., Massoulié, J. and Silman, I. (1994) Mol Pharmacol 45:335-340.

Gilson, M.K. and Honig, B.H. (1988) Proteins: Struct Funct Genetics 3:32-52.

Gilson, M.K., Sharp, K.A. and Honig, B.H. (1988) J Comput Chem 9:327-335.

Gilson, M.K., Straatsma, T.P., McCammon, J.A., Ripoll, D.R., Faerman, C.H., Axelsen, P., Silman, I. and Sussman, J.L. (1994) Science 263:1276-1278.

Harel, M., Sussman, J.L., Krejci, E., Bon, S., Chanal, P., Massoulié, J. and Silman, I. (1992) Proc Natl Acad Sci USA 89:10827-10831.

Klapper, I., Hagstrom, R., Fine, R.M., Sharp, K.A., Gilson, M.K. and Honig, B.H. (1986) Proteins: Struct Funct Genetics 1:47-59.

Ripoll, D.R., Faerman, C.H., Axelsen, P., Silman, I. and Sussman, J.L. (1993) Proc Natl Acad Sci USA 90:5128-5132.

Shafferman, A., Ordentlich, A., Barak, D., Kronman, C., Ber, R., Bino, T., Ariel, N., Osman, R. and Velan, B. (1994) EMBO J 13:3448-3455.

Sussman, J.L., Harel, M., Frolow, F., Oefner, C., Goldman, A., Toker, L. and Silman, I. (1991) Science 253:872-879.

SUBSTRATE BINDING SITE AND THE ROLE OF THE FLAP LOOP IN *CANDIDA RUGOSA* LIPASE, A CLOSE RELATIVE OF ACETYLCHOLINESTERASE[*]

Miroslaw Cygler, Pawel Grochulski, and Joseph D. Schrag

Biotechnology Research Institute
National Research Council
6100 Royalmount Avenue
Montréal, Québec H4P 2R2, Canada

INTRODUCTION

Candida rugosa lipase belongs to a large lipase/esterase family of evolutionarily related hydrolytic enzymes identified on the basis of their amino acid homology (Krejci *et al.*, 1991; Gentry *et al.*, 1991; Cygler *et al.*, 1993). Other members of this family include acetyl- and butyrylcholinesterases, carboxylesterases, cholesterol esterases, *etc.* The catalytic apparatus of these enzymes is composed of a Ser-His-Glu/Asp triad. Apart from the hydrolytic enzymes there are also other proteins with domains that clearly belong to this family but are devoid of the hydrolytic activity, eg. neurotactin, thyroglobulin, *etc.* The amino acid identity within this family varies from ~16%, for distantly related proteins, to 97% and a number of subfamilies are clearly distinguishable (Cygler *et al.*, 1993). To date, the crystal structures of three of these proteins have been determined: *Torpedo californica* acetylcholinesterase (AChE, Sussman *et al.*, 1991), *Geotrichum candidum* lipase (GCL, Schrag *et al.*, 1991) and *Candida rugosa* lipase (CRL, Grochulski *et al.*, 1993). These three enzymes display great similarities in their three-dimensional structures and provide prototypic models for other members of this large family. They belong to the α/β hydrolase fold superfamily (Ollis *et al.*, 1992). The two lipases, which share ~40% sequence identity, have also more similar structures. Approximately 90% of their Cα atoms superimpose with a root-mean-square (rms) deviation of 1.4Å. Comparison of GCL and TcAChE (~25% identity) shows that ~75% of their Cα atoms can be superimposed with a 1.9Å rms, including the catalytic apparatus composed of a Ser-His-Glu triad. Yet despite their significant structural similarity TcAChE and GCL/CRL hydrolyze very different substrates. While the

[*] NRCC publication no. 0000

Enzymes of the Cholinesterase Family, Edited by Daniel M. Quinn et al.
Plenum Press, New York, 1995

lipases break down water insoluble triglycerides, the acetylcholinesterase hydrolyzes the water soluble neurotransmitter, acetylcholine. The structure of TcAChE revealed that its active site is located at the bottom of a deep gorge and is accessible to the substrate. The consideration of the shape and type of the sidechains near the catalytic Ser 200 allowed Sussman and coworkers (1991) to propose a model of acetylcholine binding which was subsequently supported by structural data on complexes with noncovalently bound TcAChE inhibitors (Harel et al., 1993) and a covalent complex with a trifluoroketone analog of the substrate (Harel et al., this volume). TcAChE is characterized by a very high turnover number. It has been postulated that the electrostatic guidance of the substrate toward the bottom of the gorge plays an important role in the catalysis (Rippol et al., 1993). At the same time the existence of this strong electrostatic field posed a question regarding the route of escape of the products, especially a positively charged choline, from the bottom of the gorge. Recent molecular dynamics simulations (Gilson et al., 1994) provided some support for a possibility of a rearrangement of residues forming one wall of the binding site, centered around Trp 84, which would result in an opening of another access (or escape) route to the active site. These findings bring to focus the question of a role of the 67-94 loop (TcAChE numbers) in the catalytic cycle of acetylcholinesterases. This loop not only contains Trp 84 but also houses residues participating in the formation of the peripheral anionic binding site. The 67-94 loop is constrained at its base by a disulfide bridge (Cys 67 - Cys 94) and a salt bridge (Glu 92 - Arg 44). Site-directed mutagenesis indicated that the replacement of residues stabilizing the base of this loop and located far away from the substrate binding site, Glu 92 (to Gln or Leu) and Asp 93 (to Val or Asn), leads to a substantial decrease or abolition of the enzymatic activity of TcAChE (Bucht et al., 1994). This effect could result from some destabilization of the productive conformation of the 67-94 loop or from affecting its dynamic properties.

What can the investigation of structurally related lipases bring to our understanding of cholinesterase catalysis? These lipases also possess the loop equivalent to the 67-94 loop of TcAChE. As it turns out, this loop is very important for the catalytic events in lipases. It assumes multiple conformations, depending on the environment of the enzyme. The conformational rearrangement of this loop plays a crucial role in the interfacial activation of GCL and CRL lipases (Grochulski et al., 1993).

LIPASES

Although the biologically important substrates for lipases are triacylglycerides, these enzymes also hydrolyze a wide range of water soluble esters, albeit with various efficiency (eg. Jensen, Galluzo & Bush, 1990). It has been known for many years that the catalytic efficiency of lipases increases substantially in the presence of the lipid/water interface (Sarda & Desnuelle, 1956), a phenomenon known as interfacial activation. It was thought that the observed activation is associated with a conformational change in the enzyme that occurs near or at the interface. The evidence in support of this model was obtained only in recent years when the structures of the first lipases were determined. The first three-dimensional structures identified the presence of a serine protease-like catalytic triad in lipases and revealed the fact that the access to the active site is obscured by one or more loops (flap, Winkler et al., 1990; Brady et al., 1990; Schrag et al., 1991). In recent years at least three lipases were crystallized under various conditions, with and/or without inhibitors, and at least two different conformations have been observed in each case: one in which the active site triad is occluded from the solvent, and another in which the active site is available to the substrate (Brzozowski et al., 1991; Grochulski et al., 1994a; van Tilbeurgh et al., 1994).

The postulated mechanism for the lipase catalyzed reactions is rather similar to that proposed for serine proteases. In the first step the nucleophilic attack by the active site serine leads to the formation of a tetrahedral intermediate, followed by the release of the leaving alcohol group and the formation of acyl enzyme intermediate. In the second step an activated water molecule attacks the acyl enzyme, the second tetrahedral intermediate is formed, followed by a release of a free acid and a regeneration of the active enzyme (*eg.* Grochulski *et al.*, 1994b).

OVERALL STRUCTURES OF *G. candidum* and *C. rugosa* LIPASES

GCL and CRL are globular, single domain α/β type proteins with approximate dimensions of 45 x 60 x 65Å. They contain a small three stranded β-sheet at the N-terminus followed by a 11-stranded mixed β-sheet with interspersed 16 α-helices. The connections on the N-terminal side of the large β-sheet are mostly short, whereas the connections on the C-terminal side are long and form a cap over the top of the β-sheet. The fold of these proteins conforms to the α/β hydrolase fold identified in a number of hydrolases with unrelated sequences (Ollis *et al.*, 1992). The central β-sheet is significantly twisted, to the extent that the strands on the opposite ends form an angle close to 90° (Schrag & Cygler, 1993). The catalytic triad, Ser 209-His 449-Glu 341 in CRL and Ser 217-His 463-Glu 354 in GCL, is located near the center of the sheet at the C-terminal side of the strands. Following the standard nomenclature of Ollis *et al.* (1992), the catalytic serine is situated at the end of strand 5, in a tight bend between this strand and an α-helix. The acid comes from a short loop after strand 7 and the histidine is embedded in a rather long loop after strand 8.

THE FLAP

While the structures of both GCL and CRL have been determined and refined to a high resolution, only for CRL have multiple conformations been observed. Comparison of CRL structures revealed conformational flexibility of a single, extended omega loop, referred to as a flap, and located between the last strand of the small β-sheet and the second strand of the large β-sheet (Schrag & Cygler, 1993). The flap loop is anchored at its ends by a disulfide bridge, Cys 60-Cys 97, and contains 38 amino acids between the two cysteines. Additional stabilization at the bottom of the loop is provided by a salt bridge, Glu 95-Arg 37 (Grochulski *et al.*, 1994b). The flap undergoes a major conformational change which involves movement of 27 residues with clearly defined hinge points, Glu 66 and Pro 92, with the tip of the loop being shifted by as much as ~30Å. The flap *does not* move as a rigid body but instead partially refolds in the process (Grochulski *et al.*, 1994b). In the inactive conformation of CRL the flap has an elongated shape and lies flatly on the protein surface occluding the active site (Fig. 1). It encompasses a distorted helical turn and an α-helix and shows clearly an amphipathic character. The side of the flap facing the protein is hydrophobic and interacts with the hydrophobic residues surrounding the active site while the side directed toward the solvent is distinctly hydrophilic. Pro 92, located near the C-terminal end of the flap, adopts a *cis* conformation. Upon rearrangement from the closed to the open conformation the flap rotates almost 90° and becomes extended nearly perpendicularly to the protein surface (Fig. 1). This movement also involves a rotation around the C^{Ser91}-N^{Pro92} peptide bond and involves a *cis-trans* isomerization of Pro 92. Despite the positional shift and refolding from the inactive 'closed' conformation to the active 'open' conformation, the flap maintains its amphipathic character. The opening movement of the flap exposes a large hydrophobic depression with the active site at the bottom, provides an access to the catalytic Ser 209 and creates a proper binding site for the triglyceride. The depression narrows down

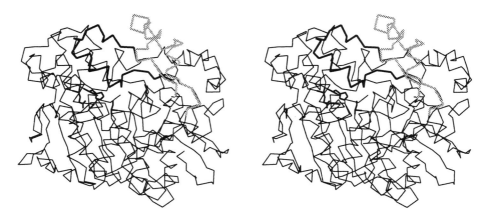

Figure 1. Stereo drawing of the Cα tracing of *Candida rugosa* lipase with open (steepled lines) and closed (thick lines) conformations shown. Residues of the catalytic triad are shown in full.

near the catalytic serine and connects to a long tunnel which leads toward the center of the protein. This tunnel exists also in the closed form of CRL but is sealed off from the solvent by the flap.

The flap of CRL has an equivalent in GCL with a corresponding disulfide bridge and a salt bridge encompasing 45 residues. Although GCL has only been observed with the flap in a closed conformation, the flap is predicted to undergo a significant conformational rearrangement upon interfacial activation (Grochulski *et al.*, 1994a). As in CRL, there is also a long tunnel in the interior of GCL but with a more complex shape. It is partly filled with ordered solvent molecules (Schrag & Cygler, 1993). A different conformation of the loop contributing to the oxyanion hole in CRL *vs* GCL suggests that the activation of GCL involves rearrangement of other loops in addition to the flap (Schrag & Cygler, 1993).

SUBSTRATE BINDING SITE

Alkyl Chain Binding

To delineate the binding site for the fatty acyl chains of the triglyceride substrate we reacted CRL with hexadecanesulfonyl chloride (HDSC), crystallized the enzyme-inhibitor complex and determined its structure (Grochulski *et al.*, 1994b). As expected, the sulphonyl group was found covalently bound to Ser 209. The alkyl chain of the inhibitor was found in the tunnel described above. This inhibitor is an analog of the second tetrahedral intermediate of the reaction catalyzed by the lipase. The alkyl chain corresponds to the position of the scissile fatty acid product. The location of the fatty acyl chain of the substrate in a tunnel inside the lipase is unique among the lipases studied to date. The residues forming the entrance to the tunnel are: Met 213, Leu 304, Phe 345 and Phe 415. The rest of the tunnel is lined with sidechains of Pro 246, Leu 302, Val 534, Phe 362, Phe 366, Ser 365 and the alkyl part of Arg 303. There is a prominent change in the tunnel direction near Leu 302. Comparison of the CRL-inhibitor complex with the unliganded open conformation shows that there are no major shifts in the backbone atoms of the protein. As mentioned above, this tunnel is also present in the closed conformation of CRL, although it is occluded from the solvent by the flap. The binding tunnel is, thus, preformed and activation requires only conformational changes which expose its entrance. Only small side chain movements of residues lining the tunnel are required to accommodate the scissile fatty acyl chain. The alkyl chain follows the shape of the tunnel by adopting a *gauche⁻* conformation at C7-C8 bonds and *gauche⁺* conformations at C10-C11 and C11-C12 bonds. As a result, C2-C8 and C11-C15 sections run in nearly perpendicular directions.

Oxyanion Hole

As is the case for serine proteases, the stabilization of the negative charge developing on the oxygen in the transition state during the reaction is aided by the hydrogen bonds from this oxygen to the mainchain NH groups of the so called oxyanion hole. These hydrogen bonds are formed to NH of Ala 210 and Gly 124. The NH of Gly 123 could also be involved in the formation of a hydrogen bond to the oxygen of the substrate. In the CRL structure there is a water molecule that competes for this hydrogen bond (Grochulski *et al.*, 1994b). Additional stabilization of the oxyanion is achieved by the electric dipole of a neighbouring α-helix (residues 210-220), which follows immediately after the nucleophilic serine (Hol *et al.*, 1978).

COMPARISON TO TcAChE

The flap described in CRL and GCL has its structural equivalent in TcAChE. This is the same loop mentioned above that contains Trp 84 and residues from the peripheral anionic binding site. The TcAChE loop, residues 67-94 (28aa), is shorter than in lipases but is stabilized by the same two features: a disulfide link and a salt bridge (Fig. 2). Interestingly,

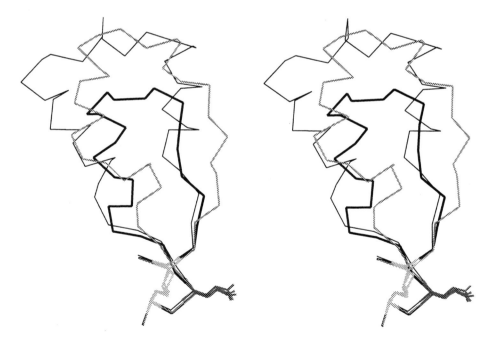

Figure 2. Flaps from GCL (thin lines), CRL (closed conformation, thin steepled) and 67-94 loop of TcAChE (thick lines) after superposition of the three structures. The common Cys-Cys bridge and Glu involved in the salt bridge are shown in full (steepled).

this loop has a much less amphipathic character than the flaps in lipases. The residues lining the face of this loop which is directed toward the rest of TcAChE include Gln 69, Tyr 70, Asp 72, Gln 74, Ser 81 and Asn 87. The loop is important for TcAChE catalysis. The Trp 84 positions the choline group of the substrate and possibly swings out to create another access to the active site (Gilson *et al.*, 1994). It was shown that mutations near the base of this loop which disrupt the interactions stabilizing the bottom of the loop are detrimental for enzymatic activity (Bucht *et al.*, 1994). In light of the importance of the flexibility of this loop in lipases it is possible that similar dynamic properties of the 67-94 loop of TcAChE may play an important role in the catalytic mechanism of TcAChE and especially in the release of the products *via* an alternate route. Systematic mutational analysis of this region should provide much needed data to understand the role of this region in detail.

REFERENCES

Brady, L., Brzozowski, A.M., Derewenda, Z.S., Dodson, E., Dodson, G., Tolley, S., Turkenburg, J.P., Christiansen, L., Huge-Jensen, B., Norskov, L., Thim, L., and Menge, U., 1990. *Nature* 343:767-770.

Brenner, S., 1980. *Nature* 34:528-530.

Brzozowski, A.M., Derewenda, U, Derewenda, Z.S., Dodson, G.G., Lawson, D.M. Turkenburg, J.P., Bjorkling, F., Huge-Jensen,B., Patkar, S., and Thim, L., 1991. *Nature* 351:491-494.

Bucht,G., Häggström, B., Radić, Z., Osterman, A., and Hjalmarsson, K., 1994. *Biochim. Biophys. Acta* in press.

Connolly, M.L. 1983. *J. Appl. Crystallogr.* 16:548-558.

Cygler, M., Schrag, J. D., Sussman, J. L., Harel, M., Silman, I., Gentry, M. K., and Doctor, B. P., 1993. *Protein Science* 2:366-382.

Cygler, M., Grochulski, P., Kazlauskas, R.J., Schrag, J.D., Bouthillier, F., Rubin, B., Serrequi, A.N., and Gupta, A.K., 1994. *J. Am. Chem. Soc.* 116:3180-3186.

Cygler, M., Schrag, J.D., and Ergan, F., 1992. *Biotechnology and Genetic Engineering Reviews* 10:143-184.

Derewenda, Z.S., and Derewenda, U., 1991. *Biochem. Cell Biol.* 69:842-851.

Gentry, M.K., & Doctor, B.P., 1991. In *Cholinesterases: Structure, Function, Mechanism, Genetics and Cell Biology* (J. Massoulié, F. Bacou, E. Barnard, A. Chatonnet, B.P. Doctor & D.M. Quinn, eds.) pp. 394-398, American Chemical Society, Washington, DC.

Gilson, M., Straatsma, T.P., McCammon, J.A., Ripoll, D.R., Faerman, C.H., Axelsen, P.H., Silman, I., and Sussman, J.L., 1994. *Science* 263:1276-1278.

Grochulski, P., Li, Y., Schrag, J.D., and Cygler, M., 1994a. *Protein Sci.* 3:82-91.

Grochulski, P. Bouthillier, F. Kazlauskas, R.J., Serreqi, A.N., Schrag, J.D., Ziomek, E., and Cygler, M., 1994b. *Biochemistry* 33:3494-3500.

Harel, M., Schalk, I., Ehret-Sabatier, L., Bouet, F., Goeldner, M., Hirth, C., Axelsen, P., Silman, I., and Sussman, J.L., 1993. *Proc. Natl. Acad. Sci. USA* 90:9031-9035.

Hol, W.G.J., van Duijnen, P.T., and Berendsen, H.J.C., 1978. *Nature* 273:443.

Jensen, R.G., Rubano Galluzo, D., and Bush, V.J. 1990. *Biocatalysis* 3:307-316.

Krejci, E., Duval, N., Chatonnet, A., Vincens, P., and Massoulié, J., 1991. *Proc. Natl. Acad. Sci. USA* 88:6647-6651.

Ollis, D.L., Cheah, E., Cygler, M., Dijkstra, B., Frolow, F., Franken, S.M., Harel, M., Remington, S.J., Silman, I., Schrag, J.D., Sussman, J.L., Verschueren, K.H.G., and Goldman A., 1992. *Protein Eng.* 5:197-211.

Rippol, D.R., Faerman, C.H., Axelsen, P.H., Silman, I., and Sussman, J.L., 1993. *Proc. Natl. Acad. Sci. USA* 90:5128-5132.

Sarda, L., and Desnuelle, P., 1958. *Biochim. Biophys. Acta* 30:513-521.

Schrag, J.D., and Cygler, M., 1993. *J. Mol. Biol.* 230:575-591.

Schrag, J.D., Li, Y., Wu, S., and Cygler, M., 1991. *Nature* 351:761-764.

Sussman, J.L., Harel, M., Frolow, F., Oefner, C., Goldman, A., Toker, L., and Silman, I. 1991. *Science* 253:872-879.

van Tilbeurgh, H., Egloff, M. P., Martinez, C., Rugani, N., Verger, R., and Cambillau, C., 1993. *Nature* 362:814-820.

Winkler, F.K., D'Arcy, A., and Hunziker, W., (1990). *Nature* 343:771-774.

STUDIES ON PARTIALLY UNFOLDED STATES OF *TORPEDO CALIFORNICA* ACETYLCHOLINESTERASE

Israel Silman,[1] David I. Kreimer,[1] Irina Shin,[1] Elena A. Dolginova,[1] Ester Roth,[1] Daniella Goldfarb,[2] Reuven Szosenfogel,[2] Mia Raves,[3] Joel L. Sussman,[3] Nina Borochov,[4] and Lev Weiner[5]

[1] Department of Neurobiology
[2] Department of Chemical Physics
[3] Department of Structural Biology
  Weizmann Institute of Science
  Rehovot, Israel
[4] The Center of Technological Education
  Holon, Israel
[5] Department of Organic Chemistry
  Weizmann Institute of Science
  Rehovot, Israel

INTRODUCTION

Chemical modification, by a repertoire of thiol reagents, of the non-conserved Cys[231] residue of *Torpedo californica* acetylcholinesterase (AChE), results in inactivation, even though Cys[231] is not involved in catalysis (Steinberg *et al.*, 1990; Dolginova *et al.*, 1992; Silman *et al.*, 1992; Salih *et al.*, 1993). Modification by disulfides and alkylating agents produces partial unfolding of native AChE (**N**) to a compact state resembling a molten globule (**MG**). The **MG** is a collapsed structure possessing much of the secondary structure of the fully folded native protein, but devoid of tertiary structure (Kuwajima, 1989; Ptitsyn, 1992); it is currently believed to serve as a folding intermediate *en route* from the nascent polypeptide chain, generated on the ribosome, to the fully folded native protein (Kim and Baldwin, 1990; Gething and Sambrook, 1992). This structural assignment for the species produced by chemical modification of *Torpedo* AChE was based upon spectroscopic evidence, including CD, intrinsic fluorescence and binding of ANS, upon hydrodynamic measurements, including sucrose gradient centrifugation and quasielastic light scattering, and upon enhanced sensitivity to proteolysis (Dolginova *et al.*, 1992). Although modification by disulfides can be rapidly reversed by exposure to reduced glutathione (GSH), the native (**N**) conformation is not restored, and no catalytic activity is recovered. AChE so demodified is a partially unfolded species whose physicochemical characteristics are virtu-

Enzymes of the Cholinesterase Family, Edited by Daniel M. Quinn et al.
Plenum Press, New York, 1995

ally identical to those of the modified enzyme, and sucrose gradient centrifugation reveals it to be stable for many hours without aggregating.

MODIFICATION BY ORGANOMERCURIALS

Chemical modification by mercurials also inactivates *Torpedo* AChE. However, such modification produces a **N**-like state, viz. a conformational state much closer to **N**, although also devoid of catalytic activity. Upon demodification with GSH of *Torpedo* AChE freshly modified with mercurials, up to 85% of the initial AChE activity is recovered, with full restoration of the spectroscopic characteristics of **N**. However, the **N**-like state produced by mercurials is not stable; it unfolds spontaneously, with a $t_{1/2}$ of *ca.* 1 hour, to a non-reactivatable, partially unfolded form which is similar to that produced by the disulfides and alkylating agents. Arrhenius plots show that the **N**-like state is separated by a low (5 kcal/mol) energy barrier from the **N** state, whereas the partially unfolded, **MG**-like state is separated from the **N**-like state by a high-energy barrier (ca. 50 kcal/mol). A schematic representation of the intraconversions between these various states is shown in Figure 1.

Comparison of the 3-D structures of native *Torpedo* AChE and of a heavy-atom derivative obtained with HgAc$_2$ (Sussman *et al.*, 1991), suggested a basis for the stabilization produced by the mercurials relative to other thiol reagents: the mercurial-modified enzyme may be stabilized by additional interactions of the mercury atom attached to Cys^{231}, specifically with Ser$^{228}O\gamma$ and with the main-chain nitrogen and carbonyl oxygen of the same serine residue (Figure 2).

The reason why AChE remains in the partially unfolded state produced by chemical modification, even after modification has been reversed, is not yet understood. Two explanations may be offered for this phenomenon. The first explanation would involve kinetic trapping. It would assume that, even though the **N** state might be at a lower free energy-level than the **MG**-like state, they are separated by a high free-energy barrier. Examples of such kinetic trapping have appeared in the literature, one of the best known being that of α-lytic protease (Baker *et al.*, 1992), for which a partially unfolded state will not fold back to the native conformation in the absence of a pro-region. AChE, however, does not contain a pro-region (Maulet *et al.*, 1990). A high energy barrier, ca. 20 kcal/mol, is observed in protein transitions involving isomerization of proline (Koide *et al.*, 1993), and there are two proline residues, Pro^{228} and Pro^{232}, adjacent to Cys^{231}, both of which are in the *trans* conformation (Sussman *et al.*, 1991). It is possible that the chemical modification of Cys^{231} 'drives' one or both of these prolines to a *cis* conformation, thus providing a high

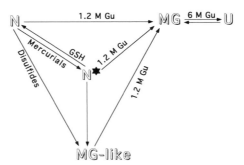

Figure 1. Schematic representation of the various conformational states of *Torpedo* AChE and their modes of interconversion.

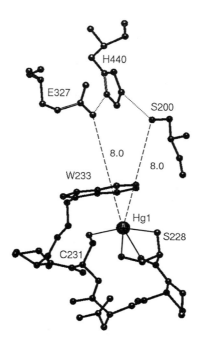

Figure 2. Crystal structure of mercurial-modified *Torpedo* AChE, showing the position of the Hg atom, relative to the catalytic triad. The distances of Oγ of Ser[200] and to Oε1 of Glu[327] are both 8.0 Å. Trp[233], one of the aromatic residues whose rings line the active-site gorge, lies below the plane formed by the catalytic triad and the Hg atom. It can be seen that this Hg atom displays close contacts not only with Sγ of Cys[231], but also with Oγ, N and O of Ser[228].

energy barrier to the reverse transition. Since *Torpedo* AChE is a large protein (a dimer of two identical subunits, with 537 residues each), it is also possible that non-native domain pairing and/or intersubunit interactions are responsible for trapping AChE in the partially unfolded state (Jaenicke, 1991).

A second explanation would be that the **N** state of *Torpedo* AChE does not correspond to the global free-energy minimum, and that the energy of the **MG**-like state is, in fact, lower than that of **N**. This seems plausible, since the quasi-**N** state produced by mercurials cannot be energetically very different from the native enzyme, yet it converts spontaneously to its partially unfolded **MG**-like counterpart. This, in turn, cannot be at an energy level very different from that of the demodified enzyme (or the analogous demodified enzyme obtained after modification with disulfides), since their spectroscopic properties are quite similar. A recent theoretical paper has indeed made the point that 'identification of native states with the most compact or minimum energy states may not strictly hold' (Bahar and Jernigan, 1994).

MONITORING AND CHARACTERIZATION OF THE MG↔U TRANSITION

Since the stable states produced by chemical modification display many of the features of the **MG**, we also produced such states of AChE by 'traditional' methods, e.g. by exposure to 1.2 M guanidinium chloride (Gu). In 1.2-2.1 M Gu, AChE is in a **MG** state, but the **N→MG** transition is irreversible; upon removing Gu, no enzymic activity is detected, and the spectroscopic characteristics of **N** are not recovered. In 6 M Gu, AChE is in an unfolded state (**U**), in reversible equilibrium with **MG** (Fig. 1). If Cys[231] is labelled selectively, with a mercury derivative of a stable nitroxyl radical, it can be seen that the EPR signal of the **MG** state is highly immobilized, whereas in the **U** state the probe is almost freely rotating. It is thus possible to observe the signals corresponding to each of these two states in the presence of the other, thus to show that they co-exist in the transition region and to measure their relative amounts. Upon elevating the Gu concentration, a decrease in the EPR signal corresponding to the **MG** state occurred concomitantly with an increase in that of the **U** state, the integral intensity of the EPR spectra remaining constant. Such behavior is characteristic of a two-state transition. The thermodynamic characteristics of the transition, whether estimated directly from the EPR data, or from both CD and fluorescence measurements, assuming a two-state scheme, are in good agreement. There has been considerable controversy, at both the theoretical (Alonso *et al.*, 1991; Finkelstein and Shakhnovich, 1991) and experimental (cf. Shimizu *et al.*, 1993; Uversky, 1993) levels, as to whether the **MG↔U** transition can, indeed, be described by a two-state model. Our data clearly demonstrate this to be the case for AChE, and show that EPR can serve as a powerful tool in monitoring such transitions.

GENERATION OF AN MG-LIKE STATE BY OXIDATIVE STRESS

Exposure of purified *Torpedo* AChE to a system generating oxygen radicals (viz. ascorbic acid/Fe(EDTA)$_2$/H$_2$O$_2$) led to inactivation. The enzyme retained its dimeric form, but electrophoresis under denaturing conditions revealed some cleavage of peptide bonds. Spectroscopic examination showed that the partially inactivated enzyme displayed spectral properties resembling those of the **MG**-like state produced by chemical modification, and that it also displayed enhanced susceptibility to proteolysis (Weiner and Silman, 1993; Weiner *et al.*, 1994). These observations may provide a model system for understanding the consequences of oxidative stress *in vivo*. We propose that partially unfolded proteins, generated by oxidative stress, may interact with molecular chaperones of the heat-shock family, thus leading to release of the heat-shock transcription factor, and to activation of heat-shock genes, as has been shown to occur in the heat-shock response (Morimoto, 1993; Matts *et al.*, 1993). The heat-shock proteins so generated could then combine with the misfolded proteins and protein fragments produced by oxidative stress, and eliminate them from the cell by transport into lysosomes, followed by degradation (Chang *et al.*, 1989). Our data thus suggest a molecular basis for the overlapping regulation of heat shock and oxidative stress which has been noted by a number of laboratories (cf. Tartaglia *et al.*, 1991; Keyse and Emslie, 1992).

INTERACTION OF THE MG-LIKE STATE OF *TORPEDO* AChE WITH LIPOSOMES

It has been proposed that the **MG** may serve not only as a folding intermediate in the biosynthesis of proteins, but also in their translocation across or insertion into plasma membranes (cf. van der Goot *et al.*, 1991). As mentioned above, the **MG**-like states of *Torpedo* AChE which we have generated are stable for many hours, under physiological conditions, without undergoing aggregation. They thus provide an experimental system for

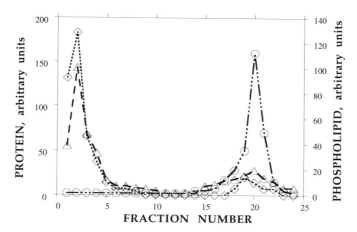

Figure 3. *Torpedo* AChE in an **MG** state interacts with the lipid bilayer. To generate an **MG** state, *Torpedo* AChE was treated with 1.2 M Gu in 50 mM Na-phosphate, pH 7.3, for 2 h at room temperature, after which the Gu was removed by gel filtration. The gel-filtered protein sample was mixed with small unilamellar vesicles of dimyristoylphosphatidylcholine obtained by ultrasonication. An aliquot of the mixture was loaded at the bottom of a sucrose step gradient (Futerman *et al.*, 1985), which was then centrifuged for 7 h at 50,000 rpm in a Beckman SW50.1 rotor. Using such a protocol, the liposomes float to the top of the gradient (◊). AChE in the **MG** state accompanies the liposomes (△), whereas native *Torpedo* AChE, which does not interact with the liposomes, remains at the bottom of a control gradient (O).

investigating **MG** interactions with lipid bilayers. Indeed, in preliminary experiments, using a flotation gradient technique, we have been able to demonstrate rapid insertion of AChE in the **MG**-like state into dimyristoylphosphatidylcholine liposomes, whereas native AChE displays no such interaction (Fig. 3). 1 M NaCl does not significantly decrease the interaction, showing that electrostatic forces do not play a major role. It was also observed that interaction of the **MG** with the liposomes causes leakage of fluorescent (calcein) or spin (TEMPO-choline) probes which had been pre-loaded into the liposomes.

CONCLUDING REMARKS

Our ability to generate partially unfolded, stable states of *Torpedo* AChE, by either chemical or physical manipulation, thus provides a model system which permits an experimental approach to several problems of current interest to protein chemists, to membrane biophysicists and to cell biologists.

REFERENCES

Alonso, D.O.V., Dill, K.A.& Stigter, D. (1991) *Biopolymers 31*, 1631-1649.
Bahar, I. & Jernigan, R.L. (1994) *Biophys. J. 66*, 454-466.
Baker, D., Sohl, J.L.& Agard, D.A.(1992) *Nature 356*, 263-265.
Chang, H.L., Teriecky, S.K., Plant, C.P.& Dice, J.P. (1989) *Science 246*, 382-385.
Dolginova, E.A., Roth, E., Silman, I.& Weiner, L.M. (1992) *Biochemistry 31*, 12248-12254.
Finkelstein, A.V.& Shakhnovich, E.I. (1989) *Biopolymers 28*, 1681-1694.
Futerman, A.H., Fiorini, R.-M., Roth, E., Low, M.G. & Silman, I. (1985) *Biochem. J. 226*, 369-377.
Gething, M.-J. & Sambrook, J. (1992) *Nature 355*, 33-45.
Jaenicke, R. (1991) *Biochemistry 30*, 3147-3161.

Keyse, S.M. & Emslie, E.A. (1992) *Nature 359,* 644-647.

Kim, P.S. & Baldwin, R.L. (1990) *Annu Rev Biochem 59,* 631-660.

Koide, S., Dyson, H.J. & Wright, P.E. (1993) *Biochemistry 32,* 12299-12310.

Kuwajima, K. (1989) *Proteins 6,* 87-103.

Matts, R.L. Hurst, R. & Xu, Z. (1993) *Biochemistry 32,* 7323-7328.

Maulet, Y., Camp, S., Gibney, G., Rachinsky, T., Ekström, T.J. & Taylor, P.(1990) *Neuron 4,* 289-301.

Morimoto, R.I. (1993)*Science 259,* 1409-1410.

Ptitsyn, O.B. (1992) in *Protein Folding* (Creighton, T.E., Ed.), pp. 243-300, W.H. Freeman, New York

Salih, E., Howard, S., Chishti, B., Cohen, S.G., Liu, W.S. & Cohen, J.B. (1993) *J Biol Chem 286,* 245-251.

Shimizu, A., Ikeguchi, M. & Sugai, S. (1993) *Biochemistry 32,* 13198-13203.

Silman, I., Krejci, E., Duval, N., Bon, S., Chanal, P., Harel, M., Sussman, J.L. & Massoulié, J. (1992) in
  Multidisciplinary Approaches to Cholinesterase Functions (Shafferman, A. & Velan, B., Eds.), pp.
  177-183, Plenum Press, New York.

Steinberg, N., Roth, E.& Silman, I. (1990) *Biochem Internat 21,* 1043-1050.

Sussman, J.L., Harel, M., Frolow, F., Oefner, C., Goldman, A., Toker, L. & Silman, I. (1991) *Science 253,*
  872-879.

Tartaglia, L.A., Storz, G., Farr, S.B. & Ames, B.N. (1991) in *Oxidative Stress: Oxidants and Antioxidants* (Sies,
  H., Ed.), pp. 155-169, Academic Press, New York.

Uversky, V.N. (1993) *Biochemistry 32,* 13288-13298.

van der Goot, F.G., Gonzáles-Mañas, J.M., Lakey, J.H. & Pattus, F. (1991) *Nature 354,* 408-410.

Weiner, L. & Silman, I. (1993) *Free Rad. Biol. Med. 15,* 524.

Weiner, L., Kreimer, D., Roth, E.& Silman, I. (1994) *Biochem. Biophys. Res. Comm. 198,* 915-922.

FTIR-SPECTROSCOPIC INVESTIGATIONS OF THE STRUCTURE AND TEMPERATURE STABILITY OF THE ACETYLCHOLINESTERASE FROM *TORPEDO CALIFORNICA*

Ferdinand Hucho[1], Dieter Naumann,[2] and Ute Görne-Tschelnokow[2]

[1] Institut für Biochemie
Freie Universität Berlin
Thielallee 63, 14195 Berlin, Germany
[2] Robert-Koch-Institut
13353 Berlin, Germany

INTRODUCTION

The acetylcholinesterase (AChE) is of interest as a key protein in chemical nerve impulse transmission at cholinergic synapses and as a model for esterases having a catalytic triad including an activated serine -OH group (Shafferman and Velan, 1992).

In the investigation presented here we used AChE as a model protein to validate the interpretation of Fourier transform infrared (FTIR) spectroscopy data which we obtained with another key protein of the cholinergic synapse, the nicotinic acetylcholine receptor (Görne-Tschelnokow et al., 1994). We show that the FTIR data concerning the secondary structure content of the enzyme are in good agreement with the three-dimensional structure of the crystalline enzyme obtained by X-ray diffraction analysis (Sussman et al., 1991). By investigating the temperature dependence of the FTIR spectra we further show, that irreversible inactivation of the enzyme occurs at temperatures where the secondary structure is not yet affected. And, finally, we describe a temperature-induced secondary structure change which we interpret as 'β-aggregation'.

MATERIALS AND METHODS

The G2-form of AChE was used exclusively in this investigation. It was released from *Torpedo californica* electric tissue membranes by cleavage of the phospholipid anchor with phosphatidylinositol-specific phospholipase C from *Bacillus thuringensis*. The enzyme

Enzymes of the Cholinesterase Family, Edited by Daniel M. Quinn et al.
Plenum Press, New York, 1995

method	α-helical structures	β-sheet-structures	β-turn-structures	undefined structures
X-ray	32 %	15 %		
CD-spectr.[1]	40 %	35 %	4 %	
CD-spectr.[2]	34 %	25 %		
CD-spectr.[3]	33 %	23 %	17 %	26 %
Raman-spectr.[4]	49 %	23 %	11 %	

Figure 1. Overview of the spectroscopically-obtained results. Estimation of secondary structure elements from the AChE. Notes: (1) tetrameric and dimeric form of the AChE from *Torpedo californica* (Wu et al., 1987); (2) A$_{12}$-form from *Torpedo californica* (Wu et al., 1987); (3) tetrameric ll-S-Form (Manavalan et al., 1985); (4) tetrameric ll-S-Form (Aslanian et al., 1987; Aslanian et al., 1991); CD-spectr.: Circular Dichroism spectroscopy; FTIR-spectr.: Fourier-Transform-Infrared-spectroscopy.

was kindly supplied by Dr. Christoph Weise. All methods and procedures have been described in Görne-Tschelnokow et al. (1993). Most of the data presented here are taken from this paper.

RESULTS AND DISCUSSION

Crystalline AChE was shown by X-ray analysis to contain roughly 33 % of the amino acids in α-helices and 15 % in β-strands (Sussman et al., 1991). CD- and Raman-spectroscopic data obtained with the enzyme in solution diverge considerably from these figures (Figure 1). Especially, they suggest a much higher β-structure content, which may be only partly due to the different esterase forms used in these investigations (tetrameric 11 S- or even the asymmetric A$_{12}$-forms).

Progress in infrared spectroscopy, providing better apparatus and improved mathematical resolution enhancement methods, promises more reliable spectroscopic secondary structure data of proteins in solution (Surewicz et al., 1993). Fig. 2 shows a survey spectrum of AChE, primarily in the so-called amide-I and amide-II absorption ranges.

The amide-I band is the most interesting part of the spectrum, because the assignment of protein secondary structure elements to components of this band is well established,

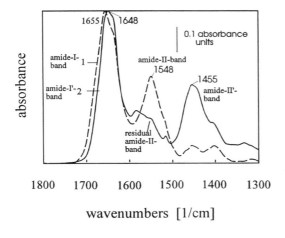

Figure 2. FTIR spectra of AChE. The FTIR spectra in H$_2$O buffer (- - -,1) and in D$_2$O buffer (——, 2) are shown after digital subtraction of the buffer spectra in the amide-I and amide-II regions.

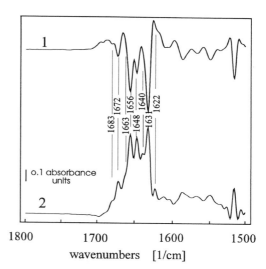

Figure 3. Resolution-enhanced FTIR spectra in the 1500-1800 cm⁻¹ region of AChE in D_2O buffer. Second-derivative spectrum (line 1) and Fourier self-deconvoluted spectrum (line 2).

whereas analysis of the amide-II band is less advanced. Unfortunately, the H_2O-absorption in the amide-I region is extremely strong. To circumvent the problem of overlapping H_2O-absorption, the deuterated protein was primarily investigated, because D_2O does not absorb in the amide-I region. H → D exchange has only little effect on the amide-I band, but shifts the amide-II band by about 100 wavenumbers. Unfortunately, with the AChE this exchange is not complete, as indicated by the residual amide-II band visible at 1548 cm⁻¹ after H → D exchange. Therefore the secondary structure content in AChE cannot be calculated from the spectra of the deuterated enzyme alone.

Mathematical resolution enhancement of the spectra of (partially) deuterated AChE by Fourier self-deconvolution and by calculating the 2nd derivative is shown in Fig. 3. The second-derivative spectra are valuable for localizing a band while the Fourier self-deconvolution allows quantitatively estimating the contribution of a particular absorption band (see below). Secondary structure assignments were made by comparing the spectra of protonated and deuterated AChE and with the help of data from the literature.

The strong band around 1656 cm⁻¹ (in D_2O) can be unambiguously assigned to α-helical structures. This assignment is based on empirical data obtained with authentic α-helical peptides and proteins (Susi, 1969; Koenig and Tabb, 1980; Byler and Susi, 1986; Susi and Byler, 1986) and on normal-mode calculations (Krimm and Bandekar, 1986). A different population of α-helices may contribute to the absorption at 1648 cm⁻¹. It has been calculated that there exists a systematic correlation between the length of an α-helix and the position of its absorption band (Nevskaya and Chirgadze, 1976): the band at the lower wavenumber (1648 cm⁻¹) accordingly, may be assigned to longer α-helices, the higher wavenumber band (1656 cm⁻¹) to shorter ones. X-ray analysis of AChE (Sussman et al., 1991) has shown two types of α-helices, long ones with fourteen residues or more, and short ones with eleven residues or less. Obviously, these two types can be discriminated by FTIR.

Analogous considerations allow the assignment of the bands at 1622 cm⁻¹, 1631 cm⁻¹, and 1683 cm⁻¹ to β-sheets and the bands at 1663 cm⁻¹ and 1672 cm⁻¹ to β-turn structures. Of special interest are the band at 1622 cm⁻¹ and the shoulder at 1690 cm⁻¹; we shall return to these absorptions below.

Based on these assignments, a quantitative evaluation of the Fourier self-deconvoluted spectra was made. Figure 4 shows the band-fitting of the deconvoluted spectra observed in H_2O and D_2O, respectively, and Figure 5 summarizes the band assignments and the

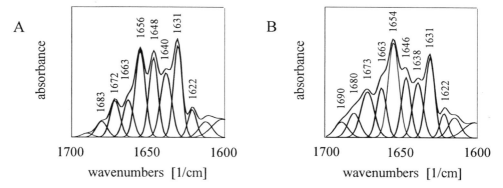

Figure 4. Deconvoluted amide-I band in D_2O buffer (A) and H_2O buffer (B) with the best fitted individual-component bands.

percentage of the contributing secondary structure elements. These FTIR data show the following secondary structure content: α-helix: 34-36 %, β-structure: 19-25 %, β-turn: 15-16%.

The temperature dependence of the FTIR spectra is shown in Figure 6. Starting at 4°C ,it is obvious that over a wide temperature range very little change occurs. Above 50°C, on the other hand, the absorptions assigned to secondary structure elements decrease and disappear around 56°C, with the only exception of the bands at 1683 cm^{-1} and 1622 cm^{-1}, which we had interpreted as being due to β-aggregation. Fig. 7 shows this in more detail: The bands assigned to α-helical and β-structures (curves 1-3 in the figure) decrease drastically (and irreversibly, not shown here) while the band at 1622 cm^{-1} increases (curve 4).

The term "β-aggregation" is based merely on the chracteristic occurence of high- and low-frequency absorption bands, which is reminiscent of the spectra of β-pleated sheets. Its meaning in structural terms is not known. One assumes that hydrogen-bonded structures are involved, with the H-bonds being intra- and/or intermolecular.

It was first described by Wu et al. (1987) that the enzyme activity decreases above 35°C and approaches zero at around 40°C. This irreversible thermal inactivation seems to not parallel changes in the secondary structure content as observed by FTIR spectroscopy.

CONCLUSIONS

Several conclusions can be drawn from these observations:

- Temperature dependent gross secondary structure changes do not occur at temperatures where thermal inactivation of AChE is observed. Inactivation seems to be due to small regional changes of the protein structure.
- Like thermal inactivation of AChE, the structural changes evidenced by FTIR spectroscopy are irreversible.
- The simultaneous increase of the bands at 1683 cm^{-1} and 1622 cm^{-1} at temperatures above 50°C supports our interpretation that these bands are due to β-aggregation.
- All information obtained by FTIR spectroscopy correlates well with the three-dimensional structure of AChE obtained by X-ray crystallography.

H$_2$O		D$_2$O		
band position cm^{-1}	band area %	band position cm^{-1}	band area %	
1690	5	1683	4	β-aggregates/ β-sheet
1680	8	1672	9	turn/ β-sheet
1673	7	1663	7	turn
1663	13	1656	20	α-helix
1654	21	1648	16	α-helix/ (random)
1646	13	1640	17	irregular structures
1638	13	-	-	β-sheet
1631	12	1631	19	β-sheet
1622	8	1622	8	β-aggregates

Figure 5. Band positions, tentative assignments and relative band areas of the amide-I component bands for AChE.

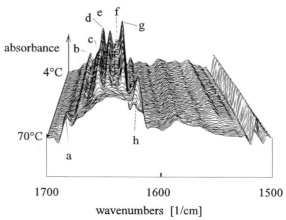

Figure 6. Temperature behavior of AChE in the amide-I and amide II regions of the spectra. The three-dimensional plot shows the deconvoluted spectra for AChE in D$_2$O buffer as a function of temperature; the bands are depicted as follows: a: 1683 cm^{-1}; b: 1672 cm^{-1}; c: 1663 cm^{-1}; d: 1656 cm^{-1}; e: 1648 cm^{-1}; f: 1640 cm^{-1}; g: 1631 cm^{-1}; h: 1622 cm^{-1}.

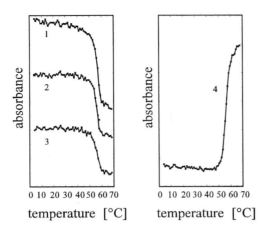

0 10 20 30 40 50 60 70

temperature [°C]

0 10 20 30 40 50 60 70

temperature [°C]

Figure 7. Temperature behavior of AChE as measured for selected amide-I-band components. Curve 1: 1631 cm⁻¹, β-structure ; curve 2: 1648 cm⁻¹, α-helix (I); curve 3; 1656 cm⁻¹, α-helix (II); curve 4: 1622 cm⁻¹, β-aggregation.

ACKNOWLEDGEMENTS

Financial support by the Deutsche Forschungsgemeinschaft (SfB 312) and the Fonds der Chemischen Industrie is gratefully acknowledged.

REFERENCES

Aslanian, D., Grof, P., Negrerie, M., Balkanski, M. and Taylor, P. (1987) FEBS Lett. 219: 202-206.
Aslanian, D., Grof, P., Bon, S., Masson, P., Negrerie, M., Chatel, M.M., Balkanski, M., Taylor, P. and Massoulié, J. (1991) Biochimie (Paris) 73: 1375-1386.
Byler, D. M. and Susi, H. (1986) Biopolymers 25, 469-487.
Görne-Tschelnokow et al. (1993) Eur. J. Biochem 213, 1235-1242.
Görne-Tschelnokow et al. (1994) EMBO J. 13, 338-341.
Koenig, J. L. and Tabb, D. L., in Analytical Applications of FTIR to Molecular and Biophysical Systems (During, J. R., ed.) D. Reidel, Boston 1980.
Krimm, S. and Bandekar, J. (1986) Adv. Protein Chem. 38, 181-364.
Manavalan, P., Taylor, P. and Johnson, W.C. (1985) Biochim. Biophys. Acta 829: 365-70.
Nevskaya, N. A. and Chirgadze, Yu. N. (1976) Biopolymers 15, 637-648.
Shafferman, A. and Velan, B., eds., Multidisciplinary Approaches to Cholinesterase Functions, Plenum Press, New York, 1992.
Surewicz et al. (1991) Biochemistry 32, 389-394.
Susi, H., in Structure and Stability of Biological Macromolecules (Timasheff, S. N. and Fasman, G. D., eds.) Marcel Dekker, New York, 1969.
Susi, H. and Byler, D. M. (1986) Methods Enzymol. 130, 290-311.
Sussman et al.(1991) Science 253, 872-879.
Wu, C.-S. C., Gan, L., and Yang, J. T. (1987) Biochem. Biophys. Acta 911, 37-41.

RESIDUES IN THE C-TERMINUS OF *TORPEDO CALIFORNICA* ACETYLCHOLINESTERASE IMPORTANT FOR MODIFICATION INTO A GLYCOPHOSPHOLIPID ANCHORED FORM

Göran Bucht, Lena Lindgren, and Karin Hjalmarsson

Department of NBC Defense
National Defense Research Establishment
S-901 82 Umeå, Sweden

INTRODUCTION

Acetylcholinesterase (AChE) is an enzyme that exists in several structurally distinct forms. Two major forms of AChE molecules are found in the electric organ of *Torpedo californica*. A hydrophilic form that is attached by a collagen-like tail to the basal lamina in the synaptic cleft, and a hydrophobic dimeric form (G2-AChE) that is attached to the cell membrane via a glycosyl-phosphatidyl inositol (GPI) anchor. These two different forms arise due to alternative splicing of two exons, exon 5 and exon 6. Exon 5 encodes the last 31 carboxy-terminal amino acids found in the hydrophobic form of AChE. This C-terminal peptide (GPIsp) contains the signal for GPI modification. The GPI-moiety is attached to a specific amino acid residue, encoded by exon 5, at the cleavage/modification site or ω-site. This residue is found about 20 amino acids upstream of the C-terminus. Studies of other GPI-modified proteins have not revealed a definitive consensus amino acid sequence in the C-terminal region, but some characteristics are found. The first 2 amino acids downstream of the ω-site, positions $\omega+1$ and $\omega+2$, are believed to interact with the active site of a putative transamidase, catalyzing the GPI-modification reaction. The $\omega+2$ position in the GPIsp is the most conserved residue. In natural proteins only a few amino acids (Gly, Ala, Ser and Thr) are found in this position (Kodukula et al., 1993). These residues are followed by a stretch of 5 to 10 small and relatively polar amino acids, the "spacer region". The "spacer region" is followed by a stretch of 10-15 hydrophobic amino acids.

We have introduced mutations in the C-terminus of the hydrophobic GPI-modified form of AChE from *Torpedo californica*. Deletions have been made to investigate the importance of the last 13 hydrophobic C-terminal amino acids, interrupted only by one serine. Furthermore, amino acid substitutions have been introduced to map the ω-site.

Enzymes of the Cholinesterase Family, Edited by Daniel M. Quinn et al.
Plenum Press, New York, 1995

MATERIAL AND METHODS

Mutagenesis and Expression Vectors

The construction of the cDNA encoding the G2-AChE form has been described previously (Bucht et al., Biochim. Biophys. Acta, accepted for publication, 1994). Mutant sequences were generated by oligonucleotide-directed mutagenesis using PCR. The correct DNA sequences were verified by PCR sequencing, (Cyclist ™ Stratagene)

Cell Culture and Transfection of Plasmid DNA

COS-1 cells (ATCC CRL 1650), were maintained in Dulbecco's modified Eagles medium (DMEM), supplemented with 10% fetal calf serum, 100 E penicillin/ml, 100 µg/ml streptomycin, 2mM L-glutamin and 1mM sodium pyruvate at 37°C in a 5% CO_2 humidified atmosphere. Cells were plated at 1.5-$2x10^6$ per 10 cm plate 16-24 h prior to electroporation. Electroporation of cells was performed with 20 µg plasmid DNA, using a BioRad Gene pulser™ at settings 200 V and 940 µFD. Transfected cells were maintained for 24 h at 37°, and then transferred to 27° for 24 h or 48 h prior to analysis or metabolic labeling.

Metabolic Labeling and Immunoprecipitation

Transfected cells were washed and incubated with a methionine/cysteine free DMEM containing 1.5% bovine serum albumin, 100 E penicillin/ml, 100 µg/ml strepto-mycin, 2mM L-glutamin and 1mM sodium pyruvate. After a 4 h Met/Cys starvation the medium was changed into 5 ml medium/10 cm plate containing 10 µCi/ml Express [^{35}S Met/^{35}S Cys] (Du Pont). The cells were incubated for additional 16 h at 27°. Labeling with ^{3}H-ethanolamine (Amersham) was performed as described by Caras et al. (1989). Labeled cells were washed with PBS (0.01 M sodium phosphate, pH 7.2, 0.145 M NaCl) and scraped off the plates. Cells were collected by centrifugation and resuspended in 200µl PBS containing 2% SDS, 5% β-mercaptoethanol. The suspension was freeze-thawed three times, and diluted 20 times in PBS containing 1% Triton X-100. Remaining particles were removed by centrifugation. Immunoprecipitation was performed with a 2000-fold dilution of a rabbit polyclonal antibody (# 81 5-21-80, personal gift from Prof. Palmer Taylor, University of Californica, San Diego) directed against *Torpedo californica* acetylcholinesterase. The cell lysate was incubated with the antibody over night at 4° before Protein A Sepharose was added. After about 4 h incubation at room temperature the sepharose beads were collected by centrifugation, washed twice with PBS containing 1% Triton X-100 and once with PBS, resuspended in 50 µl of sample buffer (0.125 M Tris-HCl, pH 6.8, 2% SDS, 25% glycerol, 5% β-mercaptoethanol) and incubated for 4 min at 95°.

SDS-Polyacrylamide Gel Electrophoresis and Autoradiography

SDS-PAGE was performed using 10% polyacrylamide gels. For autoradiography, gels containing [^{35}S Met/^{35}S Cys] were dried and exposed to Amersham Hyperfilm™ MP at room temperature. Polyacrylamide gels containing ^{3}H-ethanolamin labeled pro-teins were treated with EN^{3}HANCE™ (Du Pont) before exposure on preflashed film at -70°C.

Measurement of AChE Activity in Transfected COS-1 Cells

Cells were washed with PBS and harvested with a cell scraper. Enzyme activity was determined by the method described by Ellman et al. (1961), in microtiter plates (Labsystem iEMS reader MF, Genesis software).

Treatment with Enzymes

Phosphatidylinositol-specific phospholipase C (PI-PLC) from *Bacillus cereus* was purchased from Boehringer Mannheim. GPI-anchored proteins were released from transfected cells by resuspending cells in PBS containing 50 μg bovine serum albumin, 1U PI-PLC/ml and incubation at 25° C for 1 h. The cells were pelleted by low speed centrifugation and supernatants and pellets were used for further analysis.

RESULTS AND DISCUSSION

Mapping of the Hydrophobic Region

The GPI-anchored form of AChE from *Torpedo californica* is a protein containing 565 amino acids. The carboxy-terminal region in the precursor protein contains the Cys537 residue suggested to be involved in dimer formation and also to be the ω-site. Downstream of Cys537, a serin rich region (Ser542-Gly551) is found. This region contains many polar amino acids with small side chains, typical of a "spacer region". This "spacer region" is followed by a long stretch of hydrophobic amino acids, 13 out of 14 amino acids are hydrophobic, (see wild-type sequence in Table 1). To study the functional role of amino acids and amino acid regions in the C-terminus of *Torpedo californica* AChE, several mutants were made. The importance of the hydrophobic C-terminal region, the last 14 residues, was analyzed by deleting increasing number of residues from the C-terminus (Table 1).

A deletion of 6 C-terminal amino acids (Δ559-mutant) resulted in a total loss of enzyme activity (Fig. 1). The Δ559-mutant protein was not found in the medium (Fig 2) and could not be released by treatment with PI-PLC (data not shown), but was present in cell extracts (data not shown). This indicates that the mutant protein is retained inside the cell.

A deletion of all hydrophobic C-terminal residues (Δ550-mutant) resulted in secretion of active enzyme into the culture medium (Fig. 2). Another mutant enzyme (Δ544-mutant) with the last 21 amino acids deleted was also found to be secreted into the medium (Fig. 2). The latter mutant enzymes were found at about 5 times higher levels of total enzyme activity, compared to the wild-type enzyme (Fig. 1). A deletion of 28 amino acids (Δ538-mutant), leaving Cys537 as the C-terminal amino acid, resulted in a total loss of enzym activity

Table 1. Mutants in the hydrophobic region. C-terminal amino acid sequence of wild-type and truncated mutants of AChE from *Torpedo californica*

Construct	C-Terminal aminoacid sequence
wild-type	NATACDGELSSSGTSSSKGIIFYVLFSILYLIF
Δ559	NATACDGELSSSGTSSSKGIIFYVLFS
Δ550	NATACDGELSSSGTSSSK
Δ544	NATACDGELSSS
Δ538	NATAC

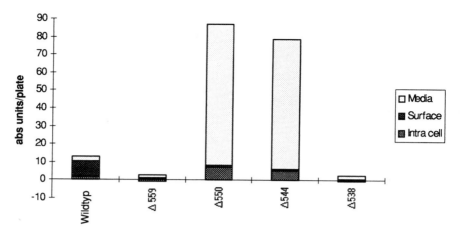

Figure 1. Enzyme activity in deletion mutants. Enzyme activity found after transfection of COS-1 cells, with DNA encoding wild-type and deletion mutants of AChE. See Table 1.

(Fig. 1). This mutant enzyme was found to be secreted in high amounts to the tissue culture medium in a monomeric form (Fig. 2). None of the deletion mutant enzymes were found localized to the cell surface (Fig. 1). These results shows that the C-terminal hydrophobic residues are absolutely required for obtaining GPI-modification of the hydrophobic G2-AChE.

Mapping of the Cleavage/Modification Site, ω-Site

We have studied the role of Cys537, Asp538 and Gly539 in intersubunit disulfide bond formation and in GPI modification by substitution with threonin (Table 2).

Cys537 was chosen for mutagenesis since the GPI-anchor has been reported to be bound to Cys537 (Gibney et. al., 1988). This residue was replaced with serin, glycin, aspartic acid and threonin. Earlier it has been shown that serins, glycines and aspartic acid are better substrates for the modification reaction than cystein, and that threonin is unable to serve as an acceptor for the GPI anchor at the ω-site (Mikanovic et al., 1990). The Gly539 residue was also replaced since a glycin in a similar position in both the mouse and human AChE

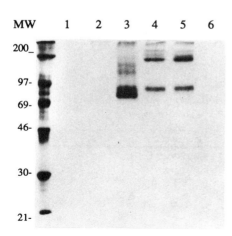

Figure 2. Immuno precipitation of deletion mutants. Transfection and metabolic labeling [^{35}S Met/^{35}S Cys] of COS-1 cells. The AChE protein was immuno precipitated from tissue culture media before SDS-polyacrylamide gel electrophoresis under nonreducing conditions and autoradiography. Lane 1: untransfected cells; lane 2: wild-type AChE; lane 3: Δ538; lane 4: Δ544; lane 5: Δ550; lane 6: Δ559.

Table 2. Mutants of the ω-site and "spacer region." C-terminal amino acid sequence of wild-type and mutants in regions affecting dimerization and GPI-modification of AChE from *Torpedo californica*

Construct	C-Terminal aminoacid sequence
wild-type	NATACDGELSSSGTSSSKGIIFYVLFSILYLIF
ALL-T	TTTATDTELTTTRTSSSKGIIFYVLFSILYLIF
C537T	NATATDGELSSSGTSSSKGIIFYVLFSILYLIF
C537S	NATASDGELSSSGTSSSKGIIFYVLFSILYLIF
TM537-539T	NATATTTELSSSGTSSSKGIIFYVLFSILYLIF
G539T	NATACDTELSSSGTSSSKGIIFYVLFSILYLIF
S542T-S544T	NATACDGELTTTGTSSSKGIIFYVLFSILYLIF

has been reported to be the ω-site amino acid residue. The mutations by which Cys537, Asp538 and Gly539 were replaced with threonin have been introduced separately (single mutants) and together (triple mutant) into the AChE cDNA to avoid unclear results if the modification site is promiscuous as has been observed in studies of u-PAR (Møller et al., 1992).

Our results clearly indicate that Cys537 is required for dimer formation. Replacement of Cys537 with any other amino acid, above-mentioned, resulted in a mutant enzyme that migrates as a monomer on SDS-PAGE (Fig. 3) under non-reducing conditions.

These monomeric mutant enzymes were fully active when anchored to the membrane, but inactive when solubilized (data not shown). This finding prompted us to analyze the stability of these monomeric enzymes. The thermal inactivation at 37°, as a function of time, followed a biphasic curve from which an initial half-life of 2.6 min could be calculated (Fig. 4). The half-life of the wild-type enzyme was calculated to be 49.8 min.

In addition to temperature sensitivity, solubilisation of the monomeric AChE from transfected cells by the addition of Triton X-100 also resulted in a rapid loss of enzyme activity (data not shown). These data might explain why the mutant having a 28 amino acid deletion (Δ538-mutant) was inactive (Fig. 1) since this mutant enzyme is secreted to the culture medium as a monomer. Unexpectedly, the triple mutant of the region 537-539 (mutant TM537-539) was not affected regarding GPI-anchoring of the enzyme in the cell membrane (Figs. 5, 6 and 7). This result can only be explained if a region different from Cys537-Gly539

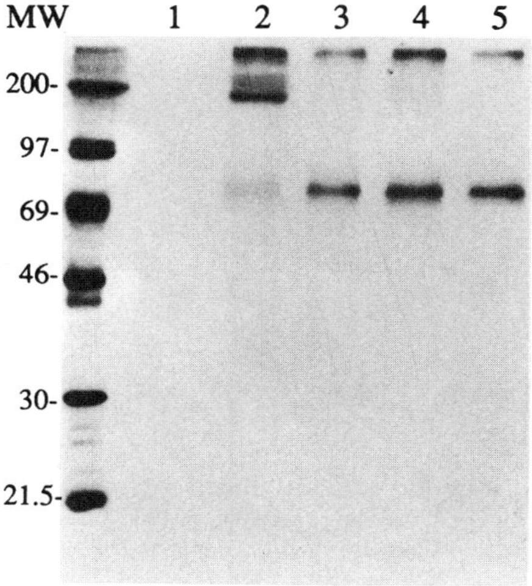

Figure 3. Immuno precipitation of mutants in residue C537. Transfection and metabolic labeling [^{35}S Met/^{35}S Cys] of COS-1 cells. The AChE protein was immuno precipitated from cell extract before SDS-polyacrylamide gel electrophoresis under nonreducing conditions and autoradiography. Lane 1: untransfected cells; lane 2: wild-type AChE; lane 3: C537D; lane 4: C537G; lane 5: C537S.

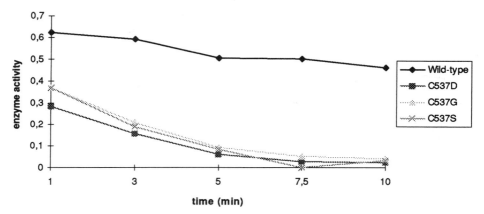

Figure 4. Heat inactivation of dimeric and monomeric AChE. Time dependent loss of AChE activity due to temperature instability at 37.2°. Enzyme activity was measured on intact COS-1 cells transfected with DNA encoding wild-type dimeric and mutant monomeric forms of AChE.

is involved in the GPI-modification reaction. In order to determine the ω-site we first constructed a mutant were all possible ω-sites were altered into threonins, ALL-T (Table 2). This mutant enzyme was found to be both catalytically inactive and retained inside the cell, indicating that the ω-site has been mutated (Figs. 5 and 6). To make a closer investigation another triple mutant was made, three serin residues (S542-S544) were substituted into three threonin (Table 2). This region was selected because our results show that the region 537-539 does not contain the w-site. If the region Ser542-Ser544 contains the attachment amino acid, the "spacer region" would be in accordance with the length of spacer regions in other GPI anchored proteins. Cells expressing this mutant enzyme contained less

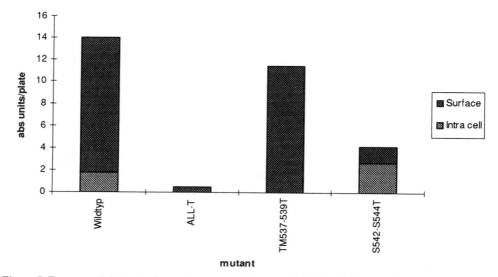

Figure 5. Enzyme activity in attachment site mutants. Enzyme activity found after transfection of COS-1 cells, with DNA encoding wild-type and mutant AChE. Mutations were introduced in regions important for dimer formation and GPI-modification. See Table 2.

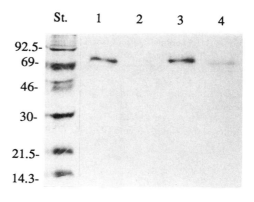

Figure 6. PI-PLC release and immuno precipitation of attachment site mutants. Transfection and metabolic labeling [^{35}S Met/^{35}S Cys] of COS-1 cells transfected with *Torpedo californica* AChE DNA. The COS cells were treated with PI-PLC 24–48 h post transfection. Immuno precipitation was performed on released protein before SDS-polyacryamide gel electrophoresis and autoradiography. Lane 1: wild-type AChE; lane 2: ALL-T; lane 3: TM537-539; lane 4: S542T-S544T,. See Table 2.

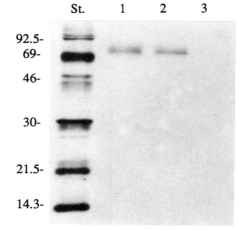

Figure 7. Etanolamin labeling and immunoprecipitation. Transfection and metabolic labeling with ^{3}H-ethanolamin of COS-1 cells transfected with *Torpedo californica* AChE DNA. The AChE protein was immuno precipitated 24–48 h post transfection with AChE DNA prior to SDS-polyacrylamide gel electrophoresis and fluorography. Lane 1: wild-type; lane 2: TM537-539; lane 3: S542T-S544T. See Table 2.

AChE activity at the cell surface, but slightly more intracellular activity (Fig. 5) compared to cells expressing wild-type enzyme. The triple mutant S542T-S544T was not GPI-anchored in the cell membranes since it was not labeled with ^{3}H-ethanolamin nor released by PI-PLC treatment (Figs. 6 and 7). This shows that the ω-site is in the region Ser542-Ser544. These results agree with the finding of a serin residue bound to the GPI-structure (Mehlert et al., 1993). In summary, our results show that the hydrophobic region in the C-terminus is absolutely required for processing into a GPI-anchored protein, and that the region Ser542-544 contains the ω-site.

ACKNOWLEDGMENTS

We thank Prof. Palmer Taylor for the kind gift of the polyclonal antibodies to *Torpedo californica* AChE. The study was supported by the National DefenseResearch Establishment, Sweden.

REFERENCES

Caras, I. W., Weddell, G. N., and Williams, S. R. (1989) J. Cell. Biol. Vol. 108, pp 1387-1396

Ellman, G. L., Courtney, K. D., Andreas, Jr, V., and Featherstone, R.M. (1961) Biochem. Pharmacol. Vol 7, pp 88-95

Gibney, G., MacPhee-Quigley, K., Thompson, B., Vedvick, T., Low, M. G., Taylor, S. S., and Taylor P. (1988) J. Biol. Chem. Vol 263, 1140-1145

Kodukula, K., Gerber, L. D., Amthauer, R., Brink L., and Udenfriend, S. (1993) J. Cell Biol., Vol 120, No 3 pp 657-664

Mehlert, A., Varon, L., Silman, I., Homans, S. W., and Ferguson, A., (1993) Biochem. J. Vol 296, pp 473-479

Micanovic, R., Gerber, L. D., Berger, J., Kodukula, K., and Udenfriend, S. (1990) Proc. Natl. Acad. Sci. USA . Vol 87, pp 157-161

Møller, L. B., Ploug, M., and Blasi. F., (1992) Eur. J. Biochem. Vol 208, pp 493-500

COMPUTER MODELING OF ACETYLCHOLINESTERASE AND ACETYLCHOLINESTERASE-LIGAND COMPLEXES

S. T. Wlodek,[1][*] J. Antosiewicz,[1][†] M. K. Gilson,[1][‡] J. A. McCammon,[1][‡] T. W. Clark,[2] and L.R. Scott[2]

[1] Chemistry Department
University of Houston
4800 Calhoun Road, Houston, Texas 77204-5641
[2] Texas Center for Advanced Molecular Computation (TCAMC)
University of Houston
Houston, Texas 77204-5641

INTRODUCTION

Site directed mutagenesis and computer simulations studies have become valuable tools for the understanding of function-structure relations of proteins. At the current level of computer technology, the quantitative agreement between experimental and molecular simulations results is not always satisfactory for proteins of acetylcholinesterase (AChE) size and complexity. However, the comparison of both types of results can increase dramatically our knowledge on molecular mechanisms of protein action.

Here we briefly present and compare with experimental data the results of four computer simulation studies of AChE: (a) determination of the enzyme pKa values from an electrostatic model, (b) Brownian dynamics simulations of the diffusional encounter rate constants between wild type and several mutants of AChE with cationic substrates, (c) quantum chemical model of the transition state for the formation of the tetrahedral intermediate and (d) thermodynamic integration simulation of the relative binding strength of tacrine (THA) and 6-chlorotacrine (ClTHA) by AChE.

---

[*] presenting author.

[†] On leave from Department of Biophysics, University of Warsaw, Poland. Current address: Department of Chemistry and Biochemistry, University of California at San Diego, La Jolla, CA 92093.

[‡] Current address: Center for Advanced Research in Biochemistry, 9600 Gudelsky Dr., Rockville, MD 20850-3479.

Enzymes of the Cholinesterase Family, Edited by Daniel M. Quinn et al.
Plenum Press, New York, 1995

Although in nature AChE occurs as a dimer, our simulations mentioned above were done using a monomer or, as in the case of thermodynamic integration, by selecting the dynamic part of the dimer and freezing the rest of it. In most cases this approach seems to be a reasonable approximation to the biological system, but, in order to study inter–monomer interactions a simulation of the complete dimer is necessary. In order to simulate the complete dimer, we utilize a parallel molecular dynamics approach described in the Methods Section. In the Results Section, we present parallel processing performance data for the simulation.

METHODS

Numerical titration procedure (Antosiewicz et al., 1994a) is based on electrostatic calculations with the use of the finite difference Poisson-Boltzmann method, which results in self and interaction energies between protein ionizable groups. Those energies are used for the calculation of mean charges of ionizable groups and their pKas.

Brownian dynamics simulation produces a large number of trajectories of the ligand around the enzyme The encounter rate constant is given by

$$k = k(b)\beta \tag{1}$$

where β is the fraction of trajectories which start on a surface of radius b around AChE and end with succesful encounter. $k(b)$ is the rate constant at which the ligand first reaches the surface of radius b, and is calculated analytically as a solution of the one-dimensional diffusion equation with spherical potential for distances $r \geq b$ (Madura et al., 1994).

Ab-initio calculations are done at SCF level, with a 3-21G basis set, with the use of the GAUSSIAN 92 program (Frisch et al., 1992). In those calculations the active site triad consisting of Ser-200, His-440 and Glu-327, is represented by methanol, imidazole and formate ion respectively, the substrate is methyl acetate and the residue Glu-199 is represented by another formate ion.

Thermodynamic integration determines the free energy difference, ΔG, between two states of the system as a sum over ensemble averages for the $\partial H/\partial \lambda$:

$$\Delta G = \sum_i \langle \frac{\partial H}{\partial \lambda} \rangle_{\lambda_i} \Delta \lambda \tag{2}$$

where λ is a perturbation parameter, $0 \leq \lambda \leq 1$, coupling both states in such a way that its integer values 0 and 1 select the Hamiltonian H for those states (Straatsma & McCammon, 1991).

The Parallel molecular dynamics program, EulerGromos (Clark et al., 1994) is used to simulate the AChE dimer placed in a box of water. Huge by conventional molecular dynamics standards, the solvated system with about 130,000 atoms outreaches conventional computer and software resources. To meet this computational demand, we run EulerGromos on 128 to 256 nodes of a massively–parallel computer. We output trajectory data every 5 steps. At this frequency, outputting the entire system generates roughly 600 Megabytes/ps, or 600 Megabytes every 1.6 hours of simulation time (see Results Section). We have reduced that output level by about a factor of 10, a manageable level, by outputting a subset of atoms using functionality provided by EulerGromos.

Most of the calculations are based on the crystal structure of AChE–THA complex (Harel et al.,1993), but a structure from the original Protein Data Bank entry, 1ACE (Sussman et al., 1991) (containing decamethonium as subsequently found (Harel et al.,1993)), was also used in numerical titration. The structures of inhibitors THA and ClTHA were found with the ab-initio calculations using 6-31G** basis set and their atomic charges were determined with the CHELPG procedure (Breneman &Wiberg, 1990). All simulations are done at 298 K. In Poisson-Boltzmann type calculations the pH was set at 7.0 and ionic strength to 85 or 150 mM. In the thermodynamic integration simulation, an OPLS AMBER (Jorgensen & Tirado-Rives, 1988) force field with explicit hydrogen atoms on aromatic side groups of Trp, Phe and Tyr was used.

RESULTS AND DISCUSSION

1. Titration Results

The first three rows of Table I present computed pKas for ionizable groups in the active site of unliganded AChE. The striking prediction is a high pKa of His-440, which exceeds the experimental value by 3 pKa units. The mechanism of AChE catalysis involves an attack of the Ser-200 hydroxyl group on the substrate ester carbon and simultaneous transfer of the hydroxyl proton to His-440 which requires that His-440 be neutral when catalysis starts. The fact that AChE functions at pH 7.0 implies that the pKa of His-440 cannot be 9.3. The same titration procedure reproduces well a pKa value of the catalytic histidine (His-57) in chymotrypsin. Why then, does our electrostatic model fail in the case of AChE? A special feature of the active site structure of AChE is the proximity of two acidic residues, Glu-199 and Glu-443, with respect to the active triad. The electrostatic potential generated by those two acidic residues could simply make His-440 more basic than the corresponding histidine in e.g. chymotrypsin. Indeed, as lines 4-6 from Table I show, the neutralization of Glu-199 and Glu-443 lowers the pKa of His-440, bringing its value to the experimental number when both of those residues become neutral. A similar effect is obtained when a monovalent cation like protonated tacrine or acetylcholine is present in the active site of AChE as illustrated by the last three lines of Table I.

It is likely then the existence of a monovalent cation in the active site of AChE is necessary to deprotonate His-440 and start catalysis. This condition is automatically met in the case of cationic substrates, like acetylcholine. For neutral substrates on the other hand, the unfavorable electrostatic potential preventing the deprotonation of His-440 might be compensated by one of the solution cations, such as Na^+.

Table I. Computed pKas of ionizable groups in the active site of *Torpedo californica* AChE (TcAChE). Protein structures are either those in the presence of bound decamethonium (I), or in the presence of bound tacrine(II). The results in line 2 are for 150 mM ionic strength; all others are for 85 mM.

Structure	State	His-440	Glu-327	Glu-199	Glu-443
I	No ligand	9.3	-1.7	1.7	5.8
I	No ligand, 150 mM	9.1	-1.2	2.0	5.8
II	No-ligand	9.2	-1.1	0.9	6.1
	Glu-199 & Glu-443 neutral	6.1	I	—	—
I	Glu-199 neutral	7.3	-1.7	—	1.8
I	Glu-443 neutral	8.5	-1.7	1.0	—
I	With acetylcholine	6.9	-1.9	-2.1	5.1
II	With tacrine	6.3	-1.2	-2.6	5.4
I	With small cation	7.4	-2.0	-1.5	5.2

2. Diffusional Encounter Kinetics

Recent experimental study of *Human* AChE (HuAChE) (Shafferman et al., 1994) shows that several mutations of acidic residues located at the surface of the enzyme into neutral groups have minor effect on the hydrolysis rate of acetylthiocholine. The authors of that study concluded that surface electrostatic properties of AChE do not contribute to the catalytic rate and that catalysis kinetics are probably not diffusion controlled. We have decided to examine this issue using Brownian dynamics simulation. Details of the simulation are described elsewhere (Antosiewicz et al., 1994b); here we want to emphasize only that the encounter criteria was defined in such a way that substrate was allowed to penetrate deeply into the reactive gorge. We have studied 5 of all 20 mutants studied experimentally by Shafferman *et al.* (Shafferman et al., 1994). They include:

1. Glu-82
2. Asp-381
3. Glu-82, Asp-285, Asp-342, Asp-351
4. Asp-342, Asp-351, Asp-380, Asp-381
5. Glu-82, Glu-278, Asp-285, Asp-342, Asp-351, Asp-380, Asp-381.

Electrostatic properties of the above mutants are provided in Figure 1.

One can see from Figure 1A that although neutralization of 7 acidic residues on the surface of TcAChE nearly annihilates the total charge of the protein at pH 7.0, it leaves a huge dipole moment of more than 1000 D. In the same mutant (mutant 5) the electrostatic potential along the reaction gorge is reduced by about 1 kcal/(e mol) but, as Figure 1B shows, its gradient in the direction of the gorge axis is only slightly changed. One may expect then, the AChE- substrate encounter rate should not differ drastically for the two enzymes. Table II compares the computed rate constants for the diffusional encounter between positively charged substrate and 6 mutants of TcAChE with experimental values for hydrolysis of acetyltiocholine by equivalent mutants of HuAChE.

It is seen that the corresponding ratios of mutant to wild type rate constants remain in reasonable agreement. The important result is that the modest changes in catalysis rate observed for the mutants could be explained in terms of a diffusional-electrostatic model of the enzyme-substrate encounter, which implies that diffusion and electrostatic steering do contribute to the catalysis kinetics.

Table II. Comparison of computed rate constants, k_{calc}, for the diffusional encounter between positively charged substrate and six mutants of TcAChE with experimental values, k_{exp}, for hydrolysis of acetylthiocholine by corresponding mutants of human AChE at 150 mM ionic strength (Shafferman et al., 1994). All rate constants are in units of 10^9 M^{-1}s^{-1}.

Mutant	k_{calc}	k_{exp}	k_{calc}/k_{calc}^{wt}	k_{exp}/k_{exp}^{wt}
wt	1.34±0.23	0.048±0.007	1.00	1.00
1	1.04±0.20	0.025±0.004	0.78	0.52
2	1.04±0.20	0.048±0.007	0.78	1.00
3	0.50±0.14	0.021±0.003	0.37	0.43
4	1.02±0.20	0.055±0.008	0.76	1.14
5	0.70±0.17	0.021±0.003	0.52	0.42

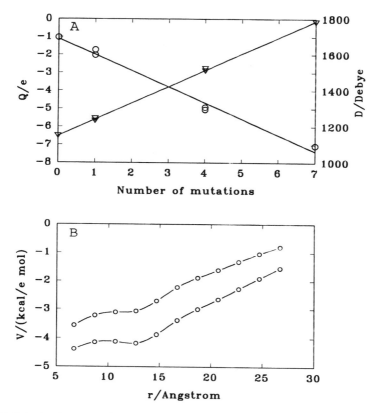

Figure 1. (A) Calculated mean total charge, Q, and dipole moment, D, of mutated TcAChE monomer as a function of the number of neutralized Asp/Glu residues in studied mutants at pH 7, 300 K and ionic strength of 150 mM. Dipole moment is calculated with respect to the center of mass of the protein. (B) Electrostatic potential along the axis of the TcAChE reaction gorge for the wild type enzyme and mutant 5 with seven acidic residues neutralized. The axis of the reaction gorge is defined as a line connecting the CD atom of Ile-444 and a geometric center of the following 4 atoms: CA of Glu-73, CB of Asn-280, CG of Asp-285 and O of Leu-333. The distance r is measured from the CD atom of Ile-444.

3. The Role of the Glu-199 Residue

A recent study by Quinn *et al.* showed that mutation of Glu-199 to Gln in TcAChE causes a 100-fold decrease in the acylation rate constant (Quinn et al., 1995). The authors concluded that the principal function of Glu-199 is specific stabilization of the chemical transition state in the acylation stage of AChE catalysis.

Below we provide the results of our preliminary quantum chemical calculations supporting that hypothesis. In our model calculations a hydroxyl oxygen from methanol attacks the planar carbon atom of methyl acetate within the gas-phase ion molecule complex where the formate ion is clustered with imidazole, methanol and methyl acetate: $(HCOO^- \cdot C_3N_2H_4 \cdot CH_3OH \cdot CH_3COOCH_3)$ in such a way that one of the formate oxygen atom forms a hydrogen bond with the NH group of imidazole, and another nitrogen atom from imidazole is hydrogen bonded to the methanol OH group. Our model reaction is:

$$HCOO^- \cdot C_3N_2H_4 \cdot CH_3OH \cdot CH_3COOCH_3 \longrightarrow$$
$$HCOO^- \cdot C_3N_2H_5^+ \cdot CH_3C(OCH_3)_2O^- \qquad (3)$$

and corresponds to the formation of the tetrahedral intermediate in the catalytic system of TcAChE, (Glu-327 ··His-440 · Ser-200 · substrate). Our calculations show that reaction (3) is endothermic with a significant activation barrier of 14 kcal/mol, and is therefore prohibited at room temperature and moderate pressure.

When another formate ion, (equivalent to Glu-199) is present in the system close to the imidazole ring, the energy barrier changes dramatically to only 1.7 kcal/mol. Although the latter energy barrier does not represent the true transition state for the system:

$$(HCOO^-)_2 \cdot C_3N_2H_4 \cdot CH_3OH \cdot CH_3COOCH_3,$$

our results clearly show that the transition state stabilization for the formation of a tetrahedral intermediate occurs by cation-anion attraction between Glu-199 and partially protonated His-440.

4. Inhibition of AChE by Tacrine and 6-Chlorotacrine

Tacrine (THA) is a well known inhibitor of AChE, which is the only drug approved to date for Alzheimer's disease treatment. Protonated THA molecule at the ring N atom, is bound in the active site of TcAChE primarily by cation–π electrons interaction with Phe-330 and Trp-84 side chains, and two hydrogen bonds: $N_{ring} \ldots O_{His-440}$ and $N_{amino} \ldots O_{water-32}$. It was recently found, that one of the THA chloro derivatives, 6-chlorotacrine (ClTHA), is about 3 times stronger inhibitor of *Electric eel* AChE than the parent THA compound (Gregor et al., 1992). Here we attempt to provide a theoretical explanation of that experimental finding, based on the calculation of the free energy difference,

$$\Delta\Delta G = \Delta G_6 - \Delta G_5 \tag{4}$$

for the reactions:

$$AChE + THA \longrightarrow AChE \cdot THA \tag{5}$$

$$AChE + ClTHA \longrightarrow AChE \cdot ClTHA \tag{6}$$

We use the thermodynamic cycle from Figure 2, where the reactions (7) and (8) are nonphysical transformations of THA into ClTHA in solution, and in the active site of the AChE, respectively.

Our earlier electrostatic calculations showed that at the conditions of the inhibition experiments, the two lowest energy ionization states of the AChE–THA complex are very close in energy (the calculated energy gap is only 0.14 kcal/mol), and correspond to ionized (state A) and neutral (state B) His-440 and Glu-443 residues. Both of those ionization states

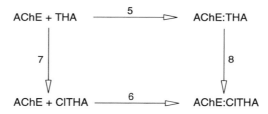

Figure 2. Thermodynamic cycle used for the calculation of $\Delta\Delta G$ given in equation (4).

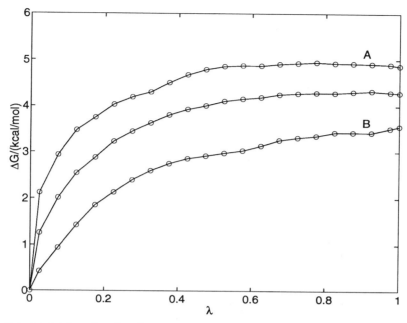

Figure 3. Calculated ΔG as a function of perturbation parameter λ coupling tacrine and 6-chlorotacrine inside the active site of TcAChE and in solution (middle curve). Curves A and B refer to two ionization states of the protein (see text).

could in principle be significantly populated, so we carried out a thermodynamic integration simulation for each of them. Figure 3 shows the variation of calculated ΔG with perturbation parameter λ for the processes (7) and (8). We find that only the ionization state B of TcAChE stabilizes the ClTHA molecule with respect to solution. From the ΔG values for $\lambda = 1$ we have:

$$\Delta\Delta G \ 0.6 \pm 0.5 \text{ kcal/mol for state A and} -0.7 \pm 0.4 \text{ kcal/mol for state B.}$$

If the strength of inhibitor binding correlates with its inhibition power, the agreement between experimental and simulated data occurs only when ionization state B is responsible for catalysis and inhibition functions. Higher strength of ClTHA binding with respect to parent THA molecule by TcAChE in ionization state B might be ascribed to the more favorable electrostatic interactions resulting from the larger dipole moment of the former.

5. EulerGromos Performance

Performance for molecular dynamics simulations of AChE dimer with the EulerGromos program might be well demonstrated by comparison with a smaller solvated myoglobin system containing 10,914 atoms. To simulate one picosecond of the solvated AChE dimer on 128 processors requires slightly less time than simulating the myoglobin system on eight processors, highlighting the benefit of massively parallel computers. Moreover, the relative motion of different globular regions has a characteristic time of from 10 ps to 100 ns (McCammon & Harvey, 1987). Since using EulerGromos to simulate the AChE dimer for

10 ps requires about 1 week of a 128 node Paragon machine, it is feasible to investigate such collective motions.

ACKNOWLEDGMENTS

We are grateful for Drs. Jim Briggs, Joel Sussman, Daniel Quinn and Palmer Taylor for helpful discussions. This work was supported in part by the N. S. F., N. I. H., A. R. P. A. and the MetaCenter Program of the N. S. F. Supercomputer Centers.

REFERENCES

Antosiewicz, J., McCammon, J.A., & Gilson, M.K. (1994a) Prediction of pH-dependent properties of proteins, *J. Mol. Biol. 238: 415-436.*

Antosiewicz, J., Wlodek, S.T., McCammon, J.A., & Gilson M.K. (1994b), Simulation of charge-mutant acetylcholinesterase, *Biochemistry*, submitted.

Breneman, C.M., & Wiberg, K.B. (1990), Determining atom-centered monopoles from molecular electrostatic potentials. The need for high sampling in formamide conformational analysis, *J. Comput.Chem.*, 11: 361-373.

Clark, T.W., Hanxleden, R.V., McCammon, J.A., & Scott L.R. (1994), *Scalable High Performance Computing Conference, Knoxville TN.* Available via anonymous ftp from softlib.rice.edu as pub/CRPC-TRs/reports/CRPC-TR93356-S.

Frisch, M.J., Trucks, G.W., Head-Gorgon, M., Gill, P.M.W., Wong, M.W., Foresman, J.B., Johnson, B.G., Schlegel, H.B., Robb, M.A., Replogle, E.S., Gomperts, R., Abdres, J.L., Raghavachari, K., Binkley, J.S., Gonzalez, C., Martin, R.L., Fox,D.J., Defrees, D.J., Baker, J., Stewart, J.J.P., & Pople, J.A., (1992), Gaussian 92, Revision E.2.

Gregor, V.E., Emmerling, M.R., Lee, C., & Moore, C.J. (1992), The synthesis and in vitro acetylcholinesterase and butyrylcholinesterase inhibition activity of tacrine (Cognex) derivatives, *Bioorg. Med. Chem. Letters,* 2: 861-864.

Harel, M., Schalk, I., Ehret-Sabatier, L., Bouet, F., Goeldner, M., Hirth, C., Axelsen, P., Silman, I., & Sussman, J.L. (1993), Quaternary ligand binding to aromatic residues in the active site gorge of acetylcholinesterase. *Proc. Natl. Acad. Sci, USA,* 90: 9031-9035.

Jorgensen, W.L., & Tirado-Rives J.,(1988), The OPLS potential function for proteins: energy minimizations for crystals of cyclic peptides and crambin, *J. Am. Chem. Soc.* 110: 1657-1666.

Madura, J.D., Davis, M.E., Gilson, M.K., Wade, R.C., Luty, B.A., & McCammon, J.A. (1994), Biological Applications of electrostatic calculations and Brownian dynamics simulations, *Rev. Comp. Chem.* 5: 229-267.

McCammon, J.A., & Harvey S.C. (1987). Dynamics of Proteins and Nucleic Acids, University Press, Cambridge, 1987.

Quinn, D., Seravalli, J., Nair, H.K., Radic, Z. Vellom, D.C., Pickering, N.A., & Taylor P. (1995), Electrostatic influence on acylation reaction dynamics of acetylcholinesterase-catalyzed hydrolysis, *Biochemistry*, in press.

Shafferman, A., Ordentlich, A., Barak, D., Kronman, C., Ber, R., Bino, T., Ariel, N., Osman, R., & Velan, B. (1994), Electrostatic attraction by surface charge does not contribute to the catalytic efficiency of acetylcholinesterase, *EMBO J.,* 13: 3448-3445.

Straatsma, T.P., & McCammon, J.A. (1991), Multiconfiguration thermodynamic integration, *J. Chem. Phys.,* 95: 1175-1188.

Susmann, J.L., Harel, M., Frolow, F., Oefner, C., Goldman, A., Toker., L., & Silman, I. (1991), Atomic structure of acetylcholinesterase from *torpedo californica*: A prototypic acetylcholine-binding protein, *Science 233: 872-879.*

STRUCTURAL ANALYSIS OF THE ASPARAGINE-LINKED OLIGOSACCHARIDES OF CHOLINESTERASES

N-linked Carbohydrates of Cholinesterases

Ashima Saxena and B. P. Doctor

Division of Biochemistry
Walter Reed Army Institute of Research
Washington, D.C. 20307-5100

INTRODUCTION

Cholinesterases are serine esterases that hydrolyse choline esters faster than other substrates. In vertebrates, two types of cholinesterases corresponding to two distinct gene products have been identified: acetylcholinesterase (AChE, E.C. 3.1.1.7) and butyrylcholinesterase (BChE, E.C. 3.1.1.8) which can be distinguished by their substrate specificity and sensitivity to various inhibitors such as BW284C51 for AChE and ethopropazine, iso-OMPA and bambuterol for BChE (Massoulié et al., 1993).

Cholinesterases are highly glycosylated proteins, with up to 24% of their molecular weight constituted of carbohydrates (Haupt et al., 1966; Liao et al., 1991; Liao et al., 1992; Trestakis et al., 1992). These carbohydrates are present primarily as asparagine-linked side chains as suggested by their susceptibility to specific glycohydrolases (Bon et al., 1987; Heider et al., 1991; Liao et al., 1991; Liao et al., 1992; Kronman et al., 1992) and their interaction with specific lectins (Gurd, 1976; Uhlenbruck et al., 1977; Raconczay et al., 1981; Bon et al., 1987, Méflah et al., 1984; Bon et al., 1988; Rotundo, 1988; Mutero and Fournier, 1992; Kerem et al., 1993). In the present study, we have attempted to make a complete characterization of the carbohydrate units of cholinesterases by investigating the carbohydrate structure of FBS AChE and horse serum (Eq) BChE. Our findings indicate that both enzymes contain only N-linked oligosaccharides. The number, structure and relative proportion of the N-linked carbohydrate unit variants were determined for both enzymes.

Enzymes of the Cholinesterase Family, Edited by Daniel M. Quinn et al.
Plenum Press, New York, 1995

MATERIALS

Electrophoretically pure AChE from FBS was purified as described (De La Hoz et al., 1986). BChE from horse serum (Eq) was purified by affinity chromatography using the procedure similar to the one described for FBS AChE. One mg of pure, native AChE and BChE contain approximately 14 and 11 nmols of active sites respectively.

CARBOHYDRATE COMPOSITION OF CHOLINESTERASES

Samples containing 50 μg of protein (in duplicate) were exhaustively dialyzed against 0.1% (v/v) trifluoroacetic acid, and subjected to hydrazinolysis using the GlycoPrep 1000 to release intact oligosaccharides from the protein. This was followed by the use of methanolic-HCl to liberate monosaccharides as the 1-O-methyl derivatives, followed by N-acetylation of any available primary amino groups and the conversion of individual monosaccharides into per-O-trimethylsilyl (TMS) methyl glycosides. The TMS-methyl glycosides were separated on a G.L.C.-M.S. system using a CP-SIL8 column and identified by comparison with standard reference TMS-methyl glycosides. Quanitation of the individual TMS-methyl glycosides was achieved using the signal from a flame ionization detector. Scylloinositol was used as an internal standard.

From the saccharide analyses, it was apparent that a substantial portion, (9-10% by weight) of FBS AChE and (20-22% by weight) of Eq BChE was in the form of carbohydrate (Table I). The relative content of mannose, galactose, N-acetylglucosamine and sialic acid was the same in both cases. Eq BChE differed from FBS AChE in that it did not contain any fucose. These results are different from those reported for *Electrophorus* AChE which lacked fucose (Bon et al., 1976) and human serum BChE which contained fucose (Haupt et al., 1966). The relatively high content of mannose suggested the presence of "N-linked" oligosaccharides and the absence of N-acetylgalactosamine indicated the absence of O-linked oligosaccharides in these enzymes. These results are in agreement with those reported for cholinesterases from other sources (Liao et al., 1991; Liao et al., 1992; Trestakis et al., 1992; Liao, et al., 1993). The total numbers of complex carbohydrate units/subunit calculated from their mannose contents were found to be 3 for FBS AChE and 11-12 for Eq BChE.

Table I. Monosaccharide composition of cholinesterases

Monosaccharide	FBS AChE (nmol / mg[a])	Eq BChE (nmol / mg[a])
Fucose	36.0	ND[b]
Mannose	133.0	383.0
Galactose	102.0	245.0
N-Acetylglucosamine	200.0	524.0
Sialic acid	49.0	137.0
Total[c]	520.0	1289.0

[a]Average of duplicate analyses.
[b]ND = not detected.

RELEASE AND RECOVERY OF OLIGOSACCHARIDES ASSOCIATED WITH CHOLINESTERASES

The oligosaccharides associated with FBS AChE (407 µg) and Eq BChE (410 µg) were quantitatively released and recovered using the GlycoPrep 1000 as described before. An aliquot of this sample was reduced with a 5-fold molar excess of 6 mM $NaBT_4$ in 50 mM NaOH, adjusted to pH 11.0 with saturated boric acid and incubated at 30° C for 4 hours. An equivalent volume of 1M $NaBH_4$ in NaOH/boric acid, pH 11.0, was then added and the incubation continued for another 2 hours. The mixture was acidified to pH 4.5 with 1M acetic acid, desalted on a column of Dowex Ag 50X12 (H^+) and subjected to descending paper chromatography for 2 days using 1-butanol/ethanol/water (4:1:1) as the solvent. The radio-chromatogram was scanned and the radioactivity remaining at the origin was subsequently eluted with water to recover radiolabeled oligosaccharide alditols.

CHARGE-DISTRIBUTION ANALYSIS OF THE OLIGOSACCHARIDE ALDITOLS RELEASED FROM CHOLINESTERASES

Aliquots of radiolabeled oligosaccharide alditols for FBS AChE and Eq BChE were subjected to high-voltage paper electrophoresis and the resulting radioelectrophoretograms are shown in Figure 1, which show that the oligosaccharides associated with both enzymes consist of neutral as well as acidic components.

The relative molar content of neutral and acidic oligosaccharides was 28% and 72% respectively for FBS AChE (panel A) as compared to 19% and 81% respectively for Eq BChE (panel B). To determine the nature of the acidic substituents, aliquots of the pool were incubated with neuraminidase from *Arthrobacter ureafaciens* before being subjected to high-voltage paper electrophoresis. The resulting electrophoretograms for FBS AChE and Eq BChE are shown in panels C and D respectively. No acidic oligosaccharides were detectable after neuraminidase treatment. Therefore, in both cases, the acidic substituent on the oligosaccharide chain was a covalently linked non-reducing terminal outer-arm sialic acid residue. These results are in agreement with previous findings in which treatment with neuraminidase induced changes in the electrophoretic mobility of various cholinesterases (Svensmark and Kristensen, 1963; Svensmark and Heilbronn, 1964; Carlsen and Svensmark, 1970; Ott et al., 1975).

HIGH pH ANION EXCHANGE CHROMATOGRAPHY-PULSED AMPEROMETRIC DETECTION (HPAEC-PAD) PROFILE OF THE TOTAL OLIGOSACCHARIDES ASSOCIATED WITH CHOLINESTERASES

The distribution of sialyated oligosaccharides in the total oligosaccharide pools obtained from FBS AChE and Eq BChE were determined by HPAEC-PAD using the following gradient:

Time (min)	[NaOH], mM	[NaOAc], mM
0	100	0
45	100	75
80	100	210

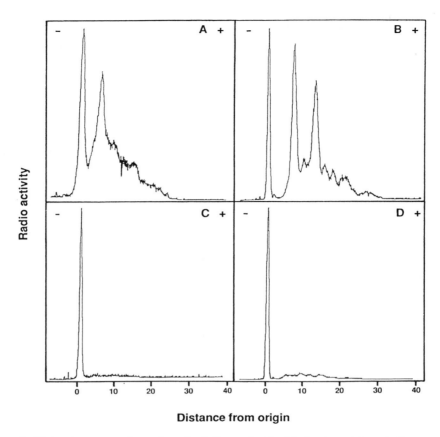

Distance from origin

Figure 1. High-voltage paper electrophoresis of oligosaccharide alditols released from cholinesterases.

The HPAEC-PAD profiles for FBS AChE and Eq BChE are shown in panels A and B respectively. The elution pattern of the total oligosaccharide pool from human serum IgG shown in panel C was used as the reference to locate the positions of neutral, asialo, mono-, di-, and tri-sialylated N-glycans. The results of this experiment demonstrate the presence of only mono- and di-sialylated N-glycans in FBS AChE. On the other hand, Eq BChE was found to contain mono-, di-, as well as tri-sialylated N-glycans.

SIZE-DISTRIBUTION ANALYSIS OF THE TOTAL POOL OF DEACIDIFIED ALDITOLS RELEASED FROM CHOLINESTERASES

An aliquot of the total pool of deacidified oligosaccharide alditols was mixed with an aqueous solution of a partial acid hydrolysate of dextran and subjected to high resolution gel permeation chromatography using the GlycoMap 1000. The column was run using water as the eluant at 55°C at a flow rate of 30 μl/min. The eluant from the column was monitored using an in-line radioactivity detector to detect radiolabeled sample as well as an in-line refractometer to detect individual glucose oligomers.

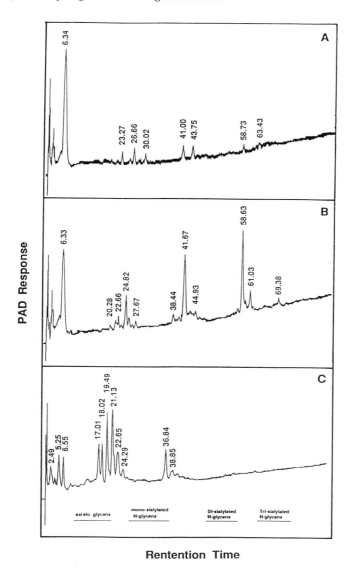

Figure 2. HPAEC-PAD profile of the total oligosaccharides associated with cholinesterases.

The gel permeation chromatograms for FBS AChE and Eq BChE are shown in panels A and B respectively. The hydrodynamic volume of individual radiolabeled oligosaccharide alditols was measured in terms of glucose units, as calculated by cubic spline interpolation between the two glucose oligomers immediately adjacent to the oligosaccharide alditol. At least two major (15.3 gu and 14.5 gu) and four minor (18.1 gu, 13.5 gu, 12.2 gu, and 11.3 gu) distinct structural components for FBS AChE and one major (13.6 gu) and six minor (17.1 gu, 15.8 gu, 12.3 gu, 11.1 gu, 9.8 gu, and 9.0 gu) distinct structural components for Eq BChE were identified.

STRUCTURAL ANALYSIS OF THE MAJOR OLIGOSACCHARIDES RELEASED FROM CHOLINESTERASES

The structures of the two major oligosaccharides released from FBS AChE (15.3 gu and 14.5 gu, Figure 3A) and one major oligosaccharide from Eq BChE (13.6 gu, Figure 3B) were obtained by sequential glycosidase digestions summarized in Table II. An aliquot of the oligosaccharide alditol was incubated with the desired amount of glycosidase at 37°C for 18 hours. At the end of the incubation period, the mixture was desalted, the glycosidase removed using ion-exchange resins, the sample evaporated to dryness, resuspended in an aqueous solution of a partial acid hydrolysate of dextran and the solution applied to a BioGel P4 (~400 mesh) gel filtration column. The eluant from the column was monitored using an

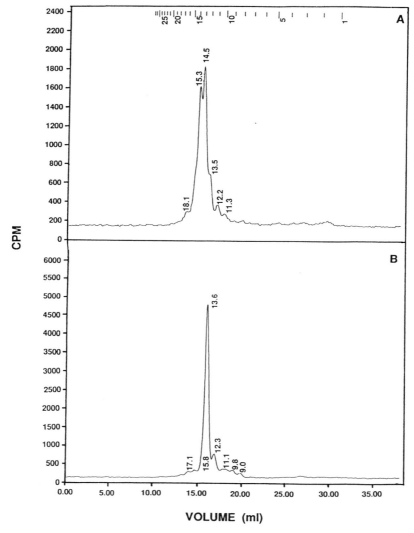

Figure 3. Gel permeation chromatography of the total pool of deacidified alditols released from cholinesterases

in-line radioactivity flow detector as well as an in-line differential refractometer. The change, if any, induced in the hydrodynamic volume of an oligosaccharide by a glycosidase of defined specificity allowed us to deduce the number and nature of the monosaccharide(s) at the non-reducing terminus of the oligosaccharide. This process was repeated sequentially using different glycosidases until the oligosaccharide was "sequenced" down to the monosaccharide alditol, or a glycosidase-resistant structure was reached.

The structures of the oligosaccharides from FBS AChE and Eq BChE are shown in Table II. The three carbohydrate structures were of the biantennary complex type but only the ones from FBS AChE were fucosylated on the innermost N-acetylglucosamine residue of the core. This result is consistent with the previously noted absence of fucose in Eq BChE

Table II. Effect of sequential glycosidase digestion on the complex N-linked oligosaccharides from cholinesterases

	FBS AChE				Eq BChE	
GLYCOSIDASE[b]	ORDER OF USE	RESIDUES[c] REMOVED	ORDER OF USE	RESIDUES[c] REMOVED	ORDER OF USE	RESIDUES[c] REMOVED
β-D-Galactosidase (ex. S. pneumoniae)	1	1	1	2	1	2
β-N-acetyl-D-hexosaminidase (ex. S. pneumoniae)	2	1	2	2	2	2
β-N-acetyl-D-hexosaminidase (ex. jack bean)	3	-	3	-	3	-
α-D-galactosidase (ex. green coffee bean)	4	1	-	-	-	-
β-D-galactosidase (ex. S. pneumoniae)	5	1	-	-	-	-
β-N-acetyl-D-hexosaminidase (ex. S. pneumoniae)	6	1	-	-	-	-
α-D-mannosidase (ex. jack bean)	7	2	4	2	4	2
β-D-mannosidase (ex. H. pomatia)	8	1	5	1	5	1
α-L-fucosidase (ex. B. epididymis)	9	1	6	1	6	-
β-N-acetyl-D-hexosaminidase (ex. jack bean)	10	1	7	1	7	1

[a]The symbols used in the structural schemes are as follows: □ GlcNAc; ■ Man; ● β-Gal; ○ α-Gal; △ Fuc. The branch location of the terminal α-Gal residues are not known.
[b]Glycosidase digestions were carried out on the two major oligosaccharides (15.3 and 14.5gu) from FBS AChE and one major oligosaccharide (13.6gu) from Eq BChE.
[c]The number and nature of the monosaccharide residues at the non-reducing terminus of the oligosaccharide were deduced from the change (if any), induced in the hydrodynamic volume of the oligosaccharide by a glycosidase of defined specificity.

(Table I). In addition to fucose, the larger oligosaccharide from FBS AChE also contained capping sugars in the form of sialyl- and α-galactosyl residues. The galactose α1-3 galactose β1-4-determinant present in FBS AChE has been identified as a potential immunogenic oligosaccharide determinant. This disaccharide sequence has been detected in secreted and cell surface glycoproteins of various non-primate mammalian species, prosimians and New World monkeys, but is absent in glycoproteins from man, apes and Old World monkeys (Thall and Galili, 1990). This species-related difference in α-galactosylation has been attributed to the absence of specific α1 → 3 galactosyl transferase activity (Galili et al., 1988). The functional significance of these structural differences in the carbohydrates of cholinesterases remains to be determined.

REFERENCES

Bon, S., Huet, M., Lemonnier, M., Reiger, F., and Massoulié, J., 1976, *Eur. J. Biochem.* 68: 523-530.

Bon, S., Méflah, K., Musset, F., Grassi, J., and Massoulié, J., 1987, *J. Neurochem.* 49: 1720-1731.

Bon, S., Toutant, J.-P., Méflah, K., and Massoulié, J., 1988, *J. Neurochem.* 51: 786-794.

Carlsen, J.B., and Svensmark, O., 1970, *Biochim. Biophys. Acta* 207: 477-484.

De La Hoz, D., Doctor, B.P., Ralston, J.S., Rush, R.S., and Wolfe, A.D., 1986, *Life Sci.* 39: 195-199.

Galili, U., Shohet, S.B., Kobin, E., Stults, C.L.M., and Macher, B.A., 1988, *J. Biol. Chem.* 26: 17755-17762.

Gurd, J.W., 1976, *J. Neurochem.* 27: 1257-1259.

Haupt, H., Heide, K., Zwisler, O., and Schwick, H.G., 1966, *Blut* 14: 65-75.

Heider, H., Meyer, P., Liao, J., Stieger, S., and Brodbeck, U., 1991, in *Cholinesterases: Structure, Function, Mechanism, Genetics and Cell Biology* (Massoulié, J., Bacou, F., Barnard, E.A., Chatonnet, A., Doctor, B.P., and Quinn, D.M., eds.), pp. 32-36, American Chemical Society, Washington, D.C.

Kerem, A., Kronman, C., Bar-Nun, S., Shafferman, A., and Velan, B., 1993, *J. Biol. Chem.* 268: 180-184.

Kronman, C., Velan, B., Gozes, Y., Leitner, M., Flashner, Y., Lazar, A., Marcus, D., Sery, T., Papier, Y., Grosfeld, H., Cohen, S., and Shafferman, A., 1992, *Gene* 121: 295-304.

Liao, J., Heider, H., Sun, M.C., Stieger, S., and Brodbeck, U., 1991, *Eur. J. Biochem.* 198: 59-65.

Liao, J., Heider, H., Sun, M.C., and Brodbeck, U., 1992, *J. Neurochem.* 58: 1230-1238.

Liao, J., Norgaard-Pedersen, B., and Brodbeck, U., 1993, *J. Neurochem.* 61: 1127-1134.

Massoulié, J., Pezzementi, L., Bon, S., Krejci, E., and Vallette, F., 1993, *Progress in Neurobiology* 41: 31-91.

Méflah, K., Bernard, S., and Massoulié, J., 1984, *Biochemie* 66: 59-69.

Mutero, A., and Fournier, D., 1992, *J. Biol. Chem.* 267: 1695-1700.

Ott, P., Jenny, B., and Brodbeck, U., 1975, *Eur. J. Biochem.* 57: 469-480.

Raconczay, Z., Mallol, J., Schenk, H., Vincendon, G., and Zanetta, J.-P., 1981, *Biochim. Biophys. Acta* 657: 243-256.

Rotundo, R.L., 1988, *J. Biol. Chem.* 263: 19398-19406.

Svensmark, O., and Kristensen, P., 1963, *Biochim. Biophys. Acta* 67: 441-452.

Svensmark, O., and Heibronn, E., 1964, *Biochim. Biophys. Acta* 92: 400-402.

Thall, A., and Galili, U., 1990, *Biochemistry* 29: 3959-3965.

Trestakis, S., Ebert, C., and Layer, P.G., 1992, *J. Neurochem.* 58: 2236-2247.

Uhlenbruck, von G., Haupt, H., Reese, I., and Steinhausen, G., 1977, *J. Clin. Chem. Clin. Biochem.* 15: 561-564.

PRESSURE EFFECTS ON STRUCTURE AND ACTIVITY OF CHOLINESTERASE

P. Masson and C. Cléry

Centre de Recherches du Service de Santé des Armées
Unité de Biochimie
24 Avenue des Maquis du Grésivaudan
38702 La Tronche Cédex, France

INTRODUCTION

The aim of this work is to present information we gained on the structure and activity of human butyrylcholinesterase (BuChE) using high pressure techniques. Also, we would like to draw attention to the usefulness of the high pressure approach for understanding catalytic mechanisms and the molecular basis for structural stability of cholinesterases.

HIGH PRESSURE EFFECTS ON PROTEINS

The volume of folded proteins is the sum of three contributions : the volume of constitutive atoms, the volume of voids (internal cavities) and a negative contribution due to hydration of peptide bonds and aminoacid side chains. Changing the physico-chemical conditions of the medium perturbs the delicate balance of intramolecular and solvent-protein interactions that maintain the native protein structure. This in turn alters protein conformation, binding properties and reactivity of enzyme active sites and may cause inactivation. Extremes of physical conditions can lead to reversible unfolding or irreversible denaturation (Jaenicke, 1991). Application of hydrostatic pressure, by changing the distance between atoms, affects weak internal interactions and controls the three-dimensional structure of proteins and enzyme activity (Gross and Jaenicke, 1994; Heremans, 1982; Morild, 1981; Mozhaev *et al.*, 1994; Silva and Weber, 1983; Weber and Drickamer, 1983).

The effects of pressure on proteins are governed by Le Chatelier's principle which states that pressure favors processes that are accompanied by reduction of the overall volume of the system. Application of high pressure to proteins allows determination of volumetric changes associated with enzyme-catalyzed reactions and conformational transitions of proteins. Thus, hydrostatic pressure is a useful physical variable to investigate protein stability and enzyme mechanisms. The pressures currently required to investigate enzyme

Enzymes of the Cholinesterase Family, Edited by Daniel M. Quinn et al.
Plenum Press, New York, 1995

mechanisms are of the order of pressure found in the deepest ocean trenches, i.e. up to 1.2 kbar; the study of conformational stability (unfolding) needs higher pressures that can be over 10 kbar.

Volume changes (ΔV) associated with equilibria are determined from the pressure (P) dependence of equilibrium constants (K). Similarly, according to the transition-state theory, activation volumes ($\Delta V^{\ddagger}$, defined as the difference between the volume of activated complex and the volume of reactants) for kinetic processes are determined from the pressure dependence of kinetic constants (k):

$$(\partial \mathrm{Ln}\ k/\partial P)_T = -\ \Delta V^{\ddagger}/RT \qquad (\partial \mathrm{Ln}\ K/\partial P)_T = -\ \Delta V/RT$$

where R is the gas constant and T the absolute temperature.

Since volume changes are related to free energy changes, they are thermodynamic parameters. Volume changes are also mechanistic criteria since their sign and magnitude depend on the nature of the reaction under study, the environment of interacting/reacting groups and reflect water structure changes (w) and protein conformation changes (conf). Thus, overall volume changes are the sum of numerous elementary contributions.

$$(\Delta V_{obs})_T = \sum_i \Delta V_{int} + \Delta V_w + \Delta V_{conf}$$

where the summation refers to weak/covalent bond formation/disruption.

These effects are strongly influenced by temperature, pH, salts, osmolytes, presence of ligands and substrate concentration. For example, differentiation of the Michaelis-Menten equation with pressure gives a relationship which expresses the variation of $\Delta V^{\ddagger}obs$ as a function of substrate concentration (Morild, 1981):

$$\Delta V^{\ddagger}_{obs} = \Delta V^{\ddagger}_{c} - Km \cdot \Delta V_b / (Km + [S])$$

At saturating, concentrations $\Delta V^{\ddagger}_{obs}$ is dominated by the volume change in the catalytic step ($\Delta V^{\ddagger}_{c}$), but at low substrate concentrations, the contribution of volume change upon binding (ΔV_b) becomes significant.

Pressure itself may affect ΔV if compressibility of the system changes. For example, downward curved logarithmic plots of rate constants (k_{obs}) as a function of pressure showing $\Delta V^{\ddagger}_{obs}$ values more positive as pressure is increased, suggest that the isothermal compressibility (β_T) is increased. That may reflect changes in hydrophobic hydration (water structuration around newly solvent-exposed apolar groups), water electrostriction around charged groups and solvation of polar groups through hydrogen bonding.

HIGH PRESSURE AS A TOOL TO STUDY CHOLINESTERASES: VOLUME CHANGES UPON LIGAND BINDING AND CATALYSIS

The first studies on the effects of pressure upon cholinesterases were undertaken to investigate adaptation of fishes to temperature and pressure by probing acetylcholinesterase (AChE) active sites of surface and abyssal fish species (Hochachka, 1974; Hochachka et al., 1975). It was found that low temperatures and high pressures induce disruption of interactions between cationic ligands/substrates and the "anionic" site. Differences between fish species were interpreted in terms of compensation between entropic and enthalpic factors. Indeed, the authors stated that binding depends upon two contributions : a hydrophobic contribution which is disrupted by low temperatures and high pressures and a coulombic contribution which is favored by high pressures. However the latter was in contradiction with model compound studies showing that ion pair formation is accompanied by volume increases and conversely that dissociation of ion pairs is associated with volume decreases due to electrostriction of water molecules around charged groups (Heremans, 1982; Gross and Jaenicke, 1994).

Studying the affinity of human BuChE for a derivative of meta amino-phenyl-trimethyl ammonium (PTMA), we found that pressure strengthens the interaction up to 1.2 kbar in water and 1.6 kbar in heavy water; above these pressures affinity dropped in a narrow pressure range (Masson and Balny, 1990). The corresponding negative binding volume change (-33 ml.mol^{-1} at 35°C) indicated that binding strength is dominated by pressure-enhanced interactions (Weber and Drickamer, 1983). As suggested from studies with fluorescent probes (Chan *et al.*, 1974), we assumed that these interactions were charge-transfer interactions between the phenyl ring of the ligand acting as an electron acceptor and an aromatic amino acid acting as an electron donor, e.g. a tryptophan residue. The importance of π electrons in stacking of aromatic rings and in interactions between positively charged groups and aromatic amino acid side chains in the active site gorge of cholinesterases is now well documented (Harel *et al.*, 1993; Barak *et al.*, 1994). However, large negative volume changes could not reflect only association of aromatic rings and hydrogen bond formation. This was confirmed by the study of BuChE car-bamylation by N-methyl-(7-dimethylcarbamoxy) quinolinium (M7C) under single-turn-over conditions in different media. We showed that negative ΔV upon binding and positive $\Delta V^{\ddagger}$ are too large to be related only to bond formation/disruption as depicted by the minimum reaction scheme for ChE carbamylation (Masson and Balny, 1988, 1990). The fact that the largest volume changes are associated with reactions that took place in buffers containing water-structure makers (Li^{+} and phosphate) suggested that extended hydration changes occur during binding and the catalytic step. In addition, deviations from linearity of Ln k_{obs} vs P for BuChE-catalyzed reactions (Masson and Balny, 1986, 1988, 1990) reflect compression phenomena that may be ascribed to pressure effects on the enzyme structure rather than on reactions.

EFFECT ON QUATERNARY STRUCTURE: MODE OF ASSOCIATION BETWEEN SUBUNITS

The effect of pressure on quaternary structure of the tetrameric form (G$_4$) of human plasma BuChE was investigated by gel electrophoresis under high pressure up to 3.5 kbar (Masson, 1995) and by conventional non-denaturing electrophoresis after exposure to pressure or ultrasound irradiation using a Branson sonicator. In parallel, the effect of cold on tetramer stability was investigated by subzero transverse temperature gradient gel electrophoresis down to -30°C (Curtil *et al.*, 1994). It should be remembered that electrostatic and hydrophobic interactions - the major forces involved in subunit asso-ciation of proteins - are disrupted by pressure (Silva and Weber, 1993). So, as a general rule, moderate pressures (0.5-2 kbar) cause dissociation of oligomeric proteins. Similarly, cold-induced dissociation reflects weakening of hydrophobic interactions at low tempera-tures (Privalov, 1990).

As regards the effect of pressure, whatever the other environmental conditions (water or heavy water as the solvent; phosphate, Tris/HCl or Tris/glycine buffers ranging from (pH)$_{P,T}$ 5.7 to 9.5; temperature ranging from 5 to 35°C; presence or absence of a) a specific ligand such as PTMA or hexyltrimethylammonium or b) a denaturing agent : propylene carbonate), no pressure-induced dissociation of G$_4$ was observed up to 3.5 kbar (Masson,

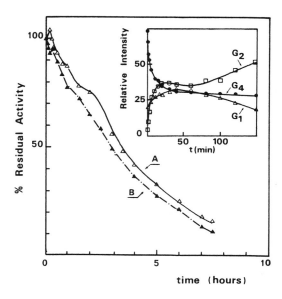

Figure 1. Effects of ultrasound on BuChE stability: Loss of activity of highly purified human BuChE (G_4) as a function of ultrasound irradiation time (B15 Branson sonicator with titanium microprobe, 20 kHz, P = 90 W) at 20°C; A, in 20 mM phosphate pH 7.0; B, in 10 mM Tris/HCl pH 7.0. Insert : ultrasound-induced partial dissociation of G_4 as determined by gel electrophoresis and densitometric analysis of samples taken at different times.

1995; Masson and Balny, 1986, 1988, 1990; Masson *et al.* 1994). Moreover, subzero electrophoresis did not show dissociation. This suggested to us that the interdimer contact area is mainly stabilized by interactions that are either pressure insensitive, e.g. H-bonds, or pressure-enhanced such as aromatic-aromatic interactions.

The effects of ultrasound were more difficult to interpret unequivocally because thermal and chemical effects due to cavitation are superposed on the specific effects of the periodic variation of the local pressure due to propagation of ultrasonic waves. Progressive loss of activity was accompanied by partial dissociation of G_4 (Fig.1). However, SDS electrophoresis revealed that only proteolytically nicked tetramers were dissociated. Inactivation and dissociation were retarded by 4 M sorbitol.

Although, the two subunits of each dimer appear to be held by helices $\alpha F'_3$ and αH (Sussman *et al.*, 1991), the mode of association between the two dimers to form the native tetrameric enzyme is not known. It has been shown that the interchain disulfide bond is important for tetramer stability, but that limited proteolysis upon the action of trypsin can lead to a tetramer in which the C-terminal hexapeptide containing the disulfide has been cleaved (Lockridge and LaDu, 1982). Unlike mutagenized human AChE (C580A) (Velan *et al.*, 1991) , mutation C571A preserves the quaternary structure of BuChE (Blong *et al.*, 1995). Moreover, the G534 stop mutant of human BuChE expressed in CHO cells is a monomer (Blong *et al.*, 1995). Thus, it appears that the peptide segment between Asn 535 and Lys 567 is directly involved in tetramer formation. This segment contains 8 aromatic amino acids regularly spaced out. Helical wheel representation of segment 538-567 shows an asymmetric distribution of hydrophobic and polar residues along the axis, suggesting that the hydrophilic side of this ideal α amphipathic helix is solvent-exposed and that the hydrophobic side interacts with a non polar surface. Prediction of the secondary structure of this segment according to the method of Chou and Fasman (Chou and Fasman, 1978) and with artificial neural network indicates that aromatic rings form a ridge on the apolar side (Fig. 2). Thus, they may face each other to form a zipper motif connecting together the C-terminal helical segments of the 4 subunits. One way to test whether these amino acids are important for tetramer stabilization would be to mutate them into helix forming aliphatic amino acids. If mutant enzymes are tetramers, application of pressure should dissociate them into dimers.

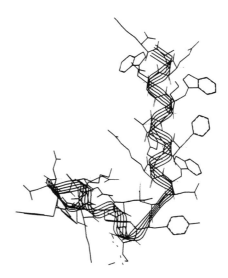

Figure 2. Secondary structure of segment 538-567 of human BuChE as predicted according to the method of Chou and Fasman.

EFFECTS ON TERTIARY AND SECONDARY STRUCTURES: CONFORMATIONAL CHANGES, INACTIVATION, UNFOLDING AND DENATURATION

The effects of pressure on three-dimensional structure of BuChE and the possibility to probe pressure-induced conformational changes were investigated by activity measurement after exposure to pressure, fourth-derivative spectroscopy under pressure in the near-ultraviolet range, high pressure electrophoresis, binding of 1-anilino-8-naphthalene sulfonate (ANS) under pressure and FT-IR spectroscopy using a diamond anvil cell operating up to 20 kbar.

The pressure inactivation study was carried out at 20°C in 10 mM Tris/HCl pH 7.0 and in 50 mM phosphate pH 7.5. Residual activity was measured immediately after pressure release. Whatever the buffer there was no irreversible inactivation up to 1.5 kbar after 12 h exposure to pressure. Activity decreased at higher pressures and the enzyme was fully inactivated beyond 4 kbar (Masson *et al.*, 1994). Increasing concentrations of propylene carbonate progressively shifted pressure "transition midpoints" to lower pressures. In contrast glycerol and ligands (10 mM procainamide, 2-PAM, edrophonium and de-camethonium) protected the enzyme against pressure inactivation . This agrees with previous findings regarding affinity for PTMA (Masson and Balny, 1990). It was observed that inactivation does not obey a first-order process, but follows apparent second-order kinetics. This suggests that a non-unimolecular process governs deactivation. Moreover, inactivation was found to be dependent on enzyme concentration. Surprisingly, after 7 h exposure to pressure, activity remained constant at least for 10 days. Actually, microhetero-geneity in the enzyme structure is expected to cause a non-first order inactivation process. Non-denaturing electrophoresis of BuChE after pressure release showed progressive decrease of the intensity of the native tetramer (N) with concomitant formation of an inactive slow-migrating form (D) and aggregates as a function of pressure (Masson *et al.*, 1994). In an attempt to interpret inactivation/denaturation of BuChE in terms of pressure-induced conformational changes, we examined the effects of pressure upon tertiary and secondary structures. First, we performed high-pressure electrophoresis in polyacrylamide gel rods of different acrylamide concentrations. Plots of log relative mobility (R_m) vs total acrylamide

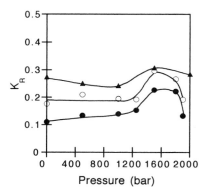

Figure 3. Ferguson plot analysis of cholinesterases: Secondary plot of K_R versus pressure for rHAChE (●, monomer; O, dimer) and human BuChE (Δ, tetramer). Electrophoresis was performed in capillary gel rods (% T : 4 to 6.5) in discontinuous buffer systems (running buffers : 0.1 M Gly/Tris pH 8.3 for rHAChE and 0.5 M Gly/Tris pH 8.3 for BuChE in the presence of 7.5 % sucrose). Migration time was 1 h under 0.3 mA/gel at 10°C.

concentration (%T), known as Ferguson plots, allow determination of molecular size of native globular proteins (Rodbard and Chrambach, 1971; Masson, 1979)

$$\log R_m = \log Y_o - K_R T$$

The ordinate intercept, Y_o, depends on charge, size and shape of the protein. Assuming sphericity, there is a linear relationship between K_R and protein size, but K_R values have been found to depend on protein conformation (Goldenberg and Creighton, 1984) and to vary with pressure (Masson, 1995). So, overall conformational changes of BuChE and rHAChE have been detected by measuring Y_o and K_R changes as a function of pressure. As shown on Fig. 3, K_R increases between 1 kbar and 1.6 kbar reflect small positive ΔV_s, suggesting a predenaturation transition toward a compact state whose hydrodynamic radius is greater than that of native enzymes. Such a transition has been observed for most of globular proteins subjected to pressure (Heremans, 1982; Morild, 1981) and may correspond to transition to a "molten globule" state. Furthermore, a continuous decrease of Y_o up to 1.5 kbar (not shown) is due to pressure-induced ionization changes of protein and buffer, but the break in $\log Y_o$-P plots beyond this pressure may reflect a change in shape and solvent-exposure of charged groups. As inferred from pressure denaturation of proteins in the presence of glycerol, molten globule states likely result from the penetration of water into protein structure (Oliveira *et al.*, 1994). In the presence of 2 M sorbitol or sucrose, we found that K_R became insensitive to pressure up to at least 2 kbar. Therefore, the effect of polyols may be interpreted in terms of preferential hydration: hydration of polyols restrains the binding of water to the enzyme (osmolyte effect), which in turn counteracts the pressure denaturing effect.

Fourth-derivative spectroscopy has been used to assess the degree of solvent exposure of aromatic residues (Padros *et al.*, 1984). Fourth-derivative spectra of BuChE recorded up to 2 kbar show a pressure dependence that reflects changes in the environment of aromatic amino acids. Indeed, red shift of peaks $\partial^4 A / \partial \lambda^4$) corresponding to absorption of tryptophan and phenylalanine and a change in the ratio R of the amplitude of the main peaks as a function of pressure (Fig. 4) indicate that aromatic amino acids undergo a transition toward more hydrophobic environments with increasing pressure. However, FT-IR-spectroscopy showed that there is no significant change in secondary structure below 3 kbar (Fig. 5); irreversible unfolding starts above this pressure and is complete at 8 kbar ($P_{1/2}$ = 5.5-6 kbar) (Cléry *et al.*, 1993). On the other hand, increased binding of ANS with pressure up to 1 kbar followed by a decrease at higher pressures (Fig. 6) indicates that low pressures promoted exposure of patches of hydrophobic residues, which is consistent with the hypothesis of "molten globule" intermediates (Semisotnov *et al.*, 1991).

So, to summarize, these results show that pressure-induced "denaturation" of BuChE -in the pressure range 10^{-3} - 3 kbar- is a multistate process including "molten globule"

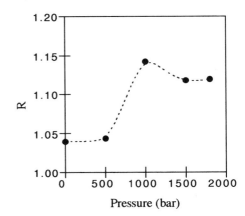

Figure 4. Fourth-derivative spectroscopy under pressure: Variation of the parameter R (ratio of the amplitude of the main tryptophan peaks) with pressure.

intermediate(s). At higher pressures, highly cooperative unfolding leads to irreversibly denatured states followed by subsequent aggregate formation. Therefore, the pressure denaturation pathway of BuChE may be depicted by the following scheme:

$P_{1/2}$ - 1.5 kbar $P_{1/2}$ ∈ 2-2.8 kbar $P_{1/2}$ - 6 kbar

N ⇄ "molten globule" and partially unfolded states ⇄ inactivated states → U (D)

Aggregates 1   Aggregates 2

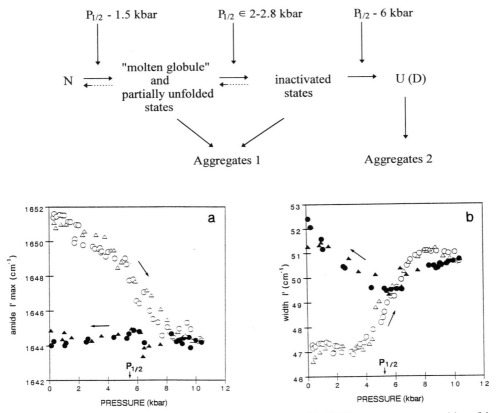

Figure 5. Pressure induced secondary structure change as monitored by FT-IR spectroscopy: a, position of the amide I' peak as a function of pressure; b, width of the amide I' band as a function of pressure. Open symbols are for spectra recorded with increasing the pressure and filled symbols for spectra recorded with decreasing the pressure.

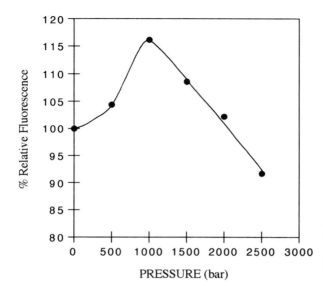

Figure 6. Binding of ANS to BuChE under pressure: Relative change of fluorescence intensity of bound ANS (λ_{ex} : 358 nm; λ_{em} max : 447 nm in 10 mM Tris/HCl pH 7.5 at 25°C.

ACKNOWLEDGMENTS

This work was supported by a grant from la Direction des Recherches Etudes et Techniques (DRET 94/5). The authors thank C. Balny and R. Lange (INSERM, Montpellier), and O. Lockridge, R. Blong and L. Kinarsky (Univ. Nebraska, Omaha) for fruitful discussions.

REFERENCES

1. Barak,D., Kronman,C., Ordentlich,A., Ariel,A., Bromberg,A., Marcus,D., Lazar,A., Velan,B. and Shafferman,A. (1994) *J. Biol. Chem., 264,* 6296-6305.
2. Blong,R.M. and Lockridge,O. (1995) *Proc. Fifth. Int. Meeting on Cholinesterases*; this volume.
3. Chan,L.M., Himel,C.M. and Main,A.R. (1974) *Biochemistry, 13,* 86-90.
4. Chou,P.X. and Fasman,G.D. (1978) *Ann. Rev. Biochem., 47,* 251-276.
5. Cléry,C., Goossens,K., Hui Bon Hoa,G., Hereman,K., Balny,C. and Masson,P. (1993) in *Stability and Stabilization of Enzymes* (van den Twell,W.J.J.,Harder,A. and Buitelaar,R.M., Eds), Elsevier, pp. 255-260.
6. Curtil,C., Channac,L., Ebel,C. and Masson,P. (1994) *Biochim. Biophys. Acta, 1208, 1-7.*
7. Goldenberg,D.P. and Creighton,T.E. (1984) *Anal. Biochem., 138,* 1-18.
8. Gross,M. and Jaenicke,R. (1994) *Eur. J. Biochem., 221,* 617-630.
9. Harel,M., Schalk,I., Ehret-Sabatier,L., Bouet,F., Goeldner,M., Hirth,C., Axelsen,P.H., Silman,I. and Sussman,J.L. (1993) *Proc. Natl. Acad. Sci. USA, 90,* 9031-9035.
10. Heremans,K. (1982) *Ann. Rev. Biophys. Bioeng., 11,* 1-21.
11. Hochachka,P.W. (1974) *Biochem. J., 143,* 535-539.
12. Hochachka,P.W., Storey,K.B. and Baldwin,J. (1975) *Comp. Biochem. Physiol.,* 52B , 13-18.
13. Jaenicke,R. (1991) *Eur. J. Biochem., 202,* 715-728.
14. Lockridge,O. and LaDu,B.N. (1982) *J. Biol. Chem., 257,* 12012-12018.
15. Masson,P. (1979) *Biochim. Biophys. Acta, 578,* 493-504.
16. Masson,P. (1995) in *High Pressure Chemistry and Physics : a practical approach* (Isaacs,N.S. Ed), Oxford Univ. Press, in press.
17. Masson,P. and Balny,C. (1986) *Biochim. Biophys. Acta, 874,* 90-98.
18. Masson,P. and Balny,C. (1988) *Biochim. Biophys. Acta, 954,* 208-215.
19. Masson,P. and Balny,C. (1990) *Biochim. Biophys. Acta, 1041,* 223-231.
20. Masson,P., Gouet,P. and Cléry,C. (1994), *J. Mol. Biol., 238,* 466-478.

21. Morild,E. (1981) *Adv. Prot. Chem.*, *34*, 93-163.
22. Mozhaev,V., Heremans,K., Frank,H., Masson,P. and Balny,C. (1994) *TIBTECH*, *12*, 493-501.
23. Oliveira,A.C., Gaspar,L.P., Da Poian,A.T. and Silva,J.L. (1994) *J. Mol. Biol, 240*, 184-187.
24. Padròs,E., Duñach,M., Morros,A., Sabés,M. and Mañosa,J. (1984) *TIBS, 9*,508-510
25. Privalov,P. (1990) *CRC Crit. Rev. Biochem. Mol. Biol.*, *25*, 281-305.
26. Rodbard,D. and Chrambach,A. (1971) *Anal. Biochem.*, *40*, 95-134.
27. Semisotnov,G.V., Rodionov,N.A., Razgulyaev,O.I., Uversky,V.N., Gripas,A.F. and Gilmanshin,R.I. (1991) *Biopolymers*, *31*, 119-128.
28. Silva,J.L. and Weber,G. (1993) *Ann. Rev. Phys. Chem.*, *44*, 89-113.
29. Sussman,J.L., Harel,M., Frolow,F., Oefner,C., Goldman,A., Toker,L. and Silman,I. (1991) *Science, 253*, 872-879.
30. Velan,B., Grosfeld,H., Kronman,C., Leitner,M., Gozes,Y., Lazar,A., Flashner,Y., Marcus,D., Cohen,S. and Shafferman,A. (1991) *J. Biol. Chem.*, *266*, 23977-23984.
31. Weber,G. and Drickamer,H.G. (1983) *Quat. Rev. Biophys., 16*, 89-112.

HYDROPHOBICITY ON ESTERASE ACTIVITY OF HUMAN SERUM CHOLINESTERASE

L. Jaganathan, K. Padmalatha, G. Revathi, and R. Boopathy[*]

Department of Bio-Technology
Bharathiar University
Coimbatore
641 046 India

The influence of hydrophobicity on the catalytic activity of acetyl cholinestrase is well documented (Wiedmer *et al.*, 1979), while that of human serum cholinesterase (BuChE) is not known. The possibility of the hydrophobicity of BuChE and its modulatary effect on its catalytic activity may exhibit the physiological role of BuChE. An attempt has been made in the present study to bring out the effect of hydrophobicity on its esterase activity.

At various stages of purification (Boopathy & Balasubramanian, 1985), BuChE was treated with Triton X-100, a non-ionic hydrophobic detergent. In general it had an inhibitory effect on esterase activity. BuChE in its native environment (plasma/serum) was inhibited to a lesser level, while on further purified enzyme the inhibitory effect of Triton X - 100 was high. This sparing effect of native environment on the Triton X - 100 inhibition on esterase activity was not observed by simple addition of bovine serum albumin to the purified enzyme. The inhibitory pattern suggests that Triton X-100 could interact with some hydrophobic region of BuChE in modulating the esterase activity. Similarly a hydrophobic detergent 3 -[(3-Cholamidopropyl)dimethylammonio]-1-propanesulfonate has been shown to affect esterase activity of BuChE (Tornel *et al.*, 1991).

The BuChE digestion with protease and its hydrophobic interaction with phenyl sepharose resin reveal that protease digestion decreased the hydrophobicity. However the recovery of esterase activity from the column increased by two folds.

Sodium metaperiodate oxidation of the glycans of BuChE increased the net hydrophobicity of the protein by 1.7 fold as measured using the fluorescent probe 8-anilino-1-napthalene sulfonic acid, concomitant to a 22% decrease in esterase activity. This increase in hydrophobicity would be attributed to the exposure of the burried hydrophobic amino acids.

Removal of sialic acid from BuChE resulted in a decrease and an increase in esterase activity at low and high concentrations of neuraminidase, respectively. These changes would

---

[*] To whom all correspondence should be addressed.

be attributed to partial and complete removal of negatively charged sialic acid. Thus esterase activity is modulated by the net negative charges on the protein. However, such changes were not associated with its hydrophobicity.

From the results it is possible to assume that hydrophobic interaction due to Triton X-100, or changes in hydrophobicity of the protein due to proteolysis or glycan oxidation and charge difference due to sialic acid distribution on the protein do have a regulatory role on the esterase activity of BuChE.

ACKNOWLEDGMENT

Supported by the Department of Science and Technology, Technology Bhavan, New Delhi, India (No. SP/SO/D-30/89 dated 24.11.89).

REFERENCES

Boopathy, R., & Balasubramanian, A.S., *Eur. J. Biochem.*, 1985, *151*, 351 - 360.
Tornel, P.L., Munoz, E., & Vidal, C.J., *Bioorg. Chem.*, 1991, *19*, 1-9.
Wiedmer, T., Di Francesco, C., & Brodbeck, U., *Eur. J. Biochem.*, 1979, *102*, 59-64.

ALTERATIONS IN THE TOPOGRAPHY OF ACETYLCHOLINESTERASE ACTIVE SITE GORGE AFTER BINDING OF PERIPHERAL ANIONIC SITE LIGANDS

Anton Štalc [1], Zoran Grubič,[1] Marjeta Šentjurc,[2] Slavko Pečar [3],
Mary K. Gentry,[4] and Bhupendra P. Doctor[4]

[1] Institute of Pathophysiology, School of Medicine
Ljubljana, Slovenia
[2] Jožef Stefan Institute
Ljubljana, Slovenia
[3] Department of Pharmacy, Faculty of Science and Technology
Ljubljana, Slovenia
[4] Division of Biochemistry, Walter Reed Army Institute of Research
Washington, D.C.

The active site of AChE is located in a pocket- or gorge-like structure, as demonstrated eighteen years ago by electron paramagnetic resonance (EPR) (Šentjurc et al., 1976a,b). EPR results were supported later by topographic studies using polyclonal and monoclonal antibodies (Ogert et al., 1990). Final confirmation came from x-ray crystallographic analysis (Sussman et al., 1991) that additionally provided a detailed three-dimensional structure of the gorge. The peripheral anionic site is located at the lip of the gorge (Shafferman et al., 1992; Radić et al., 1993). Ligands binding specifically to this site are known to inhibit AChE catalytic activity. Their mechanism of action, however, is still unknown. In order to investigate this mechanism we studied and compared the effects on the microtopography of the active site gorge of fetal bovine serum AChE of two peripheral anionic site-binding ligands, monoclonal antibody 25B1 (Fab fragments) and propidium. The EPR approach, which allowed detection of conformational alterations in the active site gorge, was used in our experiments.

As concluded from EPR spectra, the effects of the two ligands were not identical. Narrowing of the gorge, as well as stabilization against heat denaturation, was observed after incubation of spin labelled FBS AChE with Fab 25B1. On the other hand, no conformational alterations or protection against heat denaturation could be detected after propidium binding. Our results demonstrate that peripheral anionic site ligands, although directed to the same binding region on the FBS AChE molecule, may have different effects on the topography of the AChE active site. These observations suggest a complex structure of what has been called the peripheral anionic site that enables various inter-actions with different ligands.

REFERENCES

Ogert, R.A., Gentry, M.K., Richardson, E.C., Deal, C.D., Abramson, S.N., Alving, C.R., Taylor, P., and Doctor, B.P., 1990, *J. Neurochem.* 55:756-763.

Radić, Z., Pickering, N.A., Vellom, D.C., Camp, S., and Taylor, P., 1993, *Biochemistry* 32:12074-12084.

Šentjurc, M., Štalc, A., Županćić, A.O., and Schara, M., 1976, *Biochim. Biophys. Acta* 429:421-428.

Šentjurc, M., Štalc, A., and Županćić, A.O., 1976, *Biochim. Biophys. Acta* 438: 131-137.

Shafferman, A., Velan, B., Ordentlich, A., Kronman, C., Grosfeld, H., Leitner, H., Flashner, M., Cohen, Y., Barak, D., and Ariel, N., 1992, *EMBO J.* 11:3561-3568.

Sussman, J.L., Harel, M., Frolow, F., Oefner, C., Goldman, A., Toker, L., and Silman, I., 1991, *Science* 253:872-879.

ACETYLCHOLINESTERASE FROM *OCTOPUS VULGARIS* (CEPHALOPODA)

Evidence for a Specific High Salt-Soluble and Heparin-Soluble Fraction of Globular Forms

Vincenzo Talesa,[1,2] Marta Grauso,[1] Elvio Giovannini,[1] Gabriella Rosi,[1] and Jean-Pierre Toutant[2]

[1] Department of Experimental Medicine
Division of Cell and Molecular Biology
University of Perugia, Italy
[2] Différenciation cellulaire et Croissance, INRA
Montpellier, France

Octopus acetylcholinesterase (AChE) is found as two molecular forms: an amphiphilic dimeric form (G2) sensitive to phosphatidylinositol phospholipase C and a hydrophilic tetrameric (G4) form. G2 and G4 forms belong to a single pharmacological class of AChE. Thus they likely result from a post-transcriptional or post-translational processing of a single AChE gene. Sequential solubilizations reveal that a significant portion of both G2 and G4 forms can be recovered only in a High Salt-Soluble fraction (1M NaCl, no detergent). Heparin (2mg/ml) was able to solubilize G2 and G4 forms with the same efficiency than 1M NaCl. The solubilizing effect of heparin was concentration-dependent and was reduced by protamine (2mg/ml). This suggests that heparin operates through the dissociation of ionic interactions existing *in situ* between globular forms of AChE and cellular or extracellular polyanionic components.

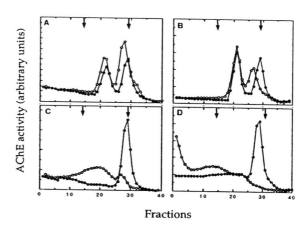

Figure 1. Sucrose gradient analysis of *Octopus* AChE present in LSS, HSS and DS fractions. In A, B and C, 250 ml of LSS, HSS and DS fractions were analyzed on HS (open circles) or HST (filled circles) gradients. D shows the effect of PI-PLC on DS fraction analyzed on HS gradient, filled circles: PI-PLC-treated, open circles: control sample. S values were determined by comparison to internal markers of sedimentation: β-galactosidase (16S, arrow on the left) and alkaline phosphatase (6.1S, arrow on the right).

Fractions

MOLECULAR POLYMORPHISM OF ACETYLCHOLINESTERASE IN *HIRUDO MEDICINALIS*

Vincenzo Talesa,[1,2] Marta Grauso,[1] Elvio Giovannini[1,] Gabriella Rosi,[1] and Jean-Pierre Toutant[2]

[1] Department of Experimental Medicine
Division of Cell and Molecular Biology
University of Perugia, Italy
[2] Différenciation cellulaire et Croissance, INRA
Montpellier, France

Two different cholinesterases were extracted from *Hirudo medicinalis* that differ in molecular forms and in substrate and inhibitor specificity. Spontaneous-soluble (SS) activity was recovered from tissue dilaceration in low salt buffer, and detergent-soluble (DS) was solubilized by LS buffer+1% Triton. Both enzymes were purified to homogeneity on edrophonium-sepharose followed by chromatography on Concanavalin A-sepharose. Purified enzymes were studied by SDS-PAGE in reducing conditions. SS enzyme gave two bands at 30 and 66 kDa, while only one band was seen at 66 kDa for DS enzyme. M_r of the purified ChEs evaluated by sephadex G 200 chromatography were about 66000 for SS enzyme and 130000 for DS. Sedimentation analysis of crude and purified enzyme preparations was performed with or without Triton in the gradient. SS ChE showed a single peak sedimenting at 5.0 S in both conditions. This result suggests that SS enzyme is a hydrophylic monomer (G1). DS enzyme showed a single peak of activity at 6.5 S and aggregated in the absence of detergent in the gradient. PI-PLC suppressed the aggregation. DS ChE is thus an amphiphilic dimer anchored to the membrane via a glycosylphosphatidylinositol. Kinetic study was carried out using p-nitrophenyl and thiocholine esters. Both enzymes appear to hydrolyze propionyl esters best but are also active on butyryl and acetyl derivatives. Both ChEs were inhibited by excess substrate and should thus be referred to as acetylcholinesterases. SS AChE displayed marked similarity between the sets of K_m values with charged and uncharged substrates, suggesting a reduced influence of electrostatic interactions in the enzyme substrate affinity. Unlike SS enzyme, ionic interactions strongly affect the formation of DS AChE substrate complex (far higher Km values with uncharged substrates). SS AChE was more sensitive to inhibition by eserine and DFP (IC50 = 10^{-7} and 10^{-8} M respectively) than DS enzyme (IC50 = 10^{-6} and 10^{-5} M).

SUBUNIT ASSOCIATION AND STABILIZATION OF BUTYRYLCHOLINESTERASE (BChE)

R. M. Blong, P. Masson, and O. Lockridge

Eppley Institute
University of Nebraska Medical Center
Omaha, Nebraska
Centre de Recherches du Service de Santé des Armées Unité de Biochimie
BP 87, 38702 La Tronche Cedex, France

ABSTRACT

The goal of this project was to determine what parts of the BChE protein are important for its tetrameric organization. In human serum the BChE molecule has a dimer of dimers structure. Dimers are covalently linked by an interchain disulfide bond between the cysteines at amino acid 571. To study the effect the disulfide bond has on tetrameric structure and stability, a recombinant mutant was engineered replacing cysteine 571 with alanine. The mutant was expressed in Chinese hamster ovary cells. Elimination of the interchain disulfide bond at amino acid 571 decreased the heat stability of the enzyme but preserved its tetrameric structure. A second mutant in which 40 amino acids were deleted from the carboxy terminus

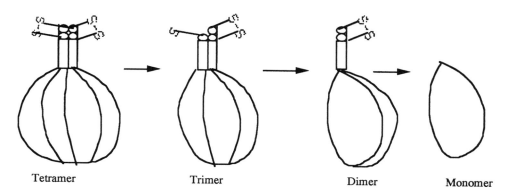

Figure 1. Model of butyrylcholinesterase (BChE) subunit dissociation.

was also expressed. This mutant gave an active monomer, thus suggesting that the region involved in dimer and tetramer contact is the carboxy terminus. Additional mutations, engineered by PCR (Polymerase Chain Reaction) site directed mutagenesis, will allow further characterization of the tetrameric stabilization of BChE.

The carboxy terminus, believed to be the contact point for dimer and tetramer formation, is depicted as a helix bundle with the interchain disulfide bond at Cys 571 stabilizing the dimers. We hypothesize that with proteolysis the carboxy terminal tail is cleaved and the subunits dissociate.

ACKNOWLEDGMENTS

Supported by U.S. Army Medical Research and Development Command Grant DAMD 17-94-J-4005

DENATURATION OF RECOMBINANT HUMAN ACETYLCHOLINESTERASE

M. Lebleu,[1] C. Cléry,[1] P. Masson,[1] S. Reuveny,[2] D. Marcus,[2] B. Velan,[2] and A. Shafferman[2]

[1] Centre de Recherches du Service de Santé des Armées
Unité de Biochimie
38702 - La Tronche Cédex - France
[2] Israel Institute for Biological Research
Ness-Ziona - 70450 Israel

The study of the unfolding process of proteins has been a subject of growing interest during recent years. The availability of recombinant cholinesterases in large amounts permits now to use biophysical tools for gaining information about folding and stability of these enzymes.

The stability of recombinant human acetylcholinesterase (rHuAChE) produced by human embryonic kidney cell line (293) in a fixed-bed reactor (Lazar *et al.*, 1993) was investigated at pH 8.0 by thermal, pressure, urea and guanidinium chloride denaturation. The thermal unfolding was studied by differential scanning calorimetry (DSC) in the temperature range from 20° to 80°C. The pressure effects were studied by following tryptophane fluorescence under pressure at 15°C up to 4 kbar, by measuring remaining enzyme activity after pressure release and by electrophoresis under pressure at 10°C (construction of Ferguson plots (Masson, 1995)). Denaturation by urea and guanidinium chloride under equilibrium conditions was studied by monitoring tryptophane fluorescence *versus* denaturant concentration at 25°C in the presence and absence of tris (2-carboxyethyl) phosphine (TCEP) as a reducing agent.

Hydrostatic pressure, up to 4 kbar, led to partially but irreversibly denatured states ($P_{1/2}$ = 2.8 kbar) followed by their subsequent aggregation. Moreover, Ferguson plots analysis showed that K_R of the dimer is almost constant up to 1 kbar. Above this pressure, K_R increased up to 1.6 kbar and then dropped, indicating conformational transition toward a compact state whose hydrodynamic radius is greater than that of native enzymes and may correspond to transition to a "molten globule" state with exposure of patches of hydrophobic residues.

The apparent conformational stability of rHuAChE, defined as the apparent ΔG_U^{H2O} calculated from urea and guanidinium chloride transition curves (Pace, 1990), was found to be 23.9 kJ.mol^{-1} and 18.4 kJ.mol^{-1}, respectively. These values are lower than values found for globular proteins, suggest that unfolding of rHAChE does not obey the two-state model. In addition, the low values of m (transition steepness) reflect multiple intermediates. This transition induced by denaturing agents under equilibrium conditions may correspond to transition from a "molten globule" state to the denaturated state. The first step of denaturation from native enzyme to "molten globule" state may occur at lower denaturant concentration, difficult to detect by monitoring tryptophan fluorescence.

Thermal denaturation cannot be described by thermodynamical equilibrium. Using DSC we showed that denaturation is an irreversible process under kinetic control. Indeed, a) the transition temperature midpoint is dependent on the scan rate (e.g. $T_m = 59.2°C$ with $v = 1°C/min$) ; b) there is no release of heat upon decreasing temperature. The ratio $\Delta H_{cal}/\Delta H_{Van't\ Hoff} = 2.5$ ($\Delta H_{cal} = 1500$ kJ.dimer mol^{-1}, as expected for a stable compact globular protein) also showed that thermal denaturation is an irreversible multistep process. The energy of activation, calculated according to several methods (Sanchez-Ruiz, 1992) is 450 kJ.mol^{-1}.

The structural basis of the stability of cholinesterases is still unresolved. What is clear from the present work is that the denaturation of rHuAChE dimer proceeds through several steps and that final denatured states depend on both the external conditions and the protein structure. Complexity may arise from the imperfect packing of aminoacids at the subunit interface and from the existence of several independent subdomains in each structural domain (monomer). The following scheme may be tentatively proposed to depict the denaturation pathway of rHuAChE.

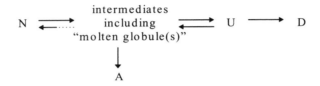

REFERENCES

Lazar, A., Reuveny, S., Kronman, C. , Velan, B. and Shafferman, A. (1993) *Cytotechnology*, *13*, 115-123.
Masson, P. (1995) in *High Pressure Chemistry and Physics : A Practical Approach* (Isaacs, N.S. Ed), Oxford Univ. Press, in press.
Pace, C.N. (1990) *TIBTECH*, *8*, 93-98
Sanchez-Ruiz, J.M. (1992) *Biophys. J.*, *61*, 921-935.

AMINO ACID RESIDUES THAT CONTROL MONO- AND BISQUATERNARY OXIME-INDUCED REACTIVATION OF O-ETHYL METHYLPHOSPHONYLATED CHOLINESTERASES

Y. Ashani,[1*] Z. Radić,[2†] I. Tsigelny,[2] D. C. Vellom,[2] N. A. Pickering,[2]
D. M. Quinn,[2‡] B. P. Doctor,[1] and P. Taylor[2]

[1] Division of Biochemistry
Walter Reed Army Institute of Research
Washington DC 20307-5100
[2] Department of Pharmacology
University of California
San Diego, La Jolla, California 92093-0636

INTRODUCTION

The limited scope of antidotal activity of commonly used reactivators of organophosphonyl (OP) conjugates of cholinesterases (ChE) such as the methanesulfonate salt of 2-PAM (P2S; Fig. 1) or TMB4, prompted the evaluation of new series of oxime reactivators (Erdman, 1969). This effort resulted in the introduction of a new series of bispyridinium monooximes. One such compound, HI-6 (Fig. 1), is among the most potent reactivating agents that serve as antidotes against organophosphate toxicity (Oldiges and Schoene, 1970).

The effectiveness of oxime reactivators as antidotes is primarily attributed to the nucleophilic displacement rate of the OP moiety from the inhibited enzyme (Fig. 2), and varies with the structure of the bound-OP, the source of the enzyme, and the oxime used. Structure-function relationships among oxime reactivators are not clearly understood.

In this report we describe studies on P2S- and HI-6-induced reactivation of wild-type recombinant mouse AChE (rMoAChE) and its mutants inhibited with 7-(methylethoxyphosphinyloxy)-1-methylquinolinium iodide (MEPQ; Fig. 2)(Levy and

* Visiting scientist from Israel Institute for Biological Research, Ness-Ziona, Israel.
† Visiting fellow from Institute for Medical Research and Occupational Health, University of Zagreb, Croatia.
‡ Visiting scientist from Department of Chemistry, University of Iowa, Iowa City, IA.

Enzymes of the Cholinesterase Family, Edited by Daniel M. Quinn et al.
Plenum Press, New York, 1995

133

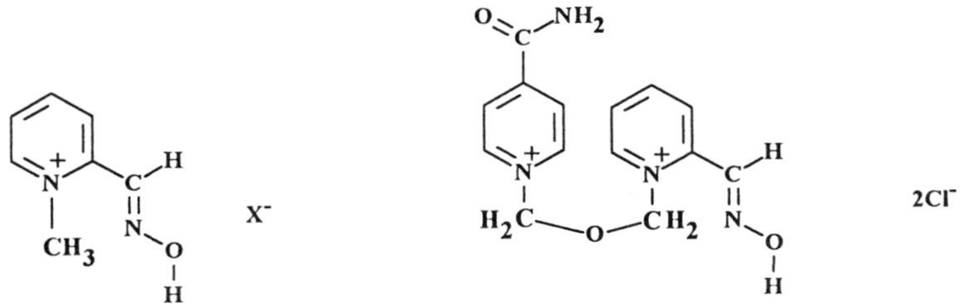

2-PAM, X=I   HI-6

P2S, X=CH$_3$SO$_3$

Figure 1. Structure of oximes.

Ashani, 1986). Biochemical constants of the reactivation of EMP-ChEs were compared to those obtained for tissue-derived TcAChE and human BChE (HuBChE). Delineation of amino acid residues that are important for reactivation highlight several aspects of the mechanism by which oximes enhance displacement of an OP from EMP-ChEs and provide an explanation for differences between the reactivation potency of HI-6 and P2S .

MATERIALS AND METHODS

Materials

MEPQ was prepared as described elsewhere (Levy and Ashani, 1986). Wild-type and mutant rMoAChEs were prepared as described previously (Vellom et al., 1993; Radić et al., 1993). Torpedo californica AChE (TcAChE), wild-type mouse AChE and some of the mutant enzymes were purified by affinity chromatography as described previously (Lee et al., 1982). HuBChE was purified by procainamide-Sepharose® 4B gel affinity chromatography. One mg of pure enzyme contained approximately 11 and 14 nmol of active sites of BChE and AChE, respectively. Inhibition and reactivation experiments were carried out in enzyme

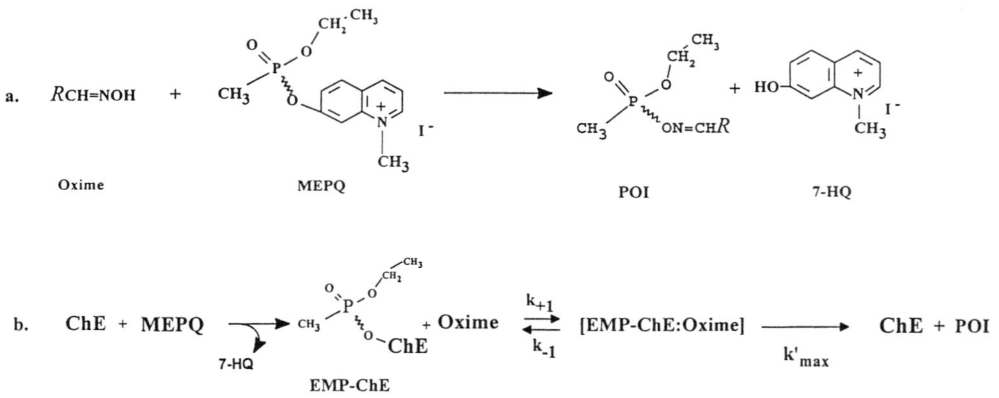

Figure 2. Chemical pathways of inhibition and reactivation.

solutions prepared in microfiltered 0.05% BSA containing 25 mM phosphate buffer, pH 7.8, at 25°C.

Enzyme Assay

AChE and BChE activities were determined by the method of Ellman et al. (1961), using 1.5 mM acetylthiocholine and butyrylthiocholine as substrates, respectively. Assays were carried out in 0.05% BSA-50 mM phosphate buffer, pH 8.0, at 25°C.

Reactivation of MEPQ-Inhibited ChEs

Reactivation was started by mixing 2-5 µL of 2-20 mM oxime stock solution in water with 0.1-0.2 ml of OP-ChE conjugate equilibrated for 5 min at 25°C. Final concentrations of oximes in the reactivation media ranged between 0.01 and 3 mM. At specified time intervals, 5-10 µL of reactivation mixture were diluted into 0.6-1 ml of assay mixture and enzyme activity was monitored as described above. Control activity was measured in the same volume ratio of oxime to nonphosphonylated enzyme. Since reactivation profiles of EMP-ChEs displayed marked deviations from a first-order approach to reactivation of a single reactivatable species, equation 1 was used to determine the best-fit values of the following parameters:

$$\%(E_{reac})_t = E_1(1-e^{-k(1)t}) + E_2(1-e^{-k(2)t}) \qquad (1)$$

where E_1 and E_2 are the percent-amplitudes of two reactivatable forms of MEPQ-inhibited ChE and the parameters k(1) and k(2) are the corresponding fast and slow pseudo first-order rate constants of the reactivation of E_1 and E_2, respectively. Ratios of E_1/E_2 ranged between 0.8 and 1.2. In those cases where nonlinear regression did not converge due to insufficient data points, curve fittings were processed assuming an E_1 to E_2 ratio of 1.

Molecular Modeling

Molecular modeling was done on a Silicon Graphics Indigo Elan using Discover 2.9, a module of InsightII 2.2.0 program (Biosym, San Diego). Coordinates from the crystal structure of TcAChE (Sussman et al., 1991) and coordinates from a model of HuBChE (Harel et al., 1992) were used in calculation of energy-minimized conformations of oximes in phosphonylated TcAChE and HuBChE. Coordinates of crystal structures of 2-PAM and HI-6 were used in docking the respective oximes in the models of EMP-ChEs.

Modeling was done in vacuo, with the dissociation state of ionizable groups set equivalent to pH 7.8. First, the O-ethyl methylphosphonyl group was covalently attached to the O^γ of the active site serine and the conformation of the conjugate minimized. By the initial placement of the oxygen of P=O bond in the oxyanion hole, the phosphonyl moiety is susceptible to an 'in line' S_N2 displacement. Oxime groups were ionized and then partial charges of oximes calculated using the MOPAC module of InsightII. Then 2-PAM and HI-6 were minimized in the model of the phosphonylated enzymes leaving specified residues to rotate freely.

RESULTS AND DISCUSSION

The dissociation constants of 2-PAM (Table 1) for the mutants tested were only moderately perturbed relative to wild-type rMoAChE. The largest destabilization was

Table 1. Biochemical constants for the reactivation MEPQ-inhibited ChEs by 2-PAM and HI-6

| Enzyme | 2-PAM[a] | | | HI-6[a] | | |
| | K_{ox}^{c} (mM) | $k_{oximate}(M^{-1}min^{-1})^{b}$ | | K_{ox}^{c} (mM) | $k_{oximate}(M^{-1}min^{-1})^{b}$ | |
		fast	slow		fast	slow
MoAChE, wt	0.049	1490	140	0.013	5570	520
W86A	ND	1200	60	ND	2510	100
W86F	0.297	1090	60	ND	ND	ND
Y337F	0.086	2170	192	ND	ND	ND
Y337A	0.357	1720	88	0.108	8200	380
E202Q	0.172	45	6	0.013	350	40
F295L	0.141	400	117	0.013	12130	2620
W286R	0.132	1340	235	ND	140	110
W286A						
Y72N						
Y124Q	0.229	920	46	0.085	79	90
TcAChE	0.089	2800	86	ND	ND	ND
HuBChE	ND	1030	20	ND	216	10

[a]Mean of 4-12 determinations. [b]Obtained by dividing k'_{max} by K'_{ox} (Eq. 2) and normalized to oximate anion concentration at pH 7.8. [c]Dissociation constant of the complex formed between nonphosphonylated enzyme and the oxime. SEM < 3 0%.

observed with Y337A and W86F as reflected by an approximately 6- to 7-fold increase in the corresponding K_{ox} compared with that of wild-type enzyme. Replacement of Y337 by phenylalanine produced a mutant that has the same sequence of aromatic amino acid residues within the active site gorge of TcAChE (Sussman et al., 1991; Cygler et al., 1993). K_{ox} values for Y337F and TcAChE were virtually equivalent. K_{ox} of [F295L:2-PAM] decreased less than 3-fold and single (W286R) and triple (W286A/Y72N/Y124Q) mutations at the entrance to the gorge increased the dissociation constants of 2-PAM 2.7- and 4.7-fold, respectively.

K_{ox} of HI-6 for wild-type rMoAChE is 4-fold lower than that of 2-PAM, suggesting that the second pyridinium ring of HI-6 contributed less than one kcal/mol to the stabilization of the bisquaternary oxime-enzyme complex relative to the monoquaternary pyridinium oxime. Replacement of Y337 by alanine increased the HI-6 dissociation constant 8-fold compared to the wild-type enzyme, whereas K_{ox} of HI-6 with E202Q, F295L and wild type rMoAChE were essentially equivalent. The three aromatic amino acid residues at the entrance to the gorge have been shown to constitute part of the peripheral anionic site for bisquaternary ligands (Radić et al., 1993; Barak et al., 1994). Their replacement with the residues found in BChE increased K_{ox} of the association of HI-6 with mutant W286A/Y72N/Y124Q only 6.5-fold relative to wild-type rMoAChE.

MEPQ-ChE conjugates are likely to be composed of a mixture of two enantiomers (Levy and Ashani, 1986; Raveh et al., 1993), designated as EMP_{R}-ChE and EMP_{S}-ChE. The symbols R and S denote absolute configuration around the phosphorus atom of the inhibited enzyme. The extent of reactivation did not differ significantly for preparations that were allowed to incubate for 1 to 24 h at 25°C before the addition of reactivator. The almost negligible spontaneous reactivation and aging reactions simplify evaluation of the kinetic profiles. The absence of competing processes, together with the high bimolecular rate constants for the inhibition of AChE and HuBChE by MEPQ (Raveh et al., 1993), resulted in rapid formation of an inhibited enzyme with defined species that could be transferred immediately to reactivation buffer. Analysis of kinetic data by curve fitting clearly showed that oxime-induced reactivation of all EMP-ChEs progressed in a manner that indicates non-homogeneity in reactivatable enzyme . The difference in component rate constants was even more pronounced for HuBChE. Semilogarithmic plots of the reactivation of MEPQ-rMoAChE by three different reactivators showed approximately an equal distribution of fast and slow components irrespective of the nucleophile used (not shown). These findings are satisfactorily explained by the presence of two kinetically distinguishable EMP-ChE enantiomers. Therefore, we categorize the two phases in Eq. 1 as fast and slow components of oxime-induced reactivation, and k(1) and k(2) are the corresponding first-order rate constants of the fast and slow components, respectively, in the presence of large stoichiometric excesses of the reactivator. The mathematical solution for the kinetic scheme of the reactivation depicted in Fig. 2b is:

$$k_{obs} = k'_{max} (1 + K'_{ox}/[oxime])^{-1} \qquad (2)$$

where k_{obs} is either k(1) or k(2), and $K'_{ox} = (k_{-1}+k'_{max})/k_{+1}$. Assuming $k_{-1} \gg k'_{max}$, K'_{ox} is approximated by k_{-1}/k_{+1} which is the corresponding dissociation constant of the complex [EMP-ChE:oxime]. The individual constants k'_{max} and K'_{ox} were determined by non-linear regression analysis of the data according to Eq. 2. The bimolecular rate constant of reactivation (k_r) was obtained by dividing k'_{max} by K'_{ox}.

Table 1 summarizes the kinetic constants of the reactivation of EMP-ChEs by 2-PAM and HI-6, respectively. The bimolecular rate constant of the reactivation by the oxime ions, RCH=N-O⁻ ($k_{oximate}$) was calculated from k_r and the pKa of the corresponding oxime. In general, dissociation constants of the reactivators for $EMP_{R,S}$-AChEs (K'_{ox}) are increased compared with non- phosphonylated enzyme (i.e., $K_{ox} < K'_{ox}$). The presence of bound acetylthiocholine or its reaction product and conjugated EMP decreased the affinity of the oximes to an equivalent extent (not shown).

Mutations at the Choline Binding Subsite

The side chains of W86, Y337 showed little involvement in the reactivation process. Replacement of E202 by glutamine ($EMP_{R,S}$-E202Q) markedly decreased the bimolecular rate constant of reactivation by 2-PAM or HI-6. Both reactivatable components were affected similarly. In the case of HI-6, the decrease in $k_{oximate}$ (13- to 16-fold) compared with wild-type enzyme almost exclusively arises from a smaller unimolecular rate constant k'_{max}. By contrast, the reduction in $k_{oximate}$ of 2-PAM (23- to 33-fold) contains contributions from both the affinity of the oxime for the phosphonylated enzyme (K'_{ox}) and k'_{max}. These findings indicate that the stability of the initial complex [$EMP_{R,S}$-AChE:HI-6] is largely controlled by residues located outside the choline-binding region.

The reactivatability of the two components of $EMP_{R,S}$-TcAChE by 2-PAM was comparable to that of mutant Y337F. Replacement of tyrosine by phenylalanine in rMoAChE produces a mutant that contains aromatic side chain residues of the choline subsite identical to TcAChE.

Mutation at the Acyl Pocket

Dimensions of the acyl pocket are likely to determine, in part, the stability of the two enantiomeric O-ethyl methylphosphonyl conjugates of AChE in manner analogous to their influence on carboxyl ester specificity. Since F295L appears to place the essential constraint in limiting butyrylthiocholine hydrolysis by AChE (Vellom et al., 1993; Ordentlich et al., 1993), it is interesting to compare $k_{oximate}$ of R_P and S_P enantiomers of EMP-F295L with wild-type $EMP_{R,S}$-rMoAChE. MEPQ-inhibited F295L reactivated only up to 50-65% of the expected activity at t_∞. By contrast, the extent of reactivation of other MEPQ-inhibited mutants, as well as that of wild-type rMoAChE and tissue derived TcAChE and HuBChE, ranged from 80 to 98%. The reactivation profiles of $EMP_{R,S}$-F295L constructed for either 2-PAM or HI-6 were fitted significantly better to a two-component model than to a single class of inhibited enzyme (not shown), but, the difference in the ratio of k_r (fast) and k_r (slow) was markedly reduced from the ratio found for the wild type enzyme.

Replacement of phenylalanine in position 295 by leucine had opposing effects on $k_{oximate}$ for 2-PAM (decreased 3.7-fold) and HI-6 (increased 2.2-fold). These findings are consistent with the relative changes observed in the affinity of the oximes for F295L. It is suggested that F295L alters the spatial constraints surrounding the O^γ S203-bound phosphonyl moiety and thereby changes the stereochemical requirements of the reactivation process.

Mutations at the Entrance to the Gorge

The potency of 2-PAM in reactivating both the fast and slow components decreased only 1.6 to 3-fold with the triple mutant involving residues at the entrance to the gorge (W286A/Y72N/Y124Q) and with W286R, relative to the wild-type enzyme. By contrast, HI-6-induced reactivation of the triple mutant decreased 70- and 6-fold for the fast and slow components, respectively. Similarly, $k_{oximate}$ of HI-6 with EMP-W286R, a mutation to the residue found in mouse BChE, was 40- and 5-fold lower for the fast and slow components respectively, compared with those observed for wild-type rMoAChE.

$k_{oximate}$ values of $EMP_{R,S}$-HuBChE that contains aliphatic amino acid residues in positions corresponding to 286, 124, and 72 of AChE, revealed that 2-PAM is superior to HI-6 in reactivating HuBChE. These observations further underscore the importance of the aromatic amino acids at the entrance to the gorge of AChE in enhancing reactivation potency of HI-6 as compared with 2-PAM.

Molecular Modeling

A stereo view of energy minimized conformations of complexes between EMP_R-AChE and EMP_R-HuBChE with 2-PAM is shown in Fig. 3. In addition the EMP_R-AChE and EMP_R-HuBChE complexes with HI-6 are shown. The overall geometries of the side chain residues that are lining the gorge of [EMP-ChE:2-PAM] complexes were similar to the starting models of $EMP_{R,S}$-ChE conjugates with one exception. The indole ring of W86 that is aligned with the gorge axis of TcAChE is slightly moved to face the gorge entrance, a rotation that appears to increase parallel contacts between the aromatic rings of 2-PAM and W86.

The oxime-containing pyridinium ring of HI-6 is oriented essentially as described above for the 2-PAM complex. The carbamoyl [$C(O)NH_2$] moiety of the distal pyridinium ring is projected toward the peripheral binding site and forms close contacts with aromatic side chains of W286, Y72, and Y124. Computer-simulated molecular dynamics of the putative transition state and the ground state of [$EMP_{R,S}$-AChE:HI-6] clearly point to the ability of the hydroxyl of Y124 to hydrogen bond to the oxygen of the bismethylene ether moiety that connects the two pyridinium rings. These interactions appear to restrict movements of HI-6 within the gorge. The proposed stabilization of HI-6 is consistent with a reported decrease in reactivation potency of a congener of HI-6 in which a three carbon methylene chain (CHS-6) is substituted for the bisoxymethylene bridge (HS-6)(de Jong and Kossen, 1985). The model of [EMP_R-AChE:HI-6] is consistent with the finding that mutations of residues W286, Y72, and Y124 decreased dramatically $k_{oximate}$ of HI-6 but not of 2-PAM. The most likely explanation for the increase in $k_{oximate}$ of HI-6 over 2-PAM stems from a better orientation of the oximate oxygen of the former oxime toward an apical

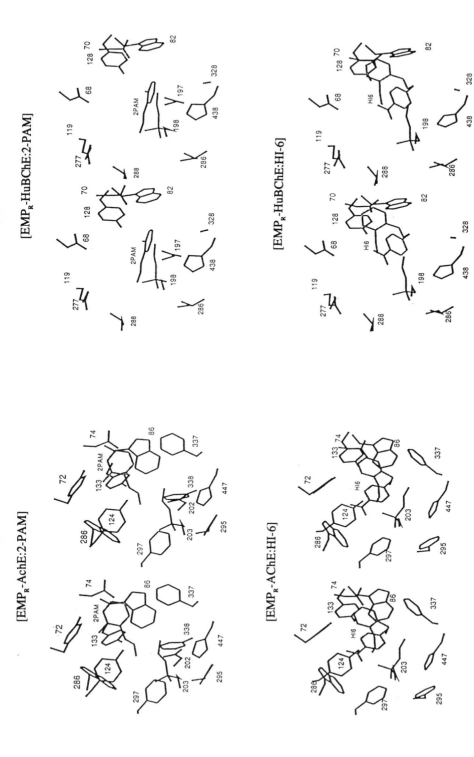

Figure 3. Stereo views of the final conformations of energy minimized 2-PAM and HI-6 in models of EMP$_R$-AChE and EMP$_R$-HuBChE. The final conformation of 2-PAM shown is one of several closely related overlapping structures with similar energy content(adapted from Ashani et al., 1995).

approach to the phosphorus atom from the face formed by three atoms and oriented perpendicular to the P-O$^\gamma$ S203 bond. Eventually, this might lead to greater stabilization of the transition state .

Using similar arguments, the smaller molecular volume of 2-PAM may confer to this reactivator sufficient flexibility to accommodate itself at various overlapping orientations within the gorge of wild-type phosphonylated AChE as well as within a gorge of diminished aromaticity seen with BChE.

ACKNOWLEDGMENT

Supported in part by U.S. Army DAMD17-91-C-1056 to P.T.

REFERENCES

Ashani, Y., Radić, Z. Tsigelny, I. Vellom, D.C., Pickering, N.A. Quinn, D.M., Doctor, B.P., and Taylor, P. (1995) *J. Biol. Chem.*, in press.

Barak, D., Kronman,C., Ordentlich, A., Ariel, N., Bromberg, A., Marcus, D., Lazar, A., Velan, B., and Shafferman, A., 1994, Acetylcholinesterase peripheral anionic site degeneracy conferred by amino acid arrays sharing a common core, J. Biol. Chem. 264: 6296-6305.

Cygler, M., Schrag, J.D., Sussman, J.L., Harel, M., Silman, I. Gentry, M.K., Doctor, B.P., 1993, Relationship between sequence conservation and three-dimensional structure in a large family of esterases, lipases, and related proteins, Protein Sci. 2:366-382.

de Jong, L.P.A., and Kossen, S.P., 1985, Stereospecific reactivation of human brain and erythrocyte acetyl-cholinesterase inhibited by soman, Biochim. Biophys. Acta., 830:345-348.

Ellman, G.L., Courtney, K.D., Andres, V., Jr., and Featherstone, R.M. 1961, A new and rapid colorimetric determination of acetylcholinesterase activity, Biochem. Pharmacol. 7: 88-95.

Erdman, W.D., 1969, A new antidote principle in alkyl-phosphate poisoning, Naunyn-Schmiedebergs Arch. für Pharmakol. Exper.Therap. 263: 61-72.

Harel, M., Sussman, J.L., Krejci, E., Bon, S., Chanal, P., Massoulie, J., and Silman, I.,1992, Conversion of acetylcholinesterase to BChE: Modelling and mutagenesis, Proc. Natl. Acad. Sci. U.S.A. 89: 10827-10831.

Lee, S. L., Camp, S., and Taylor, P.,1982, Characterization of a hydrophobic dimeric form of acetylcholi-nesterase from Torpedo californica, J. Biol. Chem. 257: 12302-12309.

Levy, D., and Ashani, Y. (1986) Biochem. Pharmacol.35: 1079-1085.

Oldiges, H., and Schoene, K., 1970, Antidotal effects of pyridinium - and imidazolium salts on soman and paraoxon poisoning in mice,. Archiv. Toxicol. 26: 293-305.

Ordentlich, A., Barak, D., Kronman, C., Flashner, Y., Leitner, M., Segall, Y., Ariel, N., Cohen, S., Velan, B., and Shafferman, A., 1993, Dissection of the human acetylcholinesterase active center determinants of substrate specificity, J. Biol. Chem. 268: 17083-17095.

Radić, Z., Pickering, N.A., Vellom, D.C., Camp, S., and Taylor, P., 1993, Three distinct domains in the cholinesterase molecule confer selectivity for acetyl- and butyrylcholinesterase inhibitors, Biochem-istry, 32:12074-12084.

Raveh, L., Grunwald, J., Marcus, D., Papier, Y., Cohen, E., and Ashani, Y. (1993) Biochem. Pharmacol. 45: 2465-2474.

Sussman, J.L., Harel, M., Frolov, F., Oefner, C., Goldman, A., Toker, L., and Silman, I., 1991, Atomic structure of acetylcholinesterase from Torpedo californica: A protoype acetyl-choline binding protein, Science, 253: 872-879.

Vellom, D.C., Radić, Z., Li, Y., Pickering, N.A., Camp, S., and Taylor, P., 1993, Amino acid residues controlling acetylcholinesterase and butyrylcholinesterase specificity, Biochemistry, 32:12-17.

MODULATION OF CATALYSIS AND INHIBITION OF FETAL BOVINE SERUM ACETYLCHOLINESTERASE BY MONOCLONAL ANTIBODIES

B. P. Doctor, Mary K. Gentry, Ashima Saxena, and Yacov Ashani

Division of Biochemistry
Walter Reed Army Institute of Research
Washington, DC 20307-5100
Israel Institute for Biological Research
Ness-Ziona, Israel

Monoclonal antibodies (MAbs) that on complexation with an enzyme molecule alter its catalytic activity can serve as useful probes to understand the mechanism of catalysis (Abe et al., 1983; Doctor et al., 1983; Brimijoin and Rakonczay, 1986; Ashani et al., 1990; Heider et al., 1991). Monoclonal antibodies have been raised against acetylcholinesterase (AChE) isolated from a variety of sources and species (Fambrough et al., 1982; Doctor et al., 1983; Brimijoin et al., 1985; Massoulié and Toutant, 1988; Heider et al., 1991). Although it is obvious that none of these MAbs bind to the esteratic site (Ogert et al., 1990), some of them appear to interact with that region of the catalytic subunit referred to as the peripheral anionic site (Ashani et al., 1991).

We describe here the production and characterization of six inhibitory MAbs against fetal bovine serum (FBS) AChE. The results show that changes in the conformation of AChE caused by interaction with MAbs at a site remote from the catalytic site result in the modulation of catalytic activity of AChE.

Four antibodies, 25B1, 4E5, 5E8, and 6H9, inhibited the catalytic activity of FBS AChE to >98% at ratios of 1-4 mole of MAb per mole of AChE. Two other anti-FBS AChE antibodies, 13D8 and 2A1, inhibited AChE approximately 70-80%. The extent of inhibition did not increase with these two antibodies even though the ratio of antibody to enzyme was increased to 32:1 (Fig. 1). None of the six antibodies bound to or inhibited *Torpedo* AChE under the same conditions and with the same concentrations of enzyme and antibody. Only MAb 5E8 inhibited recombinant human AChE at AChE-MAb ratios of 1:16 (35% inhibition) and 1:32 (75% inhibition). They neither bound to nor inhibited horse serum BChE or human serum BChE. MAb 13D8, at a very high concentration (approximately 10 μM), also recognized recombinant human AChE. Based on their inability to bind to reduced, denatured, and alkylated FBS AChE, it appears that all six MAbs have conformational epitopes.

Enzymes of the Cholinesterase Family, Edited by Daniel M. Quinn et al.
Plenum Press, New York, 1995

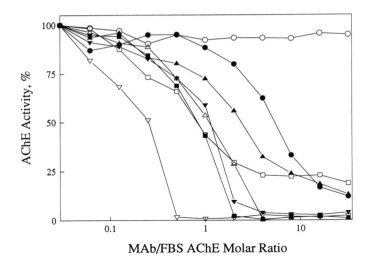

MAb/FBS AChE Molar Ratio

Figure 1. Inhibition of catalytic activity of FBS AChE by monoclonal antibodies. FBS AChE (100 pmoles/ml) was incubated at 20°C for 24 hr with varying amounts of MAbs in 50 mM phosphate, pH 8.0. Enzyme activity remaining after 24 hr incubation with MAbs was measured using the Ellman assay. O, 2C8 (control, a MAb specific for AChE but that does not inhibit catalytic activity); ●, AE2 (anti-human red cell AChE, Fambrough et al., 1982). MAbs raised against FBS AChE: ▽, 25B1; ▼, 6H9; □, 2A1; ■, 4E5; △, 5E8; and ▲, 13D8. FBS AChE, without antibody complexation but otherwise subjected to similar treatment, was also used as a control. (Awaiting permission to reproduce figure.)

Edrophonium and propidium reversibly inhibit AChE in competitive and noncompetitive manners, respectively, with the enzyme. Previously, we showed that propidium, which binds to the putative peripheral anionic site (Taylor and Lappi, 1975), but not edrophonium, which binds to the catalytic site, retarded the rate of inhibition of FBS AChE by MAb 25B1 (Ashani et al., 1990). The rate of inhibition of FBS AChE by various MAbs was followed by measuring enzyme activity. The pseudo first order plots did not produce straight lines despite the presence of a large excess of MAbs. The data were fitted to a single exponential decay equation

$$(\%AChE = A\,e^{-kt} + B)$$

where k is the inhibition rate constant. The inhibition rate constants for the six inhibitory MAbs were found to be similar. To obtain some information about the epitope for these MAbs, this inhibition reaction was carried out either in the presence of 1 µM edrophonium (which binds to the catalytic site) or 3.35 µM propidium (which binds to the peripheral anionic site). The results of this experiment shown in Figure 2 demonstrate that the rate of complex formation between the six anti-FBS AChE MAbs and the tetrameric form was unaffected by edrophonium, indicating that these MAbs did not bind at the catalytic site. A reduction in the rate of complex formation in the presence of propidium was observed in all cases, which suggests that these MAbs bound in the vicinity of the peripheral anionic site of the molecule.

Analysis of fractions from sucrose density gradient centrifugations of AChE:MAb complexes suggested that the MAbs could be categorized according to the apparent size of the complexes formed with the tetrameric form of FBS AChE. Based on these gradient data (and the stoichiometry of enzyme:antibody inhibition assays), two sizes of complexes could

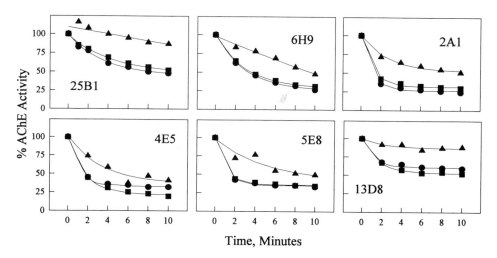

Figure 2. Effect of edrophonium and propidium on the rate of inhibition of FBS AChE by various monoclonal antibodies. Each antibody, at a concentration varying from 0.1-15 μM, was diluted at time zero into FBS AChE (0.5-1.0 nM active site) in 1 ml of 5 mM sodium phosphate, pH 8.0, containing 0.01% BSA, at 25°C. Residual enzyme activity was measured at two minute intervals by assaying 100 μl of the reaction mixture. Data for the time course of inhibition reactions were analyzed using the equation described in the text and shown as solid circles. The effect of 1 μM edrophonium (■) and 3.35 μM propidium (▲) on the rate of inhibition is also shown.

be distinguished: tetramers of enzyme cross linked by antibody molecules to form multimeric complexes (MAbs 13D8, 25B1, and 4E5) and discrete complexes of antibody molecules and single tetramers of enzyme that sedimented as single peaks (MAbs 2A1, 5E8 and 6H9). Figure 3 shows a representative sucrose density gradient centrifugation profile for one MAb from each group along with a proposed structure for the complex for that group.

FBS AChE was titrated with MAbs using two different substrates. The complexes formed with increasing AChE:MAb ratio were assayed using acetylthiocholine (ATC), a charged substrate, and indophenyl acetate (IPA), a neutral substrate, as shown in Figure 4. Complexes of MAbs 25B1, 4E5, 2A1, and 13D8 with FBS AChE produced similar inhibition curves with both substrates. In contrast, with antibodies 6H9 and 5E8 the use of IPA caused shifts in the inhibition curves. In the case of 6H9, hydrolysis of IPA appeared to increase (1.25 fold) with increasing amounts of antibody, whereas hydrolysis of ATC was inhibited. A similar increase in IPA hydrolysis was shown to occur with MAb AE-2, raised against human erythrocyte AChE (Wolfe et al., 1993). Monoclonal antibody 5E8 inhibited ATC hydrolysis but had a marginal effect on IPA hydrolysis. Data from MAbs 2A1 and 13D8 could not be analyzed by linear regression since these two antibodies did not fully inhibit FBS AChE activity.

To further characterize the differences between various FBS AChE:MAb complexes, the extent of radiolabeling of the complex by phosphorylation with the neutral covalent inhibitor [³H]DFP was determined. In all cases, sufficient amounts of MAb were used to attain maximum inhibition of catalytic activity. The extent of radiolabeling of complexes was determined following 24 hrs of incubation at 25°C with a 2.25-fold molar excess of [³H]DFP. Results, shown in Table 1, indicated that the nucleophilicity of the active-site serine of the various MAb:AChE complexes was altered.

Measurement of displacement of the organophosphate (OP) moiety from all [³H]DFP:AChE:MAb complexes by the oxime TMB$_4$ shown in Table 2, indicated that two

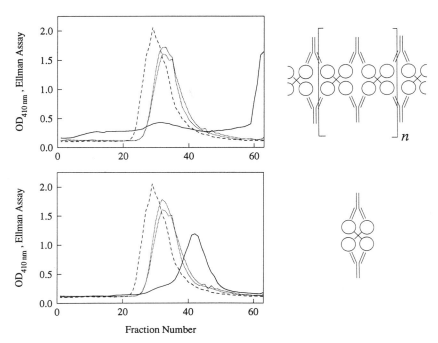

Figure 3. Sucrose density gradient centrifugation of complexes of FBS AChE and MAbs with proposed structures for the complexes. FBS AChE, approximately 40 U (0.1 nmole), was incubated for 24 hr (50 mM phosphate buffer, pH 8.0) with sufficient MAb to inhibit the enzyme > 90%. Aliquots, approximately 5 U of residual AChE activity in 200 μl with catalase as a marker, were applied to 5-20% sucrose gradients and centrifuged for 18 hr at 75,000 x g and 4°C. - - - -, FBS AChE control; ———, FBS AChE:MAb complex; ·····, catalase marker. Control and complex activities were aligned by superimposing the positions of the catalase markers. In each case, the top of the gradient is to the left. Shown are one representative gradient profile from each of the two groups of complexes and a proposed structure for the complex. Upper panel: FBS AChE:25B1 (multimeric complexes; n indicates the repeating unit). Lower panel: FBS AChE:6H9 (complexes of one tetramer of FBS AChE and two antibody molecules). (Awaiting permission to reproduce figure.)

of the MAb-bound DFP:AChE complexes, those with antibodies 25B1 and 13D8, lost nearly all radioactivity when incubated with 1 mM TMB$_4$. By contrast, the other four MAb:DFP:AChE conjugates lost smaller amounts of radioactivity, 25-50%, when incubated with 1 mM TMB$_4$. TMB$_4$, however, was able to reach the catalytic site of the enzyme in all complexes and to dephosphorylate the enzyme but to different levels.

The results also indicate that the nucleophilicity of the O$^\gamma$ atom of the catalytic serine in AChE:MAbs varies. The fact that oxime TMB$_4$ was able to reach the catalytic site and cause displacement of the organophosphoryl-bound moiety from the complexes but to different levels implies that the entrance to the active site gorge either has not been blocked (13D8 and 25B1) or is somewhat restricted. This can be attributed to conformational changes, steric hindrance, or a combination of both factors.

Trp279 (*Torpedo* AChE numbering system), which resides at the entrance to the gorge of all AChEs, has been implicated as part of the binding region of peripheral ligands (Shafferman et al., 1992; Radić et al., 1993). The fact that the peripheral anionic site ligand propidium retarded the binding of all MAbs to FBS AChE suggested that this amino acid may be a common determinant of their epitopes. However, since these MAbs do not cross

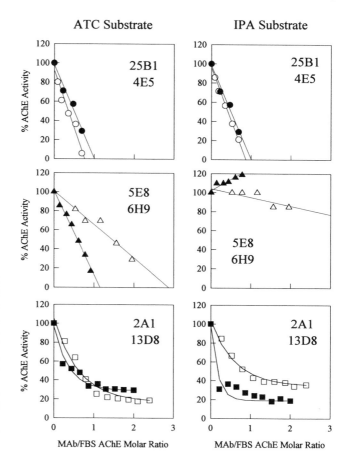

Figure 4. Titration of FBS AChE inhibition by MAbs using charged substrate ATC and neutral substrate IPA. O, 25B1; ●, 4E5; △, 5E8; ▲, 6H9; □, 2A1; and ■, 13D8. Residual enzyme activity was measured in 5 mM phosphate, pH 8.0, using the Ellman procedure in the case of ATC substrate. For the IPA substrate studies, an analogous procedure was used with measurements taken at 625 nm. (Awaiting permission to reproduce figure)

react with *Torpedo* and human AChEs (except in the case of 13D8 at high concentrations), it appears that there are other critical residues that constitute their binding sites.

Wolfe et al. (1993) have reported that MAb AE-2, directed against human RBC AChE and cross reacting with FBS AChE (Fig. 1), inhibited the catalytic activity of AChE, when the charged substrate ATC is used, but stimulated the hydrolysis of neutral substrate IPA, as

Table 1. Effect of monoclonal antibodies on binding of [^{3}H]DFP to AChE. Enzymes, except for control, were incubated with antibody for 24 hr before exposure to [^{3}H]DFP. The AChE or AChE:MAb was incubated at 25°C for 24 hr. Aliquots of AChE:MAb:[^{3}H]DFP or AChE:[^{3}H]DFP mixture were applied to BioGel P-6 columns for separation of complex from free [^{3}H]DFP. (Awaiting permission to reproduce table.)

Antibody	[^{3}H]DFP Binding % at 24 hr	Antibody	[^{3}H]DFP Binding % at 24 hr
None	100	E6-4E5	10
E4-13D8	14	E6-5E8	17
E5-25B1	7	E6-6H9	5
E6-2A1	26		

Table 2. Displacement of radioactivity by TMB_4 from MAb-bound [^{3}H]DFP:FBS:AChE. [^{3}H]DFP-treated FBS AChE was incubated for 24 hr with MAbs and then incubated with 1 mM TMB_4 at pH 8.0 for 48 hr. The protein-bound and displaced radioactivity were separated by gel filtration (BioRad P-6). (Awaiting permission to reproduce table)

Antibody	Radioactivity displaced,%	Antibody	Radioactivity displaced,%
None	100	E6-4E5	40
E4-13D8	100	E6-5E8	50
E5-25B1	95	E6-6H9	25
E6-2A1	35		

does MAb 6H9. It was suggested that AE-2 may allosterically modulate an anionic subsite in the catalytic center of FBS AChE (Trp84). Additionally, Wasserman et al., (1993), Ogert et al. (1990), and Doctor et al. (1989) have previously shown that epitopes for several MAbs generated against AChEs are located on the surface of the catalytic subunit in the region of amino acids 40-100. The role of Trp84 as an anionic subsite has been established (Harel et al., 1992; Shafferman et al., 1992; Radić et al., 1993), and it has been suggested that it and Met83 are partially exposed on the surface of the molecule (Harel et al., 1992; Gilson et al., 1994). This region appears to be highly immunogenic. The fact that MAb 6H9 and, to a lesser extent 5E8, when complexed with AChE stimulated the hydrolysis of IPA, as does AE-2, suggests that this highly immunogenic region may also be a binding site for these two MAbs. The possibility exists that these MAb combining sites can span both of these regions (B.P. Doctor and J. L. Sussman, personal communications).

The results suggest that different MAbs modulate the activity of FBS AChE via binding to a site remote from the catalytic region of the enzyme. However, the relative contribution of the MAbs to the issue of conformational changes vs. steric hindrance awaits further study.

REFERENCES

Abe, T., Sakai, M., and Saisu, H., 1983, *Neurosci. Lett.* 38:61-66.

Abramson, S., Ellisman, M., Dernick, T., Gentry, M. K., Doctor, B. P., and Taylor, P., 1989, *J. Cell Biol.* 108:2301-2311.

Ashani, Y., Gentry, M. K., and Doctor, B. P. , 1990, *Biochemistry* 29:2456-2463.

Ashani, Y., Bromberg, A., Levy, D., Gentry, M. K., Brady, D. R., and Doctor, B. P., 1991, in *Cholinesterases: Structure, Function, Mechanism, Genetics, and Cell Biology* (Massoulié, J., Bacou, F., Barnard, E., Chatonnet, A., Doctor, B. P., and Quinn, D. M., eds), pp. 235-239, American Chemical Society, Washington, DC.

Brimijoin, S. and Mintz, K. P. , 1985, *Biochim. Biophys. Acta* 828:290-297.

Brimijoin, S., Mintz, K. P., and Alley, M. C., 1985, *Mol. Pharmacol.* 24:513-520, 1985.

Brimijoin, S. and Rakonczay, Z., 1986, *Int. Rev. Neurobiol.* 28:363-410, 1986.

Doctor, B. P., Camp, S., Gentry, M. K., Taylor, S. S., and Taylor, P., 1983, *Proc. Natl. Acad. Sci. USA* 80:5767-5771.

Doctor, B. P., Smyth, K. K., Gentry, M. K., Ashani, Y., Christner, C. E., De La Hoz, D. M., Ogert, R. A., and Smith, S. W., 1989, in *Computer-Assisted Modeling of Receptor-Ligand Interactions: Theoretical Aspects and Applications to Drug Design* (Rein, R. and Golombek, A., eds), pp. 305-316, Alan R. Liss, Inc., New York.

Fambrough, D. M., Engel, A. G., and Rosenberry, T. L., 1982, *Proc. Natl. Acad. Sci. USA* 79:1078-1082.

Gilson, M. K., Straatsma, T. P., McCammon, J. A., Ripoll, D. R., Faerman, C. H., Axelsen, P. H., Silman, I., and Sussman, J. L., 1994, *Science* 263:1276-1277.

Harel, M., Sussman, J. L., Krejci, E., Bon, S., Chanal, P., Massoulié, J., Silman, I., 1992, *Proc. Natl. Acad. Sci. USA* 89:10827-10831.

Heider, H., Meyer, P., Liao, J., Stieger, S., and Brodbeck, U., 1991, in *Cholinesterases: Structure, Function, Mechanism, Genetics, and Cell Biology* (Massoulié, J., Bacou, F., Barnard, E., Chatonnet, A., Doctor, B. P., and Quinn, D. M., eds), pp. 32-36. American Chemical Society, Washington, DC.

Massoulié, J. and Toutant, J. P., 1988, in *Handbook of Experimental Pharmacology* (Whittaker V. P., ed), pp. 167-224. Springer, Berlin, 1988.

Ogert, R. A., Gentry, M. K., Richardson, E. C., Deal, C. D., Abramson, S. N., Alving, C. R., Taylor, P., and Doctor, B. P., 1990, *J. Neurochem.* 55:756-763.

Radić, Z., Pickering, N., Vellom, D. C., Camp, S., and Taylor, P., 1993, *Biochemistry* 32:12074-12084.

Shafferman, A., Velan, B., Ordentlich, A., Kronman, C., Grosfeld, H., Leitner, M., Flashner, Y., Cohen, S., Barak, D., and Ariel, N., 1992, *EMBO J.* 11:3561-3568.

Sorensen, K., Brodbeck, U., Rasmussen, A. G., and Norgaard-Pedersen, B., 1987, *Biochim. Biophys. Acta* 912:56-62.

Taylor, P. and Lappi, S., 1975, *Biochemistry* 14:1989-1997.

Wasserman, L., Doctor, B. P., Gentry, M. K., and Taylor, P., 1993, *J. Neurochem.* 61:2124-2132, 1993.

Wolfe, A. D., Chiang, P.K., Doctor, B.P., Fryar, N., Rhee, J.P. and Saeed, M.,1993, *Mol. Pharmacol.* 44:1152-1157.

INSECT ACETYLCHOLINESTERASE AND RESISTANCE TO INSECTICIDES

Didier Fournier,[1] Marie Maturano,[2] Laurent Gagnoux,[3] Philippe Ziliani,[4]
Cyril Pertuy,[3] Madeleine Pralavorio,[4] Jean-Marc Bride,[4] Leila Elmarbouh,[1]
Alain Klaebe,[2] and Patrick Masson[3]

[1] Laboratoire d'entomologie, Université Paul Sabatier
118 route de Narbonne, 31062 Toulouse, France
[2] Laboratoire IMRCP, groupe de Chimie organique biologique,
Université Paul Sabatier
118 route de Narbonne, 31062 Toulouse, France
[3] Unité de biochimie, CRSSA
24 Av. des maquis du Guésivaudan BP87 38702 La Tronche cedex, France
[4] Laboratoire de biologie des invertébrés, INRA
BP 2078, 06606 Antibes, France

Extensive utilization of pesticides against insects provides us with a good model for studying the adaptation of an eukaryotic genome submitted to a strong selective pressure. Since the early 1950s, organophosphates and carbamates have been widely used to control insect pests around the world. These insecticides are hemisubstrates that inactivate acetylcholinesterase by phosphorylating or carbamylating the active serine (Aldridge, 1950). A mechanism for insect resistance to insecticides consists of the alteration of acetylcholinesterase which is less sensitive to organophosphates and carbamates. In 1964, Smissaert first described a resistant acarina carrying a modified acetylcholinesterase. Since this report, altered acetylcholinesterases have been detected in several resistant insect species such as aphids, colorado potato beetle or mosquitoes (review in Fournier and Mutero, 1994). Resistance is very variable, from 2 to 200,000 fold depending on the species or on the strain, suggesting that several mutations may be responsible for resistance of the enzyme (Pralavorio and Fournier, 1992).

Knowledge of structure-activity relationship of wild type acetylcholinesterase being a prerequisite for studying the effect of each mutation found in resistant insects, we first studied the catalytic behavior of the enzyme with respect to substrate hydrolysis. Vertebrate acetylcholinesterase displays an inhibition by excess of substrate and vertebrate butyrylcholinesterase displays an activation at intermediate substrate concentration (Radić et al. 1993). In insects, hydrolysis of choline and thiocholine esters also deviates from the simple Michaelis-Menten model: kinetics of *Drosophila* acetylcholinesterase are triphasic, displaying complexities of both butyrylcholinesterase and acetylcholinesterase. At 25°C and pH7, K_m for acetylthiocholine was 11μM, activation at high substrate concentration ($K_{ss1} = 0.1$

Enzymes of the Cholinesterase Family, Edited by Daniel M. Quinn et al.
Plenum Press, New York, 1995

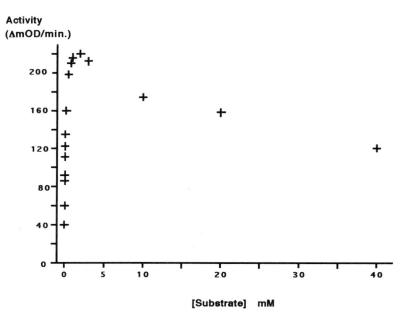

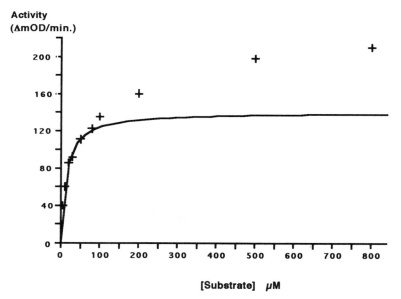

Figure 1. Substrate activity curves ranging from 0 to 40 mM acetylthiocholine (above) or from 0 to 0.8 mM acetylthiocholine (below). Crosses represent experimental data, solid line represents simple Michaelis-Menten behavior.

mM) was followed by inhibition at higher substrate concentration (K_{ss2} = 38 mM) (fig. 1). In vertebrate acetylcholinesterases, binding of the second substrate molecule responsible for inhibition, involves several aromatic residues at the mouth of the active site gorge (Schafferman et al., 1992; Radić et al., 1993). Only one of them (Tyr 408 (341)) is conserved among cholinesterase according to the allignment of Krejci et al. (1991) suggesting that other amino acids may be responsible for substrate inhibition in *Drosophila*.

The activation binding site seems to be different from the peripheral binding site since substrate activation and substrate inhibition were found in the same protein. Activation may result either from the binding of a substrate molecule on a specific site or on the choline binding site (Trp 121) of the acyl enzyme intermediate as hypothesized by Ericksson and Augustisson (1979) for horse serum butyrylcholinesterase. At low substrate concentration, the Arrhenius plot is linear and nucleophile competition using methanol shows that deacylation is the rate limiting step. On the contrary, at substrate concentration leading to activation (150 µM), acylation becomes partly limiting at 25°C and is the rate limiting step at 15°C (Table 1). Thus, activation results in a temperature induced increase of deacylation rate.

The sequence analysis of the *Ace* gene in several resistant field-strains of *Drosophila melanogaster* resulted in the identification of five point mutations associated with reduced sensitivity to insecticides (Mutero et al., 1994). These mutations are all localized near the active site gorge. Interestingly, some of the mutations were found to be identical in other insects showing that Drosophila may be a model insect to study insecticide resistance mechanisms in agricultural and medical pests. On the other hand, in some insects such as house fly, other mutations have been detected (Devonshire et al., comm. pers.) showing that there are a lot of modifications which can modify the active site conformation and/or reactivity leading to a resistant enzyme while perserving the acetylcholine hydrolase activity. Each mutation provides a weak level of resistance which is not sufficient to confer a selective advantage to the insect in a treated area. But a weak mechanism may be selected when it is in association with other resistant mechanisms such as increased degradation of insecticides by oxydases, esterases or glutathione transferases or decreased penetration of the insecticde through the insect cuticle (Raymond et al., 1989).

In most Drosophila resistant strains, several mutations were found to be present in the same protein. Combinations of several mutations in the same protein give highly resistant enzymes (Fig. 2). This result suggests that resistance originates from recombination between single mutated alleles preexisting in natural populations. This mechanism would allow insects to rapidly adapt to new selective pressures (Mutero et al., 1994).

We now have two objectives: first, to find biochemical markers which may allow controlling resistant alleles in field populations and second, to design antiresistant compounds which may be used in alternance with usual insecticides to avoid the increased frequency of resistant alleles following repetitive insecticide treatments. Preliminary results suggest that these two objectives may be reached.

Table 1. Individual rate constants of the elementary steps of acetylcholinesterase-catalyzed hydrolysis of acetylthiocholine in phosphate buffer pH7.0

| | Substrate concentration | | | |
| | low | | high | |
	15°C	25°C	15°C	25°C
k_2 (sec^{-1})	3055	4260	3050	5315
k_3 (sec^{-1})	1206	2240	>> k2	4323

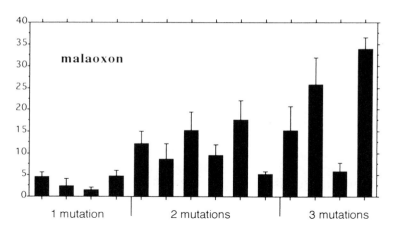

Figure 2. Effects of single and combined mutations on resistance of acetylcholinesterase to malaoxon.

Phe368(290) was found to be mutated to Tyr in some resistant strains. We mutated this residue to Tyr, Trp, Ser and Gly and expressed the protein in baculovirus infected cells. The amino acid at this position is important for affinity of the enzyme to organophosphates and hence for resistance. Harel *et al.* (1992) and Vellom *et al.* (1993) showed that Phe290 located at the bottom of the gorge may account for the specificity of substrate of the enzyme in veterbrates. As most insect AChEs metabolize butyrylcholine although less efficiently than acetylcholine (Toutant, 1989), we hypothesized that depending on the amino acid found at that position, we have either an increase or a decrease in affinity for butyrylcholine. Preliminary results seem to confirm the importance of mutations on Phe 368 in insects for substrate specificity (table 2). If other mutations do not produce this effect, we will be able to predict from biochemical data, the mutations involved in resistance with a simple test in insects from which the cholinesterase gene has not been cloned. Changes in butyrylcholine hydrolysis and affinity were observed along with resistance. In some cases, the resistant enzymes were more active or had higher affinity for butyrylcholine (Hama *et al.*, 1980; Brown and Bryson, 1992), while in other cases the affinity for butyrylcholine decreased (O'Brien et al., 1978). Thus resistance either decreases or increases the metabolization of butyrylcholine.

Among the mutations found in resistant insects, Ile199(Val129) was mutated to Val or Thr depending on the resistant strain. To study the effect of this mutation, we mutated this residues and we found a positive correlation between the bulkiness of the residue side chain and the sensitivity to insecticide. This allowed the production of mutant enzymes more sensitive to insecticide than the wild type. Since resistance originates from decrease of the side chain size, we tentatively designed antiresistant compounds by increasing the steric

Table 2. Substrate specificity and mutations at position 368 (290)

Phe (wild type)	ASCh (1) > PrSCh (0.7) > BuSCh (0.57)
Tyr	BuSCh (1.2) > PrSCh (1.2) > ASCh (1)
Trp	ASCh (1) > PrSCh (0.79) > BuSCh (0.04)
Gly	PrSCh (1.23) > ASCh (1) > BuSCh (0.77)
Ser	PrSCh (1.2) > ASCh (1) = BuSCh (1)

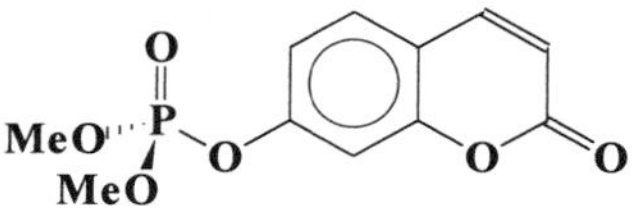

Figure 3. Example of an antiresistant compound.

hindrance of the leaving group. With some coumarin derivatives (Fig. 3), antiresistant effects have been observed.

REFERENCES

Aldridge W.N. (1950) *Biochem. J.* 46: 451-460.

Brown T.M. & Bryson P.K. (1992) *Pestic. Biochem. Physiol.* 44: 155-164.

Ericksson H. and Augustinsson K.B. (1979) *Biochem. Biophys. Acta* 567: 161-173.

Fournier D. and Mutero A. (1994) *Comp. Biochem. Physiol.* 108C: 19-31.

Hama H., Iwata T., Miyata T. and Saito T. (1980) *Appl. Ent. Zool.* 15: 249-261.

Harel M., Sussman J.L., Krejci E., Bon S., Chanal P., Massoulié J. and Silman I. (1992) *Proc. Natl. Acad. Sci. U.S.A* 89: 10827-10831.

Krejci E., Duval N., Chatonnet A., Vincens P. and Massoulié J. (1991) *Proc. Natl. Acad. Sci. USA* 88: 6647-6651.

Mutero A., Pralavorio M., Bride J.M. and Fournier D. (1994) *Proc. Natl. Acad. Sci. USA* 91: 5922-5926.

O'Brien R.D., Tripathi R.K. and Howell L.L. (1978) *Biochim. Biophys. Acta* 526: 129-134.

Pralavorio M. and Fournier D. (1992) *Biochem. Genet.* 30: 77-83.

Radić Z., Pickering N.A., Vellom D., Camp S. and Taylor P. (1993) *Biochemistry* 32: 12074-12084.

Raymond M., Heckel D.G. and Scott J.G. (1991) *Genetics* 123: 543-551.

Schafferman A., Velan B., Ordentlich A., Kronman C., Grosfeld H., Leitner M., Flashner Y., Cohen S., Barak D. and Ariel N. (1992) *EMBO J.* 11: 3561-3568.

Smissaert H.R. (1964) *Science* 143: 129-131.

Toutant J.P. (1989) *Progress in Neurobiol.* 32: 423-446.

Vellom D.C., Radić Z., Li Y., Pickering N.A., Camp S. and Taylor P. (1992) *Biochemistry* 32: 12-20.

PHOSPHONATE ESTER ACTIVE SITE PROBES OF ACETYLCHOLINESTERASE, TRYPSIN AND CHYMOTRYPSIN

Akos Bencsura, Istvan Enyedy, Carol Viragh, Rinat Akhmetshin, and
Ildiko M. Kovach

The Catholic University of America
Washington D.C. 20064

INTRODUCTION

Covalently bonded inhibitors of serine hydrolases have been considered to be good probes of the characteristics of the active site. Organophosphorus compounds (OP) and activated ketones are most prominent in this context because they should mimic the geometry of the transition state(s) of the reaction of the natural substrate.

The molecular origins of inhibition of serine hydrolase enzymes by phosphonate esters have been investigated extensively in this laboratory (Kovach et al., 1986a; Kovach, 1988a,b; Bennet et al., 1988,1989). A remarkable efficiency (60-70%) of mobilization of the catalytic power of these enzymes was observed in the P-O bond formation at the active site Ser in some cases. Small phosphonate esters use proteolytic catalysis provided by the enzyme as unnatural acyl substrates do. We have also found that the catalytic assistance is dampened or absent in the dephosphonylation of the enzymes which accounts for their often irreversible inhibition and ensuing toxicity (Kovach, 1988b).

$$SerOH + RR'OP(O)X \longrightarrow SerOP(O)RR'O + XH$$

$$SerOP(O)RR'O + H_2O \longrightarrow SerOH + RR'OP(O)OH$$

The major differences between the substrate and phosphate reactions catalyzed by AChE are: 1. the negligible leaving group dependence of the former versus the large leaving group dependence of the latter; 2. the smaller stability of the acetyl-enzyme than that of the phosphoryl-enzyme; and 3. the barrier for deacetylation is greatly reduced with respect to the nonenzymic reference, while the barrier for dephosphorylation approaches that for the nonenzymic reaction (Kovach, 1988b). The differences in leaving group effects between acylation and phosphorylation of AChE reflect on structural differences at the transition states, which must be inherent in the differences in the electronic makeup of carbon and

Enzymes of the Cholinesterase Family, Edited by Daniel M. Quinn et al.
Plenum Press, New York, 1995

phosphorus. Most of these tenets are valid for the inhibition of other serine hydrolases by OPs.

The worst aspect of OP inhibition of serine hydrolases is the occurrence of an irreversible side reaction resulting in the formation of a negatively charged monoester between Ser and phosphonate (diester of a phosphate) (Bender & Wedler, 1972; Kovach, 1988b; Kovach & Huhta, 1991; Kovach et al., 1991,1993; Ordentlich et al., 1993; Qian & Kovach, 1993). An intensely studied version of this aging reaction is dealkylation involving S_N1 or S_N2 displacement at the $C\alpha$ of an alkyl group of the phosphonate/phosphate ester of Ser (Michael et al., 1967; Schoene et al., 1980). Phenolic ester linkages, on the other hand, undergo nucleophilic displacements at P (Bender & Wedler, 1972). Each process is promoted differently by each enzyme. We have been most interested in elucidating the promotion of each of these reactions by specific active site motifs.

$$SerOP(O)RR'O + H^+ \longrightarrow SerOP(O)RO- + R'OH$$

$$SerOP(O)RArO + H_2O \longrightarrow SerOP(O)RO- + ArOH + H^+$$

The efficiency of the dealkylation reaction occurring in 2-(3,3-dimethylbutyl) methylphosphonofluoridate (soman)-inhibited AChE is unprecedented. As the crystal structure of AChE from *Torpedo californica* had become available (Sussman et al., 1991), we studied diastereomers of AChE-soman adducts by molecular mechanics (Qian & Kovach, 1993). We proposed a "push-pull" mechanism of carbonium ion formation catalyzed by Glu199 and His440 respectively. The latter also stabilizes the product, the anionic monophosphonate ester of Ser. Since then, we have extended the investigation and analysis of energy-minimized structures of diastereomeric adducts formed between soman and AChE, to trypsin and chymotrypsin. We focused attention on the availability of the acid/base catalytic apparatus of the triad, the role of the oxyanion hole, the vicinity of an electron rich residue to stabilize the incipient carbonium ion and the availability of nonpolar residues at the binding site to facilitate dealkylation. This is a report of results of computations using molecular mechanics and dynamics combined with semiempirical calculations of the Ser-OP conjugate structures. We have also constructed pH-rate profiles for the dealkylation of diastereomers of soman-inhibited AChE from *electric eel* (EE) and fetal bovine serum (FBS).

EXPERIMENTAL

Computational Work

The molecular mechanics-optimized structure reported earlier (Qian & Kovach, 1993; Kovach et al., 1994) for AChE was used in these calculations. Trypsin and chymotrypsin were optimized in an identical manner.

Minimum energy structures of the serine esters of soman were generated with MNDO as implemented in MOPAC 6.0 (Stewart, 1990). Mulliken point charges and ESP charges were also computed (calibrated to *ab initio* results) for the fragments (Kovach & Huhta, 1991). Tetravalent transients for the covalently modified Ser, to mimic the reaction of AChE with acetylcholine and that of trypsin and chymotrypsin with an amide, were generated with charges in MNDO. Each enantiomer of the serine ester of the appropriate alkoxy alkylphosphonate was then incorporated into the structure of the enzyme to be studied and the entire structure was again energy-minimized. The same protocol was applied to the tetravalent carbonyl transients.

Kinetic Studies

A 1 μM solution of AChE was prepared in a pH 10.0, 0.1 M TAPS buffer. About 0.2 mL was reserved for control. 0.5 mL of a μM solution of AChE was inhibited by 1 μL of a 10 mM solution of one of the soman enantiomers P(-)C(+) and P(-)C(-) respectively. The solutions were frozen after a few minutes and stored in the deep freezer. Control reactions were treated identically to the sample. 1. The rate constants for reactivation of each enzyme from a particular adduct with soman were measured in the presence of 5 mM HI-6 solution in 0.1 M TAPS buffer at pH 10.00 and 38 °C. The reactivation was complete in two hours with the EE AChE and within three hour with the FBS enzyme. 2. The aging reactions were carried out at μ = 0.1 M (KCl) in acetate, PIPES and HEPES buffers in the pH range 3.8-7.5 at 4.0 ± 0.5 °C. Buffers for the reactions involving FBS AChE included 0.1% Triton X-100. To obtain optimal temperature control, two mL of the buffer were preequilibrated for an hour under constant stirring in a vial. The vial was submerged in a 200 mL ethylene glycol bath which was under vigorous stirring and thermostated with a circulating ethylene glycol bath. A thermistor probe with a digital readout was used for monitoring the temperature in the reaction vessel. The enzyme-soman adduct was introduced into the reaction vessel in 5-50 μL volume and the timer was started. 50 μL aliquots were drawn into UV cells containing 10 mM HI-6 solution in 0.2 M TAPS buffers at pH 10. These cells were preequilibrated for at least 30 min. at 38 °C in a circulating water bath and the reactivation mixture was left at 38 °C for three hours before performing the assay of AChE activity by the Ellman method. AChE concentrations were 10^{-11} M or less under assay conditions.

RESULTS AND DISCUSSION

Molecular mechanics and molecular dynamics simulations were performed with AChE from *Torpedo californica*, bovine trypsin and bovine γ-chymotrypsin for purposes of comparison of the mechanisms of catalysis of the two group of enzymes. Figure 1 shows a representative comparison of the charge distribution on pinacolyl methylphosphonate ester of Ser, the corresponding dealkylated adduct and tetravalent carbonyl intermediate occurring in the AChE catalyzed reaction of an ester. Whereas the geometries of phosphonates and the tetravalent transients of carbonyl compounds are very similar, the charge accumulation on the phosphoryl and anionic oxygens in the dealkylated phosphonates are much greater than on the tetravalent intermediate formed in the natural reactions. Consequently, the phosphony-lated adducts of serine hydrolases are more stable by at least 12 kcal/mol than the tetravalent intermediates in carbonyl reactions. No doubt that transition states preceding or succeeding the covalent intermediates in acylation and deacylation of AChE have even smaller polarization and therefore bear less resemblance to the charge distribution in phosphonate ester monoanions. These major differences in charge distribution present a considerable limitation to the use of anionic phosphonate esters as "transition state analogs". In fact, the accumulation of excessive negative charge on the aged adduct, inaccessible on one side and adjacent to Glu199 on the other, further militates against the odds for successful nucleophilic displacement at P.

If the charge is solvated, a rigid network of water solvate may replace the departed alkyl chain. If the active site remains dry, in the active form of the enzyme, then a low dielectric constant promotes charge-charge repulsion.

The active site interactions between catalytic residues and the phosphonyl fragments are shown in Figure 2 and the geometric parameters for critical interactions are in Table 1.

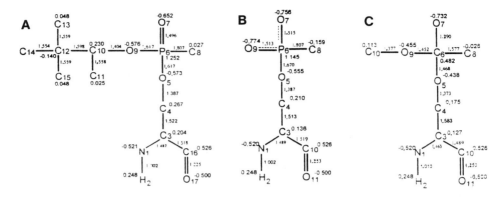

Figure 1. United atom representation of the connectivity and MNDO-generated ESP partial atomic charges for tetravalent covalent adducts of the active site Ser: A, pinacolyl methylphosphonyl-Ser; B, methylphosphonate ester anion fragment; C, methyl acetate tetravalent fragment.

1. Glu327 holds His440 in the catalytically competent position with a strong electrostatic force at 2.6 A. The interaction may also be considered a short and strong H-bond. The unusual feature of the AChE structure is that the short distance is there in the reactant state. The distance is 0.2-0.3 A longer in both the resting state and the soman adducts in the serine proteases.

2. Although mechanistic investigations strongly indicate the involvement of the protonated catalytic His in the dealkylation process, the electrostatic and H-bonding interactions both are somewhat stronger with the catalytic Ser O than with the ester O of the phosphonyl fragment in the minimum energy conformation of each adduct. The interaction is the strongest in the P_SC_S diastereomer of the soman-inhibited AChE. However, a hemisphere of positive electrostatic field, created by constituents of the oxyanion hole (Ala201) and His440, wraps around the phosphonyl fragment from the "west side" on Figure 2. This force should have a stabilizing effect on the developing negative charge on the oxygen during C-O bond cleavage.

3. The oxyanion hole interaction is very strong in the P_SC_S diastereomer of the soman-inhibited AChE. The Gly118 and Gly119, that come from a different part of the chain, are very strong H-bond donors and Ala201 provides a weak interaction to stabilize the oxyanion in AChE. The three H-bond donors form a hemisphere of positive electrostatic environment whereas there are two H-bonds known in the serine proteases. The phosphoryl group of the P_RC_S adduct of the soman-inhibited AChE cannot be accommodated in the oxyanion hole. There is one way for this diastereomer to bind, that is with the methyl group in the oxyanion hole which results in loss of many stabilizing interactions.

4. Formation of the carbonium ion seems to be promoted by a strong negative electrostatic force. In addition to a global negative electrostatic field, Glu199 in the local environment of the phosphonyl fragment at the active site of AChE, seems to provide the major driving force to the dealkylation reaction. Asp194 in the serine proteases is the counterpart of Glu199 in the sequence homology. It is 8-9 A away from the phosphonyl fragment. The remote location of negatively charged residues from the phosphonyl fragment in the serine proteases is then consistent with the much subdued tendency for dealkylation in these enzymes.

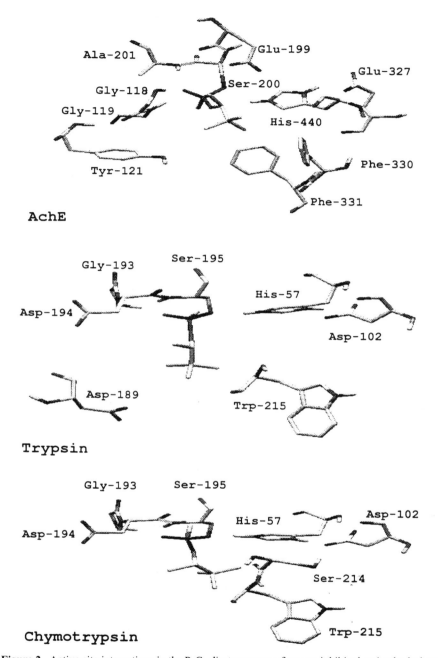

Figure 2. Active site interactions in the P_SC_S diastereomers of soman-inhibited serine hydrolases.

Table 1. Distances (A) and angles (deg, in parenthesis) in key interactions at the active site of soman-inhibited serine hydrolases (AChE/serine protease residue)

Interaction	P_SC_S			P_RC_S		
	AChE	Trp	Chy	AChE	Trp	Chy
HisNε-H- - -OESer	2.92	2.87	2.98	2.77	2.76	2.91
	(134)	(160)	(127)	(127)	(155)	(117)
HisNδ-H- - -O2Glu/Asp	2.63[a]	2.82	2.79	2.59	2.81	2.85
	(138)	(162)	**(169)**	(129)	(162)	**(174)**
HisNε-H- - -OEP	**3.61**	4.41	3.59	4.34	4.06	**2.96**
	(150)	(153)	(141)	(139)	(152)	**(167)**
INHCα- -O2Glu199/Asp194	**4.09**	7.04	8.12	4.82	8.85	9.10
INHCα- -O2Glu327/Asp102	8.16	7.82	7.69	9.38	8.73	8.71
INHCα- -O2Asp189		8.47			8.84	
GlyN-H- - -OPP	2.92	2.64	2.65		2.69	2.79
(118/193)	**(168)**	(139)	(137)		(144)	(144)
GlyN-H- - -OPP	2.77	2.73	2.79	4.76	2.66	2.71
(119/Ser195)	**(161)**	(139)	(155)	(121)	(138)	(149)
AlaN-H- - -OPP	2.69					
(201)	(120)					
INHOEP- - -Gly118	4.07					
INHC3- - -hydrophobic residues[b]	3.5-5					

[a]Strong H-bonds are indicated in boldface [b] Tyr121, Tyr84, Phe330, Phe331.

5. The pinacolyl group is stabilized by small nonspecific van der Waals interactions in the binding region of the natural substrate in each case. The greatest number of these interactions were in AChE and the fewest in trypsin.

6. The total energy of interaction was 18 kcal/mol lower for the P_SC_S diastereomer of the soman-inhibited AChE than the P_RC_S diastereomer. The protein-residual interactions were more favorable by 3-4 kcal/mol in the P_RC_S diastereomers of the soman-inhibited chymotrypsin and trypsin.

The pH-rate profiles at 4 °C for the dealkylation in the P_SC_R and P_SC_S diastereomers of soman-inhibited EE AChE and FBS AChE were bell-shaped. Table 2 shows the kinetic pKs, the maximal rate constants and solvent isotope effects for aging. The lower pKs are 3.6-3.8 for EE AChE and 4.5-4.6 for the FBS enzyme. The higher pKs are 7.6-8.0 and 6.0-6.4 respectively. The values for the FBS AChE are very close to kinetic pKs obtained with unnatural substrates. Interpretation of the lower pK values are fraught with the problem that the enzyme gradually looses activity at low pH, perhaps because of the protonation of acidic sites and the disruption of the salt bridges causing lasting conformational change. With this caveat, we assign the lower pK to Glu199 and the higher to His440 for the FBS AChE as suggested earlier (Qian & Kovach, 1993). We are not certain about the origin of the broad bell-shaped profile for the dealkylation of the soman-inhibited EE AChE.

The solvent isotope effects measured at pH 5, 5.5 and 6 were 1.1 - 1.3 in all cases (Kovach & Bennet, 1990). The lack of a larger solvent isotope effect may indicate that proton transfer is not occurring at the transition state. Instead, the electrophylic environment provided by the protonated His440 and the nearby oxyanion hole exerts an electrostatic strain on the breaking C-O bond. This is fully consistent with the "push-pull" mechanism we have proposed.

Table 2. pH-independent rate constants, solvent isotope effects and pKs for dealkylation of diastereomers of soman-inhibited AChEs

Type of soman AChE adduct	$10^3 k_{max}$, s^{-1}	k_H/k_D	pKa_1	pKa_2
$P_S C_R$ + EE	11±2	1.3±0.3	3.8±0.2	7.6±0.1
$P_S C_S$ + EE	12±2	1.1±0.2	3.6±0.2	8.0±0.2
$P_S C_R$ + FBS	46±30	1.1±0.1	4.5±0.2	6.0±0.2
$P_S C_R$ + FBS	31±14		4.6±0.2	6.4±0.2

CONCLUSIONS

Differences in the catalytic action of the serine protease and AChE group of enzymes are a more compartmentalized active site in the former than in the latter; a more elaborate apparatus in the oxyanion hole, and a potentially more versatile catalytic triad in the latter. There is a different orientation of Glu199 from Asp194 indicating a different role these evolved to fulfill. Namely, the activation of serine proteases from the zymogen, Asp194 plays a critical role, while similar development has not been observed for AChE. Serine proteases had to evolve to remove a typically long leaving group and accommodate a specific side chain. In contrast, AChE evolved to catalyze the hydrolysis of a small positively charged substrate, acetylcholine. It performs the task presumably by mobilizing a sophisticated system of negatively charged and hydrophobic residues to bring about both remarkably efficient chemistry and clearance of reactants and products. Glu199 seems to have an important role in the electrostatic catalytic mechanism. The high local negative charge density, mostly provided by Glu199 and the numerous small van der Waals interactions promote dealkylation much more effectively than any other similar enzyme. Moreover, the high concentration of negative charge created by the phosphonate ester monoanion and Glu199 adjacent to it surrounding the only accessible side of the central P, fully account for the resistance to the attack of even the strongest nucleophile applied for enzyme reactivation. Energy partitioning of the diasteromeric phosphonate esters of the enzymes showed that the potential for H-bonding to the P=O group in the oxyanion hole and the various interactions at the binding site for the evolutionarily anticipated natural substrate provide distinction between the enantiomers of an inhibitor.

REFERENCES

Bender, M. L., and Wedler, F. C., 1972, *J. Am. Chem. Soc. 94*, 2101.

Bennet, A. J., Kovach, I. M., and Schowen, R. L., 1988, *J. Am. Chem. Soc. 110*, 7892-7893.

Bennet, A. J., Kovach, I. M., and Bibbs, J. A., 1989, *J. Am. Chem. Soc. 111*, 6424-6429.

Kovach, I. M., Larson, M., and Schowen, R. L., 1986a, *J. Am. Chem. Soc. 108*, 5490.

Kovach, I. M., Larson, M., and Schowen, R. L., 1986b, *J. Am. Chem. Soc. 108*, 3054.

Kovach, I. M., 1988a, *Theochem. 170*, 159.

Kovach, I. M., 1988b, *J. Enzyme Inhib. 2*, 198-208.

Kovach, I. M., and Bennet, A. J., 1990, *Phosphorus, Sulfur and Silicon 51/52*, 51.

Kovach, I. M., Huhta, D., and Baptist, S., 1991, *Theochem. 226*, 99-100.

Kovach, I. M., and Huhta, D., 1991, *Theochem. 233*, 335-342.

Kovach, I. M., McKay, L., and Vander Velde, D., 1993, *Chirality 5*, 143-149.

Kovach, I. M., Qian, N., and Bencsura, A., 1994, *FEBS Letters 349*, 60-64.

Michael, H. O., Hackley, Jr., B. E., Berkovitz, L., List, G., Hackley, E. B., Gillilan, W., and Pankau, M., 1967, *Arch. Biochem. Biophys. 121*, 29-34.

Ordentlich, A., Kronman, C., Barak, D., Stein, D., Ariel, N., Marcus, D., Velan, B., and Shafferman, A., 1993, *FEBS Letters 334*, 215-220.

Qian, N., and Kovach, I. M., 1993, *FEBS Lett., 336*, 263.

Schoene, K, Steinhanses, J., and Wertman, A., 1980, *Biochem. Biophys. Acta 616*, 384-388.

Segall, Y., Waysbort, D., Barak, D., Ariel, N., Doctor, B. P., Grunwald, J., and Ashani, Y., 1993, *Biochemistry, 32,* 3441.

Stewart, J. J. P., 1990, QCPE # 455.

Sussman, J. L., Harel, M., Frolow, F., Oefner, C., Goldman. A., Tokder, L., and Silman, I., 1991, *Science, 253,* 872-879.

IRREVERSIBLE SITE-DIRECTED LABELING STUDIES ON CHOLINESTERASES

L. Ehret-Sabatier, I. Schalk, C. Loeb, F. Nachon, and M. Goeldner

Laboratoire de Chimie Bio-organique
Faculté de Pharmacie
Université Louis Pasteur Strasbourg
BP 24 - 67401 Illkirch Cedex, France

INTRODUCTION

Aryldiazonium salts have been demonstrated as being reactive analogues of quaternary ammonium salts and constitute therefore potential irreversible site-directed probes for the quaternary binding sites on the cholinergic proteins. The studied proteins include the cholinergic enzymes acetyl- (Goeldner & Hirth, 1980) and butyrylcholinesterase (Ehret-Sabatier et al., 1991) and the nicotinic (Langenbuch-Cachat et al., 1988) and muscarinic receptors (Ilien & Hirth, 1989). The probing of choline acetyltransferase or choline transport systems has not been achieved yet.

ARYLDIAZONIUM SALTS AS IRREVERSIBLE PROBES

Although, the chemical resemblance between quaternary ammonium and diazonium salts seems not pronounced, it remains that they both possess a diffuse positive charge in agreement with hydrophobic interactions created i.e. by π-electrons of aromatic residues from quaternary ammonium binding sites.

The chemical reactivity associated to aryldiazonium salts is very dependent on the nature of the substituents on the aromatic ring and this reactivity conditions the type of probe which will be defined. In summary, substitution by an electroattracting substituent increases the chemical reactivity and will either lead to protein modification reagents or to affinity labels. On the other hand, an electrodonating substitution stabilizes these molecules which therefore will be used as potential photoaffinity reagents. The simpliest way to establish such classification is to determine the half-life of the salts in buffered medium, which is indicative of the reactivity of these salts.

The photochemical properties of diazonium salts are remarkable for two reasons, their quantum yields are very high and the photogenerated species (arylcations) are extremely reactive. Consequently aryldiazonium salts have been described as topographical

Enzymes of the Cholinesterase Family, Edited by Daniel M. Quinn et al.
Plenum Press, New York, 1995

probes (Kotzyba-Hibert *et al.,* 1995) when used as photoaffinity reagents leading to a complete mapping of a binding site. By opposition, the labeling obtained by an aromatic diazonium salt in an affinity process will necessarily lead to a partial mapping i.e. labeling of nucleophilic residues in agreement with the predicted chemical reactivity.

SITE-DIRECTED LABELING OF TORPEDO ACETYLCHOLINESTERASE WITH ARYLDIAZONIUM SALTS: AN OVERVIEW

Figure 1 summarizes the different labeling studies accomplished on both quaternary ammonium binding sites of *Torpedo* acetylcholinesterase by using aromatic diazonium salts. These studies include the p-dimethylamino benzenediazonium (DDF) (Kieffer *et al.,* 1986; Harel *et al.,* 1993; Schalk *et al.,* 1994), the p-dibutylamino benzenediazonium (Schalk *et al.,* 1994), the p-butyroxybenzene diazonium (Ehret-Sabatier *et al.,* 1991) and the 6-coumarin diazonium salts (Schalk *et al.,* 1995). While the photoaffinity labeling study with DDF led to the identification, in a mutually exclusive manner, of either active site (Phe330 and Leu332) or peripheral site (Trp279) residues, the coumarin diazonium salt induced a selective affinity labeling at the peripheral site, predominantly on Tyr70.

6-COUMARIN DIAZONIUM SALT: A SELECTIVE PERIPHERAL SITE AFFINITY LABEL: PREDOMINANT LABELING OF TYR 70

Incubation of *Torpedo* acetylcholinesterase with 6-coumarin diazonium led to a time-, concentration- and pH dependent inactivation of the enzyme, illustrative of an affinity

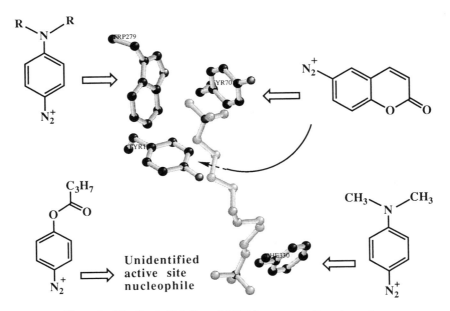

Figure 1. Site-directed labeling of AChE by aromatic diazonium salts.

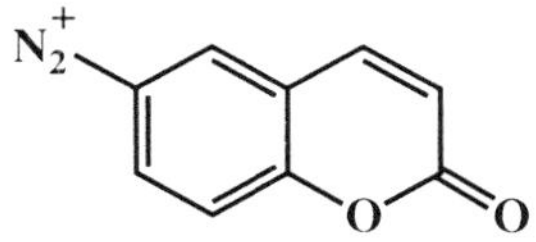

Figure 2. 6-coumarin diazonium salt.

Table 1. pH-dependence on the affinity labeling

pH	Kd (M)	k (min^{-1})
6.0	1.5 10^{-4}	0.12
7.2	2.3 10^{-4}	0.15
8.0	2.3 10^{-5}	0.47

labeling-type of inactivation. The effect of the pH (between 6 and 8) on this inactivation (dissociation constant Kd (M) and rate constant k(min-1)) is shown in Table 1.

Protection Experiments

The observed protection by NMe$_4^+$ indicates an interaction with a quaternary ammonium binding site of the enzyme which was further analyzed by using edrophonium and propidium. A concentration-dependent protection with edrophonium (Table 2) and propidium demonstrated that the labeling occured at the peripheral site. In particular, the absence of protection at lower concentration of edrophonium (1 µM), while protection by propidium was increasingly effective between 30 nM to 150 nM, was observed.

Fluorescence Measurements

The interaction of 6-coumarin diazonium with the peripheral site was directly demonstrated by measuring the decrease of propidium-induced fluorescence (Figure 3).

Mutational Analysis

(Collaboration - Y. LeFeuvre, S. Bon, J. Massoulié - ENS Paris)
Tyrosines are likely amino acid residues to be involved in this affinity labeling reaction. Within the peripheral site (Harel *et al.*, 1993; Sussman *et al.*, 1991), two tyrosine residues (Tyr70 and Tyr121) being positioned in the vicinity of Trp279, were best candidates. This hypothesis was tested by mutational analysis with respect to their reaction with the coumarin

Table 2. Edrophonium protection of AChE inactivation

Edrophonium (M)	Residual activity (%)
0	42
10^{-6}	41
5 10^{-6}	61
10^{-3}	87

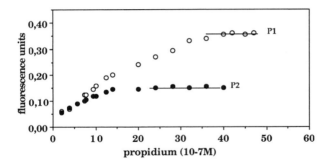

Figure 3. Titration of AChE residual peripheral sites by propidium-induced fluorescence measurement in the absence (○) or in the presence (●) of 0.2 mM of 6-coumarin diazonium.

diazonium probe. Tyr70 was successively replaced by Phe, Asn and His while Tyr121 was replaced by Phe, Gln and His. The activities of the second series were insufficient to allow accurate kinetic evaluations. The kinetic constants for acetylthiocholine hydrolysis by the recombinant enzyme and the different Tyr70 mutants (Table 3), indicated no major differences.

By opposition, the inactivation reaction by the coumarin diazonium probe showed important differences between the wild type and the different Tyr70 mutants (Figure 4). It was of particular interest to establish a correlation between the nucleophilicity of the mutated residues and the rate of inactivation: His > Tyr (wild type) > Asn > Phe. This correlation, in conjunction with the observed increase of inactivation for the Tyr70His mutant, constitute a strong argument for the involvment of this residue in the alkylation reaction and confirm therefore the belonging of Tyr70 to the peripheral site (Radić *et al.*, 1993; Barak *et al.*, 1994; Eichler *et al.*, 1994).

LABELING OF BUTYRYLCHOLINESTERASE BY DIAZONIUM SALTS

The amino acid residues labeled by [³H] DDF on human serum butyrylcholinesterase are currently investigated and might lead to an interesting comparative labeling study with AChE. For instance, the topographical mapping by DDF at the quaternary ammonium binding site of AChE active site, lacked the labeling of an important Trp84 residue (Weise *et al.*, 1990). Which residues are accessible to DDF in BuChE? Importantly, this labeling study on BuChE will allow to test directly the model structure which has been proposed for that enzyme (Harel *et al.*, 1992).

Also, the formerly described p-butyroxy benzene diazonium (Ehret-Sabatier *et al.*, 1991) led to the discovery of a new type of photosuicide inhibition by defining a substrate analogue converted to an enzymatic product which can be selectively photoactivated. The

Table 3. Kinetic constants for ATCh hydrolysis by AChE and its mutants

Enzyme	Km (μM)	kcat (nmol ATCh/ pmol active site)
Wild type	40	105
Y70H	45	127
Y70F	40	37
Y70N	35	97

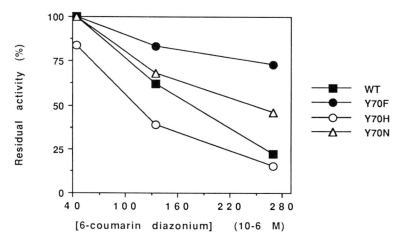

Figure 4. Inactivation of wild type (WT), and mutants: Y70F, Y70N, Y70H AChE in function of concentration of 6-coumarin diazonium salt.

expected consequence is that the photoreactive species, being derived from the enzyme product, will be generated exclusively within the active site. This inhibitor differentiates from all the others described here by the fact that it is an enzyme substrate and therefore triggers the enzymatic reaction mechanism. Consequently, the labeling of cholinesterases in different conformational states can be envisaged, by comparing the labeling obtained from a substrate to the labeling derived from an inhibitor. This work is presently undertaken.

CONCLUSION

The site directed labeling studies on AChE using aromatic diazonium salts led to successive useful structural informations on this enzyme. At first, Phe330 was identified as the first residue belonging to the quaternary ammonium binding site within the active site, and this result was obtained before the description of the 3D structure. Secondly, in conjunction with X-ray studies, Trp279 was demonstrated as being an important residue contributing to the quaternary ammonium binding at the peripheral site, and finally Tyr70 could be demonstrated as another important peripheral site residue, confirming in particular mutational studies.

What future for site-directed probes? While bringing complementary informations, in particular, to site-directed mutagenesis experiments, photoaffinity probes offer, in addition, new perspectives related to their unique property of triggering extremely rapid light-induced reactions. The trapping of transient conformations of functional proteins represents a new challenge for photoaffinity labeling for which the cholinesterases constitute an interesting target.

ACKNOWLEDGMENTS

The authors thank the Ministère de la Recherche et de la Technologie, the Centre National de la Recherche Scientifique, the Association Franco-Israelienne pour la Recherche Scientifique and the Association Française contre les Myopathies for financial support.

REFERENCES

Barak, D., Kronman, C., Ordentlich, A., Ariel, N., Bromberg, A., Marcus,.D., Lazar, A., Velan, B., & Shafferman, A., (1994) *J. Biol. Chem.* 264: 6296-6305.

Ehret-Sabatier, L., Goeldner M., & Hirth, C., (1991) *Biochim. Biophys. Acta* 1076: 137-142.

Eichler, J., Anselmet, A., Sussman, J.L., Massoulić, J., & Silman, I., (1994) *Mol. Pharmacol.* 45: 335-340.

Goeldner, M., & Hirth, C., (1980) *Proc. Natl. Acad. Sci. USA* 77: 6439-6442.

Harel, M., Schalk, I., Ehret-Sabatier, L., Bouet, F., Goeldner, M., Hirth, C., Axelsen, P., Silman, I., & Sussman, J.L., (1993) *Proc. Natl. Acad. Sci. USA* 90: 9031-9035.

Harel, M., Sussman, J.L., Krejci, E., Bon, S., Chanal, P., Massoulić, J., & Silman, I., (1992) *Proc. Natl. Acad. Sci. USA* 89: 10827-10831.

Ilien, B., and Hirth, C., (1989) *Eur. J. Biochem.* 183: 331-337.

Kieffer, B., Goeldner, M., Hirth, C., Aebersold, R., & Chang, J.Y., (1986) *FEBS Lett.* 202: 91-96.

Kotzyba-Hibert, F., Kapfer, I., & Goeldner, M., *Angew. Chemie*, In press

Langenbuch-Cachat, J., Bon, C., Mulle, C., Goeldner, M., Hirth, C., & Changeux, J-P., (1988) *Biochemistry* 27: 2337-2345.

Radić, Z., Pickering, N.A., Vellom, D.C., Camp, S., & Taylor, P., (1993) *Biochemistry* 32: 12074-12084.

Schalk, I., Ehret-Sabatier, L., Bouet, F., Goeldner, M., & Hirth,C. , (1994) *Eur. J. Biochem.* 219: 155-159.

Schalk, I., Ehret-Sabatier, L., LeFeuvre, Y., Bon, S., Massoulié, J., & Goeldner, M., Submitted

Sussman, J.L., Harel, M., Frolow, F., Oeffner, C., Goldman, A., Toker, L., & Silman, I., (1991) *Science* 253: 872-879.

Weise, C., Kreienkamp, H-J., Raba, R., Pedak, A., Aariksaar, A., & Hucho, F., (1990) *EMBO J.* 9: 3885-3888.

MUTATION OF HUMAN BUTYRYLCHOLINESTERASE GLYCINE 117 TO HISTIDINE PRESERVES ACTIVITY BUT CONFERS RESISTANCE TO ORGANOPHOSPHORUS INHIBITORS

C. A. Broomfield,[1] C. B. Millard,[1] O. Lockridge,[2] and T. L. Caviston[2]

[1] Biochemical Pharmacology Branch
U.S. Army Medical Research Institute of Chemical Defense
Aberdeen Proving Ground, Maryland 21010-5425
[2] University of Nebraska Medical Center
Eppley Cancer Institute
Omaha, Nebraska 68198-6805

INTRODUCTION

While much progress has been made in the last 50 years in the development of treatments for nerve agent intoxication, the organophosphorus anticholinesterase inhibitors (OPs) remain a chemical warfare threat. In addition, treatments for the large numbers of people are accidentally poisoned by organophosphorus insecticides each year remain inadequate. It is generally accepted that we are approaching the limiting efficacy for pharmacological treatment of this type of intoxication with drugs now in development, so any significant increase of protection without unacceptable side effects must result from new approaches. In recent years it has been shown (Raveh *et al.* 1989, Broomfield *et al.,* 1991; Broomfield, 1992) that it is possible to protect animals from OPs by prophylactic administration of scavengers that can react with and detoxify the inhibitors before they reach their critical targets. Stoichiometric scavengers (one molecule of scavenger reacts with one molecule of toxin) are effective but require a large amount of material to destroy a small amount of toxin. Catalytic scavengers (enzymes that catalyze the hydrolysis of the toxins) are also effective (Broomfield, 1992) but the enzymes that occur in nature are generally inefficient; they have high Km values and low turnover numbers. Again, a large amount of material is required to afford the desired degree of protection. To remedy this situation, we proposed to construct a more efficient organophosphorus acid anhydride hydrolase (OPAH) by site-directed mutagenesis of the butyrylcholinesterase (BuChE) gene. The rationale for this scheme was based on an hypothesis by J. Jarv (1984) that organophosphates are

Enzymes of the Cholinesterase Family, Edited by Daniel M. Quinn et al.
Plenum Press, New York, 1995

169

hemisubstrates for the cholinesterases because they form chiral phosphorylated enzymes. The histidine-bound water molecule that normally displaces the acyl group from the active site serine in the course of normal hydrolysis is prevented sterically from attacking the appropriate face of the tetrahedral phosphorus molecule, thus preventing reactivation of the enzyme. If this hypothesis is correct, we reasoned that it might be possible to introduce a second nucleophilic center into the active site in such a position that it could carry an activated water molecule to the face of the phosphorus moiety opposite the phosphorus-serine bond and thereby reactivate the enzyme. This became possible when the crystal structure of acetylcholinesterase (AcChE) was solved by Sussman, et al. (1991).

EXPERIMENTAL

For a number of reasons we felt that BuChE would be more appropriate to test this hypothesis than AcChE. BuChE has a less restricted active site and is somewhat less stereoselective than AcChE. In addition, BuChE is an intrinsic plasma protein and therefore a successful mutant would be expected to enjoy a longer biological half-life in the plasma. Therefore, we began by making a computer model of human BuChE based on the crystal structure of Torpedo AcChE, replacing those residues that are different and then minimizing the energy of the resulting structure (Millard and Broomfield, 1992). Having done that, we then measured the distance between the α-carbon of serine 198 (the "active site" serine) and the α- carbon of histidine 438 (9.62 Å on our model) and looked for residues on the opposite side of the gorge with similar spacing and with the side-chain oriented in the general direction of serine 198.   We identified three glycine residues, numbers 115, 117 and 121 with reasonable spacings and the appropriate orientation and decided to make mutants with each of residues 117 and 121 substituted by histidine as a first attempt at our goal. A model with the modification at residue 117 (G117H) is shown in Fig. 1.

The appropriate oligonucleotide probes were synthesized and purified by Dr. Charles Mountjoy in the Molecular Biology Core Facility at the University of Nebraska Medical Center and mutagenesis was carried out in M13mp19, followed by stable expression in both CHO and human embryonic kidney 293 cells. At this point in time G117H and G121H have been expressed successfully, and media containing the resulting products were screened for ability to catalyze the hydrolysis of butyrylthiocholine (BuSch), benzoylcholine (BzCh), echothiophate, soman, sarin and diisopropylphosphorofluoridate (DFP) and for inhibition by echothiophate, soman, sarin, DFP and tetraisopropyl pyrophosphoramide (iso-OMPA).

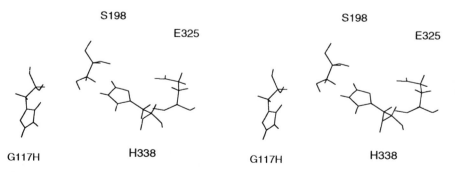

Figure 1. Computer stereo model of the active site region of BuChE with glycine residue 117 substituted by histidine.

Table 1. Summary of results, Michaelis constants

CHO cell product	Substrate	Concentrations (mM)	Km (mM)	Literature K (mM)
BuChE	benzoylcholine	0.0005 to 0.1000	0.0037 ± 0.0004	0.0045
G117H	benzoylcholine	0.0005 to 0.1500	0.0285 ± 0.0035	—
BuChE	butyrylthiocholine	0.0125 to 0.1125	0.016 ± 0.001	0.023
G117H	butyrylthiocholine	0.0125 to 0.1125	0.070 ± 0.003	—
BuChE	butyrylthiocholine	1.25 to 6.25	0.20 ± 0.016	0.260
G117H	butyrylthiocholine	1.25 to 6.25	0.23 ± 0.017	—
BuChE	acetylthiocholine	0.0125 to 0.1125	0.048 ± 0.003	0.049
G117H	acetylthiocholine	0.0125 to 0.7500	0.424 ± 0.012	—
BuChE	acetylthiocholine	1.25 to 6.25	0.516 ± 0.046	0.490
G117H	acetylthiocholine	1.25 to 6.25	0.565 ± 0.029	—
BuChE	p-nitrophenyl acetate	1.00 to 10.00	6.1	6.01
G117H	p-nitrophenyl acetate	1.00 to 10.00	12.5	—

BuChE G121H was inactive for all substrates tested, although immunoreactive material was found in the CHO cell medium.

RESULTS AND DISCUSSION

G121H appeared to have no activity at all; the hydrolysis of neither benzoylcholine nor butyrylthiocholine was catalyzed and none of the OP compounds showed evidence of hydrolysis. G117H, on the other hand, showed hydrolytic rates of both BuSCh and BzCh comparable to those of the wild type enzyme. Furthermore, it appeared to have essentially normal or only slightly altered reaction kinetics with all of the substrates tested (Table 1).

Surprisingly, it was not inhibited by echothiophate at all within the time we observed it (480 min). Further investigation showed that the rates of reaction with all of the OP inhibitors were greatly reduced. (Table 2).

It was of particular interest that, in the course of determining inhibition rates of the OPs it was found that the rate of DFP inhibition could not be measured because the inhibition rapidly reversed. Fig 2 shows a typical activity assay curve of DFP-inhibited G117H. Although the activity clearly is low at the beginning of the assay it increases with time until, after about two minutes a stable slope is achieved. The value of the stable slope finally reached decreases with time of incubation with DFP, indicating that, while the DFP-inhibited enzyme reactivates rapidly it is also aging, and the aged molecules are unable to reactivate.

Table 2. Summary of results, inhibitors

BuChE	Inhibitor	Apparent first-order K_i
WT	echothiophate	0.23 min^{-1}
	iso-OMPA	0.52 min^{-1}
	soman	> 2.8 min^{-1}
G117H	echothiophate	No inhibition at t = 480 min
	iso-OMPA	0.0043 min^{-1}
	soman	0.02 min^{-1}

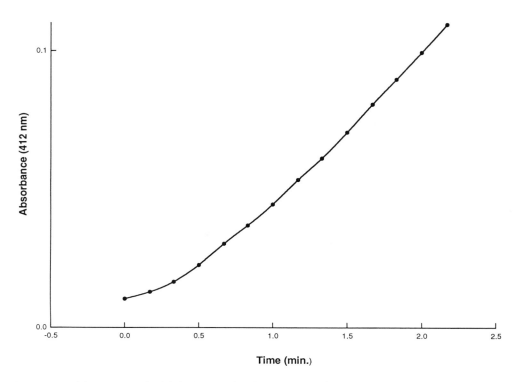

Figure 2. Activity assay result with G117H 30 min. after treatment with 1 mM DFP, using butyrylthiocholine as the substrate. This curve is typical of those observed with DFP concentrations between 0.1 and 1 mM. At concentrations of DFP 10 uM or lower little inhibition was observed.

Upon inhibition of G117H with soman no reactivation is observed and activity is not restored upon treatment with the oximes, 2PAM or TMB-4. Fortunately, however, inhibition by sarin was easily measured, reactivation was slow enough not to interfere with activity assays and there was no aging. The time course of inhibition and reactivation of G117H by sarin at different initial sarin concentrations is shown in Fig 3. Under most conditions activity was restored completely by 20 hrs. Independent determination of sarin concentration indicated that the inhibitor was completely gone at about 130 minutes , accounting for the return of activity.

Independent sarin assays indicated that there was no residual inhibitor after about 130 min.

Having shown that the inhibition by sarin is relatively rapidly reversible, an attempt was made to demonstrate sarin hydrolysis by G117H, using radiolabeled sarin as a substrate. Four parts 1mM sarin in 0.15 M saline were combined with 1 part radioactive sarin, approx. 0.5 Ci/mmole and 100 uL of the mixture added to 1 mL of culture medium containing either wild type recombinant BuChE (rBuChE) or G117H. The resulting mixtures were incubated at ambient room temperature. At predetermined intervals 100 uL aliquots were taken from each mixture and extracted with 100 uL of methylene chloride, centrifuged and 50 uL of each phase counted. The culture media contain components that catalyze the hydrolysis of sarin, so the effect of the enzyme was measured against a very high background; nevertheless, the rate of sarin hydrolysis in the presence of the mutant, G117H was significantly faster than with the rBuChE (Fig 4).

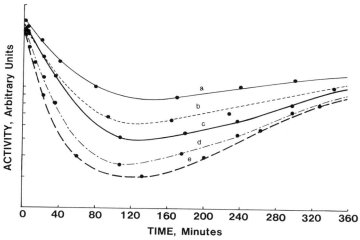

Figure 3. Time course of enzymatic activity of BuChE mutant G117H after treatment with sarin. (a) 1.64 uM, (b) 2.05 uM, (c) 2.56 uM (d) 3.20 uM (e) 4.00 uM.

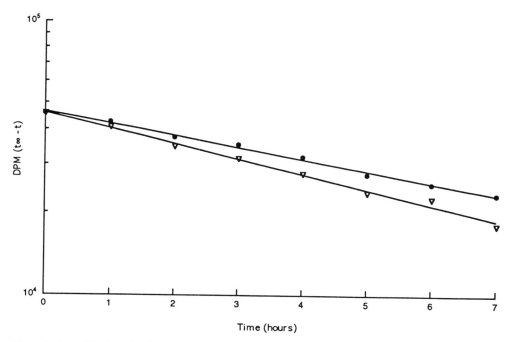

Figure 4. Rate of hydrolysis of sarin in the presence of rBuChE (●) or mutant G117H (∇) . Hydrolysis with the wild type enzyme may be regarded as background, since the wild type enzyme does not regenerate activity on the time scale of this experiment.

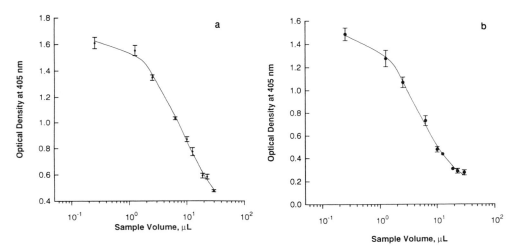

Figure 5. Competitive EIA curves for rBuChE (A) and G117H (B). The volumes at the inflection point of each curve were calibrated with respect to protein concentration by comparing with a standard curve prepared with purified, human serum BuChE.

In order to characterize fully the mutant enzyme it was necessary to determine its concentration in the culture medium. To accomplish this a quantitative, competitive enzyme-linked immunoassay (EIA) was developed, using a rabbit polyclonal antibody to denatured serum BuChE at a dilution of 1:25,000. Good four-parameter fit curves were obtained for both rBuChE and G117H (Fig 5). The sample volumes were calibrated with purified, human serum BuChE to permit the determination of enzyme protein concentrations in the samples. Specific activities with BzCh of 240 U/mg and 65 U/mg were calculated for the rBuChE and G117H, respectively.

SUMMARY

In summary, an attempt was made to produce from butyrylcholinesterase an enzyme that would catalyze the rapid hydrolysis of OPs, using computer-aided molecular modeling to identify mutation sites . The data shown here indicate that replacement of glycine 121 with histidine eliminates enzymatic activity, probably due to steric interference with substrate binding. Substitution of glycine 117 with histidine, however, produces a molecule that retains for the most part its ability to hydrolyze the normal substrates but is highly resistant to inhibition by OP anticholinesterases. Furthermore, once an OP reacts with G117H reactivation is much more rapid than is the case with the wild type BuChE, resulting in actual turnover of the OP compound. Experiments are now in progress to construct different mutants based on the same strategy used to design G117H, but using stereo optics more accurately to position the mutations. In addition, several double mutants are being constructed that are designed to overcome the slow rate of reaction with OP compounds seen with G117H. We are very encouraged by the small amount of catalysis seen with G117H and fully expect to find a highly efficient OPA hydrolase among the BuChE mutants now being made.

REFERENCES

Broomfield, C.A., Maxwell, D.M., Solana, R.P., Castro, C.A., Finger, A.V. and Lenz, D.E., 1991, Protection by Butyrylcholinesterase against Organophosphorus Poisoning in Nonhuman Primates. *J. Pharmacol. Expl. Therapeutics* 259:633-638.

Broomfield, C.A., 1992, A Purified Recombinant Organophosphorus Acid Anhydrase Protects Mice against Soman. *Pharmacol. & Toxicol.* 70:65-66.

Jarv, J., 1984, Stereochemical Aspects of Cholinesterase Catalysis, *Bioorg. Chem.* 12: 259-278.

Millard, C.B. and Broomfield, C.A., 1992, A Computer Model of Glycosylated Human Butyrylcholinesterase, *Biochem. Biophys. Res. Comm.* 183:1280-1286.

Raveh, L., Ashani, Y., Levy, D., de la Hoz, D., Wolfe, A.D. and Doctor, B..P., 1989, Acetylcholinesterase Prophylaxis against Organophosphate Poisoning: Quantitative Correlation between Protection and Blood Enzyme Level in Mice, *Biochem. Pharmacol.* 38:529-534.

Sussman, J.L., Harel, M., Frolow, F., Oefner, C., Goldman, A., Toker, L. and Silman, I., 1991, Atomic Structure of Acetylcholinesterase from Torpedo *californica:* A Prototypic Acetylcholine-binding Protein, *Science* 253:872-879.

REACTION OF ACETYLCHOLINESTERASE WITH ORGANOPHOSPHONATES

Molecular Fate at the Rim of a Gorge

Harvey Alan Berman

Department of Biochemical Pharmacology
State University of New York at Buffalo
Buffalo, New York 14260

INTRODUCTION

Irreversible inhibition of acetylcholinesterase (AchE) by organophosphorus agents is rapid, proceeding at rates that, in general, exceed 10^5-10^6 M^{-1}-min^{-1}. Numerous instances occur in which the reaction velocity, proceeding at rates greater than 10^8 M^{-1}-min^{-1}, approaches within an order of magnitude of the diffusion limitation (Berman and Leonard, 1989). The rapidity of organophosphorus reactions with AchE is striking, especially so since these molecules, having tetrahedral symmetry, occupy a measurable volume, and they engender a more complex array of steric interactions than, for example, the planar, symmetrical substrate acetylcholine. Such reactivity of organophosphonates is even more remarkable since the catalytically reactive residues exist at the base of a deep gorge, 20 Å in depth, that is deeper than it is wide, is tightly circumscribed, and offers ample opportunity for unfavorable steric interactions. Indeed, there exists within the gorge, about half-way between the enzyme surface and the base, a pronounced ledge – or restriction – that, as seen in the crystal state, would seem to restrict free diffusion of a large ligand in or out of the gorge (Sussman, et al., 1991). As such, information derived from the crystal structure provides no easy explanations for the rapid covalent reactivity of AchE toward a wide variety of organophosphorus agents. Systematic study of a defined family of structurally-related organophosphonates can provide some clues as to interactions of importance in the docking and reactivity of large molecules within the active center.

One suitable family of organophosphonates are homologous n-alkylphosphon-ofluoridates and n-alkylphosphonyl thiocholines (Figure 1). These agents contain a common cycloheptyl phosphonate ester group but differ in defined ways with respect to the chain length of the n-alkyl group linked directly to phosphorus, ranging sequentially from methyl up to the n-hexyl moiety, and the identity of the leaving group, containing a fluoride or thiocholine moiety.

Enzymes of the Cholinesterase Family, Edited by Daniel M. Quinn et al.
Plenum Press, New York, 1995

177

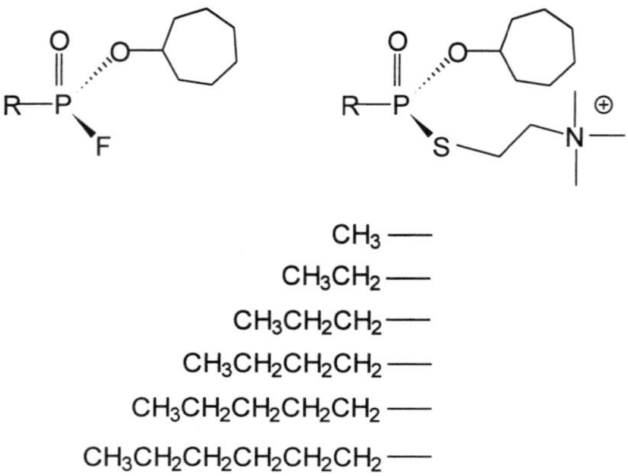

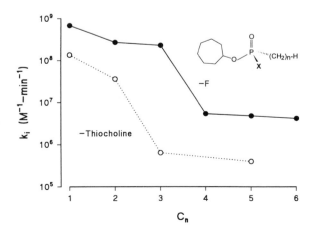

O
O
R—P
F
O
O
R—P
S
N
⊕
CH3 —
CH3CH2 —
CH3CH2CH2 —
CH3CH2CH2CH2 —
CH3CH2CH2CH2CH2 —
CH3CH2CH2CH2CH2CH2 —

Figure 1. Structures of *n*-alkylphosphonofluoridates and *n*-alkylphosphonyl thiocholines.

Figure 2 displays the relationship between inhibition constants measured for AchE from *Torpedo californica* and chain length of the *n*-alkylphosphonofluoridates and *n*-alkylphosphonyl thiocholines. Three features of these data are noteworthy. 1) The bimolecular reaction constants are uniformly rapid and, in general, exceed 10^6 M^{-1}-min^{-1}. When the *n*-alkylphosphonofluoridates contains 1, 2 or 3 carbon atoms and the *n*-alkylphosphonyl thiocholines contain 1 or 2 carbon atoms the rates of reaction are in the range 10^{-7}–10^{-8} M^{-1}-min^{-1}. 2) When the *n*-alkyl chain contains more than 3 carbon atoms for the fluoridates and more than 2 carbon atoms for the thiocholines, the inhibition rates are slower by two orders of magnitude. 3) The chain length dependence for the phosphonyl thiocholines is left-shifted relative to that for the fluoridates, and, in both cases, the chain length dependences reveal a discontinuity. For the alkylphosphonofluoridates the discontinuity occurs when the alkyl chain contains more than 3 carbon atoms; for the alkylphosphonyl thiocholines the discontinuity occurs when the alkyl chain contains more than 2 carbon atoms.

These observations prompt two general questions which make up the subject of this paper. First, what is the significance of the discontinuities in the chain length dependences and, in particular, what significance can be attached to the *position* of the discontinuity at 2 carbon atoms for the phosphonyl thiocholines and at 3 carbon atoms for the phosphonofluoridates? Second, what is the significance of the 100-fold drop in inhibitory potency that follows

10⁹
10⁸
10⁷
10⁶
10⁵
k_i (M⁻¹–min⁻¹)
1
2
3
4
5
6
C_n
–F
–Thiocholine
O
P
O
(CH2)n-H
X

Figure 2. Relationship between bimolecular inhibition constant and chain length (Cn) for cycloheptyl *n*-alkylphosphonates. (●), *n*-Alkylphophonofluoridates. (O), *n*-alkylphosphonyl thiocholines.

the discontinuity. That is, while the 100-fold reduction in reaction rate is relatively large, the reaction velocities are, on an absolute basis, uniformly rapid and comparable to rates measured for highly reactive methylphosphonates (Berman and Leonard, 1989). Asked in another way, why is there no abolition of inhibition?

Reaction Kinetics of Organophosphonates with AChE

To address the these questions is it instructive to consider the microscopic steps that underlie irreversible inhibition in the context of the operational geometry of the active center. In the case of organophosphorus reactions with AchE, as described two decades ago by R. D. O'Brien, the bimolecular reaction occurs in two steps, as a unimolecular phosphonylation that is preceded by a rapid equilibrium association (Hart and O'Brien, 1973). In the language of pharmacology, the equilibrium step represents *occupation*, the capacity of the enzyme to recognize and associate with the organophosphonate. The unimolecular phosphonylation represents *efficacy*, the capacity with which occupation is translated into covalent reaction. The two steps are distinct and can be resolved with high precision using a conventional spectrophotometer and manually-actuated stopped-flow device (Berman and Leonard, 1989).

When this reaction is obeyed, a plot of $1/k_{obs}$–*versus*–$1/$[Inhibitor] leads to a straight line that intersects the positive y-axis and the negative x-axis. The reciprocal of the slope leads to k_i, the *bimolecular reaction constant*, while the reciprocal of the y-intercept leads to a value for k_p. The dissociation constant, K_D, is obtained from knowledge of k_i and k_p. Intersection of such a plot in the positive y-axis substantiates such a two-step mechanism.

Operational Binding Domains within Active Center Gorge

The three dimensional coordinates of all atoms that make up AchE from *Torpedo* reveals the catalytic residues to exist at the base of the gorge and that the nucleophilic serine, Ser200, is situated approximately 6 Å from the base. Such an environment possesses an intrinsic asymmetry and likely represents one component governing the stereospecificity in reaction of methylphosphonate reactions (Berman and Leonard, 1989). One element important in ligand recognition is Trp84; this residue is present at the base of the gorge, makes up what was once referred to as the "anionic site", and facilitates ligand association through π-quaternary ammonium interactions. We refer to this element as representing the *leaving group domain*. An *acyl pocket* accommodates the substrate acid moiety or, analogously, the alkyl portion of the organophosphorus agent. Third, as shown through study of enantiomeric phosphonates, there exists an *alkyl binding region* that accommodates the phosphonate ester moiety and in turn promotes covalent reaction of methylphosphonates. Finally, a dipolar domain denoted as an *oxyanion hole* is thought to facilitate hydrolysis of acetate esters through polarization of the C=O bond. With respect to phosphonates, the oxyanion hole may be considered to serve as a docking or anchoring point to stabilize the already polarized dative P→O bond. It is not clear, however, whether interaction of the P→O bond in the oxyanion hole is stronger in the noncovalent complex *prior* to reaction or in the trigonal

$$E + P\text{-}X \underset{\text{Occupation}}{\overset{K_D}{\rightleftharpoons}} \{E : P\text{—}X\} \xrightarrow[\text{Efficacy}]{k_p} E\text{-}P + X^-$$

Figure 3. Kinetic scheme describing inhibition of AchE by organophosphorus agents.

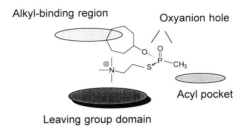

Figure 4. Operational binding domains within the active center of AchE.

bipyramidal reaction intermediate formed *during* the reaction. Figure 4 illustrates these operational relationships.

The Importance of Steric Interference and Electrostatic Steering during Irreversible Inhibition of AChE

This framework provides a basis for understanding the chain-length dependence of alkylphosphonate reactions. It is noteworthy that inhibition by *all* alkylphosphonates is rapid, irrespective of chain length. The velocity of such reactions must stand to signify the absence of steric occlusion and that any steric interference – to the extent it exists – must be of a magnitude that is insufficient to block reaction. As such, the 100-fold drop in inhibitory potency seen for the *n*-alkylphosphonyl thiocholines and the *n*-alkylphosphonofluoridates is interpreted to signify steric hindrance – rather than steric occlusion – to reaction. Such hindrance to reaction is not evident when the *n*-alkylphosphonyl thiocholines contains 2 carbon atoms and when the *n*-alkylphosphonofluoridates contain 3 carbons atoms. One source of steric interference can arise from ligand interaction with the side chains of phenylalanine residues, F295 and F297 in *Torpedo*, in spatial proximity to Ser200 (Harel, et al., 1992).

From examination of the microscopic reaction constants (Table 1) the dissociation constants for the methyl and *n*-propyl homologues are seen to be 13-30-fold lower than that for the *n*-pentyl congener. The difference in reactivity between the *n*-propyl and *n*-butyl homologues is therefore attributable primarily to a difference in K_D, that is, occupation. As such, a rotation about the P→O axis can translocate the *n*-butyl group from the unfavorable location within the acyl pocket into the alkyl-binding region, simultaneous with rotation of

Table 1. Inhibition constants for reaction of acetylcholinesterase with *n-alkyl*phosphonates

R-	k_i $(M^{-1} \cdot min^{-1})$	k_p (min^{-1})	$K_D \times 10^6$ (M)
A. *n*-Alkylphosphonofluoridates			
Methyl	6.8×10^8	540	0.82
*n*Propyl	2.3×10^8	420	1.9
*n*Pentyl	4.8×10^6	110	25
B. *n*-Alkylphosphonyl thiocholines			
Methyl	1.4×10^8	350	2.6
*n*Propyl	6.4×10^5	15	23
*n*Pentyl	4.0×10^5	6.4	16.3

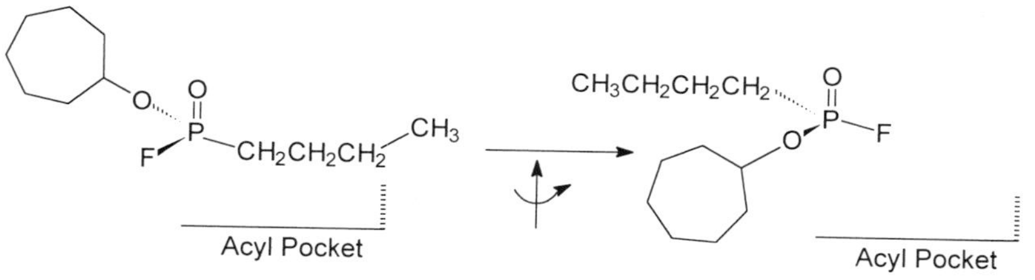

Figure 5. Intramolecular rotation of *n*-alkylphosphonofluoridate molecules in the presence of steric hindrance within the acyl pocket.

the cycloheptyl group into the leaving group domain (Figure 5). The increase in dissociation constant can be viewed as energy expended during hindered rotation within the gorge.

With increasing chain length of *n*-alkylphosphonyl thiocholines there occurs a 23-55-fold reduction in the unimolecular rate of phosphonylation; this reduction far exceeds the more modest 5-8-fold increase in K_D. In that increasing chain length leads principally to a reduction in the efficacy of covalent reactivity, the influence of chain length on reaction of the phosphonyl thiocholines arises from a different physical basis than that underlying reaction of the fluoridates. Such a conclusion implies that an increase chain length, while it doesn't preclude association, results in a displacement of the electrophilic P atom with respect to Ser200. Two explanations highlight why rotation about the P→O axis of a phosphonyl thiocholine is expected to be energetically less favorable than for a phosphonofluoridate. First, rotation of a molecule containing three sterically bulky groups is not expected to relieve steric strain; this case stands in contrast to the case for the uncharged alkylphosphonofluoridates which contain a small fluoride group of molecular volume not substantially different from that of a methyl group. Second, rotation of a thiocholine moiety out of the leaving group domain would serve to disrupt the favorable π-cation interaction between the quaternary ammonium moiety and Trp84. The reduction in k_p signifies that while the *n*-propylphosphonyl thiocholine is docked within the active center, the electrophilic P atom is not optimally positioned for nucleophilic attack by the serine hydroxyl group.

This analysis allows us to conclude that for cationic phosphonates, rotation of the organophosphonate moiety *after* entrance into the active center gorge is not particularly favorable. That is, since steric and electronic factors appear to preclude reorientation while within the gorge, cationic phosphonates must enter the gorge in close to their final orientation. In such a case, the fate of the organophosphonate complex is said to be determined at the mouth of the gorge (Figure 6). This case represents what Sussman referred to as an *electrostatic steering* mechanism (Tan, et al., 1993) in which the substantial dipole strength of the enzyme orients the alkylphosphonyl thiocholine with the cationic moiety preferentially directed into the entrance of the gorge. For the uncharged alkylphosphonofluoridates such constraints need not apply, and the molecule can enter the gorge and come to rest at its base in one of a number of possible orientations; the absence of any π-cation interactions and the presence of a small fluoride leaving group allows the noncovalent complex to relieve steric strain through hindered rotation. Uncharged phosphonates, therefore, possess a greater degree of orientational freedom in reaction with AchE than do their cationic homologues.

Docking, Stereochemistry, and Leaving Groups

Two final points are made. First, docking of the P→O moiety within the oxyanion hole would seem to provide an ideal arrangement for rotational reorientation of a reactive tetrahedral molecule within the active center. Prediction of organophosphonate stereochem-

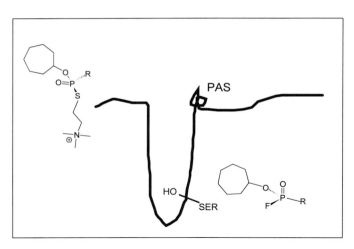

Figure 6. Orientation of *n*-alkylphosphonofluoridate and *n*-alkylphosphonyl thiocholine molecules with respect to active center gorge of AchE. Cationic phosphonates are proposed to undergo orientation at the mouth of the gorge through a mechanism of *electrostatic steering*. Such "steering" or orientation of cationic molecules is attributable to the strong electric dipole moment that emanates from the gorge. Uncharged molecules are allowed to diffuse to the base of the gorge and to undergo reorientation through hindered rotation.

istry of the final covalent conjugate is made difficult by the occurrence of *pseudo-rotation* and the subsequent scrambling of groups about the central P atom (Berman and Decker, 1989). Intramolecular rotational freedom of organophosphonates docked through P→O within the oxyanion hole contributes yet another mechanism that works to obscure configuration of the final organophosphono-enzyme adducts. That this must be so is evident since configuration of the final product is contingent on the face at which nucleophilic attack takes place and which, in turn, will be dictated by the extent of rotational reorientation. Reaction at more than one face of the tetrahedron, while resolvable in principle through analysis of the inhibition kinetics, might not be so in practice if the inhibition rates are not markedly different. Reaction at more than one face, if it occurs, might be evident from kinetic analysis of the reactivation and aging profiles of the final covalent conjugates.

Second, it is obviously not correct to view the leaving group as an inert entity that serves merely to undergo release following covalent reaction of the parent molecule. It appears quite clear that the charge and size of the leaving group play significant roles in governing the orientation, reaction pathway, and ultimate fate of organophosphonates during covalent reaction with AchE.

REFERENCES

Berman, H. A. and Leonard, K. J., 1989, Chiral reactions of acetylcholinesterase probed with enantiomeric methylphosphonothioates: Noncovalent determinants of enzyme chiral preference, *J. Biol. Chem.* 264:3942-3950.

Berman, H. A. and Decker, M. M., 1989, Chiral nature of covalent methylphosphonyl conjugates of acetylcholinesterase, *J. Biol. Chem.*, 264:3951-3956.

Harel, M., Sussman, J. L., Krejci, E., Bon, S., Chanal, P., Massoulie, J., and Silman, I., 1992, Conversion of acetylcholinesterase to butyrylcholinesterase: modeling and mutagenesis. *Proc. Nat. Acad. Sci., USA.*, 89:10827-10831.

Hart, G. J. and O'Brien, R. D., 1973, Recording spectrophotometric method for determination of dissociation and phosphorylation constants for the inhibition of acetylcholinesterase by organophosphates in the presence of substrate. *Biochemistry*, 12:2940-2945.

Sussman, J. L., Harel, M., Frolow, F., Oefner, C., Goldman, A., Toker, L., Silman, I., 1991, Atomic structure of acetylcholinesterase from *Torpedo californica*: a prototypic acetylcholine-binding protein, *Science*, 253:872-879.

Tan, R. C., Truong, T. N., McCammon, J. A., Sussman, J. L., 1993, Acetylcholinesterase: electrostatic steering increases the rate of ligand binding, *Biochemistry*, 32:401-403.

AMINO ACID RESIDUES IN ACETYLCHOLINESTERASE WHICH INFLUENCE FASCICULIN INHIBITION

Zoran Radić,[1]* Daniel M. Quinn,[2] Daniel C. Vellom,[1] Shelley Camp,[1] and Palmer Taylor[1]

[1] Department of Pharmacology
University of California San Diego
La Jolla, California 92093-0636
[2] Department of Chemistry
University of Iowa
Iowa City, Iowa 52242

INTRODUCTION

Fasciculin (FAS), a 61 amino acid peptide found in venom of green mambas, is a potent inhibitor of acetylcholinesterases from most species (EC 3.1.1.7)(AChEs) (Karlsson et al.,1984). The crystal structure of FAS (Le Du et al.,1992) reveals close structural features with a larger family of "three finger" snake toxins such as erabutoxin, cardiotoxins, bungarotoxin and cobratoxin. These toxins, however, do not inhibit AChE while FAS at low concentrations does not bind to acetylcholine receptors (Endo and Tamiya, 1987; Karlsson et al.,1984).

FAS inhibits AChEs at picomolar concentrations while closely related butyrylcholinesterases (EC 3.1.1.8) (BuChE) require FAS concentrations approaching millimolar (Karlsson et al.,1984; Radić et al.,1994). Two of the three domains of AChE and BuChE responsible for their different substrate and inhibitor specificities (Radić et al., 1993; Vellom et al., 1993; Harel et al., 1992; Loewenstein et al.,1993); the acyl pocket and choline binding site, are virtually inaccessible to FAS since they are located in a narrow active center gorge of the cholinesterases. FAS however could access a peripheral site on AChE. This site is also indicated by micromolar propidium and millimolar acetylthiocholine (ATCh) protecting the enzyme from FAS inhibition (Karlsson et al.,1984; Marchot et al.,1993). Thus FAS appears to be a highly specific peripheral site inhibitor of AChE. Since FAS binding is reversible, its high affinity is primarily a reflection of slow dissociation rates with half times in hours, while

*Visiting Fellow from the Institute for Medical Research and Occupational Health, University of Zagreb, 41000 Zagreb, Croatia.

Enzymes of the Cholinesterase Family, Edited by Daniel M. Quinn et al.
Plenum Press, New York, 1995

183

rates of association are only one to two orders of magnitude slower than the diffusion controlled rates (Marchot et al.,1993; Radić et al.,1994). The mechanism of FAS inhibition could involve physical or electrostatic obstruction of the entrance to the active center gorge and/or an allosteric influence on the active center affecting the commitment to catalysis. The capacity of FAS-AChE complex to react with DFP (Marchot et al.,1993) emphasizes importance of the latter mechanism. This presentation will consider possible mechanisms of FAS inhibition by measuring kinetic constants for inhibition of wild type and mutant cholinesterases by FAS. Secondly, the differential influence of FAS in affecting the interactions of the AChE-fasciculin complex with carboxylic acid esters, organophosphates and trifluoromethyl ketones will be examined.

MATERIAL AND METHODS

All enzymes used in this study were expressed in HEK-293 cells upon transfection of the cells with wildtype or mutant mouse AChE cDNA constructs as described earlier (Vellom et al.,1993; Radić et al.,1993). Purified and lyophilized fasciculin II was kindly provided by Dr. Carlos Cervenansky, Instituto de Investigaciones Biologicas, Montevideo, Uruguay. Concentrations of FAS stock solutions were determined by absorbance (ε_{276} = 4900 cm^{-1}). m-Tertbutyl trifluoroacetophenone (TFK) and m-trimethylammonium trifluoroacetophenone (TFK+) were synthesized as described earlier (Nair and Quinn, 1993; Nair et al., 1993).

Measurement of Enzyme Activity and Inhibition

Hydrolysis of ATCh, phenylacetate (PA) and p-nitrophenylacetate (PNPA) was measured spectrophotometrically using Ellman method (Ellman et al.,1961) for ATCh, or measurement of phenol or p-nitrophenol release at 270 nm and 405 nm, respectively. Kinetic constants for hydrolysis of the above substrates by wildtype AChE and BuChE and mutant AChEs were determined using nonlinear fit of the data to the following equation (modified from Aldridge and Reiner, 1972):

$$v = \frac{(1 + b \, S/K_{ss})}{(1 + S/K_{ss})} * \frac{V}{(1 + K_m/S)} \tag{1}$$

where S stands for substrate concentration, K_m and K_{ss} for Michaelis-Menten and substrate inhibition constants, and b for productivity of inhibitory SES complex expressed as a fraction of productivity of ES.

Reversible inhibition constants in picomolar range were determined upon overnight incubation of aliquots of enzymes with increasing concentrations of FAS, and fitting the remaining enzyme activities to the following equation:

$$v_{i\beta} = v_i[1 + \beta \, F/(K_i + v_i/k)] \tag{2}$$

where v_i is described by Ackermann and Potter equation (Ackermann and Potter, 1949) :

$$v_i = \ k/2\{(E - K_i - F) + [(K_i + F + E)^2 - 4FE]^{1/2}\}$$

and K_i is a reversible inhibition constant, equal to k_{-F}/k_F. Both equation (1) and equation (2) were derived from the following Scheme:

$$FE + S \underset{}{\overset{K_s}{\rightleftharpoons}} FES \xrightarrow{\beta\,k} FE + P$$

$$E + S \underset{}{\overset{K_s}{\rightleftharpoons}} ES \xrightarrow{k} E + P$$

$$SE + S \underset{}{\overset{K_s}{\rightleftharpoons}} SES \xrightarrow{b\,k} SE + P$$

under conditions of F=0 for equation (1) and $K_m < S < K_{ss}$ and $F \simeq E$ for equation (2).

Reversible inhibition constants in nanomolar range and higher were determined using Lineweaver-Burk or Hunter and Downs plots as described earlier (Radić et al., 1994).

Rate constants for FAS and trifluoroketone association and dissociation with enzymes were determined as described earlier (Radić et al., 1994).

RESULTS AND DISCUSSION

FAS inhibited all of the mutant enzymes tested but required a concentration range of eight orders of magnitude (Table 1), or 11 kcal, difference in stabilization energy in complexes with different mutants (Figure 1). The first five mutants listed in the table, including the ones where residues present in BuChE are introduced in AChE acyl pocket (Phe[295]Leu, Phe[297]Ile) or choline binding site (Tyr[337]Ala), have practically no effect on FAS inhibition. This provides additional evidence and therein defines several of the residues likely to constitute the peripheral anionic site, FAS being a peripheral site ligand (Karlsson et al., 1984; Marchot et al., 1993). Also visual inspection of AChE and FAS 3D structures indicates that FAS could hardly interact with residues

Table 1. Reversible inhibition constants for fasciculin 2 and wild type and mutant mouse cholinesterases[*]

Enzyme	K_i/pM
AChE wild type	2.3 ± 0.7
Phe[295]Leu	16 ± 8
Phe[297]Ile	57 ± 15
Phe[297]Tyr	7.9 ± 1.7
Phe[338]Gly	7.9 ± 1.7
Tyr[337]Ala	4.2 ± 1.7
Asp[74]Asn	43 ± 7
Tyr[124]Gln	248 ± 57
Tyr[72]Asn	$7,800 \pm 900$
Trp[286]Arg	$2,100,000 \pm 600,000$
Tyr[72]Asn/Tyr[124]Gln	$72,000 \pm 19,000$
Tyr[124]Gln/Trp[286]Arg	$8,500,000 \pm 3,100,000$
Tyr[72]Asn/Trp[286]Arg	$170,000,000 \pm 66,000$
Tyr[72]Asn/Tyr[124]Gln/Trp[286]Arg	$235,000,000 \pm 60,000,000$
BuChE wild type	$210,000,000 \pm 98,000,000$

[*]Data taken from Radić et al.,1994.

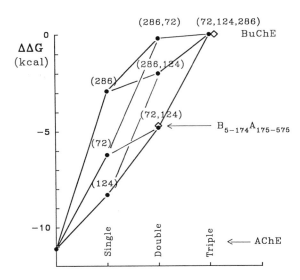

Figure 1. Diagram of the changes in ΔG for wild type, single, multiple mutant cholinesterases and AChE/BuChE chimera ($B_{5\text{-}174}A_{175\text{-}575}$). Free energy values were obtained using the formula $\Delta\Delta G = RT\ln(K_{i,mutant}/K_{i,AChEw.t.})$. Lines indicate paths of introduction of single site mutations. Diamonds indicate free energy differences for FAS complex with AChE w.t. and BuChE w.t. or $B_{5\text{-}174}A_{175\text{-}575}$ chimera.

of AChE acyl pocket or choline binding site. The elimination of the anionic residue at position 74 had surprisingly little effect on K_i keeping in mind the abundancy of cationic residues found on fasciculin and its net positive charge.

FAS is much better accommodated by the aromatic cluster of the peripheral site (Tyr[72], Tyr[124] and Trp[286]) as judged by sharp increase of FAS dissociation constants when these residues are substituted by aliphatic residues found in BuChE. Simultaneous substitution of all three peripheral site residues yields an inhibition constant virtually identical to the one found for BuChE indicating that the relatively small area enclosed by this aromatic cluster provides the majority of interaction energy in the AChE-FAS complex. Analysis of individual stabilizing contributions for the three residues (Figure 1) reveals that the stabilizing contribution least independent of other two residues comes from Tyr[124] since change in ΔG for introduction of Tyr[124]Gln mutation in Trp[286]Arg/Tyr[72]Asn double mutant is almost non-existent, and much smaller than the change in ΔG between Tyr[124]Gln and AChE w.t. Trp[286] and Tyr[72] therefore appear as major contributors in stabilizing AChE-FAS complex. The distribution of average contributions is 57%, 30% and 13% for Trp[286], Tyr[72] and Tyr[124], respectively. Substitution of Trp[286] with Ala instead of Arg, in the triple mutant, as found in human BuChE, yields FAS K_i of 7.4 uM and a more even balance of contributions of Trp[286], Tyr[72] and Tyr[124] in stabilizing the FAS complex where the first two residues provide approximately equal free energy contributions.

The AChE three dimensional structure reveals that this area is exposed on the outer surface of AChE (Sussman et al., 1991). At the entrance of the active center, a molecule of this dimension could easily interact with the peripheral site and block the gorge entrance completely; alternatively, its effect could be largely allosteric to influence the active center conformation or the charge distribution on the FAS-enzyme complex. Consistent with the DFP reactivity (Marchot et al., 1993), we find the larger phosphorates, paraoxon and echothiophate, will also react with the AChE-FAS complex, as will the trifluoroketones conjugate with the enzyme. In fact, a comparison of the rates of reaction of charged and uncharged ligands reveal an order of magnitude slower rates for charged ligands in the presence of FAS. Substitutions of Trp[86], a residue thought to provide stabilization of ATCh during catalysis (Sussman et al., 1991), by Tyr, Phe and Ala provide more insight to such a mechanism. Although these mutations do not change the affinity of fasciculin for the enzyme, fasciculin complexes with the AChE's mutated at position 86 do not show full inhibition of

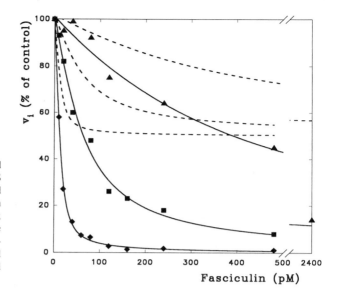

Figure 2. Inhibition of wild type and mutant AChE as a function of FAS concentration. Lines were calculated by nonlinear regression of equation (2). □ -□ wild type; ■-■ Asp[72]Asn; ▲-▲ Tyr[124]Gln. Dashed lines were generated using equation 2 and assuming a value 0.5 for β. For the solid lines β was calculated to be 0. K_i and all other constants are identical.

activity. In fact, the extent of inhibition depends on the substrate and is greater for charged, ATCh, than neutral, PhAc and PNPA, substrates. Under certain conditions with neutral substrates, fasciculin can actually activate the $W_{86}A$ enzyme.

A theoretical explanation of the data is that β in the Scheme I lies between 0 and 1 meaning that the AChE-FAS-substrate complex (FES) is in part productive. Theoretical curves in the Figure 2 are shown to illustrate a fit to experimental data where β = 0 , and dashed lines for the same K_i but where β = 0.5. This illustrates how changes in β affect the inhibition curves. These experiments indicate that substrates can also enter the gorge in the presence of FAS and suggest that the peripheral site to which fasciculin binds and Trp[86] may be allosterically linked.

Table 2 summarizes rates of association and dissociation of FAS and enzymes and substrate concentrations that affect association. It appears that mutants with lower affinity for FAS have higher dissociation rates. A similar trend was observed in Trp[86] mutants.

Table 2. Kinetic constants for rates of fasciculin association (k_F) and dissociation (k_{-F}) with acetylcholinesterases, and constants for reversible inhibition of AChEs by fasciculin (K_i) and ATCh (K_{ss})

Enzyme	k_F $10^8 M^{-1}min^{-1}$	k_{-F} $10^{-3}min^{-1}$	K_i(calc.)[a] pM	K_i(detd.)[b] pM	K_{ss}(FAS)[c] mM	K_{ss}(pS)[d] mM	b
AChE w.t.	7.0 ± 2.1	4.4 ± 2.0	6.3	2.3	25 ± 11	15 ± 2	0.23 ± 0.01
Tyr[337]Ala	6.9 ± 0.9	6.8	9.9	4.2	30 ± 10	29 ± 21	0.59 ± 0.08
Phe[338]Gly	ND	ND	—	7.9	16 ± 18	1300 ± 870	0
Phe[295]Leu	5.0 ± 1.8	3.7 ± 0.9	7.4	16	26 ± 9	62 ± 8	0
Phe[297]Ile	3.3 ± 1.2	6.0 ± 2.5	19	57	22 ± 4	43 ± 46	1.8 ± 0.3
Phe[297]Tyr	4.7 ± 2.8	3.1 ± 1.5	6.6	7.9	42 ± 14	460 ± 120	0
Asp[74]Asn	4.8 ± 1.3	40 ± 17	83	43	25 ± 7	530 ± 170	0
Tyr[124]Gln	4.9 ± 0.5	290 ± 60	590	248	21 ± 9	25 ± 13	0.35 ± 0.09

[a]$K_i= k_{-F}/k_F$.
[b]from Table 1
[c]from effect of ATCh on k_F.
[d]from nonlinear fit of enzyme activity at different ATCh concentrations to equation 1 (pS curves).

The dissociation constants for ATCh and FAS binding site resemble fairly well the K_{ss} values derived from substrate concentration dependence of enzyme hydrolysis except in four cases where the small extent of substrate inhibition precludes an accurate separation of K_{ss} and b. This leaves open the possibility that actual K_{ss} values have not changed and b values are close to 1. Since both high ATCh concentrations and FAS compete with propidium (Radić et al., 1991; Marchot et al., 1993) and with each another, one may conclude that they likely share a portion of the same binding site on AChE and that constitutes the peripheral site. However, substitution of the residues of the peripheral aromatic cluster with aliphatic residues did not result in loss of substrate inhibition (Radić et al., 1993) suggesting a relatively larger surface area at the opening of the AChE active center gorge for ligand binding to peripheral site(s).

In conclusion we note that FAS interacts with the peripheral anionic site of AChE characterized by an aromatic cluster of the three residues: Trp^{286}, Tyr^{72} and Tyr^{124}. The distribution of individual residue contributions to fasciculin binding is 57%, 30% and 13%, respectively. FAS does not completely obstruct the gorge entrance to the AChE active center, and in its mechanism of inhibition likely acts through influencing the orientation of Trp^{86}.

REFERENCES

Aldridge W.N. and Reiner E., Enzyme Inhibitors As Substrates. North-Holland, Amsterdam, 1972.

Ellman, G.L., Courtney, K.D., Andres Jr. V., & Featherstone, R.M. (1961) Biochem. Pharmacol.7, 88-95.

Endo, T., & Tamiya, N. (1987) Pharmacol. & Ther.34, 403-451.

Harel, M., Sussman, J.L., Krejci, E., Bon, S., Chanal, P., Massoulié, J., and Silman, I. (1992) Proc. Natl. Acad. Sci. U.S.A.89, 10827-10831.

Karlsson, E., Mbugua, P.M. & Rodriguez-Ithurralde, D. (1984) J.Physiol. (Paris) 79, 232-240.

Loewenstein, Y., Gnatt, A., Neville L.F., & Soreq, H. (1993) J. Mol. Biol. 234, 289-296.

le Du, M.H., Marchot, P., Bougis, P.E. & Fontecilla-Camps, J.C. (1992) J. Biol. Chem. 267, 22122-22130.

Nair, H.K., & Quinn, D.M. (1993) Bioorg. Med. Chem. Lett. 3, 2619-2622.

Nair, H.K., Lee, K., & Quinn, D.M. (1993) J. Am. Chem. Soc. 115, 9939-9941.

Marchot, P., Khelif, A., Ji, Y.H., Mansuelle, P. & Bougis, P.E. (1993) J. Biol. Chem. 268, 12458-12467.

Radić, Z., Reiner, E. & Taylor, P. (1991) Mol. Pharmacol. 39, 98-104.

Radić, Z., Pickering, N., Vellom, D.C., Camp, S. & Taylor, P. (1993) Biochemistry 32, 12074-12084.

Radić, Z., Duran, R., Vellom, D.C., Li, Y., Cervenansky, C. & Taylor, P. (1994) J. Biol. Chem. 269,11233-11239.

Sussman, J.L., Harel ,M., Frolow, F., Oefner, C., Goldman, A.,Toker, L. & Silman, I. (1991) Science 253, 872-879.

Vellom, D.C., Radić, Z., Li, Y., Pickering, N.A., Camp, S. & Taylor, P. (1993) Biochemistry 32, 12-17.

MOLECULAR ASPECTS OF CATALYSIS AND OF ALLOSTERIC REGULATION OF ACEYTLCHOLINESTERASES

A. Shafferman, A.Ordentlich, D. Barak, C. Kronman, N. Ariel, M. Leitner,
Y. Segall, A. Bromberg, S. Reuveny, D. Marcus, T. Bino, A. Lazar,
S. Cohen, and B.Velan

Israel Institute for Biological Research
Ness-Ziona, Israel

INTRODUCTION

Acetylcholinesterase is a serine hydrolase whose function at the cholinergic synapse, is the rapid hydrolysis of the neurotransmitter acetylcholine (ACh). The recently resolved 3D structure of *Torpedo californica* AChE (TcAChE) revealed a deep and narrow 'gorge', which penetrates halfway into the enzyme and contains the catalytic site at about 4Å from its base (Sussman *et al.,* 1991).The active center interacts with ACh through several subsites including the catalytic triad (Ser203*(200)*, His447*(440)* , Glu334*(327)*; Sussman *et al.*, 1991; Gibney *et al.*, 1990; Shafferman *et al.*, 1992a,b), the oxyanion hole (Gly121(*119*), Gly122(*120*), Ala204(*201*); Sussman *et al.*, 1991), the acyl pocket (Phe295 (*288*) and Phe297(*290*); Vellom *et al.*, 1993; Ordentlich *et al.*, 1993a).

The nature of the fourth element of the active center , the anionic subsite, is a matter of a longstanding controversy. One opinion, argued on the basis of the alleged presence of multiple negative charges in the active center, is that the "anionic subsite" is a true anionic locus (Quinn, 1987). The opposite view, based on the structure - activity studies with charged and noncharged substrates and inhibitors, suggested that the "anionic subsite" is in fact a trimethyl site, binding the ligands through hydrophobic interactions (Hassan *et al.*, 1980) or dispersive forces (Nair *et al.*, 1994). Herein we will provide evidence that the "anionic subsite" is not a true ionic site and that quaternary ammonium of ACh is stabilized by cation - π interactions with the indole moiety of Trp-86 at the active center.

AChE is an extremely efficient enzyme, believed to operate at, or near the diffusion control limit (Quinn, 1987). Recently, it was shown that in the structure of TcAChE an uneven spatial charge distribution results in a negative electrostatic potential that may be involved in the attraction of the positively charged substrate into the active center gorge of the enzyme (Tan *et al.,* 1993; Ripoll *et al.,* 1993). However, the bimolecular rate constants of various AChEs actually measured are about an order of magnitude lower than the expected rates of diffusion controlled reactions and furthermore they appear to be dependent on the molal

Enzymes of the Cholinesterase Family, Edited by Daniel M. Quinn et al.
Plenum Press, New York, 1995

volume of the various substrates rather than on charge (Cohen *et al.,* 1984). To further investigate the proposed contribution of electrostatic attraction to the rate of catalytic activity of AChE, we neutralized by mutagenesis up to seven of the surface negative charges, located near the rim of the active site gorge of AChE. We will show that while such substitutions have dramatic effects on the electrostatic potential, no major effects on the reactivity towards various substrates were observed.

A remarkable feature of AChE is its capacity to bind structurally diverse cationic ligands. Some of these ligands bind at the active center while others associate with a peripheral anionic site (PAS), remote from the active center (for reviews see Hucho *et al.,* 1991; Massoulié *et al.,* 1993, Taylor and Radić 1994). Allosteric modulation of AChE catalytic activity, through binding to the PAS, was first suggested by Changeux (1966). The resolution of the enzyme structure, by x-ray crystallography (Sussman *et al.,* 1991), did not yield any clues as to the mechanism of the allosteric effect but rather rendered it more enigmatic by placing the active site within a narrow gorge about 20Å away from the periphery. By employing site directed mutagenesis, molecular modeling techniques and fluorescence binding studies we generate information pertaining to the topography of the complexes of HuAChE with classical PAS ligands. We demonstrate that alternative residues participate in interaction with various PAS ligands and also provide some insight into the allosteric modulation of AChE activity.

RESULTS AND DISCUSSION

The Quaternary Ammonium Moiety of AChE Ligands Is Stabilized in the Active Center by Trp-86 through Cation-Π Interactions

Examination of the x-ray structure of TcAChE (Sussman *et al.,* 1991) and of the derived model of HuAChE (Barak *et al.,* 1992), reveals that the only negatively charged residue vicinal to the catalytic serine is Glu-202. The two other acidic residues: Asp-74 and Glu-450, located within the active site gorge, are 15.1Å and 8.9Å away from residue Ser-203 respectively and are therefore unlikely to participate in interactions of the "anionic subsite". Substitutions of Glu-202 of the human AChE (HuAChE) enzyme by neutral residues (Ala or Gln) or by Asp exerts a comparable effect on catalysis for both acetylthiocholine (ATC) and its noncharged isostere TB (Table 1). These results suggest that residue Glu-202 has no specific role in stabilizing positively charged substrates and therefore is not a part of the "anionic subsite". On the other hand, the kinetic data clearly indicate that the correct

Table 1. Kinetic constants for hydrolysis of ATC and TB by HuAChE and selected HuAChE mutants

HuAChE	Substrate	
	k^{app} (x10^{-6} M^{-1}x min^{-1})	
	ATC	TB
WT	2640	180
D74N	420	120
E202A	21	2
E202D	47	2
E202Q	273	33
E450A	27	2

Table 2. Kinetic constants for hydrolysis of ATC and TB by HuAChE and selected HuAChE mutants

	Substrate					
	K_m (mM)		k^{cat} ($\times 10^{-3} \times$ min^{-1})		k^{app} ($\times 10^{-6}$ M^{-1}xmin^{-1})	
HuAChE	ATC	TB	ATC	TB	ATC	TB
Wild type	0.14	0.30	370	55	2640	180
Y337A	0.14	0.30	100	20	710	67
F338A	0.30	0.30	170	13	570	43
W86F	0.80	0.30	200	47	260	160
W86E	90.0	0.36	40	7	0.44	20
W86A	93.3	0.44	80	18	0.86	40

positioning of the negative charge of Glu-202 plays an important role in the acylation step of the catalytic reaction as well as in phosphylation and "aging" (Shafferman 1992b, Vellom *et al.*, 1993,Ordentlich *et al.*, 1993b).

Unlike the indiscriminate effect on catalysis towards charged and noncharged substrate due to replacement of Glu-202, substitutions of Trp86 but not of Tyr337 or Phe338 affect differentially the hydrolytic activity for the two kinds of substrates. The bimolecular rate constant of the wild type enzyme for ATC is 20-fold higher than that for TB, yet in the HuAChE mutants carrying aliphatic residues at position 86 the selectivity is reversed and these enzymes show 50-fold higher reactivity for TB than that for ATC (Table 2). This reversal of selectivity towards the sterically identical noncharged substrate, and the fact that kinetic parameters for TB are only marginally affected by the various mutations, is a clear manifestation of the existence of a functional "anionic subsite" and of the role of residue Trp-86 in this subsite. Substitution of Trp-86 by the aliphatic residues alanine or glutamate, but not by phenylalanine, brings about 600-fold increase in the Michaelis Menten constant for the charged substrate ATC and an overall decrease of 3000-7000-fold in the apparent bimolecular rate constant (Table 2). These results underscore the importance of aromatic residue at position 86 in stabilizing the Michaelis-Menten complexes of HuAChE with charged substrates. Such conclusion is also supported by : a. The lack of measurable affinity of W86A and W86E mutants towards the charged active center inhibitor edrophonium; b.

Table 3. Compilation of the relative competitive inhibition constants of HuAChE and its mutants with active center and bifunctional bisquaternary ligands

	Relative Inhibition Constant (mutant/wt)			
	Ligand			
HuAChEs	Edrophonium	Decamethonium	BW284C51	Propidium
Wild Type*	1	1	1	1
W86A	>75000	15000	4000	620
W86E	>75000	8500	3900	416
W86F	63.3	83.3	70	5.3
F295A	0.5	0.1	0.4	0.6
Y337A	7.6	0.2	0.5	0.6
F338A	1	3	1	1

*Wild type HuAChE Ki values for edrophonium, decamethonium, propidium and BW284C51 are: 0.6, 6, 0.6 μM and 10 nM, respectively.

The 8500 and 15000-fold increase, relative to the wild type enzyme, in inhibition constant (Ki) value for decamethonium in the W86A and W86E enzymes respectively; c. The hundred fold higher affinity of W86F HuAChE towards edrophonium or the bisquternary inhibitors decamethonium and BW284C51, compared to either the W86A or the W86E enzymes (Table 3). In marked contrast, the nature of residue at position 86 has only a marginal contribution to the activity of HuAChE towards noncharged substrates like TB (Table 2) or noncharged inhibitors (not shown). Taken together these observations imply also that in the Michaelis Menten complexes, the orientations of the trimethyl ammonium and the 3,3-dimethylbutyl groups, of ATC and TB respectively, are not equivalent relative to Trp-86.

In conclusion, results from site directed mutagenesis studies of HuAChE and from x-ray crystallography of TcAChE (Sussman et al., 1991) and its complexes (Harel et al., 1993) provide compelling evidence for the presence of a specific site which stabilizes the quaternary ammonium groups of substrates and other ligands through cation-aromatic interactions, mainly with residue Trp-86, rather than through ionic interactions.

Electrostatic Attraction by Surface Charge Does Not Contribute to the Catalytic Efficiency of Acetylcholinesterase

The structures of acetylcholinesterases are characterized by a high net negative charge (e.g. -11e for HuAChE; -12e for bovine AChE or -14e for TcAChE) and by asymmetrical distribution of acidic and basic amino acids on the protein surface, with excess acidic residues in the "northern" hemisphere. Conservation of an excess negative charge, in the vicinity of the entrance to the active site gorge, taken together with the fact that the natural substrates bear a positive charge, may suggest that the electrostatic properties are a part of an evolutionary design for optimization of the catalytic efficiency of cholinesterases. Indeed, recent evaluation of the possible effects of the shape of TcAChE and its charge distribution, on the diffusion controlled rate of enzyme-substrate encounter by numerical brownian dynamics simulation, suggested an over 80-fold rate enhancement (Tan et al., 1993). According to this simulation, most of the enhancement (over 40 - fold) is due to electrostatic attraction, while a further minor effect could be attributed to electrostatic steering effects. Consequently, the uneven surface charge distribution and the resulting electrostatic potential, extending over the 'northern" hemisphere of AChEs, was proposed to contribute to the high catalytic efficiency of these enzymes (Ripoll et al., 1993; Tan et al., 1993).

This hypothesis was initially tested through simulated modulation of the charge distribution, in the "northern" hemisphere of HuAChE, and examination of the effects on the electrostatic potential as an indicator of the enzyme capacity for electrostatic attraction. Four out of the eleven acidic amino acids that constitute the net negative charge of the enzyme, are located within the active site gorge and do not contribute significantly to the electrostatic potential above the surface . Neutralization of remaining 7, or even 6 negative surface charges, practically abolishes the negative electrostatic potential over most of the "northern" hemisphere (Shafferman et al., 1994). Moreover, the direction of the electric field, which in the wild type HuAChE is aligned along the active site gorge axis, as observed also for TcAChE (Ripoll et al., 1993), changes by ~20° away from the z-axis. It was therefore expected that an actual neutralization of the surface negative charges should affect the bimolecular rate constant of the enzyme - substrate reaction, provided that it does indeed depend on electrostatic attraction. This was examined, through generation of 20 HuAChE enzymes, mutated in up to seven acidic amino acids, vicinal to the rim of the active site gorge. In marked contrast to the shrinking of the electrostatic potential, the kinetic constants for reactivity of the mutants towards charged substrate and inhibitor are practically invariant, indicating that the electrostatic attraction does not contribute to the reaction rates.

The values of the apparent first order rate constant of catalysis (kcat) of both ATC and TB are invariant for all the HuAChE mutants and are not sensitive to the ionic strength of the medium. This finding indicates that the active sites of the various mutants are effectively shielded from surface charges. Furthermore, it suggests that long-range electrostatic interactions, due to the surface charges of AChE, do not participate in stabilization of transition states in the catalytic process.The lack of contribution of electrostatic attraction to the catalytic rate, together with the nature of its dependence on ionic strength suggest that the rate of enzyme - substrate reaction is not diffusion controlled.

Our results, showing unequivocally the lack of contribution of electrostatic attraction to AChE catalytic properties (Shafferman et al., 1994), underscore the enigma of the uneven charge distribution, conserved throughout the cholinesterase family. It is possible that the electrostatic attraction in aqueous solutions is cryptic while in the viscous milieu of the synaptic cleft it becomes operational. Alternatively, the uneven charge distribution may be related to the noncatalytic functions of cholinesterases such as non - synaptic neuronal function, development of the nervous system or cell adhesion.

Identification of Peripheral Anionic Site (PAS) Elements and Their Involvement in Allosteric Modulation of AChE Activity

The functional significance of PAS is a controversial issue. This site was implicated in the catalytic pathway of acetylcholine (ACh) hydrolysis, in ionic strength monitoring and in substrate inhibition. A decrease in affinity for a selective PAS ligand (propidium) and a concomitant loss of substrate inhibition, were observed for certain human AChE mutants (Shafferman et al., 1992b). Determination of the structural parameters affecting ligand-affinity for the PAS can have important implications for the design of anti - AChE drugs. A number of such drugs are under clinical trials while additional AChE inhibitors are investigated as a possible treatment for Alzheimer's disease. Site directed labelling (Weise et al., 1990) and mutagenesis studies (Shafferman et al., 1992a,b; Ordentlich et al., 1993) allow for an approximate localization of PAS at or near the rim of the active center 'gorge'. Such topography is compatible with the suggestion that bisquaternary ligands, typified by decamethonium, bind to the enzyme by bridging the active and the peripheral sites (Krupka, 1966).

Several of the residues constituting the PAS in HuAChE were recently identified by kinetic studies of 19 single and multiple HuAChE mutants, fluorescence binding studies and by molecular modeling (Barak et al., 1994). Mutated enzymes were analyzed with three structurally distinct positively charged PAS ligands- propidium, decamethonium and di(p-allyl-N-dimethylaminophenyl) pentane-3-one (BW284C51), as well as with the selective active center inhibitor - edrophonium (Fig. 1). Single site mutations of residues Tyr-72, Tyr-124, Glu-285, Trp-286 and Tyr-341, resulted in up to 10-fold increase in inhibition constants for PAS ligands, whereas for multiple site mutants up to 400-fold increase was observed. The sixth PAS element residue Asp-74, is unique in its ability to affect interactions at both the active site and the PAS (Shafferman, et al., 1992b) as demonstrated by the several hundred-fold increase in Ki for D74N inhibition by the bisquaternary ligands decamethonium and BW284C51. Cooperativity of multiple mutations of PAS residues including Asp-74 may lead to over 10,000 fold increase in inhibition constants. Based on these studies, singular molecular models for the various HuAChE-inhibitor complexes were defined. Yet, for the decamethonium complex two distinct conformations were generated, accommodating the quaternary ammonium group by interactions with either Trp-286 or with Tyr-341. We propose that the PAS consists of a number of binding sites, close to the entrance of the active site gorge, sharing residues Asp-74 and Trp-286 as a common core. This

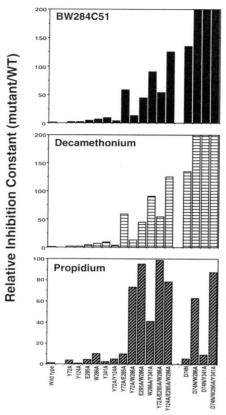

Figure 1. The relative inhibition constants of the active center inhibitor edrophonium and the three PAS inhibitors propidium, decamethonium and BW284C51 determined for single or multiple rHuAChE mutants.

functional degeneracy is a result of the ability of the Trp-286 indole moiety to interact either via stacking, aromatic-aromatic or via π-cation attractions and the involvement of the carboxylate of Asp-74 in charge-charge or H-bond interactions.

The influence of Asp-74 substitution, on inhibition by both peripheral site and active center specific ligands and in particular its striking effect on ligands bridging the two sites (Fig. 1), is consistent with its proposed role (Shafferman et al., 1992b) in the relay of allosteric signals from the periphery. Interestingly, replacement of residues Tyr-72, Tyr-124, Trp-286, Glu-285 and Tyr-341 generates AChE molecules in which substrate inhibition is affected to about the same extent as the inhibition constants for PAS ligands (up to 10-fold). Yet, unlike for inhibition by PAS ligands, no synergistic effects on substrate inhibition were observed, for the multiple mutants studied. Similar results were also reported by Radić et al (1994). This may indicate that the substrate interacts at multiple locations within the PAS array. Existence of two binding loci within this array, were suggested above for the quaternary ammonium group of decamethonium, which structurally resembles the cationic head of the substrate.

Compared to the wild type enzyme the hydrolysis of ATC by the W86A mutated enzyme is almost completely refractive to inhibition by the PAS inhibitor propidium (Table 3). These observations are quite surprising in view of the location of Trp-86 deep inside in the active site gorge, 15Å away from the surface of the enzyme and the binding site of propidium. To account for this and other results we proposed a functional cross-talk between Trp-86 and the peripheral anionic site (Shafferman et. al. 1992b,Ordentlich et al 1993a). This conclusion is consistent with spectroscopic studies by Berman *et al.*, (1981), and Berman and Nowak (1992) that demonstrated that occupation of the peripheral anionic site affects the conformation of the active center. More recent molecular models of the wild type, and of some mutated HuAChE enzymes reveal that the side chain of residue Trp-86 can occupy two conformational states. It is proposed that allosteric modulation of AChE activity, induced by binding to the peripheral anionic sites, proceeds through conformational transition of Trp-86 from the state of a functional "anionic subsite" to one that restricts access of substrates to the active center. Although the conformational flexibility of Trp-86 and its effects on the catalytic activity provide a possible mechanism for the "cross - talk" between the peripheral sites and the active center, the relay path of the allosteric signal still remains a matter for speculations.

ACKNOWLEDGMENT

This work was supported by contracts DAMD17-89-C-9117 and DAMD17-93-C-3042 from the United States Army Research and Development Command (to A.S.).

REFERENCES

Barak, D., Ariel, N., Velan, B.,and Shafferman, A. (1992). Molecular models for human AChE and its phosphonylation products. In: Multidisciplinary Approaches to Cholinesterase Functions. (Shafferman A. and Velan B. Eds). Plenum Pub. Co., London, 195-199.

Barak, D., Kronman,C., Ordentlich, A., Ariel, N., Bromberg, A., Marcus, D., Lazar, A., Velan, B., and Shafferman, A. (1994). Acetylcholinesterase peripheral anionic site degeneracy conferred by amino acid arrays sharing a common core. J.Biol.Chem. 264, 6296-6305.

Berman, H. A., Becktel, W., and Taylor, P. (1981) Spectroscopic studies on acetylcholinesterase: influence of peripheral-site occupation on active-center conformation. Biochemistry 20, 4803-4810.

Berman, H., A. and Nowak, M., W. (1992). influence of ionic composition of the medium on acetylcholinesterase conformation. In: Multidisciplinary Approaches to Cholinesterase Functions. (Shafferman A. and Velan B. Eds). Plenum Pub. Co., London . 149-156.

Changeux , J. P. (1966). Responses of acetylcholinesterase from Torpedo marmorata to salts and curarizing drugs. Mol.Pharmacol., 2, 369-392.

Cohen, S. G., Elkind, J.L., Chishti, S.B., Giner, J-L.P., Reese, H. and Cohen, J.B. (1984). Effects of volume and surface property in hydrolysis by acetylcholinesterase. The trimethyl site. J. Med. Chem. 27, 1643-1647.

Gibney, G., Camp, S., Dionne, M., MacPhee-Quigley, K. and Taylor, P. (1990). Mutagenesis of essential functional residues in acetylcholinesterase. Proc. Natl. Acad. Sci. USA. 87, 7546-7550.

Harel, M., Schalk, I., Ehret-Sabatier, L., Bouet, F., Goeldner, M., Hirth, C., Axelsen, P.H., Silman, I. and Sussman, J.L. (1993). Quaternary ligand binding to aromatic residues in the active-site gorge of acetylcholinesterase. Proc. Natl. Acad. Sci. USA 90, 9031-9035.

Hassan, F. B., Cohen, S.G. and Cohen, J.B. (1980). Hydrolysis by acetylcholinesterase. J. Biol. Chem. 255, 3898-3904.

Hucho, F., Jarv, J., and Weise, C. (1991). Substrate-binding sites in acetylcholinesterase. Trends Parmacol. Sci. 12, 422-427.

Krupka, R. M. (1966). Chemical structure and function of the active center of acetylcholinesterase. Biochemistry 5, 1988-1998.

Massoulie, J., Sussman, J., Bon, S. and Silman, I. (1993). Structure and function of acetylcholinesterase and butyrylcholinesterase. Progress in Brain Research 98, 139-146.

Nair, H. K., Seravalli, J., Arbuckle, T. and Quinn, D.M. (1994). Molecular recognition in acetylcholinesterase catalysis: free-energy correlations for substrate turnover and inhibition by trifluoroketone transition state analogs. Biochemistry 33, 8566-8576.

Ordentlich A., Barak, D., Kronman, C., Flashner, Y., Leitner, M.,Segall, Y., Ariel, N., Cohen, S., Velan, B.,and Shafferman, A. (1993a). Dissection of the human acetylcholinesterase active center - determinants of substrate specificity: Identification of residues constituting the anionic site, the hydrophobic site, and the acyl pocket. J.Biol.Chem. 268, 17083-17095.

Ordentlich A., Kronman, C., Barak, D. , Stein, D., Ariel, N., Marcus, D., Velan, B., and Shafferman, A. (1993b). Engineering resistance to 'aging' in phosphylated human acetylcholinesterase - role of hydrogen bond network in the active center. FEBS Lett. 334, 215-220.

Quinn, D. M. (1987). AChE : Enzyme structure, reaction dynamics and virtual transition states. Chem. Rev. 87, 955-979.

Radić, Z., Duran, R., Vellom, D.C., Li, Y., Cervenansky, C. and Taylor, P. (1994). Site of fasciculin interaction with acetylcholinesterase. J. Biol. Chem. 269, 11233-11239.

Ripoll, D. L., Faerman, C.H., Axelsen, P.H., Silman, I. and Sussman, J.L. (1993). An electrostatic mechanism for substrate guidance down the aromatic gorge of acetylcholinesterase. Proc. Natl. Acad. Sci. USA 90, 5128-5132. .

Shafferman, A., Kronman, C., Flashner, Y., Leitner, S., Grosfeld, H., Ordentlich, A., Gozes, Y., Cohen, S., Ariel, N., Barak, D., Harel, M., Silman, I., Sussman, J.L. and Velan, B., (1992a). Mutagenesis of human acetylcholinesterase. J. Biol. Chem. 267, 17640- 17648.

Shafferman, A., Velan, B., Ordentlich, A., Kronman, C., Grosfeld, H., Leitner, M., Flashner, Y., Cohen, S., Barak, D., and Ariel, N. (1992b). Substrate inhibition of acetylcholinesterase: residues affecting signal transduction from the surface to the catalytic center. EMBO J. 11, 3561-3568.

Shafferman, A., Ordentlich, A., Barak, D., Kronman, C., Ber, R., Bino, T., Ariel, N., Osman, R. and Velan, B. (1994). Electrostatic attraction by surface charge does not contribute to the catalytic efficiency of acetylcholinesterase. EMBO J. 13, 3448-3455.

Sussman, J.L, Harel, M., Frolow, F.,Oefner, C., Goldman, A. and Silman I. (1991). Atomic resolution of acetylcholinesterase from Torpedo californica: A prototypic acetlycholine binding protein. Science 253, 872-879.

Tan, R. C., Truong, T.N., McCammon, J.A. and Sussman, J. (1993). Acetylcholinesterase: electrostatic steering increases the rate of ligand binding. Biochemistry 32, 401-403.

Taylor, P. and Radić, Z. (1994). The cholinesterases: From genes to proteins. Annu. Rev. Pharmac. Toxicol. 34, 281-320.

Vellom, D. C., Radic, Z., Li, Y., Pickering, N.A., Camp, S. and Taylor, P. (1993). Amino acid residues controlling acetylcholinesterase and butyrylcholinesterase specificity. Biochemistry 32, 12-17.

Weise, C., Kreienkamp, H-J., Raba, R., Pedak, A., Aaviksaar, A. and Hucho, F. (1990). Anionic subsites of the acetylcholinesterase from Torpedo californica : affinity labelling with the cationic reagent N,N-di-methyl-2-phenyl-aziridinium. EMBO J. 9, 3885-3888.

STRUCTURAL DETERMINANTS OF FASCICULIN SPECIFICITY FOR ACETYLCHOLINESTERASE

Pascale Marchot,[1,2] Shelley Camp,[1] Zoran Radić,[1] Pierre E. Bougis,[2] and Palmer Taylor[1]

[1] Department of Pharmacology
School of Medicine, University of California, San Diego
La Jolla, CA 92093-0636
[2] CNRS Unité de Recherche Associée 1455
Université d'Aix-Marseille II, Faculté de Médecine Secteur Nord
13916 Marseille Cedex 20, France

INTRODUCTION

Fasciculins are the only known peptide inhibitors of acetylcholinesterase (AChE) with a high degree of selectivity. They are found in mamba snake venoms and have been shown to display powerful inhibitory activity toward mammalian and fish AChE. To date, four iso-fasciculins have been characterized: fasciculin 1 (FAS1) and fasciculin 2 (FAS2) from the venom of *Dendroaspis angusticeps* (Rodriguez-Ithurralde et al., 1983), toxin C from the venom of *D. polylepis* (Joubert and Taljaard, 1978), and fasciculin 3 (FAS3) from a venom of *D. viridis* (Marchot et al., 1993). The early pharmacological and biochemical studies of FAS2, carried out both *in vivo* and *in vitro* on various AChE-containing tissues (Karlsson et al., 1984; Lin et al., 1987), showed that i) FAS2 inhibits several (but not all) AChEs from different sources, ii) inhibition is of pseudo-irreversible type with Ki values of about 10^{-11} M, iii) FAS2 is able to displace propidium, known as a specific probe for a peripheral anionic site of AChE (See Harvey et al., 1984 for a review).

Fasciculins are 61 amino acid peptides containing four disulfide bridges. They belong to the family of structurally related three-fingered peptidic toxins from *Elapidae* venoms, which includes the nicotinic receptor blocking α-neurotoxins and the cardiotoxins. FAS1 and FAS2 differ by a single substitution, Tyr[47]Asn. Although isolated from the venom of two different *Dendroaspis* species, FAS3 and toxin C are identical. They differ from FAS1 by three substitutions: Met[2]Ile, Thr[15]Lys, and Asn[16]Asp. These substitutions are indicated by a star in the sequences aligned in Figure 1.

To examine further the fasciculin-AChE interactions at a molecular level, the following approaches have been taken.

Enzymes of the Cholinesterase Family, Edited by Daniel M. Quinn et al.
Plenum Press, New York, 1995

```
Res #: 1          1         2         3         4         5         6
                  0         0         0         0         0         0
FAS1   TMCYSHTTTSRAILTNCGENSCYRKSRRHPPKMVLGRGCGCPPGDDYLEVKCCTSPDKCNY
                                                          *
FAS2   TMCYSHTTTSRAILTNCGENSCYRKSRRHPPKMVLGRGCGCPPGDDNLEVKCCTSPDKCNY
        *         **
FAS3   TICYSHTTTSRAILKDCGENSCYRKSRRHPPKMVLGRGCGCPPGDDYLEVKCCTSPDKCNY
```

Figure 1

STRUCTURAL DETERMINANTS ON THE AChE MOLECULE RESPONSIBLE FOR SENSITIVITY TO FASCICULINS

Radiolabeling of Fasciculin and Direct Binding to Membrane Ache

Fully active, specific probes of high specific activity (1500 Ci mmole^{-1}) and exceptionally high affinity for an AChE peripheral site were obtained by enzymatic radioiodination of fasciculins (Marchot et al., 1993). FAS3 was monoiodinated on both Tyr^{47} and Tyr^{61}, while FAS2 was labeled on Tyr^{61} only. The relationships between binding site occupancy by ^{125}I-FAS3 and AChE inhibition were examined through kinetic studies conducted with ^{125}I-FAS3 and FAS3 on binding to synaptosomal rat brain AChE and on AChE inhibition, respectively. The rate of FAS3 association to AChE was found not to be modified upon iodination, and binding and inhibition were observed as concomitant processes in the time frame of the experiment. The kinetics of ^{125}I-FAS3 dissociation clearly was extremely slow ($t_{1/2}$ = 48 h), consistent with previous reports on the apparent irreversibility of the action of fasciculins on various AChE preparations (Anderson et al., 1985; Lin et al., 1987). A very low dissociation constant (Kd = 0.4 pM) for the ^{125}I-FAS3/AChE complex was calculated from the kinetic constants. The same Kd value was determined for the FAS3/AChE complex by quantifying, with ^{125}I-FAS3, the initial rate of ^{125}I-FAS3 binding over a range of reactant concentrations. It was therefore confirmed that the affinity of FAS3 for AChE was also not altered upon iodination.

In order to delineate further the topology of the fasciculin binding site on AChE, competition with the rate of fasciculin binding was measured for several AChE inhibitors or effectors. DFP and paraoxon failed to prevent ^{125}I-FAS3 from binding to its site, but inhibitors (BW284C51, d-tubocurarine, and propidium iodide) and effectors (MgCl$_2$, CaCl$_2$) presumably acting through AChE peripheral anionic sites were competitive with ^{125}I-FAS3 binding. As well, ^{125}I-FAS3 binding was prevented by the high acetylthiocholine concentrations exhibiting inhibition of the enzyme. Fasciculin, therefore, binds at a peripheral site on the enzyme distinct from the active center. This site is, at least partly, common with the binding sites of some cationic inhibitors and with the additional site at which the substrate binds at concentrations showing excess-substrate inhibition of AChE. In contrast, the initial rate of ^{125}I-FAS3 binding was found to increase up to 130% in the presence of O-ethyl-S[2-(diisopropylamino)ethyl]-methylphosphonothionate (MPT) suggesting an induced allosteric change in the conformation of the fasciculin binding site upon binding of MPT, rather than direct competition at overlapping binding sites. *The* AChE peripheral site, indeed, is likely to consist of a matrix of partially overlapping loci which can accommodate the diversity of the known peripheral ligands. Mutually exclusive binding between fasciculins and peripheral site ligands is of interest, especially since fasciculins may be expected to cover a larger area of the AChE surface than the low molecular weight ligands.

Titration of rat brain AChE was performed by direct binding of either ^{125}I-FAS3 or ^{3}H-DFP, and by recording the residual AChE activity in the presence of FAS3. The number of ^{125}I-FAS3 binding sites (equal to the number of FAS3 binding sites) was found unchanged after incubation of the AChE preparation with MPT. Also, neither the number of ^{3}H-DFP binding sites, nor the kinetics of its association to AChE were modified after incubation of the AChE preparation with FAS3. Bound fasciculin, therefore, fails to prevent subsequent occupation of the esteratic subsite of the AChE catalytic site by organophosphorus compounds. As well, the bound organophosphorus compound does not prevent occupation of the peripheral site of AChE by fasciculin. This result indicates that a ternary complex FAS3/AChE/DFP forms with a 1:1:1 ratio. Since DFP phosphorylation of the AChE catalytic serine can still occur on the FAS/AChE complex, AChE inhibition due to fasciculin binding may be thought to be allosterically linked to the anionic subsite of the AChE catalytic site (cf. the article of Radić et al., this volume).

Inhibition of Catalytic Activity by Fasciculin and Direct Binding of Radiolabeled Fasciculin to Soluble Ache

Fasciculins inhibit mammalian and fish AChE, but insect, avian, and some snake AChEs as well as the closely related butyrylcholinesterases (BuChE) from several mammalian species are largely resistant to fasciculin inhibition (Cerveñansky et al., 1991a). The availability of a large number of cholinesterase sequences (Cygler et al., 1993), the capacity to change individual residues by site-specific mutagenesis, coupled with the knowledge of the *T. californica* AChE crystal structure (Sussman et al., 1991), provide a useful framework for assigning the structural determinants of the AChE molecule responsible for sensitivity to fasciculins. The interaction of FAS2 with AChE-BuChE chimera (Vellom et al., 1993) and with several mutant forms of recombinant DNA-derived AChE from mouse, modified in the three domains of the molecule responsible for the selectivity of AChE and BuChE inhibitors (Radić et al., 1993) has been examined (Radić et al., 1994). Fasciculin appears to interact with a cluster of aromatic residues, composed of Trp[286], Tyr[72], and Tyr[124] in mammalian AChE and located near the rim of the active center gorge of the enzyme. Though one or two of these three residues are also found in some other cholinesterase species, the complete cluster is found only in the sequence of those AChEs which are susceptible to low concentrations of fasciculin. Substitution of all 3 residues with the corresponding side-chains in BuChE reduced the affinity of fasciculin (Ki) by about 8 orders of magnitude, with a single mutation, W[286]R, being itself responsible for 6 orders of magnitude decrease in affinity.

Recombinant DNA-derived AChE from mouse, as well as the derived mutant forms, are soluble species. The centrifugation technique that was used previously with synaptosomal rat brain AChE (Marchot et al., 1993), could therefore not be used for examining the direct binding of radiolabeled fasciculin to these soluble purified AChEs. A rapid filtration assay was designed to allow analysis of picomolar concentrations of soluble AChE, compatible with the low dissociation constants, and to separate efficiently the bound from free ^{125}I-FAS. A special effort has been made to optimize recovery of the ^{125}I-FAS/AChE complex on the filter. The best results were obtained with a cationic polysulfone-based membrane initially designed for oligonucleotide blots (Gelman Sciences). Equilibrium binding of ^{125}I-FAS2 was performed on both purified and non-purified recombinant AChE from mouse within a wide range of low ligand concentrations. Non-specific ^{125}I-FAS2 binding to the soluble preparations was found to be negligible when compared to the binding on the filters; it was therefore determined by omitting the enzyme from a parallel assay. For both purified and non-purified AChE, the specific ^{125}I-FAS2 binding (about 50% of the total ligand binding) was saturable and was characterized by a linear Scatchard plot, indicating a single class of non-interacting binding sites for ^{125}I-FAS2. Recovery of the enzyme complexed to ^{125}I-FAS2 (Bmax) was found to be about 50% of the initial AChE concentration, determined independently from the titration of the enzyme solution with 7-[[(methylethoxyl) phosphinyl]oxyl]-1-methylquinolinium (MEPQ). The equilibrium constant (Kd) was found to be in the 10^{-11} range, consistent with the Ki values determined previously with FAS2 (Radić et al., 1994). Fairly similar results were obtained with the purified and non-purified enzymes, indicating that the cell medium in which the enzyme is expressed does not contain other fasciculin-binding proteins. This method should therefore prove useful in the study of non-purified mutant forms of soluble AChE, devoid of catalytic activity but still recognized by fasciculins.

STRUCTURAL DETERMINANTS OF THE FASCICULIN MOLECULE RESPONSIBLE FOR SPECIFICITY TOWARD ACETYLCHOLINESTERASE

Structure of the Fasciculin Molecule

The crystal structure of fasciculins (le Du et al., 1992) is similar to those of the short α-neurotoxins and of cardiotoxins, with an eccentric dense core containing the disulfide bridges and three long loops disposed as the central fingers of a hand (referred to as loop I, II, and III, proceeding from the amino- to the carboxy-terminal ends of the peptidic chain). A fine comparison of the three prototypic toxin types, however, revealed structural features in the FAS1 molecule that distinguish it from the other two molecules. The core region, which is defined as a continuous stretch of conserved residues, is very similar to that of erabutoxin b. In contrast, the orientation of the three loops gives to FAS1 the same concavity as that of cardiotoxin $V^{II}4$, which is opposite to that of erabutoxin. Also, loops II and III of the FAS1 molecule are stabilized by a set of interactions which lack counterparts in the other two molecules.

The predominance of basic amino acids in the fasciculin molecule, together with evidence for its interaction with an anionic site on AChE, suggested an involvement of the positively charged side chains of the peptide in its interaction with the enzyme. In FAS1, several Arg and Lys side chains are clustered around the tip of loop II (Arg[27], Arg[28], and Lys[32]) and on the concave region of the molecule (Arg[24], Arg[37], and Lys[51]). Either or both of these clusters could contribute to the specificity of fasciculins. Comparison of the naturally occuring fasciculin variants for their respective affinities for AChE provides additional information on structure-function relationships of fasciculins. The Kd values of complexes formed with rat brain AChE and FAS1 (14 pM) or FAS2 (25 pM) were found to be significantly higher than that of the complex formed with FAS3 (0.4 pM) (Marchot et al., 1993). The single substitution which distinguishes FAS1 from FAS2 may account for their two-fold difference in affinity, the Tyr residue enhancing slightly the stability of the FAS1/AChE complex. In contrast, two of the three substitutions which distinguish FAS3 from FAS1 confer not only a local difference in charge but also the extended side chain of the Lys residue. These differences most likely account for the 35-fold higher affinity of FAS3. For this difference in affinity to occur irrespective of the AChE species involved would be of interest. These results, however, together with the fact that only partial loss of activity was observed upon chemical modification of Arg and Lys residues of FAS2 (Cerveñansky et al., 1991b), support the hypothesis of multipoint attachment of fasciculins on AChE, by which several residues of the fasciculin molecule simultaneously contribute to the interaction with the enzyme or/and the stability of the complex. Such a mode of attachment has already been proposed for the interaction of structurally related snake venom α-neurotoxins for their binding sites on the nicotinic receptor (Faure et al., 1983; Martin et al., 1983).

Expression of Recombinant Fasciculin in Mammalian Cells

The limited application of chemical modification and cross-linking led to considering site-specific mutagenesis as the most suitable technique for investigating further the structural determinants on the fasciculin molecule responsible for specificity for AChE. An encoding sequence for FAS2 was designed with codons of high frequency usage in mammalian systems and restriction sites were placed at convenient locations within the open reading frame and flanking regions. To insure secretion and processing, a sequence encoding the leader peptide of erabutoxin (Tamiya et al., 1985) was joined at the 5' end. The DNA was synthesized as two sets of double stranded oligonucleotides (~ 130 bp in length) ligated together, and cloned into the expression vector pGS. The pGS-FAS2 plasmid was transfected into CHO-K1 cells by standard calcium phosphate co-precipitation techniques with glycerol shock. Selection for CHO cells which had successfully integrated the fasciculin and the adjacent glutamine synthetase genes was done in glutamine-free medium (Ultraculture, Bio-Whittiker) in the presence of methionine sulfoximine. Individual colonies were selected and assayed for fasciculin activity by monitoring spectrophotometrically their inhibition capacity for mouse AChE. Clones which produced the highest fasciculin activity were then further amplified with methionine sulfoximine at higher concentrations.

Recombinant FAS2 (FAS2R), up to 250 pmol ml^{-1}, was expressed by the post-confluent cells within two weeks and remained fairly constant for many weeks. FAS2R was secreted into the culture medium which also contained, among other proteins, up to 3.6 mg of BSA ml^{-1}. The medium (10 ml per 10 cm plate) was collected and replaced every 3-4 days, extensively dialyzed against ammonium acetate, filtered to achieve sterility, and then stored at 4^{0}C in the presence of sodium azide. Purification of FAS2R was performed by FPLC using cation exchange and size exclusion chromatography. The first FAS2R-enriched fraction was concentrated from up to 500 ml of recombinant medium with more than 90% of recovery of the initial FAS2 activity, and was desorbed from the column with a salt concentration gradient. A 440-fold purification (in pmol of FAS2 *per* mg of protein) was reached at this step. After lyophilization, this fraction was subjected to size-exclusion chromatography, which led to the removal of most of the residual contaminating BSA (eluting as a major fraction) and to a 1500-fold purification. A final ion-exchange chromatography, performed under isocratic conditions, was required to remove minor contaminants. From the FAS2R-containing culture medium to the final product, the overall purification was about 2200-fold, with a final recovery of about 50% of the initial amount of FAS2R subjected to chromatography. FAS2 activity was monitored by microtitration throughout the chromatographic procedure, by recording the residual activity of recombinant wild-type mouse AChE. Homogeneity of FAS2R was examined by SDS-PAGE on a 20% acrylamide gel under reducing conditions. Specific activity of FAS2R (in pmoles FAS2 *per* pmoles FAS2R) was recorded on a sample previously quantified by UV absorbancy or/and amino acid analysis.

CONCLUSIONS

The tools are now available to shed new light on the molecular bases of the fasciculin-AChE interaction. The availability of i) numerous cholinesterase sequences with different fasciculin binding capacities, ii) the AChE and fasciculin crystal sructures, iii) site-specific mutagenesis techniques, which have been already extensively employed for recombinant AChE and can now be applied to recombinant fasciculins, and iv) a large panel of biochemical assays allowing the monitoring of AChE inhibition by fasciculins or/and direct binding of fasciculins to AChE, provide an ideal framework for investigating further the structural determinants of the fasciculin molecule responsible for its specificity, as well as the molecular mechanism of allosteric control of catalysis induced upon association of fasciculin with the peripheral site of AChE.

ACKNOWLEDGMENTS

We wish to thank Drs. Tyler White, John Lewicki, Barbara Cordell, and Andy Lin at Scios Nova Inc. for the gift of the pGS expression vector, and Mr. Kael Duprey for skillful assistance in electrophoresis and microtitration assays. Parts of this work were supported by the CNRS and NATO (to P.M.) and NIH (to P.T.).

REFERENCES

Cerveñansky, C, Dajas, F, Harvey, A.L., & Karlsson, E. (1991a) in *Snake Toxins* (Harvey, A.L., ed) pp. 303-321, Pergamon Press, Inc., New York
Cerveñansky, C., Engström, A., & Karlsson, E. (1991b) *Toxicon 29*, 1163-1164.
Cygler, M., Schrag, J.D., Sussman, J.L., Harel, M., Silman, I., Gentry, M.K., & Doctor, B.P. (1993) *Protein Science 2*, 366-382
Faure, G., Boulain,, J.C., Bouet, F., Montenay-Garestier, T., Fromageot, P., & Menez, A. (1983) *Biochemistry 22*, 2068-2076
Harvey, A.L., Anderson, A.J., Mbugua, P.M., & Karlsson, E. (1984) *J. Toxicol. Toxin Rev. 3*, 91-137
Joubert, F.J., & Taljaard, N. (1978) *South Afr. J. Chem. 31*, 107-110
Karlsson, E., Mbugua, P.M., & Rodriguez-Ithurralde, D. (1984) *J. Physiol. (Paris) 79*, 232-240
le Du, M.H., Marchot, P., Bougis, P.E., & Fontecilla-Camps, J.C. (1992) *J. Biol. Chem. 267*, 22122-22130
Lin, W.W., Lee, C.Y., Carlsson, F.H.H., & Joubert, F.J. (1987) *Asia Pac. J. Pharmacol. 2*, 79-85
Marchot, P., Khelif, A., Ji, Y.H., Mansuelle, P., & Bougis, P.E. (1993) *J. Biol. Chem. 268*, 12458-12467
Martin, B.M., Chibber, B.A., & Maelicke, A. (1983) *J. Biol. Chem. 258*, 8714-8722
Radić, Z, Pickering, N., Vellom, D.C., Camp, S., & Taylor, P. (1993) *Biochemistry 32*, 12074-12084
Radić, Z, Duran, R., Vellom, D.C., Li, Y., Cerveñansky, C., & Taylor, P. (1994) *J. Biol. Chem. 269*, 11233-11239
Rodriguez-Ithurralde, D., Silveira, R., Barbeito, L., & Dajas, F. (1983) *Neurochem. Int. 5*, 267-274
Sussman, J.L., Harel, M., Frolow, F., Oefner, C., Goldman, A., Toker, L., & Silman, I. (1991) *Science 253*, 872-879
Tamiya, T., Lamouroux, A., Julien, J.F., Grima, B., Mallet, J., Fromageot, P., & Menez, A. (1985) *Biochimie 67*, 185-189
Vellom, D.C., Radić, Z., Li, Y., Pickering, N., Camp, S., & Taylor, P. (1993) *Biochemistry 32*, 12-17

THE FUNCTION OF ELECTROSTATICS IN ACETYLCHOLINESTERASE CATALYSIS

Daniel M. Quinn,[1] Javier Seravalli,[1] Haridasan K. Nair,[1] Rohit Medhekar,[1] Basel Husseini, Zoran Radić,[2] Daniel C. Vellom,[2] Natilie Pickering,[2] and Palmer Taylor[2]

[1] Department of Chemistry
The University of Iowa
Iowa City, Iowa 52242
[2] Department of Pharmacology
University of California-San Diego
La Jolla, California 92093

INTRODUCTION

A hallmark of acetylcholinesterase (AChE) catalysis is its great speed (Quinn, 1987; Rosenberry, 1975). For example, Nolte *et al.* (1980) reported that k_{cat}/K_m for *Electrophorus electricus* AChE (EeAChE) catalyzed hydrolysis of acetylthiocholine (ATCh), extrapolated to zero ionic strength, approaches 10^{10} M^{-1} s^{-1}, a value well above the Eigen limit for bimolecular combination of ligands with enzymes (Eigen & Hammes, 1963). In addition, bimolecular rate constants that approach 10^{11} M^{-1} s^{-1} for binding of aromatic quaternary ammonium inhibitors have been observed (Nolte *et al.*, 1980). The molecular origins of these high rate constants have only recently become apparent. Ripoll *et al.* (1993) utilized the X-ray structure of *Torpedo californica* AChE (TcAChE; Sussman *et al.*, 1991) to calculate the electric field of the enzyme. They found that the hemisphere of AChE that contains the active site gorge (Sussman *et al.*, 1991) has a high negative electrical potential, and that the overall protein has a dipole moment of greater than 500 D, with the dipole moment vector aligned with the axis of the gorge. This electric field should accelerate the binding of positive charged quaternary ammonium ligands, an idea that is supported by both theoretical calculations (Tan *et al.*, 1992) and measurements of ionic strength effects (Nolte *et al.*, 1980).

The operation of the electric field of AChE seems to pose a fundamental problem. If the field accelerates quaternary ammonium ligand binding, how does the choline product of AChE catalysis get off the enzyme? This quandry has led Gilson *et al.* (1994) to propose the anthropomorphic model that AChE possesses a "back door" that may function in product release. This report describes studies of the ionic strength dependence of interaction of TcAChE with the transition state analog inhibitor *m*-(N,N,N-trimethylammonio)-tri-fluoroacetophenone, TMTFA (Brodbeck *et al.*, 1979; Nair *et al.*, 1993, 1994). The results of

Enzymes of the Cholinesterase Family, Edited by Daniel M. Quinn et al.
Plenum Press, New York, 1995

these studies obviate the need for a "back door" mechanism to account for the kinetics of product release.

EXPERIMENTAL SECTION

TMTFA is a time-dependent, reversible inhibitor that binds to AChE according to the following two-state mechanism (Nair *et al.*, 1993, 1994):

$$E + I \xrightleftharpoons[k_{off}]{k_{on}} EI$$

The corresponding first-order approach to equilibrium inhibition is described by equation 1:

$$k_{obs} = k_{on}[I] + k_{off} \tag{1}$$

Observed inhibition rate constants, k_{obs}, were determined by periodic sampling of residual activity of a buffered mixture of TcAChE and TMTFA. The resulting data were fit to equation 2 to determine k_{obs}, v_i^∞ and Δv_i^∞:

$$v_i = \Delta v_i e^{-k_{obs}t} + v_i^\infty \tag{2}$$

The parameters Δv_i and v_i^∞ are, respectively, the change in activity wrought by TMTFA binding and the residual activity of the equilibrium mixture of E and EI. From these parameters K_i, the thermodynamic dissociation constant of EI, is calculated as follows:

$$K_i = \frac{v_i^\infty}{\Delta v_i}[I] = k_{off}/k_{on} \tag{3}$$

Equations 1 and 3 both contain the unknowns k_{on} and k_{off}, and thus each time-dependent inhibition measurement provides these values and the corresponding K_i value.

Inhibition experiments were conducted in 8 mM sodium phosphate buffer at pH 7.42 and $25.0 \pm 0.1°C$. TcAChE activity was assayed in the same buffer with ATCh as the substrate (Ellman *et al.*, 1961). Since only the ketone form of TMTFA inhibits AChE, values of k_{on} and K_i were corrected for the extent of TMTFA hydration (Nair *et al.*, 1993, 1994). Ionic strength was varied by adding various concentrations of NaCl to the sodium phosphate buffers in which inhibitions were conducted.

RESULTS

The data plotted in Figures 1 and 2 indicate that both $1/K_i$, the equilibrium constant for binding of TMTFA to TcAChE, and k_{on} decrease with increasing ionic strength. These dependences can be rationalized in terms of ionic strength effects on the thermodynamic activities of the initial and final states of the processes represented by $1/K_i$ and k_{on}. The starting point for deriving the effect of ionic strength on k_{on} is the following expression:

$$k_{on} = \frac{kT}{h} \cdot \frac{a_{\ddagger}}{a_E a_I} = \frac{kT}{h} \cdot \frac{\gamma_{\ddagger}}{\gamma_E \gamma_I} \cdot \frac{[\ddagger]}{[E][I]} \tag{4}$$

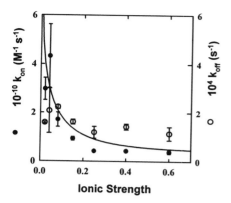

Figure 1. Dependences of k_{on} (solid circles) and k_{off} (open circles) on ionic strength for inhibition of *Torpedo californica* AChE by TMTFA. The line is a nonlinear least-squares fit of k_{on} versus I data to equation 6, as described in the text. The release rate constant k_{off} appears relatively constant over the range of ionic strengths.

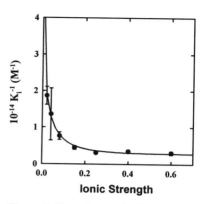

Figure 2. Dependence of $1/K_i$ on ionic strength for inhibition of *Torpedo californica* AChE by TMTFA. The line is a nonlinear least-squares fit to equation 8, as described in the text.

Equation 4 is based on transition state theory (Glasstone *et al.*, 1941) and relates k_{on} to the thermodynamic activities of the binding transition state, $a_{\ddagger}$, and of the reactants, a_E and a_I. The Debye-Huckel limiting law is used to account for the effect of ionic strength on activity coefficients:

$$\log(\gamma_i) = \log(\gamma_i^0) - 0.509 z_i^2 \sqrt{I} \tag{5}$$

In this equation z_i is the charge of the ion whose activity coefficient γ_i is affected by electrolyte, and γ_i^0 is the corresponding activity coefficient at zero ionic strength. Since the binding transition state is a complex of E and I, $z_{\ddagger}^2 = (z_E + z_I)^2$. With this in mind, equation 5 can be used to substitute for the activity coefficients in equation 4:

$$k_{on} = (k_{on}^0 - k_{on}^H) \cdot 10^{1.18 z z_E z_I \sqrt{I}} + k_{on}^H \tag{6}$$

where

$$k_{on}^0 = \frac{kT}{h} \cdot \frac{\gamma_{\ddagger}^0}{\gamma_E^0 \gamma_I^0} \cdot \frac{[\ddagger]}{[E][I]} \tag{7}$$

The charges z_E and z_I of E and I, respectively, are of opposite sign. In Figure 1 k_{on} approaches a constant value, k_{on}^H, at high ionic strength, and thus equation 6 incorporates this observation. Figure 1 shows a nonlinear-least squares fit to equation 6; the parameters determined from the fit are $k_{on}^0 = 8\pm3 \times 10^{10}$ M^{-1} s^{-1} and $z_E = -2.0\pm0.6$; k_{on}^H was constrained at 3.9×10^9 M^{-1} s^{-1}, the observed value at $I = 0.6$. Also shown in Figure 1 is the dependence of k_{off} on ionic strength. It is apparent that k_{off} varies little as ionic strength increases, even though k_{on} decreases markedly.

The treatment of $1/K_i$ in Figure 2 is similar to that for k_{on}, in that at high ionic strength the dependent variable levels off at $1/K_{i,H}$. The change in $1/K_i$ was treated as was k_{on} above to give equation 8:

$$1/K_i = (1/K_{i,0} - 1/K_{i,H}) \, 10^{1.18 z_E z_i \sqrt{I}} + 1/K_{i,H} \tag{8}$$

The data in Figure 2 were fit to equation 8, which gave $1/K_{i,0} = 6\pm2 \times 10^{14}$ M^{-1} (i.e. $K_{i,0} = 1.7$ fM), and $z_E = -3.1\pm0.8$; $1/K_{i,H}$ was constrained at 3.1×10^{13} M^{-1} (i.e. $K_{i,H} = 33$ fM), the observed value at $I = 0.6$.

DISCUSSION

The purpose of the experiments reported herein was to assess the effect of the electric field of TcAChE on reversible binding of a quaternary ammonium transition state analog to the active site. The expectation is that the electric field of the enzyme is progressively masked by increasing ionic strength, and thus that both the binding rate and equilibrium constants will decrease with increasing electrolyte concentration. These are indeed the observed dependences, which accord nicely with electrolyte effects earlier reported by Nolte *et al.* (1980) on binding of aromatic quaternary ammonium ions to EeAChE. Though Nolte *et al.* did not have a crystal structure to interpret their results, they did surmise that multiple negative charges on EeAChE were involved in accelerating binding of quaternary ammonium ligands to the active site. Their view is supported by the electric field calculations of Ripoll *et al.* (1993) on TcAChE, and by consequent calculation of the accelerative effect of this field on quaternary ammonium ligand binding (Tan *et al.*, 1993).

Because the k_{off} and k_{on} components of K_i for inhibition of TcAChE are related by microscopic reversibility, the average diffusional pathway for binding to and release from the active site must be the same, albeit in reverse sequence. That is, if TMTFA binds to the active site by diffusing through the aromatic gorge (Sussman *et al.*, 1991), release from the active site *must* involve diffusion back through the gorge. Therefore, the insensitivity of k_{off} to increasing electrolyte concentration necessitates that release of quaternary ammonium ligands from the active site is not decelerated by the electrical field of AChE. This result also accords with those of Nolte *et al.* (1980), who noted that the rate of release of N-methylacridinium ion from the active site of EeAChE only increased about two fold with increasing electrolyte concentration.

How can the electric field of AChE accelerate quaternary ammonium ligand binding but not affect release from the active site? This seemingly incongruous pair of observations can be easily rationalized by referring to the free energy diagrams in Figure 3. The equilibrium binding constant $1/K_i$ depends on the free energy difference of the E+I and EI states, while k_{on} depends on the free energy difference of the E+I state and the binding transition state, ‡. Therefore, k_{on} and $1/K_i$ share the same initial state, and their respective final states are both AChE-inhibitor complexes. The release rate constant k_{off}, on the other hand, depends on the free energy difference of two AChE-inhibitor complexes, the EI state and the binding (release) transition state. The transition state theory (Glasstone *et al.*, 1941) provides the relationship between rate constants and free energies:

$$k_{on} = (kT/h) \exp(-\Delta G_{on}^{\ddagger}/RT), \qquad \text{where } \Delta G_{on}^{\ddagger} = G_{\ddagger} - G_{E+I} \tag{9}$$

$$k_{off} = ((kT/h) \exp(-\Delta G_{off}^{\ddagger}/RT), \qquad \text{where } \Delta G_{off}^{\ddagger} = G_{\ddagger} - G_{EI} \tag{10}$$

As the free energy diagrams illustrate, if the effect of the negative electrical potential of the active site gorge on the free energies of the EI complex and the binding transition state are comparable, the free energy barrier for release will not be influenced by the electric field. Hence, masking of the field with added electrolytes will leave k_{off} relatively unaffected, as observed in Figure 1. This result is not surprising, since EI and

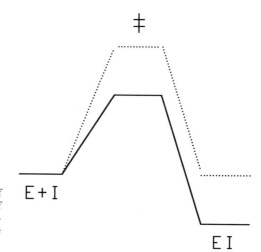

Figure 3. Free energy diagrams for interaction of quaternary ammonium ligands with the active site of AChE. Solid and dotted lines represent ligand interactions in the presence and absence, respectively, of the electric field of AChE.

the transition state are both AChE-inhibitor complexes. The lack of an electric field effect on k_{off} requires that the quaternary ammonium inhibitor interacts with negative electrical potentials of comparable magnitudes in the EI state and the binding (release) transition state. That this is so is indicated by the comparable charges on AChE calculated from the dependences of $1/K_i$ and k_{on} on ionic strength in Figures 1 and 2.

The results presented herein are pleasing from the teleological point of view. Apparently, the function of the electric field of AChE is to accelerate acetylcholine binding to the active site without retarding release of the product choline. In this way, the forward catalytic flux is maximized and product inhibition by choline is minimized. This view accords with the fact that choline is a weak AChE inhibitor, with a K_i of around 1 mM (Lee et al., 1992).

REFERENCES

Brodbeck, U., Schweikert, K., Gentinetta, R., & Rottenberg, M. (1979) *Biochim. Biophys. Acta 567*, 357-369.

Eigen, M., & Hammes, G.G. (1963) *Adv. Enzymol. Relat. Subj. Biochem. 25*, 1-38.

Ellman, G.L., Courtney, K.D., Andres, V., Jr., & Featherstone, R.M. (1961) *Biochem. Pharmacol. 7*, 88-95.

Gilson, M.K., Straatsma, T.P., McCammon, J.A., Ripoll, D.R., Faerman, C.H., Axelsen, P.H., Silman, I., & Sussman, J.L. (1994) *Science 263*, 1276-1278.

Glasstone, S., Laidler, K.J., & Eyring, H. (1941) *The Theory of Rate Processes*, McGraw-Hill, New York.

Lee, B.H., Stelly, T.C., Colucci, W.J., Garcia, G., Gandour, R.D., & Quinn, D.M. (1992) *Chem. Res. Toxicol. 5*, 411-418.

Nair, H.K., Lee, K., & Quinn, D.M. (1993) *J. Am. Chem. Soc. 115*, 9939-9941.

Nair, H.K., Seravalli, J., Arbuckle, T., & Quinn, D.M. (1994) *Biochemistry 33*, 8566-8576.

Nolte, H.-J., Rosenberry, T.L. & Neumann, E. (1980) *Biochemistry 19*, 3705-3711.

Quinn, D.M. (1987) *Chem. Rev. 87*, 955-979.

Ripoll, D.R., Faerman, C.H., Axelsen, P.H., Silman, I., & Sussman, J.L. (1993) *Proc. Natl. Acad. Sci. U.S.A. 90*, 5128-5132.

Rosenberry, T.L. (1975) *Adv. Enzymol. Relat. Areas Mol. Biol. 43*, 103-218.

Sussman, J.L., Harel, M., Frolow, F., Oefner, C., Goldman, A., Toker, L., & Silman, I. (1991) *Science 253*, 872-879.

Tan, R.C., Truong, T.N., McCammon, A., & Sussman, J.L. (1993) *Biochemistry 32*, 401-403.

FASCICULIN 2 BINDS TO THE PERIPHERAL SITE ON ACETYLCHOLINESTERASE AND INHIBITS SUBSTRATE HYDROLYSIS BY SLOWING A STEP INVOLVING PROTON TRANSFER DURING ENZYME ACYLATION

Jean Eastman,[1] Erica J. Wilson,[1] Carlos Cervenansky,[2] and Terrone L. Rosenberry[1]

[1] Department of Pharmacology
School of Medicine
Case Western Reserve University
Cleveland, OH 44106-4965
[2] Instituto de Investigaciones Biologicas
Clemente Estable
Montevideo, Uruguay 11600

INTRODUCTION

Acetylcholinesterase (AChE; EC 3.1.1.7) hydrolyzes acetylcholine at a very high catalytic rate (Rosenberry, 1975a). Several unique features that determine its catalytic power have been identified in the X-ray structure of *Torpedo* AChE (Sussman *et al.*, 1991). These include an active site gorge lined with aromatic residues that is about 20 Å deep and penetrates nearly to the center of the 70 kDa catalytic subunit. At the base of the gorge is the S200-H440-E327 catalytic triad (*Torpedo* sequence numbering), similar to triads found in other serine proteases and esterases, that is acylated and deacylated at S200 during substrate turnover. We refer to this site as the acylation site. Certain cationic inhibitors bind selectively to this site, as illustrated by the crystal structure of the AChE-edrophonium complex, in which the hydroxyl group of the inhibitor makes H-bonds to both the N^e atom of H440 and the O^γ atom of S200 (Sussman *et al.*, 1992).

AChE also possesses a peripheral site, distinct from the acylation site, that binds the cationic inhibitor propidium (Taylor & Lappi, 1975). Affinity labeling (Weise *et al.*, 1990; Schalk *et al.*, 1992) and site-directed mutagenesis (Radić *et al.*, 1993; Barak *et al.*, 1994) indicate that W279 on the rim of the gorge is a key component of the peripheral site and that residues from at least two other polypeptide loops also contribute. Recently the fasciculins, a family of snake venom neurotoxins from mambas (genus *Dendroaspis*), have emerged as

Enzymes of the Cholinesterase Family, Edited by Daniel M. Quinn et al.
Plenum Press, New York, 1995

probes of the AChE active site. Fasciculins have structures that are similar to those of the short α-neurotoxins with 4 disulfide bonds that bind to nicotinic acetylcholine receptors (le Du *et al.*, 1992). At subnanomolar concentrations, they are noncompetitive AChE inhibitors, and displace propidium from its peripheral binding site (Karlsson *et al.*, 1984; Marchot *et al.*, 1993). Furthermore, site-specific mutants reveal that residues at the gorge rim, particularly W279, are essential for high affinity fasciculin binding (Radić *et al.*, 1994). These features suggest that fasciculins bind to the same peripheral site as does propidium. We confirm this point by analyzing the kinetics of fasciculin 2 (FAS2) binding, and investigate the mechanism by which FAS2 alters AChE-catalyzed substrate hydrolysis.

EXPERIMENTAL PROCEDURES

Materials

Active site concentrations of purified human erythrocyte AChE were calculated by assuming 410 units/nmol (Rosenberry & Scoggin, 1984). Purified FAS2 concentrations were determined by absorption at 276 nm ($\varepsilon = 4900$), and stocks were mixed with BSA (to 1 mg/ml) for storage (4°C or -20°C) prior to dilution in experiments (Karlsson *et al.*, 1984).

Kinetic model of FAS2 Binding in the Presence of an Inhibitor

The interaction of FAS2 (F) and an inhibitor (I) with AChE (E) is modeled in Scheme 1.

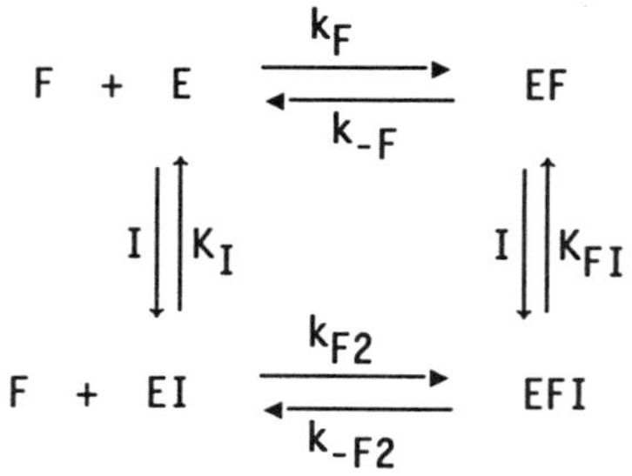

$$
\begin{array}{ccc}
F + E & \underset{k_{-F}}{\overset{k_F}{\rightleftharpoons}} & EF \\[2mm]
I \updownarrow K_I & & I \updownarrow K_{FI} \\[2mm]
F + EI & \underset{k_{-F2}}{\overset{k_{F2}}{\rightleftharpoons}} & EFI
\end{array}
$$

Scheme 1

In Scheme 1, I binding to E and EF is assumed to reach equilibrium instantaneously with dissociation constants K_I and K_{FI}, respectively. Slower binding of F to E and EI occurs with association rate constants k_F and k_{F2} and dissociation rate constants k_{-F} and k_{-F2}, respectively. Equilibrium dissociation constants are $K_F = k_{-F}/k_F$ and $K_{F2} = k_{-F2}/k_{F2}$. The observed pseudo first-order rate constant k for the approach to equilibrium is given by eq. 1.

$$k = k_{on}[F] + k_{off} \tag{1}$$

where

$$
k_{on} = \frac{k_F + k_{F2}\dfrac{[I]}{K_I}}{1 + \dfrac{[I]}{K_I}} \qquad\qquad k_{off} = \frac{k_{-F} + k_{-F2}\dfrac{[I]}{K_{FI}}}{1 + \dfrac{[I]}{K_{FI}}} \tag{2a and b}
$$

According to eqs. 2a and b, if F and I are competitive (i.e., EFI does not form), then $k_{on} = k_F/(1+[I]/K_I)$ and $k_{off} = k_{-F}$. If F and I are noncompetitive, then at saturating [I], $k_{on} = k_{F2}$ and $k_{off} = k_{-F2}$.

Analysis of Rates of Equilibrium Binding of FAS2 to AChE

Association reactions were initiated by adding FAS2 to AChE in buffer (20 mM sodium phosphate, 0.02% TX100, pH 7.0) with or without an inhibitor, at 23°C. The concentration of FAS2 was ≥ 8x that of AChE to insure pseudo first-order binding kinetics. At various times a 2.9-ml aliquot was removed to a cuvette, 120 μl of acetylthiocholine (ATCh) and 5,5'-dithiobis-(2-nitrobenzoic acid) (DTNB) were added to final concentrations of 0.5 mM and 0.33 mM, respectively, and a continuous assay trace was immediately recorded at 412 nm on a Cary 3A spectrophotometer. Dissociation reactions were measured by preincubating FAS2 with AChE at 50x the indicated final concentrations. The concentration of FAS2 was ≥ 8x that of AChE except at low concentrations of FAS2 where dissociation was virtually complete. After 1-4 hr of incubation, 58 μl was added to 2.82 ml of buffer with or without inhibitor at 23°C in a cuvette at time zero. After various times, ATCh and DTNB were added for assay as above.

Assay points v from association and dissociation reactions were fitted by the non-linear regression analysis program Fig.P (BioSoft, version 6.0) to eq. 3.

$$v = v_f + (v_i - v_f)e^{-kt} \qquad (3)$$

In eq. 3 k is the observed pseudo first-order rate constant for equilibrium FAS2 binding and v_i and v_f are the calculated values of v at time zero and at equilibrium, respectively.

Kinetic Models for Steady-State Substrate Hydrolysis in the Presence of FAS2

A general model for the interaction of substrate (S) and FAS2 (F) with AChE is given in Scheme 2, and includes an acylenzyme intermediate EA formed from ES during substrate hydrolysis. S can bind to EA to give substrate inhibition, and F can bind to E, ES, and EA.

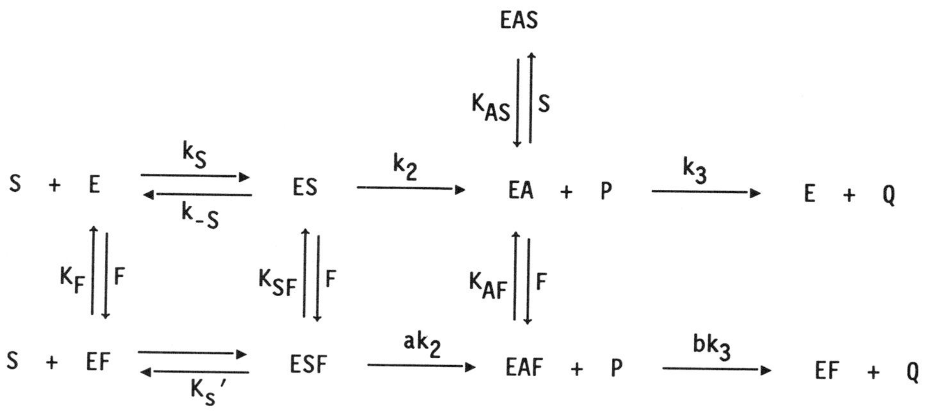

Scheme 2

In Scheme 2, the equilibrium dissociation constants $K_x = k_{-x}/k_x$, where k_x and k_{-x} are the respective association and dissociation rate constants. When [FAS2] = 0, the steady-state rate v is given by eq. 4.

$$v = -d[S]/dt = \frac{V_{max}[S]}{[S](1 + [S]/K_{SIN}) + K_{app}}$$

(4)

In eq. 4, $V_{max} = k_{cat}[E]_t$, $k_{cat} = k_2k_3/(k_2+k_3)$ and $[E]_t$ is the total enzyme concentration; $K_{app} = K_m k_{cat}/k_2$, where $K_m = (k_{-S}+k_2)/k_S$; $K_{SIN} = K_{AS}k_3/k_{cat}$. An expression for v in the presence of F when F dissociates slowly relative to k_2 and k_3 has been derived (Eastman et al., 1995). At saturating [FAS2] and $[S]/K_{SIN} \ll 1$, v reduces to the form of eq. 4, where $V_{max} = k_{cat}'[E]_t$ and $k_{cat}' = ak_2bk_3/(ak_2+bk_3)$. K_{app} corresponds to $K_{app}' = K_m'k_{cat}'/ak_2$, where $K_m' = (k_{-S}' + ak_2)/k_S'$. Primed parameters denote saturating FAS2.

Steady State Measurements of AChE-Catalyzed Substrate Hydrolysis

One unit of AChE activity corresponds to 1 μmol of ATCh hydrolyzed per min at 23°C (Rosenberry & Scoggin, 1984), as monitored by the formation of the thiolate dianion of DTNB at 412 nm ($\Delta\varepsilon = 14.15$ mM^{-1}cm^{-1}). Hydrolysis of phenyl acetate was monitored directly at 270 nm ($\Delta\varepsilon = 1.4$ mM^{-1}cm^{-1}; Rosenberry, 1975b). Hydrolysis rates were measured in 1.0 ml of 0.02% Triton X100 and sodium phosphate (0.1 M phosphate) at pH 8.0. AChE was pre-equilibrated with FAS2 for 30 min before mixing with ATCh (0.1-5 mM) and DTNB (0.1 mM) or phenyl acetate (0.1-5 mM, ≤1% MeOH final) for 0.5-5 min.

Reactions in D_2O were conducted by identical procedures except that the pH 8 buffer was adjusted to pH 8.1. For better precision in comparing rates in H_2O and D_2O, v was measured at 0.5 mM ATCh and V_{max}/K_{app} or V_{max}'/K_{app}' was determined as the constant j from the integrated form of eq 4 at low initial $[S] = [S]_o$ (12 μM ATCh, <0.2 K_{app}): $[S] = [S]_o\exp(-jt)$. V_{max} or V_{max}' was then calculated from eq. 4. To improve the precision of isotope effect measurements with phenyl acetate, multiple data sets were collected, K_{app} and K_{app}' values determined from eq. 4 were averaged, and the data sets were subjected to a second cycle of regression analysis with these average values fixed.

RESULTS AND DISCUSSION

The Kinetics of Fasciculin-2 Binding to AChE Are Consistent with a Simple Equilibrium

Equilibrium binding of fasciculins to AChE is characterized by high affinity and low dissociation rate constants (Karlsson et al., 1984; Marchot et al., 1993), and the rates of approach to equilibrium are measured on a scale of m^{-1} to h^{-1}. Examples are shown in Fig. 1A, where 36 pM FAS2 and 4 pM AChE reached equilibrium at an activity of 27% of an enzyme control without FAS2. Since FAS2 binding blocks most of the activity toward ATCh, this corresponds approximately to the percentage of enzyme that remains uncomplexed. Equilibrium was approached from initial conditions in which either no AChE was bound (an association reaction) or virtually all enzyme was bound (a dissociation reaction), and the change in enzyme activity corresponded to single exponential time courses with rate constants of about 0.07 m^{-1} for both initial conditions. Analysis of a series of rate constants at several FAS2 concentrations (Fig. 1B) supported the simple bimolecular reaction in

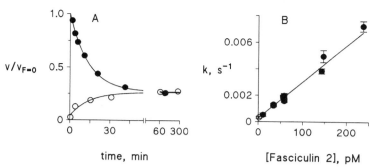

Figure 1. Kinetics of FAS2 reaction with AChE. Panel A: The association reaction (●) was initiated by mixing FAS2 and AChE and the dissociation reaction (○) by diluting these components to final concentrations of 36 pM and 4 pM, respectively. At the indicated times, aliquots were taken for assay with ATCh, as outlined in Experimental Procedures. Assay points were fitted by unweighted nonlinear regression to eq. 3 to give pseudo first-order rate constants k of 0.073 ± 0.005 m^{-1} for the association reaction and 0.07 ± 0.02 m^{-1} for the dissociation reaction. Assay points are shown after normalization to a control AChE activity without FAS2. Panel B. Values of k obtained from 11 association reactions (●) and 1 dissociation reaction (○) as shown in Panel A were plotted against FAS2 concentration according to eqs. 1 and 2. Points were weighted by the reciprocal of the observed variance of the k values and fit to a linear plot with slope $k_{on} = k_F$ and intercept $k_{off} = k_{-F}$. Values of these constants are given in Table 1.

Scheme 1, with an association rate constant $k_F = 2.7 \times 10^7$ M^{-1}s^{-1} and a dissociation rate constant $k_{-F} = 2.9 \times 10^{-4}$ s^{-1} (Table 1). The dissociation constant K_F given by k_{off}/k_{on} corresponded to 11 pM. A recent report on the interaction of FAS2 with mouse AChE found a similar value of K_F but a smaller value of k_F, perhaps because of higher ionic strength conditions (Radić *et al.*, 1994).

Fasciculin 2 Binding to AChE Is Competitively Blocked by Propidium but not by Edrophonium

To assess the site on AChE to which FAS2 binds, the effects of the peripheral site inhibitor propidium and the acylation site inhibitor edrophonium on the kinetics of FAS2 binding were analyzed. According to Scheme 1 and eqs. 1 and 2, an inhibitor that binds to the same site as FAS2 will be unable to form the ternary complex EFI. Because of this

Table 1. Ligand Competition with FAS2 Reactions[a]

Competing ligand	$k_{on} \times 10^{-7}$ M^{-1}s^{-1}	$k_{off} \times 10^4$ s^{-1}	k_{off}/k_{on}[b] pM
None	2.7 ± 0.1	2.9 ± 0.5	11 ± 2
Propidium	competitive	2.1 ± 0.3[c]	
Edrophonium	2.1 ± 0.2[d]	13 ± 2[d]	60 ± 10

[a]Values of k_{on} and k_{off} for the interaction of FAS2 with AChE were determined according to eq. 1 in the absence and the presence of the indicated ligands as shown in Fig. 1.
[b]For FAS2 alone, $k_{off}/k_{on} = K_F$, the equilibrium dissociation constant. With saturating edrophonium, $k_{off}/k_{on} = K_{F2}$, the equilibrium dissociation constant for FAS2 in the ternary complex EFI in Scheme 1.
[c]Value with 20 uM propidium.
[d]Value with 10 uM edrophonium.

competitive interaction, saturating concentrations of the inhibitor will have no effect on k_{off} = k_{-F} for FAS2 but will decrease k_{on} to an extent inversely proportional to the inhibitor concentration (eq. 2A). This type of inhibition was seen with propidium (data not shown). The value of k_{off} was unchanged with propidium (Table 1), and k_{on}^{-1} increased linearly with the propidium concentration. Also, the value of the propidium inhibition constant calculated from k_{on} agreed with the K_I determined from propidium inhibition of ATCh hydrolysis according to Scheme 2. This agreement suggests that the binding site for which propidium and FAS2 compete is the same site at which propidium inhibits substrate hydrolysis. The properties of the site-specific mutants noted in the Introduction indicate that this is the peripheral site on the rim of the active site gorge.

In contrast to propidium, saturating concentrations of edrophonium had almost no effect on k_{on} for FAS2 (Table 1). Thus the rate constant for FAS2 binding to its peripheral site is not significantly altered when edrophonium is bound to the acylation site at the bottom of the active site gorge. However, k_{off} for FAS2 dissociation from the ternary complex EFI with edrophonium and AChE did increase by 4- to 5-fold. The increase in k_{off} suggests a modest conformational interaction between the sites in the ternary complex. As noted in Table 1, the ratio of k_{off} to k_{on} is a measure of the affinity of FAS2 for the free enzyme or the ternary complex. Edrophonium decreased the affinity of FAS2 in the ternary complex about 6-fold relative to the FAS2 affinity for free AChE.

AChE-Catalyzed Hydrolysis of ATCh and Phenyl Acetate Are Inhibited by FAS2 to about Equal Extents

FAS2 inhibition of the steady-state hydrolysis of two very good substrates of AChE, ATCh and phenyl acetate, is illustrated in Figs. 2A and B. The inhibition pattern for each substrate at a saturating concentration of FAS2 is noncompetitive; that is, both k_{cat}' and the second-order acylation rate constant k_{cat}'/K_{app}' were decreased by about equal amounts relative to the corresponding parameters in the absence of FAS2. Two points deserve comment here. First, the decreases in k_{cat}' were substantial and of the same magnitude (about 100-fold) for both the cationic substrate ATCh and the neutral substrate phenyl acetate. In

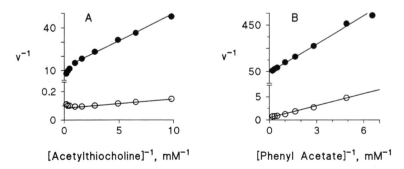

Figure 2. Steady-state inhibition of AChE-catalyzed hydrolysis of ATCh (Panel A) and phenyl acetate (Panel B) by a saturating amount of FAS2. Hydrolysis rates, v ($\Delta A/min$), were measured at pH 8.0 with ($\bullet$) or without ($\bigcirc$) 500 nM FAS2 as outlined in the Experimental Procedures. Total AChE concentrations for the four data sets ranged from 20 pM to 3 nM, and v was normalized to 2 nM AChE for these graphs. The data set for ATCh without FAS2 was fit to eq. 4 by nonlinear regression analysis to obtain V_{max}, K_{app} and K_{SIN}. For the other three data sets, $[S]/K_{SIN} \ll 1$, and reciprocal plots of eq. 4 ($1/v$ vs. $1/[S]$) were analyzed by weighted linear regression analysis to obtain V_{max} and K_{app} or V_{max}' or K_{app}' (it was assumed that v has constant percent error). Data are displayed as reciprocal plots with calculated lines. Because of apparent substrate activation in the presence of FAS2 in Panel A, the three points at the highest [S] were deleted from the regression analysis.

contrast, inhibition of AChE by the peripheral site ligands propidium and Pt(terpyridine)Cl results in a much larger decrease in k_{cat} for ATCh than for phenyl acetate (Haas *et al.*, 1992 and data not shown). [*] Thus ligands which bind competitively to the same peripheral site at the rim of the active site gorge alter reactivity at the acylation site in different ways. The conformational basis for these differences remains to be explored. Second, while the decreases from k_{cat} to k_{cat}' in Fig. 2 were substantial, the enzyme bound to FAS2 still had about 0.5 to 1% of the activity of free AChE. This residual activity was large enough to investigate by mechanistic studies, as shown below.

FAS2 Binding Inhibits Substrate Hydrolysis by Slowing Proton Transfer during the Enzyme Acylation Step

The simplest explanation for the decreased k_{cat}'/K_{app}' and k_{cat}' in Fig. 2 is that FAS2 slows formation of the acylenzyme in Scheme 2 (i.e., $a = 0.01$). This interpretation of noncompetitive inhibition also indicates that FAS2 binding has no effect on K_{app} for either ATCh or phenyl acetate, an observation that is difficult to reconcile with the decrease in affinity for edrophonium in the ternary complex with FAS2 and AChE (Table 1). To examine more closely the ways in which FAS2 binding could interfere with substrate hydrolysis, it is useful to consider additional intermediates on the catalytic pathway as shown in Scheme 3.

$$E + S \underset{k_{-1}}{\overset{k_1}{\rightleftharpoons}} ES \underset{k_{-II}}{\overset{k_{II}}{\rightleftharpoons}} ES_1 \underset{k_{-III}}{\overset{k_{III}}{\rightleftharpoons}} ES_2 \xrightarrow{k_{IV}} EA + P \xrightarrow{k_3} E + Q$$

Scheme 3

Scheme 3 involves two additional intermediates, ES_1 and ES_2. ES_1 occurs prior to the general acid-base catalysis step and may involve an induced fit conformational change of the initial ES complex, while ES_2 occurs concomitant with general acid-base catalysis and, with carboxylic acid esters, probably involves formation of a tetrahedral intermediate. Solvent isotope effects support both of these intermediates and indicate that their formation can be rate limiting with certain substrates (Rosenberry, 1975b; Rao *et al.*, 1993).

How could ligand binding to the peripheral site influence the reaction pathway in Scheme 3? A peripheral site inhibitor could sterically block access of a second ligand to the active site, but it could also alter the enzyme conformation in a way that reduces reactivity at the acylation site. To distinguish these modes of action with FAS2, it is helpful to focus on the steady-state substrate hydrolysis profiles in Fig. 2. According to Scheme 2, the second-order rate constants in the absence and presence of saturating concentrations of FAS2 are given by eqs. 5A and 5B.

$$\frac{k_{cat}}{K_{app}} = \frac{k_S k_2}{k_{-S} + k_2} \qquad \frac{k_{cat}'}{K_{app}'} = \frac{k_S' a k_2}{k_{-S}' + a k_2} \qquad \text{(5A and B)}$$

These second-order rate constants are measures of the rate-limiting step at low substrate concentrations in Scheme 3. The rate constants k_S, k_{-S}, and k_2 from Scheme 2 are combinations of the microscopic rate constants of the first four steps in Scheme 3. One useful demarcation in analyzing these combinations is the first step in which general acid-base catalysis occurs. This step involves proton transfer, and thus shows a solvent isotope effect

---

[*] E. Eckman and T. L. Rosenberry, manuscript in preparation.

(a decrease of 2- to 3-fold when D_2O replaces H_2O as the solvent) and often shows a pH dependence. In Scheme 2 and eq. 5A this step is the k_2 step, which involves general acid-base catalyzed alcohol release from the substrate, although k_2 may represent a combination of rate constants from steps III and IV in Scheme 3. Likewise eq. 5A lumps rate constants in steps I and II of Scheme 3 into the ligand association and dissociation constants k_S and k_{-S} because these steps do not involve proton transfer.

We define a steric blockade of the active site as a reduction of both k_S and k_{-S} when FAS2 is bound. Equal percent reductions in these rate constants for FAS2 binding at the peripheral site would reduce the rates at which substrate could enter and exit the active site without changing the equilibrium substrate affinity in the ternary complex. We define a conformational interaction between the AChE peripheral and active sites as a change in any rate constant not compatible with a steric blockade when FAS2 is bound (i.e., in Scheme 2, $a \neq 1$ or $b \neq 1$ or an increase in k_S or k_{-S}). Saturation of the peripheral site with FAS2 markedly decreased k_{cat} for both substrates in Fig. 2, which suggests that FAS2 binding has an effect on enzyme conformation, but it does not indicate which step in the catalytic pathway is affected. Furthermore, these definitions permit FAS2 binding at the peripheral site to have both steric and conformational effects on the AChE active site, and a clear understanding of peripheral site function requires insight into the relative contributions of these two effects.

It is difficult to measure direct steric blockade due to FAS2 binding. Most ligands that form ternary complexes at the acylation site equilibrate in less than a second even when FAS2 is bound. However, analysis of D_2O effects on the second-order rate constants in eqs. 5A and B provides insight into whether FAS2 acts primarily through a steric or a conformational blockade. Key relationships in these constants are $k_2/k_{-S} = C$ and $ak_2/k_{-S}' = C'$, termed the commitments to catalysis (see Quinn, 1987). When C is small, ES is virtually in equilibrium with E and S, and k_2 is rate limiting for acylation. Conversely, when C is large, $k_{cat}/K_{app} = k_S$ and the bimolecular reaction of E with S is rate limiting for acylation. According to eq. 5A, a small value for C implies that k_{cat}/K_{app} will include the k_2 term and show a D_2O isotope effect. This is the case for substrate hydrolysis by most enzymes. For AChE, however, k_{cat}/K_{app} values for ATCh, phenyl acetate and certain other substrates show only a small change in D_2O (Table 2 and Rosenberry, 1975b), which has led to the conclusion that k_1 or some other step prior to general acid-base catalysis is rate limiting in these cases (Rosenberry, 1975b; Quinn, 1987). In a key test of whether the decreased values of k_{cat}' and k_{cat}'/K_{app}' in the complex of FAS2 with AChE had a predominantly steric or a conformational basis, the D_2O isotope effects on these parameters for both substrates were measured. If the decreases resulted primarily from steric blockade, one would expect C' to increase and the D_2O isotope

Table 2. Deuterium oxide isotope effects on steady-state kinetic parameters for AChE and the AChE-FAS2 complex[a]

	AChE	AChE + FAS2
Kinetic constant:	k_{cat}/K_{app}	k_{cat}'/K_{app}'
ATCh	1.21 ± 0.02 (4)	1.76 ± 0.06 (6)
Phenyl acetate	1.48 ± 0.05 (5)	1.86 ± 0.06 (6)
Kinetic constant:	k_{cat}	k_{cat}'
ATCh	2.03 ± 0.05 (4)	2.45 ± 0.04 (6)
Phenyl acetate	2.45 ± 0.08 (5)	2.03 ± 0.18 (6)

[a]Kinetic constants were measured at pH 8 as outlined in Fig. 2 and the Experimental Procedures. Values shown are mean ratios of the kinetic constant obtained in H_2O to that obtained in D_2O for the number of paired data sets in parentheses.

effect on k_{cat}'/K_{app}' to become smaller than that on k_{cat}/K_{app}. Alternatively, if the decreased rates resulted primarily from a conformational interaction, C' should decrease and the D_2O effect on k_{cat}'/K_{app}' should become larger. Table 2 shows that the D_2O effects on this parameter for both substrates increased with FAS2 binding, indicating that FAS2 acts predominantly to alter the conformation of the active site in the ternary complex so that steps involving proton transfer during enzyme acylation are slowed.

Thus this test of the AChE mechanism supports the model that FAS2 blocks substrate hydrolysis by blocking a general acid-base catalyzed step in acylation. The molecular basis of this effect is being studied with substrate analogs that proceed only part of the way through the hydrolysis pathway. The most interesting data obtained so far involves reaction with organophosphate substrates. When FAS2 is bound to AChE, phosphorylation rates with either echothiophate or paraoxon are decreased $\geq$100-fold.[*] Further experiments are in progress to clarify the basis of this effect.

ACKNOWLEDGMENTS

This work was supported by grant NS-16577 from the National Institutes of Health and by grants from the Muscular Dystrophy Association.

REFERENCES

Barak, D., Kronman, C., Ordentlich, A., Ariel, N., Bromberg, A., Marcus, D., Lazar, A., Velan, B. & Shafferman, A. (1994) *J. Biol. Chem.* **264**, 6296-6305.

Eastman, J., Wilson, E.J., Cervenansky, C., & Rosenberry, T.L. (1995) *J. Biol. Chem.* **265**, in press

Haas, R., Adams, E.W., Rosenberry, M.A. & Rosenberry, T.L. (1992) In *Multidisciplinary Approaches to Cholinesterase Functions* (Shafferman, A. & Velan, B., eds), Plenum Press, New York, pp. 131-139.

Karlsson, E., Mbugua, P.M. & Rodriguez-Ithurralde, D. (1984) *J. Physiol., Paris* **79**, 232-240.

le Du, J.H., Marchot, P., Bougis, P.E. & Fontecilla-Camps, J.C. (1992) *J. Biol. Chem.* **267**, 22122-22130.

Marchot, P., Khelif, A., Ji, Y.-H., Mansuelle, P. & Bougis, P.E. (1993) *J. Biol. Chem.* **268**, 12458-12467.

Quinn, D.M. (1987) *Chem. Rev.* **87**, 955-979.

Radić, Z., Pickering, N.A., Vellom, D.C., Camp, S. & Taylor, P. (1993) *Biochemistry* **32**, 12074-12084.

Radić, Z., Duran, R., Vellom, D.C., Li, Y., Cervenansky, C. & Taylor, P. (1994) *J. Biol. Chem.* **269**, 11233-11239.

Rao, M., Barlow, P.N., Pryor, A.N., Paneth, P., O'Leary, M.H., Quinn, D.M. & Huskey, W.P. (1993) *J. Am. Chem. Soc.* **115**, 11676-11681.

Rosenberry, T.L. (1975a) *Adv. Enzymol.* **43**, 103-218.

Rosenberry, T.L. (1975b) *Proc. Natl. Acad. Sci. USA* **72**, 3834-3838.

Rosenberry, T.L. & Scoggin, D. M. (1984) *J. Biol. Chem.* **259**, 5643-5652.

Schalk, I., Ehret-Sabatier, L., Bouet, F., Goeldner, M. & Hirth, C. (1992) In *Multidisciplinary Approaches to Cholinesterase Functions* (Shafferman, A. & Velan, B., eds), Plenum Press, New York, pp. 117-120.

Sussman, J.L., Harel, M., Frolow, F., Oefner, C., Goldman, A., Toker, L. & Silman, I. (1991) *Science* **253**, 872-879.

Sussman, J.L., Harel, M. & Silman, I. (1992) In *Multidisciplinary Approaches to Cholinesterase Functions* (Shafferman, A. and Velan, B., eds), Plenum Press, New York, pp. 95-107.

Taylor, P. & Lappi, S. (1975) *Biochemistry* **14**, 1989-1997.

Weise, C., Kreienkamp, H.-J., Raba, R., Pedak, A., Aaviksaar, A. & Hucho, F. (1990) *EMBO J.* **9**, 3885-3888.

---

[*] J. Yustein & T. L. Rosenberry, unpublished observations.

NEW PHOTOLABILE INHIBITORS OF CHOLINESTERASES DESIGNED FOR RAPID PHOTOCHEMICAL RELEASE OF CHOLINE

L. Peng and M. Goeldner

Laboratoire de Chimie Bio-organique
Faculté de Pharmacie
Université Louis Pasteur Strasbourg
BP 24 - 67401 Illkirch Cedex, France

The catalytic mechanism of rapid hydrolysis of acetylcholine by acetylcholinesterase (AChE) and butyrylcholinesterase (BuChE) has gained new interest since the description of the 3D-structure of AChE (Sussman *et al.*, 1991) and the deduced model-structure of BuChE (Harel *et al.*, 1992). In particular, it is still unclear how choline is cleared from the active site which is located at the bottom of a deep, narrow "aromatic gorge". The development of the time-resolved crystallography (Cruickshank *et al.*, 1992) has made it possible to analyze the catalytic process of an enzyme at the atomic level. In order to study the dynamic process of choline escape by such method, three photolabile choline precursors (Figure) were synthesized and tested for their potential use in time-resolved crystallographic studies on cholinesterases.

All three probes showed reversible inhibitory properties on both purified *Torpedo* AChE and purified human plasma BuChE with inhibition constants in the range of 10^{-5} - 10^{-6} M (Figure). The photochemical reactions of these probes were analyzed by UV-spectroscopy, HPLC and an enzymatic assay for choline. The results demonstrated that the formation of choline was concomitant with the decay of its parent compounds during photolysis and that the conversion of choline from its precursors was stoichiometric. The kinetics of the photochemical reactions of the probes were determined by single laser flash photolysis. The rates of choline release showed a strong dependence on both pH values and the substituents at the α-benzylic position (Figure 1). Probe **A**, *O*-[1-(2-nitrophenyl)ethyl]choline iodide, is of special interest because of its high quantum yield of 0.27 and of its rapid photochemical release of choline in the microsecond range (Figure 1), which is comparable with the turnover rate of AChE (2×10^4 s^{-1}).

In conclusion, probe **A** should provide ideal experimental conditions for time-resolved crystallographic studies on release of choline from the active site of AChE.

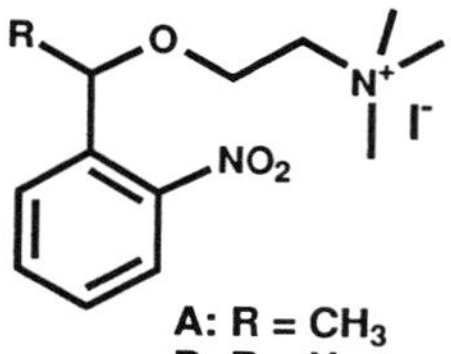

A: R = CH_3
B: R = H
C: R = CH_2NCS

probe	Ki (AChE)	Ki (BuChE)	$t_{1/2}$	Qp
A	1.3×10^{-5} M	1.1×10^{-5} M	1×10^{-5} s	0.27
B	1.0×10^{-5} M	1.9×10^{-5} M	8×10^{-2} s	0.19
C	3.7×10^{-6} M	3.1×10^{-5} M	6×10^{-4} s	0.26

Figure 1. New photolabile choline precursors: their structures, their inhibition constants on both AChE and BuChE, the kinetics ($t_{1/2}$) of their photochemical release of choline and their quantum yields (Qp).

REFERENCES

Cruickshank, D.W.J., Helliwell, J.R., and Johnson, L.N., Ed. *Phil. Trans. R. Soc. Lond. A* **1992**, *340,* 167-334.

Harel, M., Sussman, J.L., Krejci, E., Bon, S., Chanal, P., Massoulié, J. and Silman, I. *Proc. Natl. Acad. Sci. USA* **1992**, *89,* 10827-10831.

Sussman, J.L., Harel, M., Frolow, F., Oefner, C., Goldman, A., Toker, L. and Silman, I.; Science, **1991**, 872-879.

AMINO ACIDS DETERMINING SPECIFICITY TO OP-AGENTS AND FACILITATING THE "AGING" PROCESS IN HUMAN ACETYLCHOLINESTERASE

A. Ordentlich, C. Kronman, D. Stein, N. Ariel, S. Reuveny, D. Marcus, Y. Segall, D. Barak, B. Velan, and A. Shafferman

Israel Institute for Biological Research
Ness-Ziona, Israel

Cholinesterases (ChE's) are readily phosphylated at the active site serine, by a variety of organophosphorus agents (OP) (Aldrich *et al.*, 1972; Taylor, 1990). The enzyme can be reactivated by various oxime nucleophiles but in certain cases reactivation is thwarted due to a concomitant unimolecular process termed aging (Aldrich *et al.*, 1972). The aged enzyme-OP-conjugate is refractive to reactivation (Gray, 1984) and thus renders treatment, following intoxication with certain OP insecticides or nerve gas agents, extremely difficult (Glickman *et al.*, 1984). Recombinant human acetylcholinesterase (HuAChE) and several selected active site gorge mutants: W86A, E202Q, F295A, F297A, Y337A and E450A of HuAChE were studied with respect to catalytic activity towards charged and noncharged substrates, phosphylation by organophosphorus (OP) inhibitors: DFP, paraoxon and soman and subsequent aging of the resulting OP-conjugates. On the basis of these studies we have identified some of the critical elements in the active center that determine specificity to various OP-agents. Trp86, a key element of the active center "anionic subsite" that stabilizes noncovalent complexes of charged ligands by cation - π interactions (Ordentlich *et al.*, 1993a) also contributes to chemical reactivity towards various noncharged OP - inhibitors as manifested by a 5-20 fold decrease in rates of phosphorylation observed for W86A mutant. Replacement of Tyr337 by alanine had a minimal effect on the bimolecular rate constant of inhibition by the OP-molecules. The 20-fold increase in phosphorylation rates of the F295A mutated enzyme by paraoxon is attributed to the better accommodation of the ethoxy moiety of the inhibitor by the less bulky alanine residue. This is consistent with the role of Phe295 in conferring specificity to the acyl pocket (Ordentlich *et al.*, 1993a). The reduction in catalytic efficiency displayed by the E202Q and E450A mutants, and the marked decrease in the rates of both phosphylation and aging is consistent with the proposed role of these residues as key elements of the hydrogen bond network (Ordentlich *et al.*, 1993b). The role of this network is to maintain the functional architecture of the active center and proper positioning of E202, thereby stabilizing the evolving transition states of acylation and phosphylation. The impaired "aging" process displayed by the E202Q and E450A mutants provides a basis for the engineering of novel reactivatable bioscavengers.

ACKNOWLEDGMENT

This work was supported by contracts DAMD17-93-C-3042 from the United States Army Research and Development Command (to A.S.).

REFERENCES

Aldrich, W.N., and Reiner, E. (1972). Enzyme Inhibitors as Substrates, North - Holland Publishing Co., Amsterdam.
Glickman, A.H., Wing, K.D. and Casida, J.E. (1984). Toxicol. Appl. Pharmacol. 73, 16-22.
Gray, A.P. (1984). Drug Metabolism Rev. 15, 557-589.
Ordentlich, A., Barak, D., Kroman, C., Flashner, Y., Leitner, M., Segall, Y., Ariel, N., Cohen, S., Velan, B. and Shafferman, A. (1993a). J. Biol. Chem. 268, 17083-17095.
Ordentlich, A., Kronman, C., Barak, D., Stein, D., Ariel, N., Marcus, D., Velan, B. and Shafferman, A. (1993b). FEBS Lett. 334, 215-220.
Taylor, P. (1990), in: The Pharmacological Basis of Therapeutics, 8th Ed. (Gilman, A.G., Rall, T.W., Nies, A.S. and Taylor, P., Eds.), pp 131-149, Macmillan Publishing Co., New York.

ELECTROSTATIC ATTRACTION BY SURFACE CHARGE DOES NOT CONTRIBUTE TO THE CATALYTIC EFFICIENCY OF ACETYLCHOLINESTERASE

D. Barak,[1] A. Ordentlich,[1] C. Kronman,[1] R. Ber,[2] T. Bino,[1] N. Ariel,[1] R. Osman, B. Velan,[2] and A. Shafferman[1]

[1] Israel Institute for Biological Research
Ness-Ziona, 70450, Israel

[2] Mount Sinai School of Medicine
CUNY, New York, New York 10029

AChE is an extremely efficient enzyme, believed to operate at, or near the diffusion control limit (Quinn, 1987). Early kinetic estimates led to the conclusion that electrostatic attraction is a driving force for rapid binding of acetylcholine (ACh) to AChE (Quinn, 1987; Nolte *et al.,* 1980). Recently, it was shown that in the structure of *Torpedo californica* AChE (TcAChE) (Sussman *et al.*, 1991) an uneven spatial charge distribution results in a negative electrostatic potential extending roughly over half of the protein surface and a strong directionality of electric field along the axis of the active center gorge (Ripoll *et al.,* 1993). These electrostatic properties were proposed to play an important role in attracting the positively charged substrate, acetylcholine, and in steering it towards and into the active center gorge of the enzyme. To evaluate the contribution of these electrostatic properties to the catalytic efficiency, 20 single and multiple site mutants of human AChE, were generated by replacing up to 7 acidic residues, vicinal to the rim of the active center gorge by neutral amino acids.

Simulated replacement of the five residues Glu84, Glu285, Glu292, Asp349, and Glu358, which roughly encircle the gorge entrance, had a major effect on the contours of the electrostatic potential of selected mutants. For hexa or septa mutants, which include additional acidic residues (Glu389, Asp390), there is already no significant electrostatic potential (>-1kT/e) above the rim of the active site gorge area. In marked contrast to the shrinking of the electrostatic potential, the corresponding mutations had no significant effect on the apparent bimolecular rate constants of hydrolysis for charged and noncharged substrates, or on Ki value for a charged active center inhibitor. Moreover the kcat values for all the 20 mutants, are essentially identical to that of the wild type enzyme and the apparent bimolecular rate constants show a moderate dependence on the ionic strength, which is invariant for all the enzymes examined.

Based on these observations we conclude that: 1. Surface electrostatic properties of AChE do not contribute to the catalytic rate or to binding constants of charged active center ligands. 2. The catalytic rate of AChE hydrolysis is not diffusion controlled. 3. Surface charges play no role in stabilization of the transition states of the catalytic process.

ACKNOWLEDGMENT

This work was supported by contracts DAMD17-93-C-3042 from the United States Army Research and Development Command (to A.S.).

REFERENCES

Nolte, H.J, Rosenberry, T. L. and Neumann. E. (1980). Biochemistry 19, 3705-3711.
Quinn, D. M. (1987). Chem. Rev. 87, 955-979.
Ripoll, D. L., Faerman, C.H., Axelsen, P.H., Silman, I. and Sussman, J.L. (1993). Proc. Natl. Acad. Sci. USA 90, 5128-5132.
Sussman, J.L., Harel, M., Frolow, F., Oefner, C., Goldman, A., Toker, L. and Silman, I. (1991). Science 253, 872-879.

CATALYTIC PROPERTIES OF HUMAN SERUM CHOLINESTERASE PHENOTYPES IN THEIR REACTION WITH SUBSTRATES AND INHIBITORS

V. Simeon-Rudolf, M. Škrinjarić-Špoljar and E. Reiner

Institute for Medical Research and Occupational Health
Ksaverska cesta 2
41001 Zagreb, Croatia

Three samples of native sera identified as the usual UU, the fluoride resistant FS and the atypical AA human serum cholinesterase (EC 3.1.1.8) phenotypes were compared concerning their interaction with charged and uncharged ligands. The enzyme activity was measured and phenotypes identified as described earlier (Simeon et al., 1987). The interaction with reversible inhibitors HI-6, PAM-2 and 4,4'-bipyridine (BP), and progressive inhibitors tabun, sarin, soman, paraoxon, VX and phosphostigmine, and with the dimethyl-carbamate Ro 02-0683, was studied. Acetythiocholine and propionylthiocholine were used as substrates.

Activities of all three phenotypes toward both substrates deviated from Michaelis kinetics, e.g. K_m values were different when calculated from activities at low (0.02 - 0.25 mM) and at high (1 - 10 mM) substrate concentrations. K_m and Hill-coefficients were therefore calculated according to a non-linear regression procedure to fit the Hill equation. For both substrates and all three phenotypes, K_m values were between 0.6 and 1.0 mM, and Hill-coefficients between 0.5 and 0.7. The V_m for AA was about three-times, and for FS about two-times, lower than for the UU phenotype.

Reversible inhibition of phenotypes was measured in the presence of substrate. The enzyme/inhibitor dissociation constants (K_i' for the catalytic site, K_i'' for the allosteric site) were evaluated from the effect of substrate concentrations upon the degree of reversible inhibition (Reiner, 1986; Škrinjarić-Špoljar & Simeon, 1993). No obvious difference between the UU, FS and AA phenotypes was observed concerving their affinities for either charged or uncharged reversible ligands. For the three phenotypes, the mean K_i' values for HI-6, PAM-2 and BP were 0.42, 1.2 and 1.6 mM respectively, and the K_i'' value for BP was 4.6 mM.

In progressive inhibition by tabun, sarin, paraoxon and soman no difference in the rate constants of inhibition (k_i) was found between the phenotypes. However, in inhibition by VX and the positively charged progressive inhibitors phosphostigmine and Ro 02-0683,

k_1 was found to be 21, 47 and 66 times higher for the UU than for the AA, and 3, 4 and 2 times higher than for the FS phenotype.

It seems that binding of the studied compounds (expressed in terms of K_m, K_i' or K_i'') is not influenced by a postive charge, while the overall rates of phosphylation and of acylation/deacylation by the charged compounds (expressed in terms of k_i and V_m) differ for the three phenotypes.

REFERENCES

Simeon, V. Buntić, A., Šurina, B. and Flegar-Meštrić, Z., 1987, *Acta Pharm. Jugosl.* 37:107-114.
Reiner, E., 1986, *Croat. Chem. Acta* 59(4):925-931.
Škrinjarić-Špoljar, M. and Simeon, V., 1993, *J. Enz. Inhib.*, 7:169-174.

ORGANOPHOSPHATE SPECIFICITY OF ACYL POCKET CHOLINESTERASE MUTANTS

Natilie A. Pickering,[1] Palmer Taylor,[1] and Harvey Berman[2]

[1] Department of Pharmacology
University of California at San Diego
La Jolla, California 92093-0636
[2] Department of Biochemical Pharmacology
State University, New York
Buffalo, New York, 14260

Organophosphate (OP) compounds potently and irreversibly inhibit cholinesterases by phosphorylating the active site serine. Their tetrahedral geometry about the phosphate, in contrast to the trigonal carbonyl carbon of carboxyl or carbamoyl ester substrates, adds another dimension to substituent substitution. Therefore, size and geometric constraints as well as stereospecificity can be examined. As reported (Berman & Leonard, 1989), *Torpedo* acetylcholinesterase (AChE) has marked stereospecificity and substituent size specificity to OP's with the differences in reaction rates up to 200 fold between enantiomers. This specificity is suspected to be due to steric hindrance as found for substrate specificity between AChE and butyrylcholinesterase (BuChE) (Vellom et. al., 1993). The side chains of F295 and F297 have been shown in several studies to form the steric constraints on acyl pocket size, however, these residues will be shown to have distinct influences on substrate catalysis and substrate inhibition for carbonyl substrates. In the case of alkylphosphates, a more discriminating analysis of structure function is possible where we have examined acylation by the two enantiomers of a series of alkyl methyl phosphonyl thiocholines in relation to acyl pocket sustituents. Such studies provide a data base for modeling the orientation of substituent groups in the transition state and the acyl enzyme. (Supported by USHPHS grant GM 18360 & DAMD17-9-C1058).

REFERENCES

Berman, H.A., and Leonard, K., 1989, Chiral Reactions of Acetylcholinesterase Probed with Enantiomeric Methylphosphonothioates, *J. Biol. Chem.* 264(7): 3942-3950.

Vellom, D.C., Radić, Z., Li, Y., Pickering, N.A., Camp, S., and Taylor, P., 1993, Amino Acid Residues Controlling Acetylcholinesterase and Butyrylcholinesterase Specificity, *Biochemistry* 32:12-17.

LONDON DISPERSION INTERACTIONS IN MOLECULAR RECOGNITION BY ACETYLCHOLINESTERASE

Daniel M. Quinn,[1] Haridasan K. Nair,[1] Javier Seravalli,[1] Keun Lee,[1] Tomira Arbuckle,[1] Zoran Radić ,[2] Daniel C. Vellom,[2] Natilie Pickering,[2] and Palmer Taylor[2]

[1] Department of Chemistry
The University of Iowa
Iowa City, Iowa 52242
[2] Department of Pharmacology
University of California-San Diego
LaJolla, California 92093

ABSTRACT

Interaction of the aromatic residues W84, Y130 and F330 (or Y330) of *Torpedo californica* (Nair *et al.*, 1994) and mouse AChEs with the quaternary ammonium function of ligands and substrates has been probed by quantitative structure-activity relationship (QSAR) and site-directed mutagenesis studies. For a series of ten *meta*-substituted aryl trifluoroketone inhibitors, m-$XC_6H_4COCF_3$ (X = H, Me, CF_3, Et, iPr, tBu, NO_2, NH_2, NMe_2, Me_3N^+), inhibitor potency (i.e. pK_i) is not correlated with substituent hydrophobicity, but is well described by a three-dimensional correlation with the molar refractivity (MR) and σ_m of the substituents. MR depends on surface area and polarizability, and thus is a measure of London dispersion and other induced polarization interactions between ligands and the aromatic residues of the quaternary ammonium binding locus of the active site. Of the 10^7-fold range of inhibitor potency, 10^5 arises from the MR sensitivity, and the corresponding linear subcorrelation indicates that all substituents share a common binding locus and interaction mechanism. A reasonable linear correlation of pK_i for inhibition by the m-Me_3N^+ ketone versus amino acid MR values for a series of W84 mutants of mouse AChE, which includes W84Y, W84F and W84A, indicates that about a third of the binding free energy in the quaternary ammonium binding locus comes from interaction with the tryptophan indole ring. For the native mouse enzyme and these three mutants, a plot of pK_i for inhibition by the m-Me_3C ketone versus pK_i for inhibition by the m-Me_3N^+ ketone is linear with a slope of 0.7. The greater sensitivity of the charged inhibitor indicates that additional interactions, not enjoyed by the neutral inhibitor, with W84 are operating. These likely include ion-quadrupole and ion-polarizability interactions with the π-electrons of W84 (Kim *et al.*, 1994).

Similar correlations for substrate turnover support the importance of dispersion interactions in AChE catalysis. These studies indicate that the aromatic residues in the quaternary ammonium binding locus of the active site do not function as classical anionic sites.

REFERENCES

Kim, K.S., Lee, J.Y., Lee, S.J., Ha, T.-K. & Kim, D.H. (1994) *J. Am. Chem. Soc.* **116**, 7399-7400.
Nair, H.K., Seravalli, J., Arbuckle, T. & Quinn, D.M. (1994) *Biochemistry* **33**, 8566-8576.

PERIPHERAL ANIONIC SITE OF WILD-TYPE AND MUTANT HUMAN BUTYRYLCHOLINESTERASE

P. Masson,[1] M. T. Froment,[1] C. Bartels,[2] and O. Lockridge[2]

[1] Centre de Recherches du Service de Santé des Armées
Unité de Biochimie B P 87
38702 La Tronche Cédex, France
[2] Eppley Institute
University of Nebraska Medical Center
Omaha, Nebraska 68198-6805

Although the existence of a peripheral "anionic" site (PAS) on the surface of AChE has long been demonstrated, there was no direct evidence for a PAS on BuChE. However, structural similarities between the two enzymes and some kinetic complexities of BuChE suggested that the mouth of the BuChE active site gorge may possess PAS components.

Human butyrylcholinesterase (BuChE) was mutated in several residues located on the mouth of the active site gorge: A277(279)W/H, Q119(121)Y, G283(285)D and D70(72)G; the double mutant A277W,G283D was also constructed. Wild-type and mutated genes were expressed in stably transfected human embryonal 293 kidney cells. Kinetics of recombinant wild-type BuChE and its mutants was examined. Ionic strength dependence of catalytic parameters was determined for wild-type BuChE and the D70G mutant. Reversible inhibition was analyzed with two peripheral site (PAS) ligands of AChE, propidium and fasciculin-2, and with an inhibitor which binds to the active site gorge, dibucaine.

The wild-type BuChE shows a non Michaelian behavior with butyrylthiocholine (BuSCh) with activation at concentration [S] > 0.1 mM followed by an inhibition phase at [S] > 40 mM. The D70G mutant displays also a non Michaelian behavior with BuSCh with a faint substrate activation but an inhibition phase beyond 20 mM. This indicates that (D70) is implicated in substrate activation. The other mutants behave like wild-type BuChE with respect to BuSCh hydrolysis. The ionic strength dependence of the substrate activation phase for wild-type BuChE indicates that binding of the second BuSCh molecule involves a solvent-exposed negatively charged residue (D70). The fact that the D70G mutant shows a significant ionic strength dependence suggests that another negatively charged residue may also be involved. The effect of ionic strength on the substrate inhibition phase for wild-type BuChE indicates that binding of BuSCh at [S] > 40 mM does not involve charged residues. Since binding is slightly increased upon increasing the ionic strength, it may correspond to binding of a third substrate molecule on hydrophobic residues located in the active site gorge.

The biphasic ionic strength dependence of the D70G mutant suggests that at low NaCl concentrations binding of the third BuSCh molecule to this mutant may involve electrostatic interactions, but at high NaCl concentration binding involves mainly hydrophobic interactions.

The wild-type BuChE shows substrate-dependent inhibition by dibucaine (Masson *et al.*, 1993), competitive inhibition by propidium and non competitive inhibition by fasciculin at low BuSCh concentration. On the other hand, the D70G mutant shows weak competitive inhibition by dibucaine, very weak non competitive inhibition by propidium and mixed inhibition by fasciculin. Thus, residue D70 is important for binding of propidium. The other mutants behave like wild-type BuChE. Thus, except D70, residues neighboring the rim of the active site gorge do not play an important role in substrate binding and are not involved in kinetic complexities. Inhibition of the A277W mutant by propidium interacts on PAS with both D70 and W277, thus confirming results reported for AChE (Radic *et al.*, 1994). However, with wild-type BuChE propidium interacts inside the active site gorge with both D70 and residues involved in substrate binding.

It is stated that D70 is the main element of the peripheral anionic site of butyrylcholinesterase, but the contribution of another negatively charged residue (E276 ?, cf. Barak *et al.*, 1994) cannot be ruled out.

REFERENCES

Barak, D., Kronman, C., Ordentlich, A., Ariel, N., Bromberg, A., Marcus, D., Lazar, A., Velan, B. and Shafferman, A. (1994) *J. Biol. Chem.*, *264*, 6296-6305.
Masson, P., Adkins, S., Gouet, P. and Lockridge, O. (1993) *J. Biol. Chem.*, *268*, 14329-14341.
Radic, Z., Duran, R., Vellom, D.C., Li, Y., Cervenansky, C. and Taylor, P. (1994) *J. Biol. Chem.*, *269*, 11233-11239.

DUAL CONTROL OF ACETYLCHOLINESTERASE CONTENT OF MATURE INNERVATED FAST MUSCLES

Victor Gisiger

Départment d'Anatomie
Université de Montréal
CP 6128, Succursale Centre-Ville
Montréal H3C 3J7, Canada

INTRODUCTION

A series of studies has established that enhancement of neuromuscular activity, as achieved by exercise training programs, results in very selective G_4 adaptations in fast muscles of the rat, while the other forms, in particular the junction-associated A_{12}, remain essentially unchanged (Fernandez & Donoso, 1988; Jasmin & Gisiger, 1990; Gisiger et al., 1991). Actually, fast muscles exhibit G_4 increases or decreases according to whether their action during training is predominantly dynamic or tonic (Jasmin & Gisiger, 1990). As for the slow-twitch soleus, in spite of being heavily recruited during exercise (Gardiner et al., 1982; Roy et al., 1985, 1991), it invariably responds to training by only minor and non-selective AChE changes. The very selectivity of the AChE adaptations suggests that G_4 and the other molecular forms, including the A_{12} form, constitute separate compartments in fast muscles which are subject to distinct regulations. Studies performed recently, in collaboration with the group of Dr. Phillip Gardiner at University of Montréal and Dr. Bernard Jasmin at University of Ottawa, support this conclusion and provide additional information on the factors controlling these two AChE compartments.

COMPENSATORY HYPERTROPHY (JASMIN ET AL., 1991)

First, we examined the impact on AChE molecular forms of compensatory overload which is an alternative model of enhanced tonic neuromuscular activity. In this approach, the workload of several synergists is transfered by tenotomy to one muscle. In contrast to the training programs used previously, this classical model generates a significant hypertrophy of the overloaded muscle which exhibits a concomitant increase in total AChE activity (Guth et al., 1966; Snyder et al., 1973). In our case, overload of the medial gastrocnemius was induced by way of bilateral section of the distal tendons of soleus, plantaris and lateral

Enzymes of the Cholinesterase Family, Edited by Daniel M. Quinn et al.
Plenum Press, New York, 1995

gastrocnemius. In this condition, the weight of the medial gastrocnemius increased by 25 %, and total AChE activity by 28 %. However, the increase in enzyme content was not confined to G_4 but affected almost equally all molecular forms, resulting in a general increase of the forms. Nonetheless, the specific activities of the molecular forms were essentially unchanged, with the exception of G_1, a fact which is consistent with the similar increases in weight and total AChE activity displayed by the overloaded muscle. It has been shown that, in compensatory hypertrophy, enlargement of the muscles fibers is accompanied by a corresponding expansion of the size of the motor endplates (Granbacher, 1971). Thus, it appears that the general increase in the content of all molecular forms compensated for the increase in size of the endplates, therefore maintaining the concentration and, probably, the spatial distribution of the various molecular forms. This suggests that the size of the endplates constitutes one factor regulating the amount of A_{12} present in innervated muscles.

TETRODOTOXIN-GENERATED MUSCLE PARALYSIS (JASMIN ET AL., 1994)

Another aspect of AChE regulation we were interested in is the molecular form content of mature innervated, but inactive muscles. Activity of the hindlimb muscles was prevented by blocking the conduction of action potentials in the sciatic nerve with tetrodotoxin (TTX). This was achieved by placing around the sciatic nerve of rats a silastic cuff to which TTX is continuously delivered via an osmotic pump. This method which results in total conduction blockade for two weeks, has several advantages of importance to us. First, consistent with the fact that TTX blockade does not impair axonal transport (Lavoie et al., 1976) and the spontaneous release of transmitter (Drachman et al., 1982), the TTX-generated paralysis is easily reversible: rats already begin to move their hindlimbs less than 12 hours after removal of the cuff, and anatomical as well as functional recuperation occurs within 4 weeks (St-Pierre, 1987). Second, the morphological changes of the muscles are limited to a marked reduction in the size of both their fibers and motor endplates with no significant concurrent damage of the fibers (Gardiner et al., 1991).

After 10 days of TTX blockade, three hindlimb muscles, EDL, plantaris and soleus, were taken for analysis of their AChE content. These muscles displayed a marked atrophy, with a weight loss of about 40-50 % compared to controls. As expected, their AChE activity was sharply reduced, too, by about 50-70 %. Interestingly, analysis of the molecular forms revealed a marked convergence in the AChE content of the muscles. The profiles of both EDL and plantaris were transformed into ones resembling that of the slow-twitch soleus (Gisiger & Stephens, 1982-83; Massoulié et al., 1993), the latter showing essentially no change. This transformation resulted mainly from a preferential reduction of G_4, superimposed over a general decrease affecting all molecular forms in both fast and slow-twitch muscles. What was especially striking is that the TTX-generated paralysis reproduced the effect of murine muscular dystrophy (Gisiger & Stephens, 1983) and other neuromuscular diseases (Massoulié et al., 1993) on AChE molecular forms, which, too, is characterized by a shift of the fast muscles' AChE content toward that typical of their slow-twitch counterparts. It should be mentioned here that, in a recent publication describing the impact of botulinum toxin-induced paralysis on the AChE content of EDL, Sketelj and collaborators (1993) reported, two weeks after toxin application, molecular form profiles resembling those produced by TTX.

Together, these facts suggest the following conclusions. Once innervation is physically established, muscles contain a basic equipment of AChE molecular forms of the type displayed by the soleus muscle. The size of this set of molecular forms would depend primarily, though not exclusively, on the state of innervation, including the dimension of the motor endplates. In turn, the G_4 pool, characteristically found in fast muscles, constitutes an additional compartment of AChE, separate from the basic set of molecular forms in both its localization and regulation. Indeed, in addition to be predominantly located around the motor endplates, this G_4 pool seems to exist only as far as fast muscles exhibit an activity of the dynamic type (Jasmin & Gisiger, 1990; Massoulié et al., 1993).

VOLUNTARY WHEEL RUNNING (GISIGER ET AL., 1994)

One of the unanswered questions regarding the G_4 pool characterizing fast muscles is how the size of this pool relates to the amount of dynamic muscle activity. We addressed this problem by examining G_4 plasticity resulting from voluntary activity as evoked by housing rats in live-in wheel cages. Indeed, in this model, rats spontaneously run for great and also variable distances (Rodnick et al., 1989; Lambert & Noakes, 1990). Upon introduction in these wheel-cages, each rat begins to run at its own individual pace for distances which while varying widely from day to day as well as from one rat to another, progressively increase up to a plateau reached after 3 to 4 weeks (Rodnick et al., 1989). The rats exhibited considerable amounts of activity, completing at the end of the 4th week about 12,000 revolutions per day, on the average, which corresponds to a distance of about 18 kilometers per day.

We measured the effect of spontaneous activity on AChE after 5 days, and 4 weeks of wheel running in the same 5 hindlimb muscles used in the training experiments, namely 4 fast muscles and the slow-twitch soleus. All fast muscles exhibited massive increases of their G_4 form. Both the extent of the G_4 adaptations and the level of voluntary activity displayed by the individual rats varied over a wide range. However, the muscles of the most active rats showed the greatest increases: more than 4 times after 4 weeks of voluntary exercise resulting in the total AChE activity to more than double. The G_4 adaptations were very selective, with only minor changes, if any at all, affecting the other forms. Comparison of the sedimentation of the molecular forms in the presence of Triton X-100 and Brij 96 revealed that the bulk of the additional G_4 molecules was amphiphilic. The similarity in amphiphilic properties exhibited by muscle G_4 in exercising and sedentary rats suggests that the tetramer has the same intramuscular distribution in both situations, namely a preferential perijunctional localization (Gisiger & Stephens, 1988; Massoulié et al., 1993).

Taking advantage of the wide range of both wheel running and G_4 adaptations exhibited by individual rats, we searched for a correlation between these two parameters. The best fit was found by relating the G_4 content with the logarithm of the number of wheel revolutions completed during the 5 days before sacrifice. The correlations were highly significant suggesting that the amount of G_4 present in fast muscle depends on the level of recent activity, and that G_4 of fast muscles continuously adapts to the level of activity. This fact together with the large magnitude of the G_4 adaptations strongly support our hypothesis that G_4 of fast muscles fulfils an important function of its own, distinct from that of the junction-associated A_{12} (Gisiger & Stephens, 1988).

We have proposed earlier that the mainly perijunctional G_4 controls endplate excitability by eliminating acetylcholine molecules diffusing out of the synaptic cleft (Jasmin & Gisiger, 1990). This would prevent elevation of acetylcholine background level during high-frequency trains of impulses and the ensuing desensitization of nicotinic receptors. Interestingly, electrophysiological data obtained in mice, a species expected to display G_4 responses similar to those in rats (Gisiger & Stephens, 1982-83, 1988), seem to support our proposal. Indeed, Wernig and collaborators (Dorlöchter et al., 1991) have found in mouse EDL that voluntary wheel running increases EPP amplitude as well as the resistance of neuromuscular transmission to blocking by curare or Mg^{++}, two adaptations quite consistent with an increased endplate excitability.

ARE ALL MUSCLE FIBERS EQUALLY DEPENDENT ON AChE FOR THEIR FUNCTION?

The marked G_4 plasticity induced by voluntary activity would appear to directly contradict the view that, in the neuromuscular junctions of non-exercising animals, AChE sites are already present in excess of the quantity required for normal function, a notion reiterated recently by Anglister and collaborators (1994). The source of this seeming contradiction probably resides in our present concept of how the neuromuscular junction works. Indeed, in the light of accumulating evidence including the massive G_4 adaptations to voluntary activity, this concept appears more and more as an oversimplification, an approximation of reality. This statement is further supported by two additional observations.

The first one is the AChE adaptations to wheel running shown by soleus (Gisiger et al., 1994). This slow-twitch muscle, which is heavily recruited during running (Gardiner et al., 1982; Roy et al., 1985, 1991), responded by a marked selective reduction of its asymmetric forms, up to 45 % in the case of A_{12}. In other words, while being faced with increased trans-synaptic activation, soleus actually reduced its already low AChE content by selectively diminishing its junction-associated AChE molecules.

The second example comes from a recent study (Panenic et al., 1994), in which we examined the impact of AChE inhibition on the isometric tension developed by gastroc-nemius medialis in response to stimulation by the nerve at increasing frequencies. AChE was irreversibly inhibited to unmeasurable levels by in situ application of MSF (methane-sulfonyl fluoride) followed by abundant rinsing. Classically, the untreated muscles respond to increasing stimulation frequencies by increased tensions, up to a maximum reached when tetanus occurs. The response of inhibited muscle to indirect stimulation at 25 Hz is also well established (Hobbiger, 1976): there is, at first, a potentiation of the tension followed by a plateau of lower value which is maintained for the remainder of the stimulation (i.e., 800 msec in our conditions). What has not been reported before are the tensions developed at higher frequencies, 50 and 200 Hz in this case. Strikingly and opposite to what the present concept of neuromuscular transmission predicts, the level of tension in the plateau increased with the stimulation frequency. Also, the oscillations of tension remained well synchronized with the stimulation impulses indicating that a significant portion of fibers continued to respond to each stimulus in spite of the absence of active AChE. This seems to suggest that gastrocnemius of the rat contains a subpopulation of muscle fibers which depend very little on AChE for their function, are less prone to desensitization. Characterization of these particular muscle fibers, on which we are working, should be a key element in clarifying the functional meaning of the AChE adaptations to exercise, in particular the intriguing decreases induced in fast muscles by tonic activity (Jasmin & Gisiger, 1990) and in soleus by wheel running (Gisiger et al., 1994).

REFERENCES

Anglister, L., Stiles, J.R., and Salpeter, M.M., 1994, Acetylcholinesterase density and turnover number at frog neuromuscular junctions, with modeling of their role in synaptic function, *Neuron* 12:783-794.

Dorlöchter, M., Irintchev, A., Brinkers, M., and Wernig, A., 1991, Effects of enhanced activity on synaptic transmission in mouse extensor digitorum longus muscle, *J. Physiol.* 436:283-292.

Drachman, D.B., Stanley, E.F., Pestronk, A., Griffin, J.W., and Price, D.L., 1982, Neurotrophic regulation of two properties of skeletal muscle by impulse-dependent and spontaneous acetylcholine transmission, *J. Neurosci.* 2:232-243.

Fernandez, H.L., and Donoso, J.A., 1988, Exercise selectively increases G_4 AChE activity in fast-twitch muscle, *J. Appl. Physiol.* 65:2245-2252.

Gardiner, K.R., Gardiner, P.F., and Edgerton, V.R., 1982, Guinea pig soleus and gastrocnemius electromyograms at varying speeds, grades, and loads, *J. Appl. Physiol.* 52:451-457.

Gardiner, P., Favron, M., and Corriveau, P., Contractile and histochemical responses of rat gastrocnemius to TTX-disuse, *Can. J. Physiol. Pharmacol.* 70:1075-1081.

Gisiger, V., and Stephens, H.R., 1982-83, Correlation between the acetylcholinesterase content in motor nerves and their muscles, Gif Lectures in Neurobiology, *J. Physiol.(Paris)* 78:720-728.

Gisiger, V. and Stephens, H.R., 1983, Asymmetric and globular forms of AChE in slow and fast muscles of 129/ReJ normal and dystrophic mice, *J. Neurochem.* 41:919-929.

Gisiger, V., and Stephens, H.R., 1988, Localization of the pool of G_4 acetylcholinesterase characterizing fast muscles and its alteration in murine muscular dystrophy, J. Neurosci. Res. 19:62-78.

Gisiger, V., Sherker, S., and Gardiner, P.F., 1991, Swimming training increases the G_4 acetylcholinesterase content of both fast ankle extensors and flexors, *FEBS Lett.* 278:271-273.

Gisiger, V., Bélisle, M., and Gardiner, P.F., 1994, Acetylcholinesterase adaptation to voluntary wheel running is proportional to the volume of activity in fast, but not slow, rat hindlimb muscles, *Europ. J. Neurosci.* 6:673-680.

Granbacher, N., 1971, Relation between the size of muscle fibers, motor endplates and nerve fibers during hypertrophy and atrophy, *Z. Anat. Entwicklungsgesch.* 135:76-87.

Guth, L., Brown, W.C., and Ziemnowicz, J.D., 1966, Changes in cholinesterase activity of rat muscle during growth and hypertrophy, *Am. J. Physiol.* 211:1113-1116.

Hobbiger, F., 1976, Pharmacology of anticholinesterase drugs, In: Neuromuscular Junction, *Handb. Exp. Pharmacol.* 42:486-581.

Jasmin, B.J., and Gisiger, V., 1990, Regulation by exercise of the pool of G_4 acetylcholinesterase characterizing fast muscles: opposite effect of running training in antagonist muscles, *J. Neurosci.* 10:1444-1454.

Jasmin, B.J., Gisiger, V., and Michel, R.N., 1994, Fast muscles display an acetylcholinesterase molecular form profile typical of the slow-twitch soleus following tetrodotoxin-induced paralysis, submitted.

Jasmin, B.J., Gardiner, P.F., and Gisiger, V., 1991, Muscle acetylcholinesterase adapts to compensatory overload by a general increase in its molecular forms, *J. Appl. Physiol.* 70(6):2485-2489.

Lambert, M.I., and Noakes, T.D., 1990, Spontaneous running increases $VO_{2\ max}$ and running performance in rats, *J. Appl. Physiol.* 68:400-403.

Massoulié, J., Pezzementi, L., Bon, S., Krejci, E., and Vallette, F.M., 1993, Molecular and cellular biology of cholinesterases, *Prog. Neurobiol.* 41:31-91.

Panenic, R., Gardiner, P.F., and Gisiger, V., 1995, Effect of acute irreversible inhibition of acetylcholinesterase on nerve stimulation-evoked contractile properties of rat medial gastrocnemius, (Abstr.) Proc. Fifth International Meeting on Cholinesterases, in: *Enzymes of the Cholinesterase Family*, D.M. Quinn et al. eds., Plenum Press, New York (this volume) page 287.

Rodnick, K.J., Reaven, G.M., Haskell, W.L., Sims, C.R., and Mondon, C.E., 1989, Variations in running activity and enzymatic adaptations in voluntary running rats, *J. Appl. Physiol.* 66:1250-1257.

Roy, R.R., Hirota, W.K., Kuehl, M., and Edgerton, V.R., 1985, Recruitment patterns in the rat hindlimb muscle during swimming, *Brain Res.* 337:175-178.

Roy, R.R., Hutchison, D.L., Pierotti, D.J., Hodgson, J.A., and Edgerton, V.R., 1991, EMG patterns of rat ankle extensors and flexors during treadmill locomotion and swimming, *J. Appl. Physiol.* 70:2522-2529.

Sketelj, J., Crne-Finderle, N., Sket, D., Dettbarn, W.D., and Brzin, M., 1993, Comparison between the effects of botulinum toxin-induced paralysis and denervation on molecular forms of acetylcholinesterase in muscles, *J. Neurochem.* 61:501-508.

Snyder, D.H., Rifenberick, D.H., and Max, S.R., 1973, Effects of neuromuscular activity on choline acetyltransferase and acetylcholinesterase, *Exp. Neurol.* 40:36-42.

St-Pierre, D., 1987, Recovery of muscle TTX-induced disuse and the influence of daily exercise. 1. Contractile properties, *Exp. Neurol.* 98(3):472-488.

RESTRICTED REGULATION OF AChE TRANSCRIPTION, TRANSLATION, AND LOCALIZATION

Individual Nuclei Respond to Locally Generated Signals in Multinucleated Skeletal Muscle Fibers

Richard L. Rotundo, Susana G. Rossi, Rosely O. Godinho,
Ana E. Vazquez, and Bhavya Trivedi

Department of Cell Biology and Anatomy
University of Miami School of Medicine
Miami, Florida 33136

INTRODUCTION

The highly localized accumulation of acetylcholinesterase (AChE) at vertebrate neuromuscular junctions, and the paucity of this enzyme in extrasynaptic regions, requires that expression and/or retention be differentially regulated in innervated and non-innervated segments of the muscle fibers. The endpoint of this differential regulation is the increased density of AChE molecules attached to the synaptic basal lamina interposed between the nerve terminal and the postsynaptic specializations of the neuromuscular junction. At the molecular level, regulation can occur at many different points during biogenesis of the several AChE oligomeric forms including transcription, post-transcriptional processing of AChE mRNAs, translation of individual catalytic subunits, and assembly and processing into homo- and hetero-oligomeric forms. Finally, the assembled molecules must be transported, targeted, and retained at their appropriate extracellular sites. Studies from many laboratories, most of them reporting their recent findings during this conference, indicate that AChE expression is regulated at virtually all stages of biogenesis from transcription initiation to final association with the synaptic basal lamina. It therefore follows that all of these regulatory events must themselves be differentially controlled in innervated and non-innervated regions of muscle, and that the sources of the regulatory signals themselves must be highly compartmentalized within the fibers. Our current studies, the emphasis of this chapter, focus on the compartmentalization of AChE regulation in muscle, *in vivo* and in tissue-cultured myotubes, and the possible signaling systems involved in transducing membrane depolarization and muscle contraction into messages capable of modulating AChE biogenesis.

Enzymes of the Cholinesterase Family, Edited by Daniel M. Quinn et al.
Plenum Press, New York, 1995

COMPARTMENTALIZATION IN SKELETAL MUSCLE

Compartmentalization of AChE Synthesis, Assembly, and Localization in Myotubes

Evidence from several laboratories has shown that expression of many, if not most, of the proteins in skeletal muscle is highly compartmentalized (reviewed in Hall and Ralston, 1991). Once expressed, many classes of transcripts do not diffuse far from the nuclei transcribing them thereby increasing the probability that they will be locally translated and their products locally assembled and utilized. The biogenesis of AChE follows a similar pattern where locally-translated mRNAs encode the enzyme molecules that are subsequently attached to the overlying region of the muscle fiber (Rotundo, 1990; Rossi and Rotundo, 1992, Jasmin et al., 1993).

Compartmentalization of AChE mRNA in Myotubes

Studies from our laboratory have shown that primary quail myoblasts initially express low levels of AChE transcripts and protein which subsequently increase prior to the time of myoblast fusion. The transcripts accumulate adjacent to the myoblast nuclei, detected by *in situ* hybridization, and the vast majority of nuclei in newly formed myotubes are bracketed by "patches" containing high concentrations of AChE mRNA (Rossi et al., 1994, and in preparation). As differentiation proceeds, and the myotubes begin to spontaneously contract, a dramatic decrease in the number of nuclei exhibiting accumulations of AChE mRNA occurs, in parallel with the lower mRNA levels measured by Northern blot analysis. Thus it appears that the decrease in AChE mRNA results at least in part from a decrease in the fraction of nuclei expressing this gene rather than a synchronous decrease by all myotube nuclei.

LOCALIZED EXPRESSION OF AChE mRNA AND PROTEIN AT THE ADULT NEUROMUSCULAR JUNCTION

The compartmentalization of AChE expression observed in tissue-cultured myotubes is even more conspicuous in mature fibers isolated from adult muscle. In skeletal muscle fibers, AChE is concentrated at sites of nerve muscle contact (reviewed in Salpeter, 1987) where a portion of the expressed AChE is attached to the synaptic basal lamina (McMahan et al., 1979). Likewise, we have shown that AChE transcripts are more highly concentrated at sites of nerve-muscle contact (Jasmin et al., 1993). In contrast, AChE transcripts are almost undetectable in non-innervated regions where enzyme levels are less than 3-5% of those found at the synapse (Figures 1 and 2).

Although the mechanisms underlying the increased expression of AChE transcripts and enzyme at the neuromuscular junction are unknown, sufficient evidence exists to postulate that locally-generated signals are responsible for increasing AChE expression at sites of nerve-muscle contact. Membrane depolarization, whether spontaneous or nerve-evoked, appears to be necessary for suppressing AChE expression in skeletal muscle (Randall et al., 1988; Randall, 1994), as has been shown for regulation of the nicotinic acetylcholine receptor protein and mRNAs. These observations, along with the apparent lack of diffusion of AChE transcripts in myotubes (Rotundo, 1990) make it unlikely that accumulation of AChE mRNA and protein at the neuromuscular synapse could occur via recruitment of these molecules from neighboring extrajunctional nuclei. It follows, then, that a signaling system or systems responsible for increasing and regulating AChE expression at

Figure 1. Acetylcholinesterase Activity in Innervated and Adjacent Non-innervated Segments of Avian Skeletal Muscle Fibers. Single skeletal muscle fibers were dissected from adult quail posterior latissimus dorsi muscles, stained briefly using a histochemical technique for visualizing AChE at the neuromuscular junction, and cut into single junctional and adjacent non-innervated extrajunctional segments. Each segment was approximately 150 μm in length. Enzyme activity was measured using a micro radiometric assay.

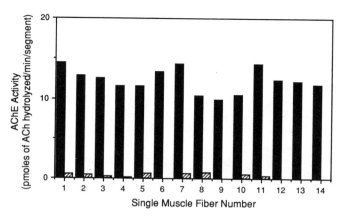

Samples 1-11 are sets of adjacent junctional and extrajunctional segments, while samples 12-14 are extrajunctional segments obtained from more distal non-innervated regions of adult fibers taken for comparison. The mean AChE activity in the non-innervated segments of the fibers was 0.36 ± 0.07 pmol/min whereas the mean activity was 12.24 ± 0.40 pmol/min in the innervated portions, a difference of over thirty fold. Reproduced from Jasmin et al., Neuron 11: 467-477 (1993), with permission.

the transcriptional as well as translational levels must be localized at the postsynaptic region of the neuromuscular junction. Although several candidate signaling systems have emerged, the nature of the physiologically relevant systems functioning in the post-junctional sarcoplasm are unknown.

ATTACHMENT OF AChE TO THE NEUROMUSCULAR JUNCTION AND POSSIBLE ROLE OF HEPARAN SULFATE-LIKE PROTEOGLYCANS

It appears that regardless of the extraction procedures used to solubilized AChE from adult skeletal muscle, short of enzymatic digestion with proteases (Hall and Kelly, 1971;

Figure 2. Relative Acetylcholinesterase Transcript Levels in Innervated and Non-innervated Regions of Adult Muscle Fibers. The AChE transcript levels were determined using a sensitive quantitative reverse transcriptase polymerase chain reaction technique which incorporated quantitation of an internal synthetic AChE RNA standard and endogenous muscle actin transcripts. The average number of AChE transcripts per junctional nucleus is about 800, however there is considerable variability between individual junctions with over half having undetectable levels. (A) Relative transcript levels in small pools (2-10 fiber segments each) consisting of junctional (J; n = 23), adjacent extrajunctional (P; n = 21), and more distal extrajunctional (D; n = 9) segments of adult muscle fibers isolated from the quail posterior latissimus dorsi muscle. (B) Relative AChE transcript levels in single neuromuscular junctions and paired extrajunctional segments. Of the 52 pairs of segments assayed, only 17, or one third, had detectable AChE mRNA. The sensitivity of the assay procedure is sufficient to detect as few as 5 mRNA copies per nucleus, and in all cases the internal standard and the actin transcripts were readily detected. Reproduced from Jasmin et al., Neuron 11: 467-477 (1993), with permission.

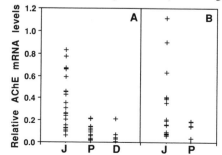

Betz and Sakmann, 1973), a significant portion of the enzyme always remains insoluble. If all of the junctional and extrajunctional AChE molecules were associated with the basal lamina via ionic interactions, then extraction using high ionic strength buffers, ionic detergents, or chaotropic agents, would be expected to remove most if not all of the extracellularly disposed enzyme. This, in fact, appears not to be the case. Recent immunofluorescence localization studies have shown that the synaptically-localized AChE molecules cannot be removed following homogenization of skeletal muscle in high salt and detergent containing buffers. Analysis of the insoluble pelleted material following centrifugation showed that the neuromuscular junctions and their associated AChE remained essentially intact (Figure 3). Even extraction with chaotropic agents such as 8 M urea or 4 M guanidine HCl failed to remove the junctional AChE molecules (Rossi and Rotundo, 1993). Therefore, stronger than ionic interactions must be postulated to account for the these observations. The most likely possibility is that the enzyme is covalently linked to the synaptic basal lamina.

On the other hand, these observations are not inconsistent with the large number of studies indicating that heparin and heparan sulfate-like proteoglycans (HSPs) solubilize asymmetric AChE forms from nerves and muscle (Brandan et al., 1985 and references therein). First, the fraction of total muscle AChE that is actually concentrated at the neuromuscular junction is only a small fraction of the total enzyme. Second, there is clearly a substantial amount of extracellular AChE which is ionically associated with the extracel-

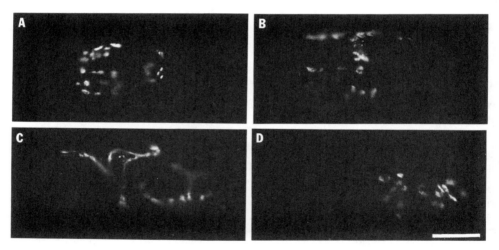

Figure 3. Retention of AChE on the synaptic basal lamina following homogenization in high and low ionic strength buffers. Adult quail anterior latissimus dorsi (ALD) or posterior latissimus dorsi (PLD) muscles were dissected, washed, and homogenized using a polytron in 20 mM borate extraction buffer, pH 9.0, containing 1% Triton X-100, 5 mM EDTA, 5 mg/ml bovine serum albumin, protease inhibitors, and either 1 M NaCl or 150 mM NaCl plus 2 mg/ml heparin. Following centrifugation the supernatants were analyzed by velocity sedimentation whereas the pellets were washed and stained for AChE using double antibody immunofluorescence. The gradient profiles showed that all oligomeric forms of AChE were extracted under these conditions. In contrast, immunofluorescence localization of the remaining non-extractable AChE indicate that it was localized at the sites of nerve muscle contact. (A) ALD fragment extracted with high salt buffer. (B) ALD muscle fragment extracted with low salt buffer and heparin. (C) PLD fragment extracted with high salt buffer. (D) PLD muscle fragment extracted with low salt buffer and heparin. In all cases there was no reduction in fluorescence intensity of AChE staining at the neuromuscular junctions compared to controls indicating that the junctional enzyme was not extracted. Reproduced from Rossi and Rotundo, J. Biol. Chem. 268: 19152-1915 (1993), with permission.

lular matrix and capable of being solubilized using high salt buffers or polyanions. Third, it is likely that a significant fraction of the polyanion-solubilized AChE derives from intracellular pools of newly-synthesized molecules. Finally, it should be emphasized that the "non-extractable" AChE accounts for only a small portion of the total active enzyme molecules and therefore quantitatively its significance was minimized. It just happens to account for the vast majority of the active enzyme concentrated at the neuromuscular synapse (Rossi and Rotundo, 1993).

At the same time, the possible importance of a role for HSP and HSP-like molecules in localizing AChE to the muscle cell surface should not be dismissed. Recent studies from our laboratory have shown that incubation of quail myotubes in culture medium containing heparin prevents the accumulation of asymmetric AChE at cell surface clusters. There is no effect on the synthesis nor assembly of the asymmetric AChE molecules, however their attachment to the cell surface is blocked. The heparin in the medium can be shown to attach to the collagen-like tail which results in secretion of this AChE form rather than localization to the cell surface. Thus one possible role for the HSP molecules may be to serve as an initial attachment or targeting site, with a second step, possibly involving enzymatically catalyzed crosslinking, serving to cement the collagen-tailed AChE to the synaptic basal lamina. The nature of the chemical bonds attaching AChE to the extracellular matrix remain to be determined. One interesting possibility has recently been suggested by Haynes et al. who showed that inhibitors of glutamine transaminase blocked accumulation of surface AChE molecules in tissue-cultured quail myotubes (Hand and Haynes, 1992). Alternatively, the large number of cysteins present in the carboxyl terminus domain of the collagen-like tail polypeptide chains (Krejci et al., 1991) could also participate in attaching AChE to the extracellular matrix.

REGULATION OF AChE BY SECOND MESSENGERS

Regulation of AChE Forms and Transcripts by Membrane Depolarization

Spontaneous contractile activity, or membrane depolarization, is necessary for the expression of the collagen-tailed AChE form in muscle both *in vivo* and in tissue culture (reviewed in Massoulié et al., 1993). Antagonists of voltage-dependent sodium channels decrease expression of the collagen-tailed AChE whereas channel agonists increase its expression. However this regulation must occur at the level of assembly, either through the availability of the tail subunit or some other assembly-related event, since agents that affect membrane depolarization do not affect the rates of translation of the catalytic subunit (Fernandez-Valle and Rotundo, 1989). In contrast to the expression of the AChE protein, the transcripts appear to be regulated in an opposite manner where they more closely parallel the levels of transcripts encoding acetylcholine receptor subunits. Preliminary studies in our laboratory using tetrodotoxin-treated (TTX) myotubes indicate that blocking spontaneous contraction increases total AChE mRNA with most nuclei being surrounded by transcripts, whereas sodium channel agonists such as veratridine or scorpion toxin (ScTx) reduce transcript expression and the number of nuclei with detectable AChE mRNA. These observations are consistent with studies of AChE mRNA levels in innervated and non-innervated segments of adult muscle, as well as studies indicating that AChE transcripts are elevated in denervated avian muscle (Randall, 1994), which suggest that muscle activity plays a role in regulating AChE at the transcriptional level.

Regulation of Cell Surface AChE Clustering

Skeletal muscle fibers synthesize AChE molecules for secretion into the surrounding environment, for localization on the surface plasma membrane, and for attachment to the extracellular matrix enveloping each fiber. In mature adult muscle, the collagen-tailed AChE molecules are more highly concentrated on the synaptic basal lamina, whereas in non-innervated tissue-cultured myotubes many of the cell surface collagen-tailed AChE molecules are localized in discrete clusters or patches attached to the extracellular matrix (Rossi and Rotundo, 1992). These patches may also contain other extracellular matrix proteins such as heparan sulfate proteoglycans, membrane proteins such as acetylcholine receptors and integrins, and perhaps even some molecules derived from the culture medium, although the ratios of these molecules vary considerably from cluster to cluster and not all are detectable in each cluster. However, the similarity in molecular composition between clusters of cell surface acetylcholine receptor and AChE on the one hand, and the neuromuscular junction on the other, lends support to the widely held belief that similar molecular mechanisms are involved in assembling both of these structures.

Since membrane depolarization is necessary for assembly of the collagen-tailed AChE form, and this form is the predominant, if not unique, AChE oligomer present in the cell surface clusters, it is therefore not surprising that antagonists of voltage-dependent sodium channels also prevent accumulation of cell surface AChE clusters whereas sodium channel agonists can increase them. Using indirect immunofluorescence to localize and quantitate surface AChE clusters, we can show that incubation of immature myotubes with sodium channel antagonists prevents formation of AChE clusters, while a similar treatment of fully differentiated myotubes prevents additional accumulation of clusters. Untreated myotubes normally have 0.2-0.3 clusters per nucleus, whereas inactive myotubes have about ten-fold lower cluster densities. In contrast, treatment of myotubes with sodium channel agonists for 24-48 hours raises the number of clusters to around 0.7-0.8 per nucleus, about a 300-400% increase over controls. These studies show that not only is AChE expression and assembly regulated by muscle activity, the mechanisms responsible for its localization are as well.

Down Regulation of AChE by Electrical Stimulation

If nerve-evoked or spontaneous membrane depolarization can downregulate AChE expression in skeletal muscle and increase the relative abundance of the collagen-tailed AChE form, then coordinated membrane depolarization by electrical stimulation of non-innervated myotubes in culture should mimic this effect. To examine this possibility we have electrically stimulated tissue cultured myotubes at a frequency and pattern designed to emulate the firing patterns of motoneurons innervating fast twitch muscle fibers *in vivo*. Electrical impulses of alternating polarity were delivered for 24-48 hours at a frequency of 100 Hz for one second every 100 seconds via platinum electrodes immersed in the culture medium. Control unstimulated cultures were maintained together with the stimulated ones. Preliminary results indicate that electrical stimulation decreases the levels of newly-synthesized globular AChE forms while increasing the total percentage of the collagen-tailed AChE. This system now permits more controlled *in vitro* studies on the mechanism(s) linking membrane depolarization to the transcriptional and translational control of AChE in skeletal muscle.

Generation of Second Messengers Capable of Regulating AChE in Culture

At this point an almost bewildering array of pharmacological agents and chemical compounds have been shown to affect AChE expression in tissue cultured myotubes, which in turn tempers any efforts aimed at deciphering the sequence of events beginning with the release of the neurotransmitter acetylcholine and other molecules from motoneurons to the increased accumulation of AChE at sites of nerve-muscle contact and its decrease in non-innervated regions of the adult muscle fiber. This confusion stems in part from the extensive cross-talk between signaling systems known to occur in all cells, the physiological immaturity of the tissue-cultured muscle cells used in many of the studies compared to fully differentiated fibers in adult organisms, and also to the lack of complete compartmentalization of the signal transduction systems expressed in myotubes compared to their organization *in vivo* where the nerve can induce additional specializations of the post-synaptic membrane. Thus, at present, our best approaches can only serve to indicate which signaling systems might be capable of regulating AChE in mature skeletal muscle fibers rather than identifying those responsible for regulating AChE under normal physiological conditions. With these qualifications in mind, recent observations from our laboratory are summarized.

Previous studies from our laboratory indicated that activators of protein kinase C (PKC) could reverse the effects of muscle paralysis induced by sodium channel antagonists (Fernandez-Valle and Rotundo, 1988) suggesting one possible pathway linking membrane depolarization to changes in AChE expression. More recent studies from our laboratory suggest that diacylglycerol (DAG), an endogenous activator of protein kinase C, can be generated in response to application of cholinergic agonists and, to a lesser extent, membrane depolarization in response to electrical stimulation. The cholinergic agonists act via muscarinic receptors expressed on tissue-cultured myotubes, while the mechanism responsible for the increase in response to electrical stimulation is not known. Cholinergic agonists can increase expression of the collagen-tailed AChE form when applied in culture, which is at least consistent with the observations relating activation of PKC to AChE expression. Whether this same pathway is present in adult skeletal muscle fibers, and more specifically at the neuromuscular junction, is currently under investigation. A second pathway, capable of down regulating AChE expression, appears to involve generation of cAMP. This pathway is of great interest because it could be responsible for suppressing AChE expression in extrajunctional regions of the muscle fibers. In addition to these two major pathways, several additional pathways include the release of Ca^{++} from intracellular and extracellular stores and an unknown pathway responsible for generating intranuclear DAG with a concomitant activation of nuclear PKC. However, the cross talk between these pathways is extensive and the activation of one can affect the functions of others resulting in parallel agonistic and antagonistic effects, even under the most controlled conditions. Therefore, sorting out the various regulatory pathways remains a daunting challenge indeed.

COMPARTMENTALIZATION OF REGULATION: EACH NUCLEUS RESPONDS TO LOCALLY-GENERATED SIGNALS FROM THE PLASMA MEMBRANE

If skeletal muscle fibers are compartmentalized, and AChE is regulated by signals generated at the plasma membrane, then it should follow that AChE expression at the level of individual nuclei should be regulated via signals generated locally on the overlying region of the plasma membrane. To test this hypothesis we used modified microchambers to physically isolate segments of individual myotubes and exposed them to either TTX or ScTx. The results show that localized membrane depolarization increases AChE expression, whereas local inhibition of membrane depolarization decreases it, indicating that each nucleus responds to signals generated on the overlying plasma membrane. These signals are generated by combinations of membrane depolarization and production of second messengers such as DAG which increases A_{12} AChE expression by activating protein kinase C, or increases in cAMP which specifically blocks assembly of A_{12} AChE. Together, these molecular events could provide the foundation for local control of AChE expression in synaptic and extra-synaptic regions of skeletal muscle fibers.

SUMMARY

Multinucleated skeletal muscle fibers are highly compartmentalized cells where each nucleus can receive independent signals specifying regulatory events. Thus the molecular events regulating AChE expression are themselves compartmentalized both *in vivo* and in tissue-cultured myotubes. In adult muscle, AChE oligomers and their transcripts are higher in innervated versus non-innervated regions of the fibers. In tissue-cultured myotubes, AChE transcripts are localized around individual nuclei where their levels are regulated by signals originating on the overlying plasma membrane. The translated AChE polypeptides are then transported to the overlying plasma membrane where the collagen-tailed form is localized into clusters. The presence or absence of clusters over individual nuclei is determined in part by local membrane depolarization, which, together with other signals regulates local expression of this enzyme. These studies provide a structural framework in which to understand the local regulation of AChE expression in mature skeletal muscle together with the necessary compartmentalization of signal transduction systems responsible for increasing AChE at the neuromuscular junction and suppressing its expression in non-innervated portions of the fibers.

ACKNOWLEDGMENTS

This research was supported by grants from the National Institutes of Health to R. Rotundo and A.E. Vazquez and a fellowship from FAPESP to R.O. Godinho. B. Trivedi was supported by the MD/PhD program of the University of Miami School of Medicine.

REFERENCES

Brandan, E., Maldonado, M., Garrido, J., and Inestrosa, N. (1985) J. Cell Biol. 101: 985-992.
Betz, W. and Sakmann, B. (1973) J. Physiol. (Lond.) 230: 673-688.
Fernandez-Valle, C. and Rotundo, R.L. (1989) J. Biol. Chem. 264: 14043-14049.
Godinho, R.O., Trivedi, B., and Rotundo, R.L. (1994) Soc. Neurosci. Abstr. 20.
Hall, Z.W. and Kelly, R.B. (1971) Nature [New Biol.] 232: 62-63.
Hall, Z.W. and Ralston, E. (1989) Cell. 59: 771-772.
Hand, D. and Haynes, L.W. (1992) Biochem. Soc. Trans. 20: 158S.
Jasmin, B.J., Lee, R.K., and Rotundo, R.L. (1993) Neuron 11: 467-477.
Krejci, E., Duval, N., Chatel, J.M., Legay, C., Puype, M., Vandekerckhove, J., Cartaud, J., Bon, S., and Massoulie, J. (1991) The EMBO Journal 10: 1285-1293.
Massoulié, J., Pezzementi, L., Bon, S., Krejci, E., and Vallette, F.-M. (1993) Progress in Neurobiology 41:31-91.
McMahan, U.J., Sanes, J.R. and Marshall, L.M. (1978) Nature 271: 172-174.
Randall, W.R., Fernandez-Valle, C. and Rotundo, R.L. (1988) Soc. for Neurosci. Abstr. 14: 1163.
Randall, W.R. (1994) Third Intern. Symposium on the Cholinergic Synapse. Abst.
Rossi, S.G. and Rotundo, R.L. (1992) J. Cell Biol. 119: 1657-1667.
Rossi, S.G. and Rotundo, R.L. (1993) J. Biol. Chem. 268: 19152-19159.
Rossi, S.G., Vazquez, A.E., and Rotundo, R.L. (1994) Soc. Neurosci. Abstr. 20
Rotundo, R.L. (1990) J. Cell Biol. 110: 715-719.
Salpeter, M.M.(Ed.)(1987) The Vertebrate Neuromuscular Junction. Alan R. Liss, N.Y.

MECHANISM AND IMPLICATIONS OF SELECTIVE NEURAL LESIONS BY ACETYLCHOLINESTERASE ANTIBODIES IN THE CENTRAL AND PERIPHERAL NERVOUS SYSTEMS

AChE Autoimmunity

S. Brimijoin, P. Hammond, and Z. Rakonczay

Department of Pharmacology
Mayo Clinic
Rochester, Minnesota 55905

INTRODUCTION

We are exploring the use of monoclonal antibodies to acetylcholinesterase (AChE) as a means of targeting cholinergic neurons for destruction. By administering such antibodies to experimental animals one can create highly selective cholinergic deficits in brain and spinal cord as well as in the peripheral nervous system (Brimijoin and Lennon, 1990; Bean et al., 1991; Brimijoin et al., 1993; Rakonczay et al., 1993). The immunologically induced deficits show promise as models of natural neurologic disorders in human patients and as tools for neurobiology.

LESIONS IN THE PREGANGLIONIC SYMPATHETIC SYSTEM

Interestingly, discrete subsets of the population of AChE-bearing neurons are selectively vulnerable to damage by AChE antibodies. Other neurons, equally rich in the target antigen, are largely resistant. Thus, it is now well established that AChE antibodies injected systemically into rats destroy preferentially the cholinergic terminals of preganglionic sympathetic neurons. At the same time, these antibodies have only minimal and transient effects on other AChE-bearing cells such as somatic motor neurons, adrenergic neurons, adrenal chromaffin cells, and skeletal muscle. The factors that determine vulnerability or resistance to the antibodies are still unknown and remain under active investigation. In

Enzymes of the Cholinesterase Family, Edited by Daniel M. Quinn et al.
Plenum Press, New York, 1995

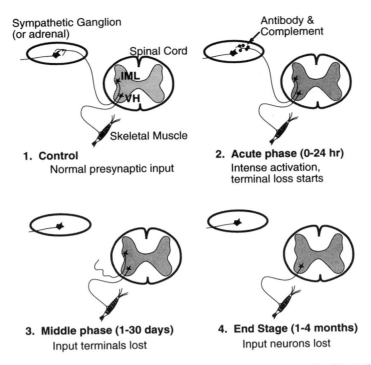

Figure 1. Schematic illustration of preganglionic sympathetic damage by AChE antibodies. 1. Control state. Healthy preganglionic sympathetic neurons are located in the intermediolateral cell column (IML) of the thoracolumbar spinal cord. Preganglionic axons run from these cells to sympathetic ganglia and adrenal medulla via the sympathetic chain, splanchnic nerves and other pathways. 2. Acute phase of lesion (0-24 hr after antibody injection in the tail vein): Anti-AChE IgG is deposited on AChE-bearing nerve terminals throughout the peripheral sympathetic system; complement is activated locally; presynaptic fibers and terminals rapidly disappear from the adrenal medulla and sympathetic ganglia. 3. Middle phase (1-30 days): Antibody and complement deposits are gradually cleared, but presynaptic fibers and terminals recover poorly or not at all. 4) End phase (1-4 months): Up to 70% of preganglionic sympathetic perikarya disappear from the spinal cord IML but cholinergic motor neurons in the ventral horn (VH) are spared. Regeneration or collateral sprouting from surviving sympathetic neurons leads to slight return of presynaptic input.

contrast, the localization and structural characteristics of selective sympathetic lesioning are well established.

General aspects of the experimental dysautonomia induced by AChE antibodies have been described in detail in several recent publications. Key features are: 1) rapid onset of sympathetic dysfunction, evidenced within hours of injection by eyelid drooping, low blood pressure and slowed heart rate (Brimijoin and Lennon, 1990); 2) failure of ganglionic neurons to respond to preganglionic nerve stimulation, as shown by intracellular microelectrodes (Szurszewski and Miller, unpublished observations) 3) simultaneous disappearance of presynaptic fibers and terminals from the adrenal glands and most sympathetic ganglia (Brimijoin et al., 1993); 4) global deficits of ganglionic transmission, reflected by a reduced catecholamine release in response to sympathoadrenal stress (Brimijoin et al., 1994); 5) Persistent impairment of sympathetic function paralleled by permanent biochemical and structural damage in the preganglionic system. Overall, this pattern resembles the pathophysiology in patients with a spontaneous autonomic disorder known as Shy-Drager syndrome or Multiple System Atrophy.

A few of the more important questions surrounding the dysautonomia induced by AChE antibodies are as follows:

- What is the immunologic mechanism of neural damage?
- Why does the damage last so long?
- Is it related to autonomic nerve disease in humans?

These questions form the central focus of our recent work.

Mechanism of Peripheral Lesions by AChE Antibodies

Two independent lines of evidence suggest that local activation of complement is the principal mechanism by which AChE antibodies trigger damage to preganglionic sympathetic nerve terminals. First of all, immunofluorescence microscopy reveals deposition of murine IgG in sympathetic ganglia within hours after injection of antibodies into a peripheral vein (Brimijoin et al., 1993). Specific immunofluorescence is localized in a manner consistent with a preferential binding of AChE antibodies at cholinergic ganglionic synapses. Soon after antibody binding, complement fixation is detected as deposits of immunoreactive C3, the component common to both the classical and alternative complement pathways. The structural evidence for complement activation is supported by the effects of the anti-complement agent, cobra venom factor. This factor, which irreversibly activates component C3 and depletes it from plasma, prevents activation of complement by immune complexes deposited on cells and tissues. It is therefore significant that pretreatment of rats with cobra venom factor blocks the signs of sympathetic damage by AChE antibodies, including eyelid drooping and enzyme loss in ganglia (Brimijoin and Lennon, 1990).

All information thus points toward antibody deposition and complement activation as critical steps toward neural lesions in our animal model. However, these steps are not sufficient, because both AChE antibody and complement component C3 are also deposited at sites which resist lesioning (Brimijoin et al., 1993). Examples of such resistant sites are the motor endplate of the diaphragm and other skeletal muscles, the neuropil of parasympathetic ganglia, and the cholinergic parasympathetic terminals in the heart. We have not yet determined why these structures resist lesioning, but the topology of their AChE, their sensitivity to lysis by complement, and their content of other mediators are factors worth considering.

Reason for Persistent or Permanent Sympathetic Deficits

Normally one would expect autonomic neurons to recover quickly from damage to their nerve terminals. For example, it is well established that preganglionic neurons will regenerate after mechanical lesions to their axons in the cervical sympathetic chain. Therefore, the long lasting or even permanent loss of preganglionic terminals at sites like the superior cervical ganglion suggests that AChE antibodies damage the parent neuron irreversibly. This suggestion was confirmed by neuropathological observations showing that AChE antibodies cause a delayed but substantial loss of preganglionic sympathetic perikarya in the spinal cord (Brimijoin et al., 1993). The neuronal loss occurs without IgG accumulation in the cord or other signs that the blood-cord barrier is penetrated by the antibodies. Furthermore, the neuronal loss does not reach its peak for more than a month after antibody injection, when circulating murine IgG is no longer detectable. At this time, counts of cells with immunoreactive choline acetyltransferase indicate that 70% of the cholinergic preganglionic neurons in the intermediolateral column of the thoracic spinal cord have disappeared. It seems likely that this change reflects delayed neuronal death in reaction to loss of trophic support from the periphery. We are now addressing the possibility that apoptotic mechanisms are involved.

Lessons Concerning Human Dysautonomia

Experimental dysautonomia induced by AChE antibodies is so similar to Multiple System Atrophy that one might ask whether the human disorder reflects spontaneous autoimmunity to AChE. We have not been able to answer this question definitively, but our observations to date do not support the idea of a common etiology. We addressed the issue by looking for circulating anti-AChE antibodies, either IgG or IgM, in patients with various neurological or autoimmune disorders including myasthenia gravis, Lambert-Eaton syndrome, amyotrophic lateral sclerosis, multiple sclerosis, Graves disease, miscellaneous cases of myositis and myopathies, and 50 cases of autonomic neuropathies, including Multiple System Atrophy. Over 350 plasma samples were analyzed. Two patients had IgM antibodies that bound butyrylcholinesterase, but no conclusive evidence for pathogenic anti-AChE antibodies was found. It is still wise to reserve judgment on the possibility of spontaneous AChE autoimmunity, since "active" or acute cases of dysautonomia were not studied. For the meanwhile, however, the causal mechanism of Multiple System Atrophy appears unlikely to involve AChE antibodies. Even so, experimental AChE autoimmunity remains in our view an excellent model of the human disorder.

DIRECT LESIONS IN THE CENTRAL NERVOUS SYSTEM

In addition to an indirect loss of central sympathetic neurons when AChE antibodies are injected in the periphery, central lesions can be produced directly by AChE antibodies introduced into brain or spinal cord. For example, injection of the antibodies into the caudate nucleus of adult rats causes profound local destruction of AChE-positive fibers (Bean et al., 1991). On the other hand, the AChE-rich cell bodies of the cholinergic interneurons that give rise to the damaged fibers appear unscathed by this treatment, at least initially. A variety of peptidergic fibers in the same region are also spared. The mechanism for the immediate and evidently selective central lesioning has not been fully defined but may, as in the periphery, involve activation of complement. It is interesting that the caudate lesions induced by AChE antibodies seem to be at least partly reversible, with significant recovery over a period of two to four weeks.

CNS lesions can be produced in newborn rats by giving AChE antibodies intraperitoneally (Rakonczay et al., 1993). This effect requires injection during the first few days after birth, when the blood brain barrier is relatively permeable to IgG. In comparison with the effects of intracerebral injection in adults, the lesions in newborn brain are more widespread and include almost complete disappearance of the AChE-positive fibers throughout the cerebral cortex, together with more modest lesions of the striatum, medulla, and cerebellum. AChE-rich nerve terminals in the spinal cord are also selectively destroyed (Li et al., 1994). Structural damage in the newborn brains remains extensive for up to two weeks. Beyond this point, however, a surprising degree of plasticity takes hold. By three weeks after treatment, the local density of AChE positive fibers and of other cholinergic markers (e.g., ChAT activity) recovers completely in most areas of the brain. More study is needed to determine why the fibers of central neurons can regenerate or recover while the peripheral processes of preganglionic sympathetic neurons seldom do.

CONCLUSION

It should be worthwhile to exploit AChE antibodies further for probing the function of AChE-rich neural pathways, both in the peripheral autonomic system and in the brain and spinal cord. In the periphery, we have already begun to use the antibody induced lesions as a way to study the transsynaptic regulation of neuropeptides in sympathetic ganglia and adrenal chromaffin cells (Dagerlind et al., 1994a; Dagerlind et al., 1994b; Brimijoin et al., 1995). Antibody effects in newborn brain and spinal cord are potentially of even greater interest for what they imply about the plasticity of the developing central nervous system, and for the possibility of deriving clues to the function of AChE-expressing neural systems. For example, the selective destruction of cholinergic "C-terminal" fibers in the spinal cord (Li et al., 1994) provides a new basis for investigating the physiological role of these specialized inputs to motor neurons in the ventral horn. Other applications will present themselves as the fascinating effects of AChE antibodies in vivo continue to unfold.

ACKNOWLEDGMENTS

This work was supported by the National Institutes of Health under grant NS29646.

REFERENCES

Bean, A. J., Xu, Z., Chai, S. Y., Brimijoin, S. and Hökfelt, T., 1991, Effect of intracerebral injection of monoclonal acetylcholinesterase antibodies on cholinergic nerve terminals in the rat central nervous system, Neurosci. Lett. 133:145-149.

Brimijoin, S., Dagerlind, Å., Rao, R., McKinzie, S. and Hammond, P., 1995, Accumulation of enkephalin, proenkephalin mRNA, and neuropeptide Y in immunological denervated rat adrenal glands: evidence for divergent peptide regulation, J. Neurochem. In Press.

Brimijoin, S., Hammond, P., Khraibi, A. A. and Tyce, G. M., 1994, Catecholamine-release and excretion in rats with immunologically induced preganglionic sympathectomy, J. Neurochem. 62:2195-2204.

Brimijoin, S., Hammond, P., Moser, V. and Lennon, V., 1993, Death of Intermediolateral spinal cord neurons follows selective, complement-mediated destruction of their preganglionic sympathetic terminals by cholinesterase antibodies, Neuroscience 54:201-223.

Brimijoin, S. and Lennon, V. A., 1990, Autoimmune preganglionic sympathectomy induced by acetylcholinesterase antibodies, PNAS 87:9630-9634.

Dagerlind, Å., Pelto-Huikko, M., Lundberg, J. M., Ubink, R., Verhofstad, A., Brimijoin, S. and Hökfelt, T., 1994a, Immunologically induced sympathectomy of preganglionic nerves by antibodies against acetylcholinesterase: increased levels of peptides and their mRNAs in rat adrenal chromaffin cells, Neuroscience in press

Dagerlind, Å., Zhang, X., Brimijoin, S., Lindh, B. and Hökfelt, T., 1994b, Effects of preganglionic sympathectomy on peptides in the rat superior cervical ganglion., Neuroreport 5:909-912.

Li, W., Ochalski, P., Brimijoin, S., Jordan, L. M. and Nagy, J. I., 1994, C-Terminals on motoneurons: EM localization of cholinergic markers in adult rats and antibody-induced depletion in neonates., Neuroscience In Press

Rakonczay, Z., Hammond, P. and Brimijoin, S., 1993, Lesion of central cholinergic systems by systemically administered acetylcholinesterase antibodies in newborn rats., Neuroscience 54:225-238.

NEURAL REGULATION OF ACETYLCHOLINESTERASE EXPRESSION IN SLOW AND FAST MUSCLES OF THE RAT

J. Sketelj[1] and B. Črešnar[2]

[1] Institute of Pathophysiology
[2] Institute of Biochemistry
School of Medicine
University of Ljubljana
61105 Ljubljana, Slovenia

INTRODUCTION

Although acetylcholinesterase (AChE, EC 3.1.1.7) is present all along the muscle fibres, it is highly concentrated in the neuromuscular junction (nmj), especially its asymmetric A_{12} molecular form (Marnay and Nachmansohn, 1938; Hall, 1973; Sketelj and Brzin, 1985). This molecular form of AChE, however, is not motor endplate-specific. Its focalization in the nmj is a developmentally regulated phenomenon: the A_{12} AChE form is very pronounced extrajunctionally in immature rat muscles during early postnatal period, and becomes restricted to the nmjs by the end of the first month after birth (Sketelj and Brzin, 1980).

Several lines of evidence indicate that AChE activity in muscles is regulated by the motor nerve. First, AChE focalization in the nmjs starts shortly after establishment of nerve-muscle contacts and is limited strictly to the region underneath the nerve ending (Brzin et al., 1981). Second, AChE regulation in fast muscles is significantly different from that in slow ones (Bacou et al., 1982; Groswald and Dettbarn, 1983, Lomo et al., 1985; Sketelj et al., 1991). Motor units in the rat slow muscle are stimulated tonically with long trains of low-frequency impulses, whereas those in the fast are excited phasically with short high frequency bursts occuring rarely, so that the total number of impulses per day is much higher in the slow than in the fast motor units (Hennig and Lomo, 1985).Third, AChE activity in muscles changes rapidly after denervation. In rat muscles it is reduced to about 30% of control value during the first week after denervation (Guth et al., 1964).

The aim of our work was to test the hypothesis that neural regulation of muscle AChE is exerted primarily on the level of mRNA of the catalytic subunit of AChE. The steady-state level of AChE mRNA in muscles was correlated with activity of molecular forms of AChE under different experimental conditions in which muscle AChE activity is affected by changing neural influences. In all examined cases the observed changes of AChE activity,

Enzymes of the Cholinesterase Family, Edited by Daniel M. Quinn et al.
Plenum Press, New York, 1995

most notably those of the monomeric AChE form, can be explained in terms of neurally induced changes of AChE mRNA level.

METHODS

Adult male Wistar rats (200-250 g) were used in the experiments. Ketalar (60 mg/kg), Rompun (8 mg/kg) and atropine (0.6 mg/kg) were applied IP for surgical anaesthesia. Soleus (SOL), extensor digitorum longus (EDL), and the white portion of the sternomastoid muscle (STM) were examined.

Denervation of SOL and EDL muscles was performed by transection of the sciatic nerve in the thigh. Transient denervation and subsequent reinnervation of the two muscles was achieved by crushing the sciatic nerve. Disuse of the SOL muscle was produced by hind limb immobilization (Fischbach and Robbins, 1969). Regeneration of the SOL muscle followed muscle injury due to ischemia and injection of a myotoxic local anaesthetic bupivacaine. The motor nerve to the muscle was left intact to enable immediate reinnervation (Carlson et al., 1981).

Regions of muscles containing no neuromuscular junctions (the extrajunctional regions) were isolated, and molecular forms of AChE and AChE mRNA level in these regions were analysed by velocity sedimentation and Northern blotting as described earlier (Sketelj et al., 1992; Crešnar et al., 1994). The cDNA probe for AChE mRNA analysis was a gift by Dr. C. Legay and Dr. J. Massoulié from Paris, the actin probe was from Dr. W. Mages from Regensburg.

RESULTS AND DISCUSSION

Slow vs. fast muscles

Extrajunctionally, AChE activity per unit of muscle weight is much higher in the fast than in the slow muscle, which is due to much higher activity of the globular forms of AChE (G_1, G_2 and G_4) in the former (Fig. 1 A). Northern blotting revealed two transcripts of the catalytic subunit of AChE (2.4 and 3.2 kb) in the extrajunctional regions of fast STM muscles in mature rats (Fig. 1 B; see also Legay et al., 1993).

The level of AChE transcripts in the junctional region of the STM muscle, encompassing about 3 mm broad band of muscle around motor nerve branches, was about 20% higher than that in the extrajunctional region (Crešnar et al., 1994). This indicates that AChE expression in the junctional nuclei of rat muscles is enhanced in regard to the extrajunctional ones.

The extrajunctional steady-state level of AChE mRNA in the slow SOL muscle was substantially lower than that in the fast muscle (Fig. 1 B). This explains relatively low activity of AChE in the SOL muscle, especially of the globular forms. We propose that high amount of neural impulses per day, characteristic for a slow muscle in contrast to a fast one, actually suppress AChE mRNA level in the SOL muscle. This hypothesis was tested during SOL muscle disuse and reinnervation, two conditions in which neural stimulation pattern and AChE activity are affected (Groswald and Dettbarn, 1983; Dettbarn et al., 1991).

Disuse of SOL Muscle

Hind limb immobilization in the rat causes SOL muscle disuse because SOL can no longer perform its load-bearing function as an antigravity muscle. The neural stimulation pattern in disused SOL muscles becomes phasic instead of tonic, and the number of impulses

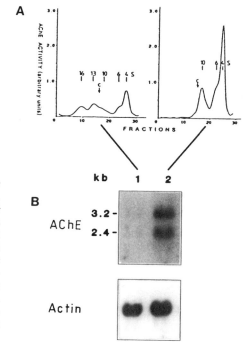

Figure 1. AChE in the extrajunctional regions of slow soleus (SOL) and fast sternomastoid (STM) rat muscles. A: velocity sedimentation analysis of AChE molecular forms. AChE activity per unit of muscle wet weight is given in arbitrary units comparable in all presented gradients. The top of the gradients is on the right hand side. Molecular forms of AChE are identified by their respective sedimentation coefficients. The arrow indicates position of catalase (11.3 S) in a gradient. B: Northern blot analysis of the level of AChE mRNA. Lane 1 - soleus muscle, lane 2 - sternomastoid muscle. Actin mRNA was analyzed to control for nonuniform sample loading on the gel.

per day is reduced to 15% of normal (Fischbach and Robbins, 1969). After ten days of leg immobilization, activity of the monomeric G_1 AChE form in the SOL muscle increased substantially (Fig. 2 A, samples 1 and 5). Accordingly, the steady-state level of AChE mRNA was found to be much higher in disused than in normal SOL muscles (Fig. 2 B, lines 1 and 5). The decrease of neural stimulation of the SOL muscle, therefore, enhanced AChE expression. This result corroborates our hypothesis that a high amount of neural stimulation decreases AChE mRNA level in muscle.

Reinnervation of Previously Denervated SOL Muscle

After reinnervation of denervated rat SOL muscle, AChE activity in this muscle increases transiently to several times its normal value (Groswald and Dettbarn, 1983). Extrajunctionally, activity of both the globular and asymmetric AChE forms changed in parallel (Figure 2 A, samples 1-4). Changes in AChE activity correlated precisely with a transient increase and subsequent decrease of the steady-state level of AChE mRNA in the extrajunctional regions of the SOL muscle (Fig. 2 B, lines 1-4). It can be assumed that, due to interrupted afferent neural pathways, the typical stimulation pattern of the SOL muscle is probably not reestablished immediately after reinnervation. Therefore, full expression of AChE occurs at that period, to be suppresed later when the normal tonic pattern of SOL muscle stimulation is established after complete recovery.

Denervation and Reinnervation of Fast Muscles

In a few days after denervation, activity of the G_1 AChE form in the extrajunctional regions of the rat fast muscles drops to less than 10% of its normal value (Groswald and Dettbarn, 1983; Sketelj et al., 1992). In parallel, a rapid decrease of the steady state level of

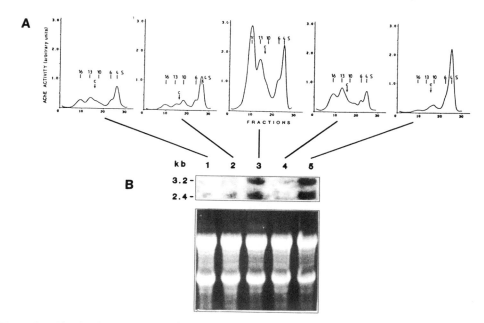

Figure 2. AChE in the extrajunctional regions of the slow soleus muscle under different experimental conditions: 1 - normal muscle, 2 - muscle denervated for 21 days, 3 - reinnervated muscle 21 days after sciatic nerve crush, 4 - reinnervated muscle 35 days after sciatic nerve crush, 5 - muscle disused for 10 days. A: velocity sedimentation analysis of AChE molecular forms, for details see the legend to Fig.1. B: Northern blot analysis of AChE mRNA level. An ethidium bromide-stained gel was used to control for nonuniformity in relative amounts of total RNA loaded on the gel.

the AChE mRNA was observed (Crešnar et al., 1994). After reinnervation of the previously denervated EDL muscle, both the extrajunctional activity of the G_1 AChE form (Groswald and Dettbarn, 1983; Fig. 3 A) and the steady-state level of AChE mRNA (Fig. 3 B) increased again, approaching normal levels 35 days after sciatic nerve crush, i.e. about three weeks after the first signs of EDL muscle reinnervation.

In conclusion, a good correlation between AChE activity, especially activity of the monomeric G_1 form, and the level of AChE mRNA in muscles was observed in all experimental manipulations of muscle innervation examined so far. This corroborates our assumption that neural influence on muscle AChE is largely exerted through regulation of the level of mRNA of the AChE catalytic subunit, either by determining the transcription rate of the AChE gene or by modifying stability and, thereby, degradation rate of AChE transcripts (see Fuentes and Taylor, 1993). A certain level of muscle electromechanical activity, such as that provided by the 'fast' motoneurons, seems to be required for optimal expression of AChE gene (or highest transcript stability) in the extrajunctional muscle regions in the rat. Either inactivity or very high muscle activity may actually suppress AChE gene expression (or decrease its transcript stability), at least in the SOL muscle.

Developmental Aspect of Nerve Dependence of Muscle AChE

In contrast to mature muscles, AChE expression in immature rat muscles in culture or during regeneration is nerve-independent in the sense that it is high even in the absence of the nerve (Sugiyama, 1977; Sketelj et al., 1987 - Fig. 4 g, h).

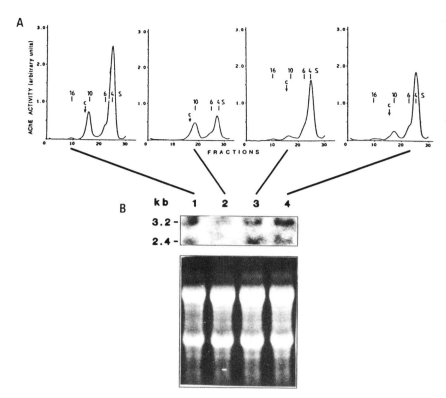

Figure 3. AChE in the fast extensor digitorum longus (EDL) muscle under different experimental conditions: 1 - normal muscle, 2 - muscle denervated for 21 days, 3 - reinnervated muscle 21 days after sciatic nerve crush, 4 - reinnervated muscle 35 days after sciatic nerve crush. A: velocity sedimentation analysis of AChE molecular forms in the extrajunctional regions of the EDL muscles, for details see the legend to Fig. 1. B: Northern blot analysis of AChE mRNA levels in whole EDL muscles. An ethidium bromide-stained gel was used to control for nonuniformity in relative amounts of total RNA loaded on the gel.

If the regenerating SOL muscle was immediately reinnervated (Carlson et al., 1981), just a few days of innervation should be enough for the G_1 form activity in the regenerating muscle to decline after subsequent denervation, although not to such low levels as observed in mature SOL muscle (Fig. 4 c,d). The AChE pattern did not return to that in non-innervated regenerating muscles. If the regenerating SOL was denervated a week later, AChE activity dropped nearly to levels characteristic for denervated mature SOL muscle (Fig. 4 e,f). It seems that in just a few days after innervation the motor nerve starts to take over the regulation of AChE in immature regenerating rat muscles. This process is completed in about a week and muscle AChE becomes as nerve-dependent as in mature innervated rat muscle.

ACKNOWLEDGMENTS

The authors acknowledge skillful technical assistance of Mr. Boris Pečenko and Mr. Marjan Kužnik. We are grateful to Dr. C. Legay and Dr. J. Massoulié for giving us the AChE cDNA probe, and to Dr. W. Mages for the actin cDNA probe. This study was supported by the Ministry of Science and Technology of the Republic of Slovenia.

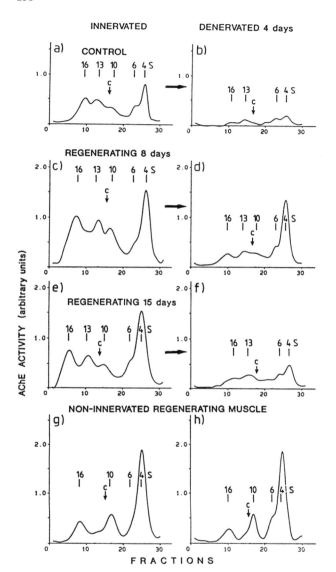

Figure 4. Effect of denervation on AChE in regenerating SOL muscles at different time points during early stages of regeneration and maturation. Denervated muscles were analysed four days after nerve transection. a,b - control SOL muscle before and after denervation; c,d - 8-day old regenerating muscle before and after denervation; e,f - 15-day old regenerating muscle before and after denervation; g,h - 8-day and 15-day old regenerating muscles that had never been innervated. For details see the legend to Fig. 1.

REFERENCES

Bacou, F., Vigneron, P., and Massoulié J., 1982, Acetylcholinesterase forms in fast and slow rabbit muscle, *Nature* 296:661-664.

Brzin, M., Sketelj, J., Tennyson, V.M., Kiauta, T., and Budininkas-Schoenebeck, M., 1981, Activity, molecular forms and cytochemistry of cholinesterases in developing rat diaphragm, *Muscle & Nerve* 4:505-513.

Carlson, B.M., Hnik, P., Tuček, S., Vejsada, R., Bader, D.M., and Faulkner, J.A., 1981, Comparison between grafts with intact nerves and standard free grafts of the rat extensor digitorum longus muscle, *Physiol. Bohemoslov.* 30:505-513.

Črešnar, B., Crne-Finderle, N., Breskvar, K., and Sketelj, J., 1994, Neural regulation of muscle acetylcholinesterase is exerted on the level of its mRNA, *J. Neurosci. Res.* 38:294-299.

Dettbarn, W.-D., Groswald, D.E., Gupta, R.C., Misulis, K.E., and Patterson, G.T., 1991, In vivo regulation of acetylcholinesterase in slow and fast muscle of rat, in *Cholinesterases* (Massoulié. J., Bacou, F.,

Barnard, E., Chatonnet, A., Doctor, B.P., and Quinn D.M., eds.), pp. 71-75, American Chemical Society, Washington.

Fischbach, G.D., and Robbins, N., 1969, Changes in contractile properties of disused soleus muscles, *J. Physiol.* 201P:305-320.

Fuentes, M.A., and Taylor, P., 1993, Control of acetylcholinesterase gene expression during myogenesis, *Neuron* 10:679-687.

Groswald, D.E., and Dettbarn, W.-D., 1983, Nerve crush induced changes in molecular forms of acetylcholinesterase in soleus and extensor digitorum muscles, *Exp. Neurol.* 79:519- 531.

Guth, L., Albers, R.W., and Brown, W.C., 1964, Quantative changes in cholinesterase activity of denervated muscle fibers sole plates, *Exp. Neurol.* 10:236-250.

Hall, Z.W., 1973, Multiple forms of acetylcholinesterase and their distribution in endplate and non- endplate regions of rat diaphragm muscle, *J. Neurobiol.* 4:343- 361.

Hennig, R., and Lomo, T., 1985, Firing pattern of motor units in normal rats, *Nature* 314:164-166.

Legay, C., Bon, S., Vernier, P., Coussen, F., Massoulié, J. 1993, Cloning and expression of a rat acetylcholinesterase subunit: generation of multiple molecular forms and complementarity with a Torpedo collagenic subunit, *J. Neurochem.* 60:337-346.

Lomo, T., Massoulié, J., and Vigny, M., 1985, Stimulation of denervated rat soleus muscle with fast and slow activity patterns induces different expression of acetylcholinesterase molecular forms, *J. Neurosci.* 5:1180-1187.

Marnay, A., and Nachmansohn, D., 1938, Choline esterase in voluntary muscle, *J. Physiol.* 92:37-47.

Sketelj, J., and Brzin, M., 1980, 16 S acetylcholinesterase in endplate-free regions of developing rat diaphragm, Neurochem. Res. 5:653-658.

Sketelj, J., and Brzin, M., 1985, Asymmetric molecular forms of acetylcholinesterase in mammmalian skeletal muscles, *J. Neurosci. Res.* 14:95-103.

Sketelj, J., Crne, N., and Brzin, M., 1987, Molecular forms and localization of acetylcholinesterase and nonspecific cholinesterase in regenerating skeletal muscles, *Neurochem. Res.* 12:159-165.

Sketelj, J., Crne-Finderle, N., and Brzin, M., 1992, Influence of denervation on the molecular forms of junctional and extrajunctional acetylcholinesterase in fast and slow muscles of the rat, *Neurochem. Int.* 21:415-421.

Sketelj, J., Crne-Finderle, N., Ribariš, S., and Brzin, M., 1991, Interaction between intrinsic regulation and neural modulation of acetylcholinesterase in fast and slow skeletal muscles, *Cell. Mol. Neurobiol.* 11:35-54.

Sugiyama, H., 1977, Multiple forms of acetylcholinesterase in clonal muscle cells, *FEBS Lett.* 84:257- 260.

NEUROMUSCULAR FACTORS INFLUENCING ACETYLCHOLINESTERASE GENE EXPRESSION IN SKELETAL MUSCLE FIBERS

B. J. Jasmin, C. Boudreau-Larivière, R. Chan, D. A. Hubatsch, and
H. Sveistrup

Department of Physiology
Faculty of Medicine
University of Ottawa
Ottawa, Ontario
Canada K1H 8M5

INTRODUCTION

Acetylcholinesterase (AChE) is of particular interest with regards to muscle plasticity since levels of AChE molecular forms are known to be highly sensitive to neural influences. For example, muscle paralysis induced via surgical denervation results in general in the rapid disappearance of the synaptic collagen-like tailed AChE forms (Massoulié et al., 1993). Alternatively, enhanced neuromuscular activation achieved by exercise training programs and compensatory hypertrophy lead to significant increases in whole muscle AChE activity which are reflected by specific and prominent changes in the levels of the various molecular forms (Fernandez and Donoso, 1988; Jasmin and Gisiger, 1990; Gisiger et al., 1991; Jasmin et al., 1991; Sveistrup et al., 1994). Despite the wealth of information available on the plasticity of AChE molecular forms confronted with altered levels of neuromuscular activation, our knowledge of the cellular and molecular basis underlying the activity-linked regulation of AChE in muscle is still rudimentary. In this context, several levels of regulatory mechanisms including transcriptional, post-transcriptional as well as post-translational may be envisaged. Within the last few years, several laboratories have succeeded in isolating cDNA and genomic clones encoding AChE in a variety of species (Schumacher et al., 1986; Sikorav et al., 1987; Rotundo et al., 1988; Maulet et al., 1990; Rachinsky et al., 1990; Li et al., 1991; Legay et al., 1993a) thus allowing for the study of the cellular and molecular mechanisms involved in the regulation and localization of AChE at the nucleic acid level.

Enzymes of the Cholinesterase Family, Edited by Daniel M. Quinn et al.
Plenum Press, New York, 1995

COMPARTMENTALIZATION OF AChE mRNA EXPRESSION

In an initial series of studies, we examined the molecular basis underlying the accumulation of AChE at avian neuromuscular synapses (Jasmin et al., 1993). At first, we compared levels of AChE transcripts in junctional versus extrajunctional regions of Quail muscle fibers. Using quantitative RT-PCR, we showed that AChE mRNA is a moderately expressed transcript in junctional regions of skeletal muscle fibers (~800 copies per junctional myonuclei) whereas in extrajunctional segments, it is either undetectable or a rare transcript. Unexpectedly, we also observed in these studies that the levels of AChE transcripts were highly variable between junctional samples and that they were detected in less than 50% of the junctional samples assayed. This was in contrast to both AChE enzyme activity and actin mRNA levels which were remarkably constant in these samples. Converging lines of evidence suggested that the observed variability in AChE mRNA levels was in fact genuine. On the basis of these results, we proposed a model whereby transcription of the AChE gene occurs intermittently, i.e. rhythmically (Sassone-Corsi, 1994) rather than constitutively. This would provide muscle cells with a regulated transcriptional control mechanism influenced by neuromuscular activation.

Since considerable differences exist between mammalian and avian species with regard to the regulation of AChE (for review see Massoulié et al., 1993), we were interested in determining whether rat skeletal muscle fibers would also display selective accumulation of AChE mRNAs within the postsynaptic sarcoplasm. Thus, we determined AChE transcript levels in junctional and extrajunctional regions of rat muscle fibers by quantitative RT-PCR (Michel et al., 1994). In these assays, AChE primers that amplified a fragment corresponding to the rat AChE T subunit were used (Legay et al., 1993b) and sample size was monitored by measuring the abundance of dystrophin transcripts since dystrophin is a cytoskeletal protein evenly distributed throughout the length of individual muscle fibers (Matsumura and Campbell, 1994). Following AChE histochemical staining of extensor digitorum longus (EDL) muscles, bundles of fibers containing 10 to 15 neuromuscular junctions were teased from the muscle and separated into junctional and extrajunctional regions. Subsequent to RNA extraction, an aliquot of each sample was reverse transcribed and the resulting AChE and dystrophin cDNAs amplified. Results of our experiments showed that AChE mRNA levels were considerably higher in junctional regions as compared to extrajunctional regions while dystrophin transcripts were relatively equal in both samples. Quantitative analysis revealed that on average, AChE mRNA levels were approximately 5- to 14-fold higher in junctional versus extrajunctional regions of EDL muscle fibers; values that correspond closely to our earlier findings in avian muscle (Jasmin et al., 1993). Our observations in rat muscle were confirmed by a series of in situ hybridization experiments using a synthetic ^{35}S-labelled oligonucleotide complementary to the common coding region of AChE transcripts. Our experiments performed on cryostat sections from both EDL and soleus muscles disclosed accumulations of AChE mRNAs within the sarcoplasm immediately beneath the postsynaptic membrane of neuromuscular synapses. Interestingly, as predicted by our intermittent model of transcription, we failed to observe accumulations of silver grains at several neuromuscular synapses in spite of the presence of a strong signal at neighbouring junctions. Thus, our recent studies using two different approaches showed that indeed AChE transcripts selectively accumulate within the junctional sarcoplasm of mammalian muscle fibers and further provided evidence for the model of intermittent transcription of the AChE gene (Michel et al., 1994).

The mechanisms responsible for the preferential accumulation of AChE mRNAs have not yet been elucidated. However, results from several laboratories have shown the selective accumulation of mRNAs encoding the subunits of the acetylcholine receptor (AChR) (Merlie and Sanes, 1985; Fontaine and Changeux, 1989; Goldman and Staple, 1989; Brenner et al., 1990) and recent promoter analyses demonstrated that this compartmentalization results primarily from localized gene transcription (Klarsfeld et al., 1991; Sanes et al., 1992; Simon et al., 1992; Duclert et al., 1993; Bessereau et al., 1994). In addition, numerous specialized cellular structures and organelles including mitochondria (see Jasmin et al., 1994b and Refs therein), a synapse-specific Golgi apparatus (Jasmin et al., 1989; Jasmin, et al., 1994a) and a stable array of microtubules (Jasmin et al., 1990) are contained within the postsynaptic sarcoplasm. Thus, it may be hypothesized that selective expression of the AChE gene by the junctional myonuclei may itself explain the observed compartmentalization of AChE transcripts. This proposed mechanism is in fact coherent with the notion that renewal of postsynaptic membrane proteins is achieved by a mechanism involving localized transcription of genes encoding synaptic proteins, with subsequent local translation and post-translational processing as well as focal insertion of newly synthesized molecules at the level of the motor endplate.

NEURAL FACTORS REGULATING AChE GENE EXPRESSION

Because the presence of the motor nerve significantly influences AChE expression in muscle, we were also interested in identifying the neural factors responsible for the selective accumulation of AChE transcripts within the postsynaptic sarcoplasm of neuromuscular synapses. In a recent study (Michel et al., 1994), we examined this by using two well-characterized models of muscle paralysis namely, surgical denervation and chronic superfusion of the sodium channel blocker tetrodotoxin (TTX) onto the sciatic nerve. With denervation both the propagation of action potentials along motor axons as well as axonal transport of various substances are eliminated. Although chronic TTX delivery also abolishes neuromuscular activation, it offers the distinct advantages of preserving the integrity of nerve-muscle contacts and allowing axonal transport to proceed normally (Lavoie et al., 1976). This experimental approach is therefore useful to partition the impact of nerve-evoked electrical activity versus the release of putative trophic factors in the regulation and selective accumulation of AChE transcripts within the junctional sarcoplasm.

Although the loss in muscle mass was similar following 10 days of paralysis, denervation and chronic TTX superfusion affected AChE mRNA levels to different extent. For instance, AChE transcripts in EDL muscles were decreased by over 10-fold following denervation but were much less reduced in TTX-paralyzed muscles. These pronounced decreases in AChE transcript levels following either denervation or TTX inactivation were confirmed with a series of in situ hybridization experiments and, accordingly, accumulations of AChE transcripts within the postsynaptic sarcoplasm of EDL and soleus muscle fibers could no longer be observed following the period of inactivity. Taken together, these findings indicate that although nerve-evoked activity per se appears a key regulator of AChE mRNA levels in vivo, other factors such as the integrity of the synaptic structure and transport of molecules down the axons are capable of maintaining AChE transcript levels closer to normal. One implication is that the constitutive release of nerve-derived trophic factors may exert a pronounced influence on the local expression of AChE transcripts within the postsynaptic membrane domain of neuromuscular synapses. However, the identity of the putative trophic factors involved in this regulation still remains elusive. Potential candidates for such a role include the ciliary neurotrophic factor (CNTF) which was recently shown to have in addition to its neurotrophic actions, profound myotrophic effects in preventing morphological and physiological changes that accompany muscle denervation (Helgren et al., 1994). We are currently testing whether CNTF indeed affects expression of the AChE gene in rat muscle.

It is well established that denervation and TTX-induced paralysis lead to significant increase in AChR transcripts as a result of transcriptional activation of AChR genes in nuclei located in the extrasynaptic sarcoplasm (Merlie et al., 1984; Fontaine and Changeux, 1989; Goldman and Staple, 1989; Tsay and Schmidt, 1989; Witzemann et al., 1991). Our observation that the levels of AChE mRNA are markedly reduced in paralyzed muscles stands in sharp contrast to the findings on the expression of AChR. It strongly indicates that under these conditions expression of these transcripts encoding synaptic proteins is independently controlled. Nonetheless, it still remains to be determined whether the AChE and AChR genes are coordinately expressed during assembly of postsynaptic membrane domains. During early stages of myogenesis, expression of both AChE and AChR increase sharply during the transitional phase from myoblasts to myotubes. As a result of exploratory motor axons reaching the surface of the muscle fibers, these two proteins become concentrated within the synaptic region of muscle fibers. Thus, monitoring transcript levels during embryogenesis may provide another opportunity to test whether expression of the AChE gene is co-regulated with that of the AChR subunit genes. For example, a first indication that expression of these genes is co-regulated would be provided by a similar and concomitant pattern of progressive restriction of these mRNAs as the neuromuscular junction develops and is stabilized on the muscle surface. If this hypothesis is confirmed, a picture could emerge whereby expression of these genes is co-regulated during development of muscle fibers and assembly of postsynaptic membrane domains but is independently regulated in adult muscle as we have recently demonstrated (Michel et al., 1994).

PASSIVE AND ACTIVE MECHANICAL FACTORS

From our studies, it is clear that expression of AChE mRNAs in skeletal muscle fibers is significantly modulated by nerve-evoked electrical activity and putative nerve-derived trophic factors. The possibility exists however that passive and active mechanical factors associated with cycles of muscle contraction and relaxation also influence AChE gene expression. This hypothesis is based on the observations that mechanical forces modulate the growth and differentiation of various tissues including for instance, cardiac and smooth muscles as well as bone and lung (Vandenburgh, 1992; Ingber, 1993). In addition, skeletal muscle fibers are known to be particularly sensitive to mechanical stimuli since in response to passive forces, these cells readily adapt by synthesizing large amounts of contractile proteins (Goldspink and Booth, 1992).

In the course of our on-going studies on the regulation of AChE gene expression in muscle, we examined levels of AChE mRNA in rat hemidiaphragm muscle fibers denervated for 2, 5, 10 and 20 days (Hubatsch et al., 1994). For these experiments, denervation was performed by cutting and removing a 5 mm portion of the phrenic nerve via an intrathoracic incision. As observed in hindlimb muscles, levels of AChE transcripts were markedly affected in denervated hemidiaphragms. In these muscles, a rapid and pronounced decrease in AChE mRNAs was evident as early as 2 days after denervation. Five days following denervation however, we noted that levels of AChE transcripts began to increase and appeared to reach a plateau between 10-20 days post-denervation. At these latter time points, AChE mRNA levels were significantly less reduced in the hemidiaphragm as compared to levels observed 2 days after sectioning of the motor nerve. The up-regulation of AChE transcripts that we observed coincides in fact with the reappearance of normal breathing patterns following thoracotomy (Torres et al., 1989). As a result of this recovery, denervated hemidiaphragms undergo mechanical events associated with respiratory movements. Recent studies have shown using sonomicrometry crystals, that denervated hemidiaphragms indeed undergo passive stretching during contraction of the intact contralateral side (Zhan et al.,

1992). Thus, our results are consistent with the notion that passive stretching of skeletal muscle fibers in vivo regulates expression of the AChE gene. This hypothesis fits well with converging lines of evidence indicating that imposition of a mechanical stretch stimulus dramatically affects expression of several muscle genes encoding specialized contractile proteins. It will be important to determine whether the up-regulation in mRNA levels observed in denervated hemidiaphragms occurs within the junctional region since our in situ hybridization experiments revealed, as for hindlimb muscles, the selective accumulation of AChE mRNA within the postsynaptic sarcoplasm of control diaphragm muscle fibers.

CONCLUSION AND PERSPECTIVE

The successful isolation of cDNA and genomic clones encoding AChE in a variety of species has initiated the launching of numerous studies on the molecular mechanisms presiding over the regulation and localization of AChE in skeletal muscle fibers. Our results have indicated thus far that regulation of the AChE gene in adult skeletal muscle is a multifactorial process that may involve in addition to neural activation, nerve-derived trophic substances as well as passive and active mechanical factors. Naturally, the results discussed here along with those obtained in other laboratories (for example Fuentes and Taylor, 1993) represent only the beginning of an exciting era which will ultimately lead to the identification of the molecular pathways regulating AChE gene expression. Together, these studies should not only increase our knowledge of the mechanisms regulating AChE expression but they should also contribute to our basic understanding of how interactions between muscle and nerve result in the modulation of physiological functions through complex cellular and molecular events.

ACKNOWLEDGEMENTS

Work in the laboratory is supported by grants from the Medical Research Council (MRC) and the Muscular Dystrophy Association of Canada to BJJ. BJJ is a Scholar of the MRC.

REFERENCES

Bessereau, J.L., Stratford-Perricaudet, L.D., Piette, J., Le Poupon, C., and Changeux, J.-P., 1994, In vivo and in vitro analysis of electrical activity-dependent expression of muscle acetylcholine receptor genes using adenovirus, *Proc. Natl. Acad. Sci. USA* 91:1304-1308.

Brenner, H.R., Witzemann, V., and Sakmann, B., 1990, Imprinting of acetylcholine receptor messenger RNA accumulation in mammalian neuromuscular synapses, *Nature* 344:544-547.

Duclert, A., Savatier, N., and Changeux, J.-P., 1993, An 83-nucleotide promoter of the acetylcholine receptor ε-subunit gene confers preferential synaptic expression in mouse muscle, *Proc. Nat. Acad. Sci. USA* 90:3043-3047.

Fernandez, H.L, and Donoso, J.A., 1988, Exercise selectively increases G4 Ache activity in fast-twitch muscle, *J. Appl. Physiol.* 65:2245-2252.

Fontaine, B., and Changeux, J.-P., 1989, Localization of nicotinic acetylcholine receptor α-subunit transcripts during myogenesis and motor endplate development in chick, *J. Cell Biol.* 101:1025-1037.

Fuentes, M.E., and Taylor, P., 1993, Control of acetylcholinesterase gene expression during myogenesis, *Neuron* 10:679-687.

Gisiger, V., Sherker, S., and Gardiner, P.F., 1991, Swimming training increases the G_4 acetylcholinesterase content of both fast ankle extensors and flexors, *FEBS Lett.* 278:271-273.

Goldman, D., and Staple, J., 1989, Spatial and temporal expression of acetylcholine receptor RNAs in innervated and denervated rat soleus muscle, *Neuron* 3:219-228.

Goldspink, G., and Booth, F., 1992, Mechanical signals and gene expression in muscle. *Am. J. Physiol.* 262:R327-R328.

Helgren, M.E., Squinto, S.P., Davis, H.L., Parry, D.J., Boulton, T.G., Heck, C.S., Zhu, Y., Yancopoulos, G.D., Lindsay, R.M., and DiStefano, P.S., 1994, Trophic effect of ciliary neurotrophic factor on denervated skeletal muscle, *Cell* 76:493-504.

Hubatsch, D.A., Comtois, A.S., Vu, C.Q., and Jasmin, B.J., 1994, Does passive mechanical stretch of skeletal muscle fibers influence acetylcholinesterase gene expression? *Mol. Biol. Cell* (abstract) in press.

Ingber, D.E., 1993, The riddle of morphogenesis: a question of solution chemistry or molecular cell engineering? *Cell* 75:1249-1252.

Jasmin, B.J., Antony, C., Changeux, J.P., and Cartaud, J., 1994a, Nerve-dependent plasticity of the Golgi complex in skeletal muscle fibers: compartmentalization within the subneural sarcoplasm, *Eur. J. Neurosci.* in press.

Jasmin, B.J., Campbell, R.J., and Michel, R.N., 1994b, Nerve-dependent regulation of succinate dehydrogenase in junctional and extrajunctional compartments of rat muscle fibers, *J. Physiol. (Lond.)* in press.

Jasmin, B.J., Cartaud, J., Bornens, M., and Changeux, J.-P., 1989, Golgi apparatus in chick skeletal muscle: changes in its distribution during end plate development and after denervation, *Proc. Natl. Acad. Sci. USA* 87:7218-7222.

Jasmin, B.J., Changeux, J.-P., and Cartaud, J., 1990, Compartmentalization of cold-stable and acetylated microtubules in the subsynaptic domain of chick skeletal muscle fibres, *Nature* 344:673-675.

Jasmin, B.J., Gardiner, P.F., and Gisiger, V., 1991, Muscle acetylcholinesterase adapts to compensatory overload by a general increase in its molecular forms, *J. Appl. Physiol.* 70:245-2489.

Jasmin, B.J., and Gisiger, V., 1990, Regulation by exercise of the pool of G4 acetylcholinesterase characterizing fast muscles: opposite effect of running training in antagonist muscles, *J. Neurosci.* 10:1444-1454.

Jasmin, B.J., Lee, R.K., and Rotundo, R.L., 1993, Compartmentalization of acetylcholinesterase mRNA and enzyme at the vertebrate neuromuscular junction, *Neuron* 11:467-477.

Klarsfeld, A., Bessereau, J.L., Salmon, A.M., Triller, A., Babinet, C., and Changeux, J.-P., 1991, An acetylcholine receptor α-subunit promoter conferring preferential synaptic expression in muscle of transgenic mice, *EMBO J.* 10:625-632.

Lavoie, P.-A., Collier, B., and Tennenhouse, A., 1976, Comparison of α-bungarotoxin binding to skeletal muscles after inactivity or denervation, *Nature* 260:349-350.

Legay, C., Bon, S., Vernier, P., Coussen, F., and Massoulié, J., 1993a, Cloning and expression of a rat acetylcholinesterase subunit: generation of multiple molecular forms and complementarity with a Torpedo collagenic subunit, *J. Neurochem.* 60:337-346.

Legay, C., Bon, S., and Massoulié, J., 1993b, Expression of a cDNA encoding the glycolipid-anchored form of rat acetylcholinesterase, *Febs Lett.* 315:163-166.

Li, Y., Camp, S., Rachinsky, T.L., Getman, D., and Taylor, P., 1991, Gene structure of mammalian acetylcholinesterase: alternative exons dictate tissue-specific expression, *J. Biol. Chem.* 266:23083-23090.

Massoulié, J., Pezzementi, L., Bon, S., Krejci, E., and Vallette, F.M., 1993, Molecular and cellular biology of cholinesterases, *Prog. Neurobiol.* 13:31-91.

Matsumura, K., and Campbell, K.P., 1994, Dystrophin-glycoprotein complex: its role in the molecular pathogenesis of muscular dystrophies, *Muscle & Nerve* 17:2-15.

Maulet, Y., Camp, S., Gibney, G., Rachinsky, T., Ekstrom, T.J., and Taylor, P., 1990, Single gene encodes glycophospholipid-anchored and asymmetric acetylcholinesterase forms: alternative coding exons contain inverted repeat sequences, *Neuron* 4:289-301.

Merlie, J.P., Isenberg, K.E., Russell, S.D., and Sanes, J.R., 1984, Denervation supersensitivity in skeletal muscle: analysis with a cloned cDNA probe, *J. Cell Biol.* 99:332-335.

Merlie, J.P., and Sanes, J.R., 1985, Concentration of acetylcholine receptor mRNA in synaptic regions of adult muscle fibers, *Nature* 317:66-68.

Michel, R.N., Vu, C., Tetzlaff, W., and Jasmin, B.J., 1994, Neural regulation of acetylcholinesterase mRNAs at mammalian neuromuscular synapses, *J. Cell Biol.* in press.

Rachinsky, T.L., Camp, S., Li, Y., Ekstrom, T.J., Newton, M., and Taylor, P., 1990, Molecular cloning of mouse acetylcholinesterase: tissue distribution of alternatively spliced mRNA species, *Neuron* 5:317-327.

Rotundo, R.L., Gomez, A.M., Fernadez-Valle, C., and Randall, W.R., 1988, Allelic variants of acetylcholinesterase: genetic evidence that all acetylcholinesterase forms in avian nerves and muscle are encoded by a single gene, *Proc. Natl. Acad. Sci. USA* 85:7805-7809.

Sanes, J.R, Johnson, Y.R., Kotzbauer, P.T., Mudd, J., Hanley, T., Martinou, J.C., and Merlie, J.P., 1992, Selective expression of an ACh receptor-LacZ transgene in synaptic nuclei of adult mouse muscle fibres, *Development* 113:1181-1191.

Sassone-Corsi, P., 1994, Rhythmic transcription and autoregulatory loops: winding up the biological clock, *Cell* 78:361-364.

Schumacher, M., Camp, S., Maulet, Y., Newton, M., MacPhee-Quigley, K., Taylor, S.S., Friedmann, T., and Taylor, P., 1986, Primary structure of Torpedo californica acetylcholinesterase deduced from a cDNA sequence, *Nature* 319:407-409.

Sikorav, J.L., Krejci, E., and Massoulié, J., 1987, cDNA sequences of Torpedo marmorata acetylcholinesterase: primary structure of the precursor of a catalytic subunit; existence of multiple 5' untranslated regions, *EMBO J.* 6:1865-1873.

Simon, A.M., Hoppe, P., and Burden, S.J., 1992, Spatial restriction of AChR gene expression to subsynaptic nuclei, *Development* 114:545-553.

Sveistrup, H., Chan, R., and Jasmin, B.J., 1994, Chronic enhancement of neuromuscular activity increases acetylcholinesterase mRNA levels in rat hindlimb muscles, submitted.

Torres, A., Kinball, W.R., Qvist, J., Stanek, K., Kacmarek, R.M., Whyte, R.I., Montalescot, G., and Zapol, W.M., 1989, Sonomicrometric regional diaphragmatic shortening in awake sheep after thoracic surgery, *J. Appli. Physiol.* 67:2357-2368.

Tsay, H.J., and Schmidt, J., 1989, Skeletal muscle denervation activates acetylcholine receptor genes, *J. Cell Biol.* 108:1523-1526.

Vandenburgh, H.H., 1992, Mechanical forces and their second messengers in stimulating cell growth in vitro, *Am. J. Physiol.* 262:R350-R355.

Witzemann, V., Brenner, H.R., and Sakmann, B., 1991, Neural factors regulate AChR subunit mRNAs at rat neuromuscular synapses, *J. Cell Biol.* 114:125-141.

Zhan, W.Z., Farkas, G.A., and Sieck, G.C., 1992, Passive stretch does not affect denervation-induced muscle fiber adaptations, *FASEB J.* 6:A2025.

POST-TRANSLATION PROCESSING OF ACETYLCHOLINESTERASE

Cellular Control of Biogenesis and Secretion

Baruch Velan, Chanoch Kronman, Arie Ordentlich, Yehuda Flashner, Raphael Ber, Sara Cohen, and Avigdor Shafferman

Israel Institute for Biological Research
Ness-Ziona, P.O. Box 19
70450, Israel

INTRODUCTION

Acetylcholinesterase (AChE) occurs in multiple molecular forms in different tissues of vertebrates and invertebrates (for recent reviews see Taylor, 1991; Massoulie et al 1993). This heterogeneity is generated through associations of various catalytic subunits (T or H) and structural subunits. The T subunit is involved in formation of several multimeric AChE ectoenzyme configurations: Secreted AChE is usually composed of soluble homooligomers, whereas cell-bound AChEs consist of tetramers attached to cell-associated structural subunits. The biogenesis of these various molecular forms is believed to be a regulated process involving various control mechanisms.

Studies on the assembly of AChE have shown that intracellular AChE monomers serve as precursors for the multimeric AChE forms (Rotundo, 1984; Lazar et al., 1984) and that subunit dimerization through the C-terminal cysteine (Roberts et al., 1991; Velan et al., 1991b) is an intermediate step in formation of the various complex enzyme configurations. As a first step in the elucidation of the mechanisms involved in generation of the varied forms of AChE, we examined in detail the cellular processes related to the biogenesis of the T-subunit homodimer. To focus on this step, we have used a recombinant expression system which under specified conditions (Velan et al., 1992; Lazar et al., 1993) gives rise to soluble, secreted, dimeric human AChE (huAChE) molecules. This expession system consists of 293 human embryonal kidney cells transfected by multipartite expression vectors (Velan et al 1991a; Kronman et al., 1992; Shafferman et al., 1992a) which carry the coding sequences of the human AChE T-subunit (Soreq et al., 1990) under control of the cytomegalovirus immediate early promoter enhancer. The recombinant 293 expression system proved to be instrumental in identifying amino acid residues involved in the catalytic machinary of AChE (reviewed in Shafferman et al., 1992b and Shafferman et al., 1995) as well as in the establishment of a biotechnological process for the generation of large quantities of the

Enzymes of the Cholinesterase Family, Edited by Daniel M. Quinn et al.
Plenum Press, New York, 1995

human enzyme (Lazar *et al.,* 1993). In this study we use the recombinant 293 expression system to examine the events related to the formation of a folded, glycosylated catalytic subunit, the assembly of these subunits into a dimeric form and the transport of this dimer through the secretory pathway out of the cell.

RESULTS AND DISCUSSION

Processing of rHuAChE - Sequestering in Cells en Route to Secretion

Assembly and secretion of rHuAChE were examined in pulse labeled 293 cells. The newly formed intracellular monomers are gradually assembled into dimers and these are then secreted efficiently into the medium (Velan *et al.,* 1992). Secretion appears to be preferential for assembled forms, even though a minor fraction of monomeric forms is detected occasionally in the medium. As expected, all secreted rHuAChE molecules were endoglycosidase H (endo H) resistant, due to their terminal glycosylation. In contrast, most of the intracellular rHuAChE was endo H sensitive. Endo H resistance of cellular forms was revealed only in oligomers and not in monomers (Kerem *et al.,* 1993). The cellular compartment in which rHuAChE is oligomerized appears to be the endoplasmic reticulum (ER). This is indicated by the fact that brefeldin A (BFA) which inhibits vesicular export of proteins from ER (Klausner *et al.,* 1992) fails to impair oligomerization (Kerem *et al.,*1993).

Transport of rHuAChE from the ER through the central vacuolar system of 293 cells was monitored by lectin blot analysis. Steady-state pools of intracellular and extracellular rHuAChE were probed with various lectins as well as specific antibodies (Fig 1). Visualization with antibodies revealed the presence of significant pools of dimers as well as unassembled AChE monomers in the cell. Both cellular forms (monomers and dimers) are stained by Con-A, a lectin which binds terminal mannose and is a marker for residence in the ER and early Golgi compartments. On the other hand, WGA, which binds terminal N-acetylglucosamine and is a marker for residence in the med-Golgi, stains faintly the intracellular dimers but not the monomers. Ricin, a lectin marker for residence in the trans-Golgi does not identify any intracellular AChE pool, yet stains the extracellular pool. Both the lectin binding studies and the endo H resistance studies indicate that transport and processing prior to cis-Golgi glycosylation appears to be rate limiting, while the following passage to cell

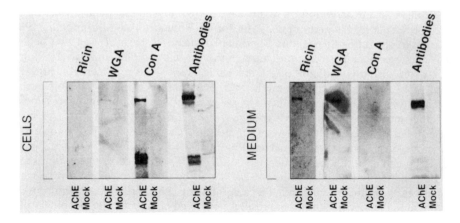

Figure 1. Lectin-staining of intracellular and secreted rHuAChE.

surface is rapid. This could suggest a transient retention or sequestration of AChE in the ER related to the operation of 'various quality' control processes in this compartment (Hurtly and Helenius, 1989). These could involve the control of folding, assembly or export of AChE from the ER.

AChE Polypeptide Folding - Role of N-Glycosylation

The role of N-glycosylation in biogenesis of HuAChE was examined by site directed mutagenesis (Asn to Gln substitution) of the three potential N-glycosylation sites, Asn-265, Asn-350 and Asn-464. Analysis of HuAChE mutants defective in single or multiple N-glycosylation sites, by expression in 293 cells suggested that glycosylation at all sites is important for effective secretion (Velan *et al.*, 1993). Extracellular AChE levels in mutants defective in one, two or all three sites amounted to 20-30%, 2-4% and about 0.5% of wild-type level, respectively. Nevertheless, all the secreted glycosylation mutants, including the triple-mutant, hydrolysed efficiently acetylthiocholine, displaying kinetic constants similar to those of the wild-type enzyme (Velan *et al.*, 1992). It therefore appears that all secreted molecules, even those secreted at a 0.5% efficiency, attained conformational maturity, prior to secretion.

Analysis of secretion kinetics in 293 cells expressing glycosylation-defective AChE revealed that the inefficient biogenesis of these mutants is related to accumulation of non active, malfolded molecules in the cell (Velan *et al.*, 1993). N-glycosylation appears to be required for efficient folding of the AChE polypeptide into a functional, secretion-competent molecular form. This could be effectuated through the involvement of glycan-binding chaperones such as calnexin (Hammond *et al.*, 1994) in AChE folding. Nevertheless, the rudimental secretion of a functional non-glycosylated AChE subunit into the medium, clearly indicates that an alternative, less efficient pathway, which is not dependent on glycans can operate in the cell. The existence of the two folding pathways, an efficient glycosylation-dependent route and an inefficient glycosylation independent route can explain some of the conflicting observations on AChE processing in different experimental systems. It could be possible that different cells utilise these two pathways to different extents. Preferential utilization of the non-efficient pathway may have led to formation of the retained, inactive molecules identified in specific tissues (Rotundo *et al.*, 1989; Chatel *et al.*, 1993).

AChE Subunit Assembly - Effect on Folding and Secretion

The interrelations between AChE assembly and processing was analysed by mutating the free Cys-580 to Ala (Velan et al, 1991b). Cells expressing this mutant secreted monomeric AChE molecules at efficiencies similar to those of wild type enzyme (Table 1). Detailed analysis of the secretion kinetics (Kerem et al, 1993) did not reveal a significant difference between the intracellular transport rates of rHuAChE wild-type oligomers and mutated C580A monomers.

Analysis of the cell-associated AChE pools in cells expressing the monomeric mutant and the wild-type enzyme reveals the presence of buffer-soluble and detergent-soluble AChE fractions both composed mainly of endo H - sensitive molecular species (not shown). These two fractions appear to contain intracellular AChE forms en-route to secretion, and the phase partition could be related to the amphiphilicity of the T-subunit (Massoulie *et al.*, 1993). Quantitation of the intracellular AChE-polypetide by ELISA reveals a similar ratio between activity and antigenic mass (Shafferman *et al.*, 1992c) in the secreted and in the cellular fraction (Table 1). The intracellular steady-state AChE pools do not contain, therefore, detectable amounts of non-folded enzyme molecules. This can be observed in cells expressing the wild-type enzyme as well as in cells expressing the dimerization-impaired mutant.

Table 1. Production of AChE by cells expressing wild-type and dimerization-impaired HuAChE and their KDEL-appended derivatives

HuAChE	C-terminus type	Quantitation method	Amounts of HuAChE (µg/plate)		
			Secreted	Cellular	Total
Wild Type	CSDL	Activity	21.5	2	23.5
		ELISA	20	2	22
C580A	ASCL	Activity	19.5	3	22.5
		ELISA	21	3	24
C580A-KDEL	ASKDEL	Activity	2.5	22.5	25
		ELISA	2	21	23
AChE-KDEL	CSKDEL	Activity	17	5.5	22.5
		ELISA	18	6	24

It therefore appears that the folding process is not linked to the dimerization process in AChE biogenesis.

AChE Accumulation in the Cell - Effect on Production and Processing

To examine the possible regulatory role of AChE cellular retention, we have engineered a mutant that is impaired in its ability to exit the cell. This was achieved by appending the well-characterized KDEL retention-signal tetrapeptide (reviewed by Pelham , 1990) at the C-terminus of the C580A AChE monomeric subunit (Velan et al., 1994). As expected the C580A-KDEL mutant was fully retained in the transfected cells (Table 1). The ratio between activity and antigenic mass of the retained enzyme is similar to that of the secreted enzyme suggesting that accumulation does not affect folding. Moreover, the total amount of enzyme produced by cells expressing the retained and non-retained molecular species remains constant (Table 1), indicating that enzyme production is not affected by impairing the secretion process. Pulse labeling experiments performed in cells expressing the C580A-KDEL mutant substantiate this observation and indicate that the enzyme molecules accumulated at concentrations as high as 1mM are not cleared from the ER through proteolytic degradation (Velan et al., 1994). It should be mentioned that degradation of AChE molecules in the cell can occur, yet it appears to be restricted to molecules defective in their folding properties (Shafferman et al., 1992a). Taken together, the AChE retention experiments suggest that concentration per-se in the ER does not serve as a signal in the control of AChE biogenesis.

AChE Retention - Interrelationship with Subunit Dimerization

The preferential secretion of dimerized wild-type HuAChE molecules in the recombinant cell system (Kerem et al., 1993) suggests a mechanism which restricts secretion of unassembled subunits containing a free cysteine at their C-terminus. To test whether this control can be mediated through the selective ER retention of wild-type monomers, the KDEL tetrapeptide was appended to an AChE subunit in which Cys-580 is conserved. As opposed to the KDEL appended monomeric mutant, the KDEL-appended dimerization-competent AChE is not retained in the cell and is secreted to the medium almost as efficiently as the wild-type enzyme (Table 1). Thus, subunit dimerization can effectively reverse the effect of an ER retention signal.

It, therefore, appears that by juxtaposing the signal for ER retention and the signal for subunit assembly one can generate a control system for the selective export of assembled forms and retention for monomers, not yet assembled. The C-terminus of native AChE subunits does not contain the canonical KDEL retention signal yet it carries the tetrapeptide CSDL which conserves some of the structural features involved in KDEL-retention (The C-terminal Glu/Asp-Leu configuration; Pelham et al 1990). Nevertheless, a comparative analysis of the structural components of KDEL and CSDL by site directed mutagenesis (Kronman *et al.,* 1995) failed to prove that CSDL can act as a dual signal for retention and assembly of AChE.

Transport Through the Golgi - Disassembly of Subunits

In specified tissues AChE assembly is not restricted to homooligomerization and involves interaction with structural subunits as well. In avian muscle cells homooligomerization in the ER is believed to be followed by assembly with the collagen-like subunits in the Golgi (Rotundo *et al.,* 1984). The fact that the same C-terminal cysteine residues are involved in homooligomerization as well as heterooligomerization (Roberts *et al.,* 1991) should invoke disassociation of disulfide-bonded AChE molecule in a post-ER milieu.

Recent observations on the processing of the KDEL- appended AChE (Velan *et al.,* 1994), provide some support to this unorthodox approach to processing of multi-subunit protein. Endo H analysis of the KDEL-appended AChE has revealed a delay in the export of enzyme from the cell caused by a KDEL-mediated sequestering in the distal Golgi compartments. This sequestering would involve an equilibrium between KDEL-AChE dimers, KDEL-AChE monomers and the KDEL receptor in the distal compartment of the Golgi-apparatus, which obviously implies the dissociation of the dimerized molecule in this compartment.

CONCLUSIONS

By examining rHuAChE biogenesis (Fig 2) in a non specialized recombinant cell-line that produces the least complex multisubunit forms of AChE we made the following observations:

1. Newly formed AChE molecules can be folded into an active subunit through either a glycosylation independent or a glycosylation dependent mechanism, the latter being much more efficient than the former.
2. Once folded monomers have been formed, they can accumulate in the ER at practically unlimited quantities. Non-folded molecules on the other hand are prone to rapid degradation.
3. Folding of AChE monomers is not coupled to subunit dimerization. Assembly is not a prerequisite for exit from the ER, yet there appears to be a mechanism which selects for preferential exit of dimerized forms.
4. An engineered control system that couples between assembly and retention can be generated. Such a system could operate in the selective retrieval of unassembled molecule, yet we were not able to prove its operation in AChE processing.
5. Molecules are transported to secretion in their dimerized form yet disassembly in the Golgi compartments may be possible. This could play a role in the generation of more complex AChE molecular forms.

In summary, the recombinant 293 cells expressing HuAChE provide an advantageous system for studying the general aspects of secretory proteins biogenesis, mainly due to the

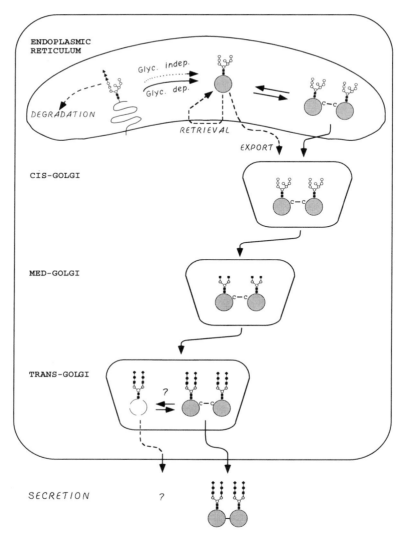

Figure 2. Schematic presentation of AChE processing in recombinant 293 cells.

effective analytical tools available for AChE. In addition, this system has revealed some of the more specific aspects of AChE biogenesis related to the relaxed control of processing and the utilization of alternative pathways. These features probably reflect the fact that manufacturing of AChE in the cell must be compatible with more than a single assembly pathway to allow for the generation of the various molecular forms of the enzyme.

ACKNOWLEDGEMENT

This work was supported by contracts DAMD17-89-C-9117 and DAMD17-93-C-3042 from the United States Army Research and Development Command (to A.S.).

REFERENCES

Chatel, J.M., Grassi, J., Frobert, Y., Massoulie, J. and Vallette, F.M (1993) Existence of an inactive pool of acetylcholinesterase in chicken brain. Proc. Natl. Acad. Sci. USA. 90, 2476 - 2480.

Hammond, C., Braakman, I., and Helenius, A. (1994) Role of N-linked oligosaccharide recognition, glucose trimming, and calnexin in glycoprotein folding and quality control. Proc. Natl. Acad. Sci. USA. 91, 913 - 917.

Hurtly S.M., and Helenius, A. (1989). Protein oligomerization in the endoplasmic reticulum. Annu. Rev. Cell. Biol. 5, 277 - 307.

Kerem, A., Kronman, C., Bar-Nun, S., Shafferman, A., and Velan, B. (1993) Interrelation between assembly and secretion of recombinant human acetylcholinesterase. J. Biol. Chem. 268, 180 - 184.

Klausner, R.D., Danaldson, J.G. and Lippincott-Schwartz J. (1992) Brefeldin A: Insights into the control of membrane traffic and organelle structure. J. Cell Biol. 116, 1071 - 1080.

Kronman, C.,Velan, B., Gozes, Y., Leitner, M., Flashner, Y., Lazar, A., Marcus, D., Sery, T., Papier, Y., Grosfeld, H., Cohen, S., and Shafferman, A. (1992) Production and secretion of high levels of recombinant human acetylcholinesterase in cultured cell lines. Gene, 121, 295 - 304.

Kronman, C., Flashner, Y., Shafferman, A. and Velan, B. (1995) Signal-mediated cellular retention and subunit assembly of human acetylcholinesterase. This volume.

Lazar, M., Salmeron, E., Vigny, M. and Massoulie, J. (1984) Heavy isotope labelling study of the metabolism of monomeric and tetrameric acetylcholinesterase forms in murine neuronal like T28 hybrid cell line. J. Biol Chem. 259, 3703 - 3713.

Lazar A., Reuveny, S., Kronman, C., Velan, B. and Shafferman, A. (1993) Evaluation of anchorage-dependent cell propagation system for production of human acetylcholinesterase by recombinant 293 cells. Cytobiology. 13, 115 - 123.

Massoulie, J., Pezzementi, L., Bon, S., Krejci, E., Vallette, F.M. (1993) Molecular and cellular biology of cholinesterases. Prog. Neurobiol. 41, 31 - 91.

Pelham, H.R.B. (1990) The retention signal for soluble proteins of the endoplasmic reticulum.Trends Biochem. Sci. 15, 483 - 486.

Roberts, W.L., Doctor, B.P., Foster, J.D., and Rosenberry T.L. (1991) Bovine brain acetylcholinesterase primary sequence involved in intersubunit disulfide linkages. J. Biol. Chem. 266, 7481 - 7487.

Rotundo, R.L. (1984) Asymmetric acetylcholinesterase is assembled in the Golgi apparatus. Proc. Natl Acad. Sci. USA 81, 479 - 483.

Rotundo, R.L. Thomas, K., Porter-Jordan, K., Benson, R.J. Fernandez-Valle, C., and Fine, R.E. (1989) Intracellular transport, sorting and turnover of acetylcholinesterase. J.Biol. Chem. 264, 3146 - 3152.

Shafferman, A., Kronman, C., Flashner,Y., Leitner, M., Grosfeld, H., Ordentlich, A., Gozes, Y., Cohen,S., Ariel,N., Barak, D., Harel, M., Silman,I., Sussman,J.L., and Velan, B. (1992a). Mutagenesis of human acetylcholinesterase. Identification of residues involved in catalytic activity and in polypeptide folding. J. Biol. Chem. 267, 17640 - 17648.

Shafferman, A., Velan, B., Ordentlich, A., Kronman, C., Grosfeld, H., Leitner, M., Flashner, Y., Cohen, S., Barak, D., and Ariel, N. (1992b) Acetylcholinesterase catalysis- protein engineering studies. In: Multidisciplinary Approaches to Cholinesterase Functions. (Shafferman A. and Velan B. Eds). pp. 39 - 47. Plenum Press, New York 1992.

Shafferman, A., Velan, B., Ordentlich, A., Kronman, C., Grosfeld, H., Leitner, M., Flashner, Y., Cohen,S., Barak, D., and Ariel, N. (1992c) Substrate inhibition of acetylcholinesterase: residues affecting signal transduction from the surface to the catalytic center. EMBO J. 11, 3561-3568

Shafferman. A., Ordentlich, A., Barak,D., Kronman,C., Ariel,N., Leitner, M., Segall,Y., Bromberg, A.,.Reuveny, S., Marcus, D., Bino, T., Lazar, A., Cohen, S. and Velan, B. (1995) Molecular aspects of catalysis and of allosteric regulation of acetylcholinesterases. This volume.

Soreq, H., Ben-Aziz, R., Prody, C.A., Gnatt, A., Neville, A., Lieman-Hurwitz, J., Lev-Lehman, E., Ginzberg, D., Seidman, S., Lapidot-Lifson, Y. and Zakut, H. (1990) Molecular cloning and construction of the coding region for human acetylcholinesterase reveals a G+C rich attenuating structure. Proc. Natl. Acad. Sci. USA 87, 9688 - 9692.

Taylor, P. (1991)The cholinesterases. J. Biol. Chem. 266, 4025 - 4028.

Velan, B., Kronman, C., Grosfeld, H., Leitner, M., Gozes, Y., Flashner, Y., Sery, T., Cohen, S., Ben-Aziz, R., Seidman, S., Shafferman, A., and Soreq, H. (1991a) Recombinant human acetylcholinesterase is secreted from transiently transfected 293 cells as a soluble globular enzyme. Cell. Mol. Neurobiol. 11, 143 - 156.

Velan, B., Grosfeld, H., Kronman, C., Leitner, M., Gozes, Y., Lazar, A., Flashner, Y., Marcus. D., Cohen, S., and Shafferman, A. (1991b) The effect of elimination of intersubunit disulfide bonds on the activity, assembly and secretion of recombinant human acetylcholinesterase. J. Biol Chem. 266, 23977 - 23984.

Velan, B., Kronman, C., Leitner, M., Grosfeld, H., Flashner, Y., Marcus, D., Lazar, A., Kerem, A., Bar-Nun, S., Cohen, S. and Shafferman, A. (1992) Molecular organization of recombinant human acetylcholinesterase. In: *Multidisciplinary Approaches to Cholinesterase Functions.* (Shafferman A. and Velan B. Eds). pp. 165 - 175. Plenum Press, New ~York.

Velan, B., Kronman, C., Ordentlich, A., Flashner, Y., Leitner, M., Cohen, S., and Shafferman, A. (1993) N-glycosylation of human acetylcholinesterase: effects on activity, stability and biosynthesis. Biochem. J. 296, 649 - 656.

Velan, B., Kronman, C., Flashner, Y. and Shafferman, A. (1994) Reversal of signal-mediated cellular retention by subunit assembly of human acetylcholinesterase. J. Biol. Chem. 269, 22719 - 22725.

ACETYLCHOLINESTERASE AT NEUROMUSCULAR JUNCTIONS

Density, Catalytic Turnover, and Molecular Forms with Modeling and Physiological Implications

L. Anglister,[1] J. R. Stiles,[2] B. Haesaert,[1] J. Eichler,[1] and M. M. Salpeter[2]

[1] Department of Anatomy and Embryology
Hebrew University—Hadassah Medical School
Jerusalem 91120, Israel
[2] Section of Neurobiology and Behavior
Division of Biological Sciences
Cornell University
Ithaca, New York 14853-2702

INTRODUCTION

Acetylcholinesterase (AChE) which is concentrated at vertebrate neuromuscular junctions between the presynaptic release sites and the postsynaptic receptors (AChR), hydrolyzes acetylcholine (ACh) and thereby assists in termination of neuromuscular transmission. Knowledge of junctional AChE and AChR concentrations is crucial for understanding transmission at this synapse. However, although AChR density has been measured in endplates of several species, and shows little variability (Fertuck and Salpeter, 1976; Matthews-Bellinger and Salpeter, 1978; Land et al., 1980), AChE density has been determined only in mouse endplates (Rogers et al., 1969; Salpeter et al., 1972, 1978).

Based on our recent findings (Anglister et al., 1994a, 1994b), we hereby show that the density of AChE (σ_E) measured by EM-autoradiography and biochemistry at frog neuromuscular junctions, which contain both globular and asymmetric molecular forms, is ~4-fold lower than reported for mouse endplates. On the other hand the rate of ACh hydrolysis by frog AChE active site (turnover number, k_{cat}) was high but well within the relatively large range reported for other enzyme sources (e.g. Rosenberry, 1975; Vigny et al., 1978). The physiological implications of these experimental results, were evaluated using Monte Carlo computer simulations of miniature endplate currents (MEPCs) for frog and lizard neuromuscular junction (NMJ) geometries. Our simulation results show that a wide range of density and k_{cat} values would be compatible with normal neuromuscular transmis-

Enzymes of the Cholinesterase Family, Edited by Daniel M. Quinn et al.
Plenum Press, New York, 1995

sion with either endplate geometry, and that even a density of only ~400 sites/μm^2 and k_{cat} of only ~1,000 s^{-1} are sufficient for normal quantal currents. Thus, the relative insensitivity of MEPC amplitude, rise and decay time, to both σ_E and k_{cat} provides an additional component to the high safety factor known for this synapse.

DETERMINATION OF AChE Site Density By EM-AUTORADIOGRAPHY

Labeling with ^{3}H-Diisopropylfluorophosphate (^{3}H-DFP)

Cutaneous pectoris muscles of frogs were dissected and their AChE labeled with ^{3}H-DFP, an organophosphorus reagent which inhibits AChE, as well as other serine-hydrolases, by covalent binding to the active sites (McIsaac and Koelle, 1959). Thus, to label selectively extracellular AChE active sites muscles were treated by a 3-step procedure (Rogers et al., 1969; Anglister et al., 1994b), with initial exposure to non-radioactive DFP, which is lipid soluble, permeates membranes rapidly and thus saturates all intra- and extracellular DFP-binding sites. The extracellular AChE sites were then reactivated by exposure to the membrane-impermeable quaternary oxime, pyridine-2-aldoxime methiodide (2-PAM) (Hobbiger, 1963), and subsequently labeled with ^{3}H-DFP.

Examination of EM autoradiograms showed that the developed grains corresponding to ^{3}H-DFP binding to extracellular true AChE (2-PAM reactivated) sites were concentrated around NMJs, as illustrated in Figure 1. The number of synapse specific grains (in 275 NMJ-sections of 4 frogs) was used to calculate the extracellular AChE site density (σ_E). A value of 560 ± 60 AChE sites μm^{-2} was determined if the post-junctional membrane (or the adjacent basal lamina sheath) was assumed to be the surface along which the label was distributed (Table 1) (Anglister et al., 1994b). Since the nerve terminals can provide AChE to the synaptic cleft (Anglister, 1991), another value could be calculated for σ_E (~400 ± 40 μm^{-2}, Table 1) if one assumes that the AChE in the primary cleft is distributed equally along all synaptic membrane, i.e. along both the post-junctional and pre-synaptic (axonal) membrane surface. Thus the σ_E for frog (400 μm^{-2}) is very low relative to that at mouse (or lizard, Fig. 1b) NMJs (~2500 μm^{-2}).

Labeling with ^{125}I-Fasciculin

Determination of AChE site density by EM-autoradiography using ^{3}H-DFP is difficult for several reasons: First, the lipid soluble DFP may bind not only to AChE but also to other extra- and intracellular serine-hydrolases. Thus, to label specifically the extracellular AChE, a complex sequence of inhibition/reactivation steps (as described above) is required. Second, the specific activity of the tritiated compound is low. Consequently, the exposure time of EM-autoradiograms required to measure AChE density even at normal NMJs is very long (3-12 months). This limits the ability to study the regulation of AChE, where measurements of low site densities are required. To overcome these difficulties we have introduced the use of the radio-iodinated anticholinesterase toxin, fasciculin, for EM-autoradiographic determination of AChE sites. A purified preparation of fasciculin II from the venom of the Mamba Dendroaspis angusticeps (Karlsson et al., 1985) was labeled. The ^{125}I-fasciculin, as the original unlabeled toxin, strongly inhibits AChE from several species by binding with very high affinity to the peripheral-anionic site on the AChE catalytic subunit (Marchot et al., 1994; Radic et al., 1994). The

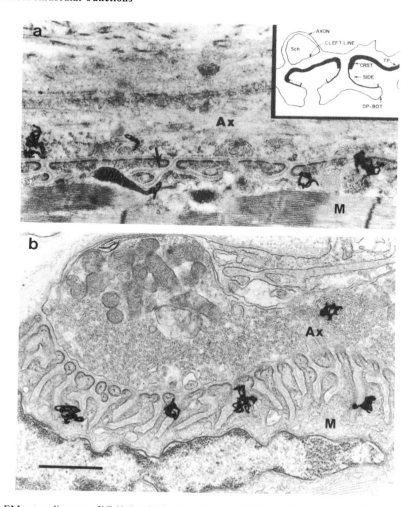

Figure 1. EM autoradiogram of NMJs in which extracellular (2-PAM reactivated) true AChE sites were labeled with ^{3}H-DFP. Developed grains are typically localized over region of primary and secondary clefts. Basal lamina follows contours of junctional folds (illustrated best at arrow). Magnification bar = 1 μm; M = muscle; Ax = axon. a) Frog cutaneous pectoris NMJ. Insert shows a schematic representation of post-synaptic membrane regions in frog elongated NMJ [top (TP), crest (CRST), side, and deep bottom (DP-BOT)] where apparent AChR site densities are 10000 (TP and CRST), 5000 (SIDE), or 0 μm^{-2} (DP-BOT), Sch = Schwann process. For miniature endplate current computer simulations, the lateral branches at the bottom of frog junctional folds were pulled down to make a model fold of uniform width extending straight down into the muscle cell. b) Lizard intercostal muscle NMJ, unlike frog NMJ, has a typical endplate structure as in mammals, with junctional folds that are mostly straight and show relatively uniform width, and can thus be well represented by a simplified model fold design. Synaptic grain counts indicate that lizard σ_E is somewhat lower but similar to mouse value. For modeling, we used the experimental values of frog and lizard NMJs, for width of primary cleft, spacing of folds, width of mouth of secondary cleft and overall ratio of pre- to postsynaptic membranes, as well as AChR and AChE densities and distributions.

number of toxin molecules bound is, therefore, equal to the number of AChE active sites (or DFP-binding sites). Binding of ^{125}I-fasciculin to frog muscles resulted in a very specific binding to the NMJs (Fig. 2). After the relatively short exposure required to obtain sufficient grain counts, analysis of EM-autoradiograms revealed a value for AChE site density at frog NMJs similar to the density determined using ^{3}H-DFP.

Table 1. Synaptic AChE site density (σ_E) in frog muscle

	σ_E Value (AChE sites μm^{-2})	
	Autoradiographic determination	Biochemical determination
Postjunctional membrane	560 ± 60	540 ± 104
Postjunctional plus axonal membrane	400 ± 40	386 ± 74

The density was determined for frog cutaneous pectoris muscle using two methods, quantitative EM autoradiography and biochemical assay. Values were expressed relative to either post-junctional membrane area or the sum of post-junctional plus axonal membrane areas. For autoradiographic determinations, the values were obtained separately for each muscle and then averaged. This gives a mean SEM averaged over 4 animals. For biochemical determinations, the values were calculated based on: synaptic AChE activity, measurements of AChE turnover number (k_{cat}), the number of myofibers per muscle and mean NMJ surface area (see Text). The error ranges shown were obtained from the square root of the sum of the squares of SE for each of the above terms in the calculation. Thus, the error in the final biochemical value is larger than that for the EM autoradiography, which is based on fewer independent variables.

BIOCHEMICAL DETERMINATION OF SYNAPTIC AChE SITE DENSITY

To compare the EM autoradiographic results with those obtained by a radically different technique, we used biochemical measurements for σ_E determination. Accordingly, it was necessary to: 1) estimate the amount of AChE activity present in frog NMJs (extracellular, synaptic AChE); 2) determine k_{cat} for the frog muscle enzyme and thus convert synaptic AChE activity to a number of active sites; and 3) relate the number of active sites to NMJ geometry. Synaptic AChE activity was estimated from separate measurements of extracellular activity for junctional (NMJ-containing) and non-junctional muscle regions: Dissected muscles were treated using a sequence of DFP and 2-PAM as outlined above for extracellular AChE labeling, but omitting ^{3}H-DFP step. This produced muscle preparations in which the intracellular AChE was covalently inhibited but the extracellular AChE was active and could be directly examined. Junctional and nonjunctional muscle regions were separated and all their AChE content was extracted. "Synapse-specific" AChE activity was assessed by subtracting from the activity measured in the junctional region the value measured for an equal amount of protein in the nonjunctional regions, assuming a uniform distribution of non-synaptic extracellular AChE along the length of each muscle fiber. Accordingly, the synapse-specific AChE activity calculated was 12.68 ± 1.28 pmol ACh s^{-1} (n = 27). We also found that about 83% of the extracellular activity of the entire muscle was present at the NMJ, which by our measurements constitute less than 0.3% of myofiber surface area.

In order to calculate σ_E (i.e. as sites μm^{-2} of junctional membrane) from our biochemical estimate of synaptic AChE activity, the following calculation is carried out: When the muscle's synaptic AChE activity (12.68 ± 1.28 pmol s^{-1}) is multiplied by Avogadro's number per pmol, we obtain a value of $7.64 \pm 0.77 \times 10^{12}$ ACh molecules hydrolyzed per second. Division of this value by k_{cat} value for the frog (9500 ± 750 ACh molecules s^{-1} site^{-1}; Anglister et al., 1994b) gives $8.1 \pm 1.0 \times 10^8$ synapse-specific AChE sites in the entire muscle. Since there are about 500 singly innervated muscle fibers in the cutaneous pectoris muscle (Rotshenker and McMahan, 1976), and the average post-junc-

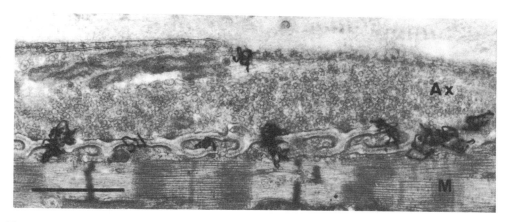

Figure 2. EM autoradiogram of frog NMJ labeled with ^{125}I-fasciculin. Cutaneous pectoris muscles were first incubated with ^{125}I-fasciculin (.5 μM) to inhibit AChE activity, and then fixed and processed for EM-autoradiography (as in Anglister et al., 1994b). Developed grains appeared almost exclusively at the NMJs. Bar, 1 μm.

tional membrane area is about 3,000 μm^2 (Matthews-Bellinger and Salpeter, 1978), calculation of σ$_E$ based on our biochemical measurements yields 540 ± 104 μm^{-2}. This value, which as described is based on determinations of several parameters, is less accurate but in excellent agreement with our σ$_E$ determination based on EM autoradiography (560 ± 60 μm^{-2}, Table 1).

MOLECULAR FORMS OF FROG SYNAPTIC AChE

Muscle contains asymmetric and globular isoforms of AChE (for reviews see Massoulié and Bon, 1982; Anglister and McMahan, 1984; Silman and Futerman, 1987; Rotundo, 1987; Toutant and Massoulié, 1988). The following experiments were aimed to determine the molecular forms of the extracellular AChE at the neuromuscular junctions that are associated with the extracellular matrix at the synapse. The 2-PAM-reactivated AChE in junctional regions of frog muscles (prepared as in previous section) was extracted at high salt concentration but no detergents, so as to minimize any solubilization of plasma-membrane-bound AChE that might occur. The high ionic strength buffer alone could solubilize 84% ± 3 (SD, n = 3) of the total pool of extracellular enzyme in the junctional region extracted by high salt and detergent. As illustrated in Fig. 3, sucrose-gradient-sedimentation analysis showed that the high salt extract of the extracellular junctional enzyme contained both asymmetric (17.5 S and 13.8 S) and small globular AChE forms (4-6 S, with a maximum at ~5 S). Moreover, the small globular AChE accounted for >50% of the total activity, and was solubilized in Ringer's solution while the rest of the enzyme remained insoluble. Thus, the 4-6 S AChE comprises more than half of the extracellular AChE at frog NMJs, and is extracted, together with the asymmetric forms, in high salt in the absence of detergent, but unlike the asymmetric forms, it is soluble also in isotonic buffers. Further analysis of the 4-6 S species revealed that it consists of a major 5.3 S non-amphiphilic form (75% of the activity) and a minor 6.9 S amphiphilic component (Anglister et al., 1994a).

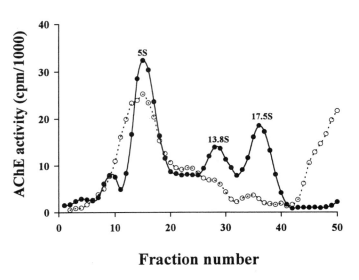

Figure 3. Molecular forms of extracellular AChE in the synaptic region of frog muscle. Muscles were first incubated with DFP (1mM, 1h, RT) to block all AChE activity, and then reactivated with 2-PAM (0.1 mM, 45 min, RT) to unmask the activity of external AChE. Molecular forms of AChE in muscle extracts were resolved by sucrose gradient sedimentation and assayed radiometrically (as in Anglister et al., 1994a). Extraction and sedimentation were done in the absence of detergents either in high salt (1M NaCl)(l) or in low salt (0.1M NaCl)(ü) buffers, revealing both asymmetric (17.5 S and 13.8 S) and smaller globular molecular forms of external AChE, with a major 4-6 S component (maximum at ~5 S) solubilized in both high salt and low salt buffers.

EFFECT OF σ_E AND k_{cat} ON MEPCs; MONTE CARLO COMPUTER SIMULATIONS

Many of the essential features of neuromuscular transmission may be studied at the level of single quantal currents (i.e., miniature endplate currents, MEPCs). Inhibition of AChE has been shown to affect the MEPC and the multi-quantal endplate current (EPC), by increasing the amplitude (Ac), rise time and decay time (t_f) (e.g. Katz & Miledi, 1973; Anderson & Stevens, 1973; Gage & McBurney, 1975; Hartzell et al., 1975; Land et al., 1980, 1981, 1984). To evaluate the effects of AChE density (σ_E) and hydrolytic rate (k_{cat}) on neuromuscular physiology, and specifically on MEPC Ac and t_f, we have used Monte Carlo computer simulations of MEPC with σ_E and k_{cat} as the variable parameters. Briefly, in the Monte Carlo algorithms individual ACh molecules from a released quantum are examined as they diffuse through the primary and secondary synaptic cleft spaces and interact with AChR or AChE molecules. The primary simulation output is the membrane conductance at voltage clamp conditions, from which the Ac and t_f are obtained (Bartol et al., 1991).

The modeling required specified values for two types of input parameters: 1) Synaptic geometry parameters - primary and secondary cleft dimensions, the distance between junctional folds and the location and distribution of AChRs and AChE. To compare the effect of geometry features on simulation results, modeling was done for the two drastically different NMJ morphologies of frog, which are elongated and lie longitudinally to the axis of the myofiber, and of lizard, with a coiled compact configuration as in mammalian endplates (see Fig. 1a and b, data reviewed in Salpeter, 1987). 2) Chemical kinetic parameters - eg. the rate constants for the interaction of ACh with AChR and AChE. We used two different sets of values previously determined by using simulations to fit published values and experimental lizard MEPC data under different conditions, including various degrees of AChR inactivation (see, Land et al., 1980, 1981, 1984) .

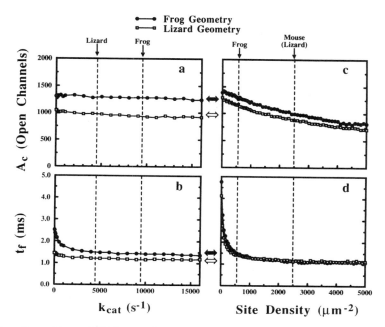

Figure 4. Effect of σ_E and k_{cat} on MEPC properties of frog and lizard NMJs. Monte Carlo computer simulations were used to illustrate dependence of MEPC amplitude (Ac, expressed as peak number of bi-liganded open AChR channels) and e-fold fall time (t_f, ms) on AChE turnover number (k_{cat}, s^{-1}; panels a ,b) or site density (σ_E, μm^{-2} post-synaptic membrane area; panels c,d). Each point represents the average result of 3 independent simulations. Dashed lines indicate experimental values for frog k_{cat} and σ_E, and for lizard k_{cat} as well as the preliminary lizard σ_E (see, Fig. 1b). Arrows on ordinate axes indicate predicted frog (solid) or lizard (open) Ac or t_f corresponding to their respective experimental σ_E or k_{cat} values.

Figure 4 shows MEPC simulation results as a function of variations in σ_E or k_{cat} for frog and lizard model geometries. For both geometries, the MEPC Ac (Fig. 4 a) and rise time (not shown) decrease very slightly (%) as k_{cat} increases, over the entire range tested (0 - 16,000 s^{-1}). The dependence of t_f on k_{cat} (Fig. 4b) is highly nonlinear and more complex for both geometries: Between 1,000 and 16,000 s^{-1}, t_f is less dependent on k_{cat} than Ac, reaching an asymptote equal to the AChR mean burst duration (1.1 ms, see below) at $k_{cat} \approx 5,000$ s^{-1}. Below 1,000 s^{-1}, however, t_f increases significantly, especially for the frog NMJ. For example, at $k_{cat} = 500$ s^{-1}, t_f for frog geometry has increased by 36%, whereas for lizard geometry, it has increased only by 10%.

The effect of σ_E on MEPC Ac is stronger (steeper slope) than that of k_{cat} for both model geometries (Fig. 4c). Even so, Ac decreases only by a factor of ~2 as σ_E increases from 0 to 5,000 sites μm^{-2}, with most of the change occurring below the experimental mouse value of 2,500 μm^{-2}. The dependence of t_f on σ_E, however, is dramatically different (Fig. 4d): t_f is markedly prolonged for σ_E below ~ 400 μm^{-2}, (e.g., by 200 μm^{-2} it has increased by almost 50%). Above the critical value of 400 μm^{-2}, t_f curves approach an apparent asymptote (which is equal to the mean burst duration, Fig. 4d).

Figure 4 also gives the predicted values of Ac and t_f for frog and lizard junctional geometries (open and closed arrows on ordinates) at their respective experimental σ_E and k_{cat} values. The predicted Ac and t_f are 1300 open channels and 1.5 ms for the frog and 1,000 open channels and 1.2 ms, respectively, for the lizard conditions. These predicted MEPC values are close to the experimental values (1,000-2,000 open channels and 1.5 ms, respectively; Dionne and Parsons, 1981; Coquhoun and Sakmann, 1985). Furthermore, the model predicted changes in Ac after complete AChE inhibition ($\sigma_E = 0$) were ~10% for frog and ~40% for lizard (Fig. 4c), whereas for t_f, the predicted change was ~3- to 4- fold for both junctions (Fig. 4d), are also in agreement with experimental data (Miledi et al., 1984; Land et al., 1984; Bartol et al., 1991). We also found that if AChE sites were related to both post-synaptic and pre-synaptic surfaces (e.g. frog σ_E ~400 μm^{-2}; Table 1) the simulation results did not change appreciably. One feature of our results is the minor but systematically higher predicted value of Ac and t_f for frog versus lizard model conditions, which reflects the fact that there is a lower ratio of AChE to AChR binding sites in the frog.

SUMMARY AND CONCLUSIONS

AChE site density in frog neuromuscular junctions was determined by quantitative EM-autoradiography and by biochemical assays. EM-autoradiography included two kinds of measurements: The first involved AChE labeling with ^{3}H-DFP, and the second introduced the application of ^{125}I-fasciculin in AChE quantitations in situ. All three studies resulted in frog σ_E ~600 μm^{-2}, ~4-fold lower than at mouse endplates. Further biochemical analysis showed that frog synaptic AChE contained both asymmetric and globular isoforms. To evaluate the compatibility of our results with neuromuscular physiology we used Monte Carlo computer modeling. In total, our modeling studies showed that: 1) A wide range of densities and k_{cat} would be compatible with normal transmission. 2) The effects of σ_E and k_{cat} on MEPC Ac and t_f do not depend on either gross endplate geometry or any particular set of valid simulation conditions. 3) There are well-defined basic minimal requirements of AChE site density and k_{cat} (of only ~400 μm^{-2} and ~1,000 s^{-1}, respectively) for normal synaptic function. We have thus demonstrated that some NMJs have AChE densities and k_{cat} well in excess of these critical lower limits.

ACKNOWLEDGMENTS

We thank Rachel Cohen and Maria Szabo for excellent technical assistance. Supported by US-Israel Binational Science Foundation BSF89-00502 (LA and MMS) and Israel Science Foundation of Israel Academy of Sciences and Humanities - Charles H. Revson Foundation 675/94 (LA), NIH NS09315 and GM10422 (MMS), NIH Fellowship NS09126 (JS). Computer simulations were conducted at the Cornell National Supercomputer Facility, a resource of the Center for Theory and Simulation in Science and Engineering, Cornell University, Ithaca, NY.

REFERENCES

Anderson, C.R. & Stevens, C.F. (1973) *J. Physiol. (Lond) 235,*655

Anglister, L. (1991) *J. Cell Biol. 115,*755.

Anglister, L. & U. J. McMahan (1984) In *Basement Membranes and Cell Movement* (Porter, R. & Whelan, J., Eds.) pp. 163-178, Pitman Books Ltd., London.

Anglister, L., Haesaert, B., & McMahan U.J. (1994a) *J. Cell Biol. 125,* 183.

Anglister, L., Stiles, J.R., & Salpeter, M.M. (1994b) *Neuron 12,* 783.

Bartol Jr., T.M., Land, B.R. Salpeter, E.E. & Salpeter, M.M. (1991) *Biophysical J. 59,* 1290.

Colquhoun, D., & Sakmann, B. (1985) *J. Physiol. 369,*501.

Dionne, V.E., & Parsons, R.L. (1981) *J. Physiol. 310,* 145.

Fertuck, H.C. & Salpeter, M.M. (1976) *J. Cell Biol. 69,* 144.

Gage, P.W., & McBurney, R.N. (1975) *J. Physiol. 244*, 385.

Hartzell, H.C., Kuffler, S.W., & Yoshikami, D. (1975) *J. Physiol. 251*, 427.

Hobbiger, F. (1963) In *Cholinesterase and Anticholinesterase Agents*, (Koelle, G.B. Ed.) Handbook Exp. Pharmacol. Vol. XV, pp. 921-988, Springer-Verlag, Berlin-Gotingen-Heidelberg.

Karlsson, E., Mbugua, P.M., & Rodriguez-Ithurralde, D. (1985) *Pharmac. Ther. 30*, 259.

Katz, B. & Miledi, R. (1973) *J. Physiol. (Lond) 231*, 549.

Land, B.R., Salpeter, E.E., & Salpeter, M.M. (1980) *Proc. Nat. Acad. Sci. USA 77*, 3736.

Land, B.R., Salpeter, E.E., & Salpeter, M.M. (1981) *Proc. Nat. Acad. Sci. USA 78*, 7200.

Land, B.R., W.V. Harris, W.V., Salpeter, E.E., & Salpeter, M.M. (1984) *Proc. Natl. Acad. Sci. USA 81*, 1594.

Letinsky, M.S., Fischbeck, K.H., & McMahan, U.J. (1976) *J. Neurocytol. 5*, 691.

Marchot, P., Khelif, A., Ji., Y.-H., Mansuelle, P. & Bougis, P.E. (1993) *J. Biol. Chem. 17*, 12458.

Massoulié, J. & Bon, S. (1982) *Ann. Rev. Neurosci. 5*, 57.

Matthews-Bellinger, J. & Salpeter, M.M. (1978) *J. Physiol. 279*, 197.

McIsaac, R. S. & G. B. Koelle. (1959) *J. Pharmacol. Ther. 126*, 9.

Radic, Z., Duran, R., Vellom, D.C., Li, Y., Cervenansky, C., & Taylor., P. (1994) *J. Biol. Chem. 260*, 1.

Rogers, A.W., Derzynkiewicz, Z., Salpeter, M.M., Ostrowski, K., & Barnard, E.A. (1969) *J. Cell Biol. 41*, 665.

Rosenberry, T.L. (1975) Acetylcholinesterase. *Adv. Enzymol. 43*, 103.

Rotshenker, S., & McMahan, U.J. (1976) *J. Neurocytol. 5*, 719.

Rotundo, R. L. (1987) In *The Vertebrate Neuromuscular Junction. Neurology and Neurobiology* (Salpeter, M. M., Ed.) Vol. 23, pp. 247-284, Alan R. Liss, Inc., NY. .

Salpeter, M.M. (1987) In *The Vertebrate neuromuscular Junction. Neurology and Neurobiology* (Salpeter, M.M. Ed.), pp. 1-54, Alan R. Liss, Inc., New York.

Salpeter, M.M., Plattner, H., & Rogers, A.W. (1972) *J. Histochem. Cytochem. 20*, 1059.

Salpeter, M.M., Rogers, A.W., Kasprzak, H., & McHenry, F.A. (1978) *J. Cell Biol. 78*, 274.

Silman, I. & Futerman, A. H. (1987) *Eur. J. Biochem. 170*, 11.

Toutant, J.-P. & Massoulié, J. (1988) In *Handbook of Exp. Pharmacol.* (Whittaker, V.P., Ed.), Vol. 86. pp. 225-265, Springer-Verlag, Germany.

Vigny, M., Bon, S., Massoulié, J., & Letterier, F. (1978) *Eur. J. Biochem. 85*, 317.

EFFECT OF ACUTE IRREVERSIBLE INHIBITION OF ACETYLCHOLINESTERASE ON NERVE STIMULATION-EVOKED CONTRACTILE PROPERTIES OF RAT MEDIAL GASTROCNEMIUS

R. Panenic,[1] P. Gardiner,[1] and V. Gisiger[2]

[1] Département d'Education Physique
[2] Département d'Anatomie
  Université de Montréal
  Montréal, Québec, Canada, H3C 3J7

This study was conducted to determine the effect of decreased muscle acetylcholinesterase (AChE) on the expression of muscle contractile properties during indirect stimulation in situ. In anesthetized Sprague-Dawley rats, the medial gastrocnemius was surgically isolated, and isometric forces and muscle EMG were measured in response to a protocol of sciatic nerve stimulation. Contractions examined included twitches, and responses to trains (800 ms) at 25, 50, and 200 Hz. Fatigue resistance was also determined by stimulating with trains (100 ms) at 75 Hz, once every 1.5 seconds, for 3 minutes. In one group, AChE was irreversibly inhibited before the protocol by bathing the muscle for 25 minutes with a solution of methanesulfonyl fluoride (MSF, 1 mg/ml in saline) followed by repetitive rinsing with saline. This resulted in at least 90% inhibition of AChE, as measured in muscle homogenates following experiments. In MSF-treated muscles, peak forces for twitch and 25 Hz contractions were potentiated (77% and 40%, respectively), while force at 200 Hz was depressed (by 36%), compared to controls. Thus, the ratio of twitch/maximum tetanic force was 50% in MSF-treated muscles, compared to 16% in controls. The pattern of force decline during the fatigue protocol was similar in control and MSF preparations, although the latter generated significantly less contractile force throughout the 3 minutes. During trains, forces declined dramatically from peak force early in the train to a force plateau in MSF-treated muscles. Interestingly, these force plateaus increased with increasing frequencies of stimulation in MSF-treated muscles. This finding, as well as the lack of effect of MSF-treatment on the pattern of force decrement during intermittent trains, would appear inconsistent with an increased desensitization of ACh receptors due to elevated ACh accumulation at higher frequencies of stimulation. Funded by NSERC Canada.

DEVELOPMENTAL REGULATIONS OF ACETYLCHOLINESTERASE AT THE MOLECULAR LEVEL IN THE CENTRAL NERVOUS SYSTEM OF THE QUAIL

Alain Anselmet, Jean-Marc Chatel, and Jean Massoulié

Neurobiologie Cellulaire et Moleculaire
CNRS URA 1857
Ecole Normale Supérieure
46 rue d'Ulm
75005 Paris, France

ABSTRACT

We have analyzed the evolution of acetylcholinesterase (AChE) transcripts and molecular forms during development in the central nervous system of the quail. We found that several aspects are developmentally regulated: 1) the production of multiple mRNA species; 2) the production of enzymatically active and inactive AChE molecules; 3) the production of amphiphilic and nonamphiphilic AChE forms; 4) the proportions of tetrameric G4, dimeric G2 and monomeric G1 forms (Anselmet et al., 1994).

REFERENCES

Anselmet, A., Fauquet, M., Chatel, J.-M., Maulet, Y., Massoulie, J. & Vallette, F.-M. (1994) J. Neurochem. **62**, 2158-2165.

REVERSAL OF GLYCEROL ETHER-STIMULATED ACETYLCHOLINESTERASE ACTIVITY BY *c-fos* ANTISENSE OLIGONUCLEOTIDE IN PRIMARY NEURONAL CULTURES

J. R. Dave, E. Fasnacht, F. C. Tortella, J. M. Best, B. P. Doctor, and H. S. Ved

Divisions of Neuropsychiatry and Biochemistry
Walter Reed Army Institute of Research
Washington, D.C. 20307

We have previously reported that dodecylglycerol (DDG), an alkyl glycerol lipid similar to platelet activating factor (PAF), stimulates differentiation in primary neuronal cultures obtained from fetal rat cerebral cortex, hippocampus and cerebellum (Ved *et al*, 1994). The neuronal differentiation was evident both morphologically as development of axon-like extensions and biochemically as an increase in neuron specific enzyme activity (Ved *et al*, 1991). We also reported that DDG mediated neuronal differentiation resulted in a transient increase in *c-fos* mRNA levels (Ved *et al*, 1993). The objectives of present study were to determine if *c-fos* antisense oligonucleotide (ASO) would reverse DDG-stimulated acetylcholinesterase (AChE) and choline acetyltransferase (ChAT) activities, and neuronal differentiation. Primary enriched neurons derived from embryonic (E-17) rat cerebral cortex, cerebellum or hippocampus were treated in a serum-free media with either vehicle, DDG (4 µM) or DDG with either *c-fos* antisense oligonucleotide (ASO) (5-20 µM), *c-myc* ASO or non-sense ASO for 24 hrs. Treatment with DDG produced a maximal stimulation in AChE and ChAT activities, and outgrowth of neuronal processes in cultures obtained from cerebellum and a minimum effects in those obtained from cerebral cortex. The percent stimulation in AChE activity following DDG treatment in these neuronal cells were approximately 200%, 150% and 65% above the control levels in cerebellum, hippocampus and cortex, respectively. The percent stimulation in ChAT activity following DDG treatment in neuronal cells were approximately 260%, 220% and 45% above the control levels in cerebellum, hippocampus and cortex, respectively. Pretreatment with *c-fos* ASO partially inhibited DDG-stimulated AChE and ChAT activities and outgrowth of the neuronal processes, however, *c-myc* ASO and nonsense ASO had no significant effect. The inhibitory effect of *c-fos* ASO on DDG-stimulated AChE and ChAT activities and neuronal differentiation was greatest in neurons obtained from cerebellum and hippocampus (100%-52% inhibition). These results suggest a causative role of *c-fos* proto-oncogene in DDG-mediated neuronal differentiation and further suggest a direct role of c-fos gene product on modulation of AChE and ChAT enzyme activities in the neuronal cells.

REFERENCES

Ved, H.S., Fasnacht, E.A., Knight, E.S., Doctor, B.P. and Dave, J.R. 1994, *FASEB J.* (Abstract)
Ved, H.S., Gustow, E. and Pieringer, R.A. 1991, *J. Neurosci. Res.* 37: 156-159.
Ved, H.S., Doctor, B.P., Genovese, R.F., Tortella, F.C. and Dave, J.R. 1993 (Abstract) *Soc Neurosci Annual Meeting.*

ACETYLCHOLINESTERASE ACTIVITY IN PLANTS

S. Madhavan, Gautam Sarath, and Patricia L. Herman

Department of Biochemistry
University of Nebraska
Lincoln, Nebraska 68583

INTRODUCTION

Acetylcholinesterase, AChE, (E.C.3.1.1.7), has been reported to be present along with acetylcholine, ACh, in many plant species (Hartmann and Gupta, 1989; Tretyn and Kendrick, 1991). A defined role for ACh in plant cellular processes is still unclear but suggestions have been made that it may act as a second messenger in various processes (Hartmann and Gupta, 1989). We have presented in this study data supporting: a) the presence and enrichment of AChE in guard cells of stomata in three different plant species, and b) the possible involvement of ACh in the regulation of stomatal movements.

ACETYLCHOLINESTERASE ACTIVITY IN GUARD CELLS

AChE activity has been detected in extracts of guard cell protoplasts (GCP) of *Vicia faba* (10.3±5.2 nmols min^{-1} mg^{-1} protein) *Nicotiana glauca* (19.2±0.4 nmols min^{-1} mg^{-1} protein) and *Kalanchoe* (47.3±3.6 nmols min^{-1} mg^{-1} protein). When compared to the mesophyll cell protoplasts (MCP) or whole leaf extracts, GCP from *V. faba*, *N. glauca* and *Kalanchoe* showed a 50-90% enrichment in AChE activity. The enzyme from GCP extracts of the above three species also displayed significant substrate specificity towards acetylthiocholine than either propionylthiocholine or butyrylthiocholine. Inhibition studies with animal AChE inhibitors, neostigmine, tacrine and eserine, on *V. faba* GCP-AChE suggested that both neostigmine ($I_{0.5}$, 0.1 mM) and tacrine ($I_{0.5}$, 0.14 mM) inhibit GCP-AChE more effectively than eserine. Polyclonal antibodies raised against a purified AChE from fetal bovine serum cross reacted with proteins present in *V. faba* GCP extracts and the apparent molecular weight of the cross reacting band was 63 kD. Thus, enrichment of AChE activity, substrate specificity, inhibition by neostigmine, tacrine and eserine and the detection of GCP proteins that immunoreact with antibodies raised to bovine AChE strongly suggest that the enzyme in GCP extracts is a true plant AChE.

ACH AND STOMATAL OPENING

Exogenous application of ACh, PrCh and BuCh on stomatal movement indicate that *V. faba* leaf stomata from isolated epidermal peels respond to these choline esters similar to the substrate specificity displayed by their AChE. ACh causes maximal stomatal closing, PrCh induces some closure after 30 minutes and BuCh is minimally effective. AChE inhibitors, tacrine and eserine, also promoted stomatal closure, perhaps by increasing intracellular ACh concentration, although their effect was not as pronounced as that of ACh application.

DETECTION OF AChE-LIKE SEQUENCES BY SOUTHERN ANALYSIS IN *ARABIDOPSIS* GENOMIC DNA

Southern analysis showed that a probe prepared from a Torpedo cDNA clone that encodes AChE hybridizes to sequences in *Arabidopsis* genomic DNA under permissive hybridization conditions. The pattern of hybridization suggested the presence of an AChE homologue as a single copy in the *Arabidopsis* genome.

CONCLUSIONS

Guard cells constitute about 2% or less of the epidermal cell population, and constitute an even smaller percentage of the total leaf cells. Thus the enrichment of ChE activity in guard cells over other cell types of the leaf indicates a physiological significance for ACh and AChE in these cells. The presence of a true AChE in GCP extracts and the effects of exogenously applied ACh and AChE inhibitors on stomatal movements suggest that the cholinergic system could function as another potential signal-transduction pathway in guard cells.

REFERENCES

Hartmann, E. & Gupta, R. (1989) in *Second Messengers in Plant Growth and Development* (Boss, W.F., Moore, D.J., Eds.) pp 257-287, Alan R. Liss, Inc., New York.
Tretyn, A., & Kendrick, R.E., (1991) *Bot. Rev.* 57, 33-73.

SIGNAL-MEDIATED CELLULAR RETENTION AND SUBUNIT ASSEMBLY OF HUMAN ACETYLCHOLINESTERASE

C. Kronman, Y. Flashner, A. Shafferman, and B. Velan

Israel Institute Biological Research
Ness-Ziona, Israel

Kinetic analysis of human AChE T- subunit secretion by 293-HEK cells revealed preferential secretion of assembled dimers (Kerem *et al.*, 1993). Secretion of ER resident enzymes is precluded through a signal mediated process involving a membrane-bound receptor (Munro and Pelham, 1987) which recognizes a specific signal at the COOH-terminus of the lumenal proteins (Pelham, 1990) of which the Acid-Leu terminus appears to be the most critical structural element in the interaction between the ER lumenal proteins and their cognate receptor (Pelham, 1990; Andres *et al.*, 1991). The presence of a CSDL motif at the C-terminus of the AChE polypeptide could suggest the participation of a selective signal-mediated ER retrieval process in the preferential export of assembled AChE catalytic subunits. To examine this possibility, secreted and cell-associated enzymatic activities in transiently transfected cells expressing HuAChE carboxyl-terminus mutants were determined (See Table 1).

Neither mutations that increase similarity to the canonical retention signal (substitution of CSDL by KSDL) nor those that deviate from if it (substitution to CSAV) altered the ratio of secreted to intracellular enzyme, yet the KDEL appended dimerization-impaired enzyme was retained in a cell associated fraction. The retained product

Table 1.

HuAChE type	C-terminal tetrapeptide	HuAChE Medium[*]	Cells[*]	AChE ratio (cells/medium)
Wild type	CSDL	760±67	88±20	0.11
D582A/L583V	CSAV	650±45	68±25	0.1
C580A	ASDL	680±32	86±17	0.13
C580K	KSDL	750±90	60±10	0.08
C580A-KDEL	KDEL	20±4	855±54	42.5

[*]Milliunits/plate

displayed kinetic constants similar to those of wild-type enzyme and pulse-chase experiments did not indicate its intracellular degradation. Part of the retained enzyme molecules acquired endoH resistance suggesting that the KDEL receptor is recycled between the various compartments of the secretory pathway. These results suggest that: a. AChE-monomers accumulating in 293 cells in large quantities by signal mediated retention acquire full enzymatic activity and are probably folded into the authentic functional configuration.b. Dimerization of AChE subunits through Cys-580 overrides KDEL mediated intracellular retention. c. The native C-terminal tetrapeptide CSDL of HuAChE provides a dimerization signal but does not serve as a variant of the KDEL retention signal. d. Conjunction of an ER oligomerization process with a consecutive step of retrieval of non-oligomerized subunits can provide, in principle, a 'quality control' system for the proper sorting of assembled subunits from the ER.

ACKNOWLEDGMENT

This work was supported in part by the U.S. Army Research and Development Command, Contract DAMD17-93-C-3042 (to A.S.).

REFERENCES

Andres, D.A., Rhodes, J.D., Meisel, R.L., and Dixon, J.E. (1991). J. Biol. Chem. 266, 14277-14282.
Kerem, A., Kronman, C., Bar-Nun, S., Shafferman, A. and Velan, B. (1993). J. Biol. Chem. 268, 180-184.
Munro, S. and Pelham, H.R.B. (1987). Cell 48, 899-907.
Pelham, H.R.B. (1990). Trends Biochem. Sci. 15, 483-486.

THE GRAIN APHID CHOLINESTERASES

B. Leszczynski,[1] T. Bakowski,[1] A.F.G. Dixon,[2] and H. Matok[1]

[1] Agricultural and Pedagogic University
Department of Biochemistry
ul. Prusa 12, PL-08110 Siedlce, Poland
[2] University of East Anglia
School of Biological Sciences
Norwich NR4 7TJ, United Kingdom

SUMMARY

Cholinesterases are key enzymes in transmission of nervous impulses across insect synapses. In the present paper we report on activity of acetylcholinesterase (AChE) [E.C. 3.1.1.7] and butyrylcholinesterase (BuChE) [E.C. 3.1.1.8] in grain aphid *Sitobion avenae* (Fabr.) homogenates, determined according to Ellman at al. (1961). In addition, the effect of cereal allelochemicals on activity of these enzymes has been studied.

The activity of both cholinesterases was found in homogenates from the grain aphid. The acetylcholinesterase was mostly associated with mitochondria and microsome membranes, instead about 55% of the activity of the butyrylcholinesterase occured in the cytosolic fraction. Within the studied morphs of the aphid, winged apterous adults had higher activity of acetylcholinesterase but larvae have higher activity of the butyrylcholinesterase. The EDTA extracts from whole seedlings of the resistant and susceptible wheat cultivars showed much stronger inhibitory effect on activity of the AChE and BuChE than EDTA phloem exudates. With a few exceptions, *in vitro* cereal allelochemicals inhibited the activity of the aphid cholinesterases even at the lowest studied concentrations. Therefore it is clear that the cholinesterases play an important role in chemical interactions between cereals and the grain aphid, especially during the host-plant colonisation stage.

REFERENCES

Ellman, G.L., Coutney, G.L., Andres, V., Jr. and Featherstone, R.M. 1961. Biochem. Pharmacol. 7: 88-95.

REVIEW OF NERVE AGENT INHIBITORS AND REACTIVATORS OF ACETYLCHOLINESTERASE

David H. Moore[1], Charles B. Clifford[1], Isabelle T. Crawford[2], Greg M. Cole[2], and Jack M. Baggett[2]

[1] U.S. Army Medical Research and Material Command
Fort Detrick, Frederick, Maryland 21702-5012
[2] Science Applications International Corporation
Joppa, Maryland 21085

BACKGROUND

Organophosphorus (OP) compounds have agricultural, military, and medical applications. Historically, the discovery of these compounds date back to 1854 when Clermont first synthesized an OP (Taylor, 1990). The establishment of the insecticidal properties of OPs by the German chemist Gerhard Schrader in 1937 (cited in Taylor, 1990) led to the development of anticholinesterase pesticides with potent selective insect toxicity. The highly toxic OP agents soman, tabun, and sarin were developed as chemical warfare agents in this same laboratory during World War II. These nerve "gases" were not used at that time; however, the potential threat that they posed then and now has sustained research interest in the pathophysiology of and protection from these agents. Anticholinesterase agents have therapeutic uses as well, including the treatment of glaucoma, myasthenia gravis, and Alzheimer's disease (Taylor, 1990).

Over 50,000 OP compounds have been synthesized and screened for insecticidal potency, and nearly 200 of them are in the marketplace. Currently, OP esters account for about 40% of the registered pesticides (Salem and Olajos, 1988). Insecticide poisonings comprise 3% of all poisonings reported to US poison centers, and of the 36,541 pesticide poisonings reported in 1986, 12,142 involved OPs (Litovitz et al., 1987). It has been estimated that on a worldwide basis, 500,000 illnesses (and a corresponding 20,000 deaths) are the result of pesticide poisoning each year.

Nerve agents have been a chemical warfare threat since their development during World War II. They are easily produced from ordinary industrial chemicals, and large stocks of these agents are known to exist, even in smaller countries. Up until about 10 years ago, there was no documented use of nerve agents. However, in 1983 the United Nations confirmed allegations that Iraq had employed tabun against Iran (Somani et al., 1992).

Enzymes of the Cholinesterase Family, Edited by Daniel M. Quinn et al.
Plenum Press, New York, 1995

Although organophosphorus esters are among the few classes of poisons for which there are specific antidotes, the need for improved treatment regimens for anticholinesterase poisoning still remains.

HISTORICAL PERSPECTIVES FOR REACTIVATORS

Studies initiated by Wilson and Bergmann in the early 1950s established that the active moiety of acetylcholinesterase (AChE) consists of both a choline binding site and an esteratic site (Taylor, 1990). The positively charged choline moiety of acetylcholine (ACh) is attracted to the choline binding site by electrostatic forces. X-ray crystallography has shown the choline binding site to be lined with aromatic groups and a single, probably protonated glutamate side chain (Taylor, 1990). The imidazole group of the esteratic site enhances the nucleophilic activity of the serine hydroxyl group (the acidic component of the esteratic site), thus enabling it to form a covalent bond with the carbonyl C atom of ACh. An acetylated enzyme intermediate is formed with the release of choline. Upon nucleophilic attack by the electronegative oxygen atom of a water molecule, the acetylated enzyme is hydrolyzed, yielding the regenerated enzyme and acetic acid. This scheme is diagrammed in Fig. 1 below.

Anti-AChEs act by generating a carbamylated or phosphorylated enzyme, instead of the acetylated enzyme generated during ACh hydrolysis. These adducts prevent substrate binding and are hydrolyzed more slowly than the acetyl enzyme. With some inhibitors, the enzyme-inhibitor complex can undergo what is thought to be a conformational change, rendering the enzyme-inhibitor complex extremely resistant to lysis by water or reactivators. This phenomenon is known as "aging" (Wilson *et al.,* 1992). The half-life of phosphoryl-inhibited AChE ranges from hours to months; in effect, permanently inactivating the enzyme. In 1951, Wilson found that hydroxylamine would reactivate phosphorylated AChE by releasing the phosphoryl group attached to the serine residue more rapidly than water. However, hydroxylamine was toxic at the concentrations required for reactivation (Wilson,

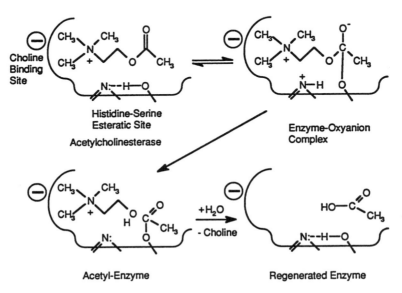

Figure 1. Hydrolysis of acetylcholine by acetylcholinesterase.

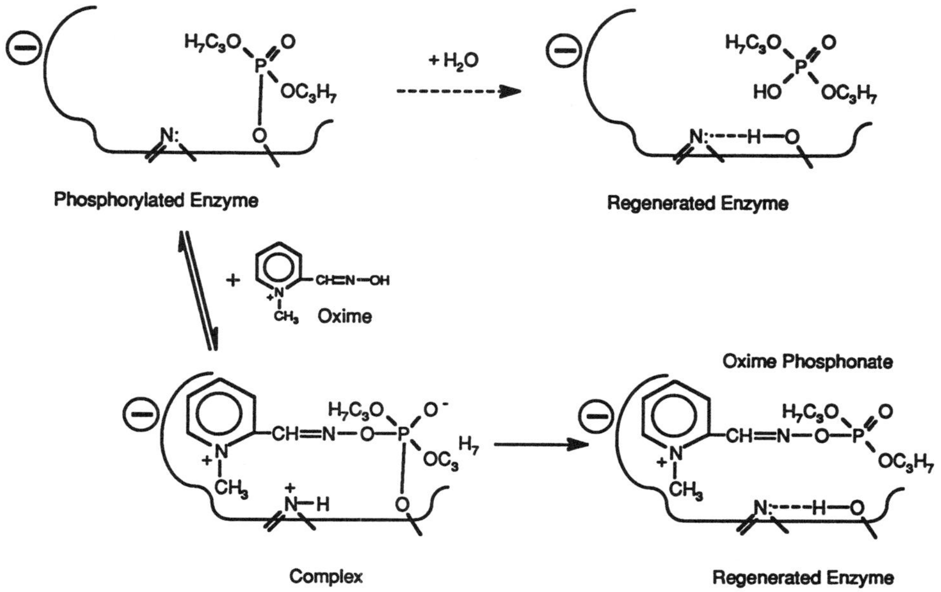

Figure 2. Regeneration of inhibited acetylcholinesterase.

1951). A diagrammatic example of the reactivation of phosphorylated AChE following normal hydrolysis or treatment with oxime is shown in Fig. 2.

The search for more effective reactivators unearthed many hydroxamic acids (RCONHOH) and oximes (RCH=NOH) of interest, the hallmark of which was the discovery of 2-PAM (pyridine 2-aldoxime, pralidoxime) by Wilson and Ginsburg (1955). The quaternary ammonium group in the pyridine ring of 2-PAM orients it in such a way as to exert a nucleophilic attack on the phosphorus of the inhibited enzyme, transferring the phosphate from the active site of the enzyme to the oxime. Pralidoxime reactivates OP-inhibited AChE at one million times the rate of hydroxylamine (Heilbronn-Wikström, 1965). However, its therapeutic effectiveness has been limited because 1) it cannot cross the blood-brain barrier, and hence cannot reverse the inhibition of brain AChE; and 2) it is not an efficient reactivator of soman-inhibited AChE.

Certain symmetric bisquaternary compounds were subsequently shown to be more potent reactivators of AChE. Obidoxime (Toxogonin; 1,1'-oxydimethylene bis-[4-formyl pyridinium bromide] dioxime), TMB-4 (Trimedoxime; 1,1'-trimethyl bis-[4-formyl pyridinium chloride] dioxime), and MMB-4 (Methoxime; 1,1'-methylene bis [4-[(hydroxyimino) methyl] pyridinium] dibromide) are examples of compounds having the advantage of being effective in the central nervous system (Das Gupta, 1982; Ellin, 1982; Harris, 1989). In spite of their improved potency, these first-generation reactivators were more toxic than 2-PAM, and none proved effective against the rapidly aging chemical warfare agent, soman.

During the late 1970s, a series of bis-pyridinium mono-oximes were synthesized by Hagedorn and coworkers (1978). One of the most potent members of this class of Hagedorn oximes, HI-6, has shown efficacy against soman-induced toxicity in several different animal species, in a multitude of in vitro and in vivo studies (Bošković, 1981; Clement, 1981; Lenz and Maxwell, 1981; Četkovic, 1984; de Jong and Wolring, 1984). The efficacy of HI-6 against soman did not appear to be solely due to reactivation of soman-inhibited unaged AChE in peripheral tissues (Goldman, 1990). Direct effects of HI-6 in the central nervous

Table 1. NATO and other fielded drug regimens

Country	Pretreatment	Treatment
Belgium	Pyridostigmine	Atropine
		Toxogonin
Canada	Pyridostigmine	Atropine
		Toxogonin/HI-6
		Prodiazepam/
		Diazepam
France	Pyridostigmine	Atropine
		P2S
		Diazepam
Israel	Pyridostigmine	Trimedoxime
		Atropine
		Benactyzine
Netherlands	Pyridostigmine	Atropine
		Toxogonin
Norway	Pyridostigmine	Atropine
		Toxogonin
Sweden	Pyridostigmine	Atropine
		HI-6
		Diazepam
United Kingdom	Pyridostigmine	Atropine
		P2S
		Diazepam
United States	Pyridostigmine	Atropine
		2-PAM
West Germany	Pyridostigmine	Atropine
		Toxogonin

system have been postulated to include 1) antimuscarinic and antinicotinic activity (Amitai et al., 1980; Kuhnen-Clausen et al., 1983; Szinicz et al., 1988); 2) ganglionic blockade (Kirsch and Weger, 1979; Lundy and Tremblay, 1979); and 3) neuromuscular junction effects (Clement, 1979; Amitai et al., 1980). HI-6 has not been shown to be effective in reactivating tabun-inhibited brain AChE in vivo (Clement, 1982) or tabun-inhibited human erythrocyte AChE in vitro (Galosi et al., 1988). There have also been problems with the stability of HI-6 in solution at therapeutic concentrations (Fyhr et al., 1987). Another compound in the Hagedorn series, HLö7 (a bispyridinium dioxime), has displayed potent reactivation of tabun-inhibited electric eel AChE and human erythrocyte AChE, as well as reactivation of soman-inhibited enzyme (de Jong et al., 1989). However, HLö7, like HI-6, is of questionable solubility at therapeutic concentrations in aqueous solution (Eyer et al., 1989).

There is no consensus among NATO and other countries on the oxime of choice for reactivation of nerve agent-inhibited AChE. The fielded drug regimens of selected countries are shown in Table 1. Although the Hagedorn oximes represent a significant advance in anti-AChE treatment, the pursuit of a universal reactivator is continuing.

CURRENT RESEARCH EFFORTS

As the understanding of structure-function relationships between the AChE enzyme, OP anti-AChEs, and reactivators matures, treatment compounds that better suit our needs can be designed. The most recent generation of reactivators includes such examples as the imidazolium oximes and the AB-oximes.

Several quaternary imidazolium salts have displayed improved efficacy against soman and tabun in early animal testing (Harris III *et al.*, 1990). The absence of a correlation between the *in vitro* and the *in vivo* activity of these compounds suggests modes of AChE protection other than reactivation (Goff *et al.*, 1991; Koolpe *et al.*, 1991). Analyses of the imidazolium oximes have been performed by Musallam *et al.*, 1993 using the Quantitative Structure-Activity Relationships (QSAR) technique. The Hansch approach was applied to the QSAR mathematical analysis. These data were compared with actual *in vivo* data to assess the predictive value of the technique. A novel imidazolium compound was predicted using this approach. The compound was synthesized and found to perform at least as well as all other synthesized compounds of its class in primary treatment assays.

The "AB-oximes" are another promising class of compounds designed to combine the structural aspects of oximes that confer ACh reactivation potency with those that give anticholinergic properties. Several of the synthesized compounds have demonstrated marked efficacy against soman and tabun poisoning in mice and guinea pigs, and preliminary results of the therapeutic value of these oximes in higher animal species (dogs and monkeys) are promising (Amitai *et al.*, 1993).

The dimethanesulfonate salt of HLö7 has generated recent interest because it is the first water-soluble salt of HLö7 (Eyer *et al.*, 1992). It is stable in aqueous solution and is potentially suitable for use in the autoinjector. In addition, this compound was synthesized using the nonmutagenic *bis*(methylsulfonoxymethyl)ether, rather than the standard carcinogenic *bis*(chloromethyl)ether. HLö7 dimethanesulfonate has been shown to be superior to HI-6 in reactivating soman-, tabun-, and sarin-inhibited erythrocyte AChE *in vitro*. It is more effective than HI-6 in atropine-protected, soman- and sarin-poisoned mice, but less effective in higher animal species (atropine-protected guinea pigs).

Recent research has also confirmed the pretreatment efficacy of exogenously added AChE. Maxwell and others (1992; Wolfe *et al.*, 1992) demonstrated that pretreatment of rhesus monkeys with fetal bovine serum (FBS) AChE provided complete protection against 5 LD_{50}s of soman. This pretreatment strategy was augmented by the coadministration of FBS AChE and HI-6 in sarin-exposed mice, wherein the AChE is continuously reactivated by the oxime (Caranto *et al.*, 1994).

FUTURE PROSPECTS

The future outlook for this research is both exciting and promising. The 3-dimensional structure of AChE has been determined by X-ray crystallographic analysis, and the catalytic site of the enzyme has been characterized at the atomic level (Sussman *et al.*, 1991; Shafferman *et al.*, 1993). Site-directed mutagenesis is being used to elucidate the role of individual amino acid residues in enzyme-substrate specificity. The knowledge gained will allow for the production of specifically targeted bioengineered molecules for use as nerve agent scavengers. Computer modeling procedures will allow for rapid structural evaluation of candidate compounds, and may play an important predictive role in directed drug design. Further elucidation of the mechanisms of action of reactivators may aid in the development of alternative approaches to the treatment of OP intoxication.

In this chapter, some of the most current findings in the research of reactivation of AChE and the treatment of nerve agent poisoning are presented. Four of the papers (Dr. Amitai, Dr. Das Gupta, Dr. Dube and Mr. Maxwell) discuss the efficacy of combination therapies—to include combinations of both well-known and novel broad-spectrum reactivators, carbamates, and enzyme scavengers. Dr. Benschop describes the toxicokinetics of inhaled nerve agent and a suitable substitute research model for inhalation exposure. The potential role of OP-sensitive muscarinic receptors in the clinical expression of AChE inhibition is reviewed by Dr. Pope. Dr. Adler gives evidence for the direct actions of HI-6 in abrogating one effect of soman-induced toxicity. Finally, Dr. Somani presents research on the effect of physostigmine and/or exercise stress on AChE and cholineacetyltransferase levels in rat brain.

REFERENCES

Amitai, G., Kloog, Y., Balderman, D., and Sokolovsky, M., 1980, The interaction of bis-pyridimium oximes with mouse brain muscarinic receptor, *Biochem. Pharmacol.* 29:483-488.

Amitai, G., Rabinovitz, I., Chen, R., Cohen, G., Zomber, G., Adani, R., Manistersky, B., Leader, H., and Raveh, L., 1993, Antidotal efficacy of bisquaternary oximes against soman and tabun poisoning in various species, in *1993 Medical Defense Bioscience Review*, Proceedings, Vol. 2, Baltimore Convention Center, Baltimore, Maryland, May 10-13, U.S. Army Medical Research and Materiel Command, Fort Detrick, Frederick, Maryland.

Baggot, J.D., Buckpitt, A., Johnson, D., Brennan, P., Cone, N., and Goldman, M., 1990, Bioavailability, disposition and pharmacokinetic study of WR249,655 (HI-6) in Beagle Dogs, U.S. Army Medical Research and Materiel Command, Washington, DC, Contract No. DAMD17-86-C-6177, Task Order UCD#9, Draft Study Report.

Boeke, A.W., 1978, The stability of pralidoxime salts in solution - a review, *Pharm. Weekb.* 113:713-718.

Bošković, B., 1981, The treatment of soman poisoning and its perspectives, *Fundam. Appl. Toxicol.* 1:203-213.

Caranto, G.R., Waibel, K.H., Asher, J.M., Larrison, R.W., Brecht, K.M., Schutz, M.B., Raveh, L., Ashani, Y., Wolfe, A.D., Maxwell, D.M., and Doctor, B.P., 1994, Amplificaton of the effectiveness of acetylcholinesterase for detoxification of organophosphorus compounds by bis-quaternary oximes, *Biochem. Pharmacol.* 47(2):347-357.

Četkovic, S., Četkovic, M., Jandri, D.,Osi, M., and Bošković, B., 1984, Effect of PAM-2 cl, Hl-6, and HGG-12 in poisoning by tabun and its thiocholine-like analog in the rat. *Fundam. Appl. Toxicol.* 4:S116-S123.

Clement, J.G., 1979, Pharmacological actions of HI-6, an oxime, on the neuromuscular junction, *Eur. J. Pharmacol.* 60:135-141.

Clement, J.G., 1981, Toxicology and pharmacology of bispyridinium oximes - Insight into the mechanism of action vs soman poisoning *in vivo, Fundam. Appl. Toxicol.* 1:193-200.

Clement, J.G., 1982, HI-6: Reactivation of central and peripheral acetylcholinesterase following inhibition by soman, sarin and tabun *in vivo* in the rat, *Biochem. Pharmacol.* 31(7):1283-1287.

Das Gupta, S., Ghosh, A.K., Moorthy, M.V., Jaiswal, D.K., Chowdhri, B.L., Purnand, and Pant, B.P., 1982, Comparative studies of pralidoxime, trimedoxime, obidoxime and diethyxime in acute fluostigmine poisoning in rats, *Pharmazie* 37:605.

de Jong, L.P.A., Verhagen, M.A.A., Langenberg, J.P., Hagedorn, I., Löffler, M., 1989, The bispyridinium-dioxime HLö-7: A potent reactivator for acetylcholinesterase inhibited by the stereoisomers of tabun and soman, *Biochem. Pharmacol.* 38(4):633-640.

de Jong, L.P.A., and Wolring, G.Z., 1984, Stereospecific reactivation by some Hagedorn-oximes of acetylcholinesterases from various species including man, inhibited by soman, *Biochem. Pharmacol.* 33(7):1119-1125.

Ellin, R.I., 1981, Anomalies in theories and therapy of intoxication by potent organophosphorus anticholinesterase compounds, aberdeen proving ground, Maryland, April, U.S. Army Biomedical Laboratory, Biomedical Laboratory Special Publication, USABML-SP-81-003, DTIC No. AD A101364.

Ellin, R.I., 1982, Anomalies in theories and therapy of intoxication by potent organophosphorus anticholinesterase compounds, *Gen. Pharmacol.* 13:457-466.

Eyer, P., Hagedorn, I., Klimmek, R., Lippstreu, P., Löffler, M., Oldiges, H., Spöhrer, U., Steidl, I., Szinicz, L., and Worek, F., 1992, HLö7 dimethanesulfonate, a potent bispyridinium-dioxime against anticholinesterases, *Arch. Toxicol.* 66(9):603-621.

Eyer, P., Ladstetter, B., Schäfer, W., and Sonnenbichler, J., 1989, Studies on the stability and decomposition of Hagedorn-oxime HLö7 in aqueous solution, *Arch. Toxicol.* 63:59-67.

Galosi, A., Deljac, A., Deljac, V., and Binenfeld, Z., 1988, Reactivators of acetylcholinesterase inhibited by organophosphorus compounds. Imidazole derivatives, *Acta Pharm. Jugost.* 38:23-29.

Goff, D.A., Koolpe, G.A., Kelson, A.B., Vu, H.M., Taylor, D.L., Bedford, C.D., Musallam, H.A., Koplovitz, I., and Harris, R.N., III, 1991, Quaternary salts of 2-[(hydroxyimino)methyl]imidazole. 4. Effect of various side-chain substituents on therapeutic activity against anticholinesterase intoxicaton, *J. Med. Chem.* 34:1363-1368.

Hagedorn, I., Stark, I., Schoene, K., and Schenkel, H., 1978, Reaktivierung phosphorylierter acetylcholin-esterase: Insomere bisquartäre salze von pyridin-aldoximen. [Reactivation of phosphorylated pyri-dine-aldoximes. Isomeric bis-quaternary pyridine-aldoxime salts. [In German; summary in English], *Arzneim.-Forsch.* (Drug Res.) 28:2055-2057.

Harris, L.W., Anderson, D.R., Lennox, W.J., Woodard, C.L., Pastelak, A.M., and Vanderpool, B.A., 1989, Evaluation of HI-6, MMB-4, 2-PAM and ICD 467 as reactivators of unaged soman- inhibited whole blood acetylcholinesterase in rabbits. u.s. army medical research institute of chemical defense, aberdeen proving ground, Maryland, USAMRICD-TR-89-13, Technical Report.

Heilbronn-Wikstrosm, E., 1965, Phosphorylated cholinesterases, their formation, reactions and induced hydrolysis, *Svensk kem. Tidskr.* 77:11-43.

Kirsch, D., and Weger, N., 1979, Effects of soman and bispyridiniumoximes on the isolated, circumfused cervical ganglion of the rat, *Naunyn-Schmiedebergs Arch. Pharmacol.* 307(94):R24.

Koelle, G.B., 1981, Anticholinesterase Agents, in *The Pharmacological Basis of Therapeutics*, 5th Ed., Goodman, L.S. and Gilman, L, eds., MacMillan Publishing Co., Inc., New York.

Koolpe, G.A., Lovejoy, S.M., Goff, D.A., Lin, K.-Y., Leung, D.S., Bedford, C.D., Musallam, H.A., Koplovitz, L., and Harris, R.N., III, 1991, Quaternary salts of 2-[hydroxyimino)methyl]imidazole. 5. Structure-activity relationships for side-chain nitro-, sulfone-, amino-, and aminosulfonyl-substituted analogues for therapy against anticholinesterase, *J. Med. Chem.* 34:1368-1376.

Lenz, D.E., and Maxwell, D.M., 1981, Protection afforded by the h oximes, in *evaluation of h-series oximes: proceedings of a symposium*, Aberdeen Proving Ground, Maryland, April, U.S. Army Medical Research and Materiel Command, Biomedical Laboratory Technical Report, USABML-SP-81-001.

Litovitz, T.L., Martin, T.G., and Schmitz, B., 1987, 1986 Annual report of the American Association of Poison Control Centers National Data Collection System, *Am. J. Emerg. Med.* 5:405-445.

Lundy, P.M., and Tremblay, K.P., 1979, Ganglion blocking properties of some bispyridinium soman antago-nists, *Eur. J. Pharmacol.* 60:47-53.

Maxwell, D.M., Castro, C.A., De La Hoz, D.M., Gentry, M.K., Gold, M.B., Solana, R.P., Wolfe, A.D., and Doctor, B.P., 1992, Protection of rhesus monkeys against soman and prevention of performance decrement by pretreatment with acetylcholinesterase, *Toxicol. Appl. Pharmacol.* 115:44-49.

Musallam, H.A., Foye, W.O., Hansch, C., Harris, R.N., III, and Engle, R.R., 1993, Quantitative structure-ac-tivity relationships of imidazoluim oximes as nerve agent antidotes, in *1993 Medical Defense Bioscience Review*, Proceedings, Vol. 2, Baltimore Convention Center, Baltimore, Maryland, May 10-13, U.S. Army Medical Research and Materiel Command, Fort Detrick, Frederick, Maryland.

Salem, H., and Olajos, E.J., 1988, Review of pesticides: Chemistry uses and toxicology, *Toxicol. Ind. Health* 4(3):291-321.

Shafferman, A., Velan, B., Barak, D., Kronman, C., Ordentlich, A., Flashner, Y., Leitner, M., Segall, Y., Grosfeld, H., Stein, D., and Ariel, N., 1993, in *1993 Medical Defense Bioscience Review*, Proceedings, Vol. 3, Baltimore Convention Center, Baltimore, Maryland, May 10-13, U.S. Army Medical Research and Materiel Command, Fort Detrick, Frederick, Maryland.

Somani, S.M., Solana, R.P., and Dube, S.N., 1992, Toxicodynamics of nerve agents, in *Chemical Warfare Agents*, Academic Press, Inc.

Sussman, J.L., Harel, M., Frolow, F., Oefner, C., Goldman, A., Toker, L., and Silman, I., 1991, Atomic structure of acetylcholinesterase from *Torpedo californica*: A prototypic acetylcholine-binding protein, *Science* 253:872-879.

Szinicz, L., Arndt, H., and Arbogast, H., 1988, Effect of soman, atropine and hi 6 on various behavioural parameters in mice, in *NATO Research Study Group Panel* III/RSG-3, Washington, DC, September, U.S. Army Medical Research Institute of Chemical Defense, Aberdeen Proving Ground, Maryland.

Taylor, P., Anticholinesterase agents, in The Pharmacological Basis of Therapeutics 8th Ed., Gilman, A.G., Rall, T.W., Nies, A.S. and Taylor, P. eds. Pergamon, New York. pp. 131-149.

Wilson, B.W., Hooper, M.J., Hansen, M.E., and Nieberg, P.S., 1992, Reactivation of Organophosphorus Inhibited AChE with Oximes, in *Organophosphates: Chemistry, Fate, and Effects*, J.E. Chambers and P.E. Levi, eds., Academic Press, Inc.

Wilson, I.B., 1951, Acetylcholinesterase. XI. Reversibility of tetraethylpyrophosphate inhibition, *J. Biol. Chem.* 190:111-117.

Wilson, I.B., and Bergmann, F., 1950, Acetylcholinesterase. VIII. Dissociation constants of the active groups, *J. Biol. Chem.* 186:683-692.

Wilson, I.B., and Ginsburg, S., 1955, A powerful reactivator of alkylphosphate-inhibited acetylcholinesterase, *Biochim. Biophys. Acta* 18:168-170.
Wolfe, A.D., Blick, D.W., Murphy, M.R., Miller, S.A., Gentry, M.K., Hartgraves, S.L., and Doctor, B.P., 1992, Use of cholinesterases as pretreatment drugs for the protection of rhesus monkeys against toxicity, *Toxicol. Appl. Pharmacol. 117:189-193.*

ORGANOPHOSPHATE-SENSITIVE CHOLINERGIC RECEPTORS

Possible Role in Modulation of Anticholinesterase-Induced Toxicity

C. N. Pope, J. Chaudhuri, and T. K. Chakraborti

Toxicology Program
College of Pharmacy and Health Sciences
Northeast Louisiana University
Monroe, Louisiana 71209-0470

INTRODUCTION

Organophosphorus (OP) pesticides are commonly used throughout the world (Marquis, 1986). These pesticides work by inhibiting the enzyme acetylcholinesterase (AChE), allowing acetylcholine to accumulate in central and peripheral synapses and overstimulate postsynaptic cells (Ecobichon, 1991). Recent reports indicate that some of these OP compounds can also bind directly to cholinergic receptors and these direct OP-receptor interactions may participate in the expression of toxicity (Katz and Marquis, 1989; Jett et al., 1991; Chaudhuri et al., 1993; Huff et al., 1994).

Parathion (O,O'-diethyl-p-nitrophenyl phosphorothioate) and chlorpyrifos (O,O'-diethyl-O-3,5,6-trichloro-2-pyridinyl phosphorothioate) are common OP pesticides. Following phosphorylation of acetylcholinesterase, the enzyme is modified with a diethylphosphoryl moiety in both cases, thus one would expect similar biochemical changes (e.g.,reactivation, aging, recovery) following enzyme inactivation with these compounds, when similar levels of inhibition occurred. Previous studies in our laboratory, however, (Chaudhuri et al., 1993; Chakraborti et al., 1994) have demonstrated marked differences in signs of toxicity following such exposures. Essentially, parathion exposure produces more extensive signs of toxicity than does CPF following respective doses which produce similar rates and levels of maximal brain AChE inhibition.

Compensatory changes in postsynaptic cholinergic receptors (i.e., down-regulation) is thought to constitute a primary mechanism of tolerance to AChE inhibition (Costa et al., 1982; Russell and Overstreet, 1987). We have observed that while total muscarinic receptor density is reduced in a time-dependent manner in forebrain regions after high acute doses of either PS or CPF, a subset of muscarinic receptors (quantitated by measuring paraoxon-sensitive binding to [^{3}H]quinuclidinyl benzilate) is differentially down-regulated and up-regu-

Enzymes of the Cholinesterase Family, Edited by Daniel M. Quinn et al.
Plenum Press, New York, 1995

lated by parathion and chlorpyrifos, respectively (Chaudhuri *et al.*, 1993). The high affinity muscarinic agonist [³H]*cis*-methyldioxolane) appears to label selectively a subset of muscarinic receptors sensitive to some OPs (Ward *et al.*, 1993) and these appear to represent muscarinic m2 receptors (Huff and Abou-Donia, 1994). We propose that these OP-sensitive muscarinic receptors possess autoreceptor function (i.e., they regulate acetylcholine release *via* feedback inhibition) and that their differential up- and down-regulation following chlorpyrifos and parathion exposures modulates the toxic effects of AChE inhibition.

The present study examined cholinergic neurochemical markers and functional changes in rat striatum after either parathion or chlorpyrifos exposures causing similar levels of maximal AChE inhibition (Chaudhuri *et al.*, 1993). In addition, *ex vivo* acetylcholine release was examined in parallel groups of rats to study the effects of OP exposure on autoreceptor function. The results indicate that these OP-sensitive muscarinic receptors are autoreceptors and that differential changes in this subset of cholinergic receptors may play a role in the relative toxicity of AChE inhibitors.

METHODS

Chemicals

Parathion (O,O-diethyl-O-4-nitrophenyl phosphorothioate, 98% purity) and chlorpyrifos (O,O-diethyl-O-3,5,6-trichloro-2-pyridinyl phosphorothioate, 99% purity) were purchased from Chem Service, West Chester, PA. Paraoxon (O,O-diethyl-O-4-nitrophenyl phosphate, 95% purity) was purchased from Sigma Chemical Co., St. Louis, MO. Chlorpyrifos oxon (O,O-diethyl O-3,5,6-trichloro-2-pyridinyl phosphate, 98% purity) was obtained from the central repository of the U.S. EPA Facility, Research Triangle Park, NC, USA. Quinuclidinyl benzilate , L-[benzilic-4,4'-³H] (QNB) , specific activity 43.0 Ci/mmol, dioxolane, L (+)-cis-[2-methyl-³H] (CD) , specific activity 64.5 Ci/mmol and acetylcholine iodide, [acetyl-³H] (ACh), specific activity 120 Ci/mmol, and [³H]choline chloride (specific activity 86.7 Ci/mmol) were purchased from New England Nuclear, Boston, MA. All other chemicals were purchased from Sigma Chemical Co., St. Louis, MO.

Animals and Treatments

Adult, male Sprague-Dawley rats (3 months of age, average body weight about 375g) were used throughout. Chlorpyrifos and parathion were administered subcutaneously in peanut oil. Previous studies (Pope *et al.*, 1991; Chaudhuri *et al.*, 1993) have established acute maximum tolerated doses (MTDs) for chlorpyrifos (279 mg/kg, sc) and parathion (18 mg/kg, sc). Control animals received peanut oil only (2 ml/kg). Animals were housed in individual steel mesh cages and were maintained on a 12L:12D illumination cycle. Food and water were given *ad libitum*.

Acute Toxicity

A set of functional observations was used for examining signs of acute toxicity on day 2 following treatment. The tests included signs for the following: a) salivation, lacrimation, urination, diarrhea (SLUD); b) motor activity c) body temperature and d) arousal and were derived from Moser *et al.* (1988). Motor activity was measured by the method of Finn and coworkers (1990) using five 2-channel Electronic Activity Monitors (No. 31404, Stoelting Co., Wood Dale, IL, USA).

Tissue Preparation and Biochemical Assays

Groups of rats were decapitated 2, 7 and 14 days after treatment and striatum was dissected as described by Glowinski and Iversen (1966). For acetylcholine release studies, the striatum was dissected and sliced (400 micron, unidirectional) with a McIlwain tissue slicer (Weiler, 1984) and transferred to 2.0 ml of Krebs Ringer Bicarbonate buffer (KRB) previously equilibrated with $O_2:CO_2$ (95:5). The slices were triturated, buffer was replaced and slices were preincubated at 37°C for 60 minutes under a constant stream of $O_2:CO_2$. The slices were then washed and incubated with [^{3}H]choline (2.5 uCi/ml) at 37°C for 40 minutes under $O_2:CO_2$. After radiolabel incorporation, the buffer was removed, tissues were rinsed once with fresh buffer, and then fresh KRB containing 10 uM hemicholinium-3 (to prevent high affinity choline uptake) and (in some cases) 20 uM physostigmine (to inhibit acetyl-cholinesterase) was added. The slices were then transferred to a superfusion apparatus (Brandel SF12, Gaithersburg, MD) and perfused (0.5 ml/min) with KRB for 60 minutes containing hemicholinium-3. Twenty 5-minute fractions were then collected and the effect of either one (first stimulation, S1) or two (second stimulation, S2) pulses of buffer containing high potassium chloride (20 mM) on acetylcholine release was examined. When exogenous agents were added *in vitro*, they were included in the superfusion buffer 10 minutes before S2. The effect of exogenous agonist/antagonist on release was determined by changes in the S2/S1 ratio. Samples of each fraction as well as alkaline hydrolysates of the tissue samples following superfusion were assayed by liquid scintillation counting and data were normalized according to total radioactivity.

Data Analysis

All statistical analyses were performed using the SAS package for personal computers (SAS, 1988). Median values for arousal and SLUD signs were tested for significance by the Kruskall-Wallis test and/or Mann Whitney U test, whereas body temperature, motor activity, enzyme and receptor data were tested for significance by ANOVA followed by Duncan's multiple range test. Striatal acetylcholine release data were compared between treatments and timepoints by two-way ANOVA and between different *in vitro* agonist/antagonist concentrations by ANOVA using the SAS GLM procedure followed by Duncan's test.

RESULTS

Acute Toxicity

Figure 1 shows the effects of single doses of parathion (18 mg/kg, sc) or chlorpyrifos (280 mg/kg, sc) on functional endpoints in adult male rats measured 2 days after treatment. Parathion at the dose administered (18 mg/kg, sc) had a significant ($p < 0.05$) effect on arousal (Fig. 1A), body temperature (Fig. 1B), SLUD signs (Fig. 1C), and motor activity (Fig. 1D). Chlorpyrifos at the dose used (280 mg/kg, sc) had no effect on arousal but affected motor activity and SLUD signs. The degree of effect of chlorpyrifos on motor activity and SLUD signs was significantly lower than that of parathion.

AChE Activity and Muscarinic Receptor Binding

The effect of parathion or chlorpyrifos on AChE activity, atropine-sensitive QNB (i.e., total muscarinic receptors) and paraoxon-sensitive binding (i.e., OP-sensitive mus-

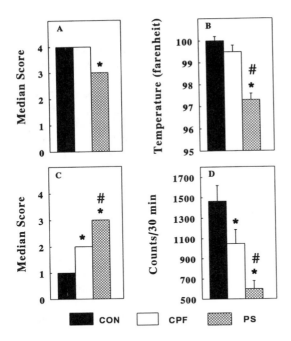

Figure 1. Changes in functional endpoints two days after treatment with either parathion (18 mg/kg) or chlorpyrifos (280 mg/kg). Parathion or chlorpyrifos was administered subcutaneously to 90 day old male Sprague Dawley rats and observed ("blind") for A) arousal scores, B) body temperature, C) SLUD signs and D) motor activity essentially as described in Moser *et al.*, (1988). Asterisks indicate significant (p < 0.05) differences between treated and control values. Pound signs indicate significant (p < 0.05) differences between OP treatment groups.

carinic receptors) in striatum 2, 7 and 14 days after treatment is shown in Figure 2 (data derived from Chaudhuri *et al.*, 1993). In general, parathion (18 mg/kg) and chlorpyrifos (280 mg/kg) caused similar degrees of maximal AChE inhibition in striatal tissues (Fig. 2A) but more persistent AChE inhibition was noted after chlorpyrifos exposure. Fig. 2B shows the effects of parathion or chlorpyrifos on total muscarinic receptor binding in striatum 2-14 days after treatment. Whereas both treatments reduced total muscarinic receptor binding, more extensive maximal reductions were seen after chlorpyrifos treatment (56% vs 32%). Parathion also reduced the apparent density of QNB binding sites that were blocked by

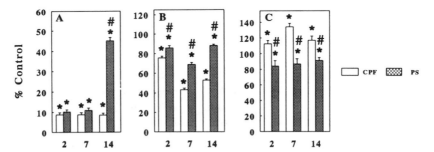

Figure 2. Effect of parathion or chlorpyrifos on striatal neurochemistry. Parathion or chlorpyrifos was administered as in Figure 1 and rats were sacrificed at either 2, 7 or 14 days after treatment for measurement of A) AChE activity, B) atropine-sensitive [³H]QNB binding and C) paraoxon-sensitive [³H]QNB binding. Data represent percent (mean ± SE) of control values. Combined control values for AChE activity = 214 ± 3 nmoles ACh hydrolyzed/min/mg protein; Atropine-sensitive QNB binding = 2.12 ± 0.05 pmoles QNB bound/mg protein; Paraoxon-sensitive QNB binding = 0.19 ± 0.02 pmoles QNB bound/mg protein. Asterisks indicate significant (p < 0.05) differences between treated and control values. Pound signs indicate significant (p < 0.05) differences between OP treatment groups.

paraoxon preincubation (10 uM final concentration, Fig. 2C). In contrast, chlorpyrifos exposure increased paraoxon-sensitive QNB binding up to 37% in striatum 7 days after treatment (Fig. 2C). A similar OP-specific change in receptor binding using the radioligand [³H]cis-methyldioxolane was also noted seven days following chlorpyrifos or parathion exposure, i.e., chlorpyrifos caused an up-regulation of these binding sites while parathion caused a decrease (Chaudhuri et al., 1993).

Acetylcholine Release

Figure 3A shows the effects of physostigmine on acetylcholine release from striatal slices. As reported by others (D'Agostino et al., 1986; Feuerstein et al., 1992), physostigmine (20 uM) decreased net acetylcholine release, presumably via indirect activation of muscarinic autoreceptors. This effect of physostigmine could be antagonized in a dose-dependent manner by atropine (Fig. 3B).

The nonspecific muscarinic agonist carbachol also decreased net acetylcholine release (Fig. 4A).

The high affinity muscarinic agonist cis-dioxolane also decreased acetylcholine release in a concentration-dependent (Fig. 4B) and atropine-sensitive (Fig. 5) manner.

In vivo treatment with either parathion or chlorpyrifos caused a time-dependent change in the response to physostigmine, i.e., the ability of physostigmine to reduce

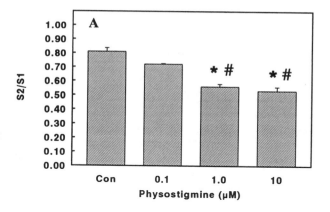

Figure 3. Concentration-dependent effects of physostigmine and atropine on K⁺-evoked ACh release from striatal slices. A) Striatal slices were prelabeled with [³H]choline and superfused in the presence of hemicholinium-3. Tissue slices were stimulated twice (S1 and S2) with high potassium (20 mM) buffer. Physostigmine was added to the superfusion buffer 10 minutes prior to S2 and results were expressed as S2/S1 ratios. B) Striatal slices were prelabeled as above and superfused in the presence of both hemicholinium-3 and physostigmine (20 uM). Atropine was added 10 minutes prior to S2 and release evaluated as above. Asterisks indicate significant (p < 0.05) differences between control and drug treated tissues. Pound signs indicate significant (p < 0.05) differences between drug concentrations. Values represent means (± SE) of 3-5 different animals.

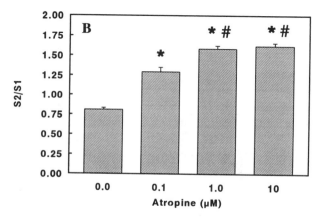

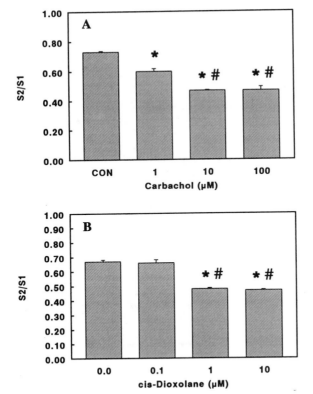

Figure 4. Concentration-dependent effects of carbachol (A) and *cis*-dioxolane (B) on K⁺-evoked ACh release from striatal slices in the absence of physostigmine. Agonist-mediated changes in release were examined as described in Figure 3. Asterisks indicate significant (p < 0.05) differences between control and drug treated tissues. Pound signs indicate significant (p < 0.05) differences between drug concentrations. Values represent means (± SE) of 3-5 different animals.

acetylcholine release was attenuated after OP exposure (Fig. 6). The onset and recovery in this change in response to physostigmine were different between treatments, however. Two days after parathion exposure, striatal acetylcholine release was markedly less sensitive to physostigmine but sensitivity returned to near control levels by 14 days after treatment. On the other hand, no difference in response to physostigmine (relative to control tissues) was noted two days after chlorpyrifos exposure, but the ability of physostigmine to reduce acetylcholine release was diminished from 4-14 days after chlorpyrifos.

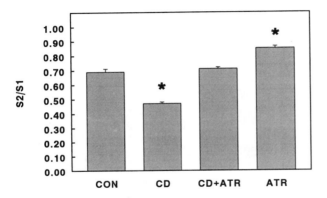

Figure 5. Effect of *cis*-dioxolane (CD, 1 uM) and atropine (ATR, 0.1 uM) on striatal acetylcholine release in the absence of physostigmine. Release was examined as described in Figure 3 with either CD, ATR or both being added 10 minutes prior to S2. Asterisks indicate significant (p < 0.05) differences between control and drug treated tissues. Values represent averages of 2-6 different animals.

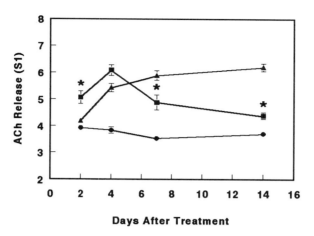

Days After Treatment

Figure 6. Time-dependent changes in autoreceptor function following chlorpyrifos or parathion exposure. Rats were treated with OP and sacrificed either 2, 4, 7 or 14 days after treatment. Slices were prepared and stimulated with potassium as in Figure 3. Physostigmine (20 uM) was included in the superfusion buffer and the peak in potassium-evoked acetylcholine release (S1) was measured. Note that an increase in S1 over control values indicates a decreased sensitivity to the physostigmine-mediated reduction in acetylcholine release. Asterisks indicate significant ($p < 0.05$) differences between OP treatment groups. The S1 from both OP treatment groups was significantly different from control at all time points except 2 days (for chlorpyrifos) and 14 days (for parathion). Values represent means ($\pm$ SE) of 4-6 different animals.

DISCUSSION

Parathion and chlorpyrifos are both diethylorganophosphorothioate insecticides for which the mode of toxicity is generally agreed to be AChE inhibition (Ecobichon, 1991). We previously established the maximum tolerated doses for these compounds (Pope *et. al.*, 1991). In the present study, rats treated with the maximum tolerated dose of parathion (18 mg/kg, sc) exhibited a higher degree and greater incidence of acute toxicity than animals treated with the maximum tolerated dose of chlorpyrifos (280 mg/kg, sc, Fig. 1). Previous studies (Pope *et al.*, 1991; Chaudhuri *et al.*, 1993) have shown that similar maximal changes in AChE activity occur with these treatments. The initial rates of inhibition in brain AChE activity are also similar after these doses of chlorpyrifos and parathion (Pope *et. al.*, 1991). Differences in AChE inhibition thus do not appear to be involved in the differential expression of acute toxicity with parathion and chlorpyrifos exposures.

We reported (Chaudhuri *et al.*, 1993) that exposure to doses of parathion or chlorpyrifos causing similar maximal changes in striatal AChE activity were associated with down-regulation and up-regulation, respectively, of muscarinic receptors labeled by the high affinity muscarinic agonist [^{3}H]*cis*-methyldioxolane (CD). Specific binding to CD in brain has been reported by several groups to be blocked by the active metabolites of parathion and chlorpyrifos, i.e., paraoxon (Jett *et al.*, 1991; Ward *et al.*, 1993) and chlorpyrifos-oxon (Huff *et al.*, 1994).

We hypothesized that these OP-sensitive CD binding sites were presynaptic autoreceptors involved in feedback inhibition of acetylcholine release (Chaudhuri *et al.*, 1993). Muscarinic presynaptic autoreceptors regulate acetylcholine release *in vitro* in both human and rat forebrain (Weiler, 1989; Marchi *et al.*, 1990; Feuerstein *et al*, 1992). A change in the density of autoreceptors could affect the release of acetylcholine and thereby alter the consequences of AChE inhibition.

Our data provide support for the notion that the subset of OP-sensitive striatal muscarinic receptors differentially modulated by parathion and chlorpyrifos possesses autoreceptor activity (Figures 4b, 5). In addition, it appears that autoreceptor function (interpreted by sensitivity to a physostigmine-mediated decrease in acetylcholine release) may remain viable longer with persistent AChE inhibition following chlorpyrifos treatment than after parathion exposure. Specifically, loss of feedback inhibition of acetylcholine release appears to occur two days after parathion treatment but appears intact two days after chlorpyrifos exposure, i.e., the time of maximal incidence of signs of toxicity with either treatment (Chaudhuri et al., 1993). By four days after treatment, autoreceptor function appears altered in chlorpyrifos treated tissues as well, but by this time, alternate compensatory processes may occur which allow tolerance to extensive AChE inhibition. Further studies should clarify the role of these OP-sensitive cholinergic receptors in the expression of anticholinesterase-induced toxicity.

ACKNOWLEDGEMENTS

This research was supported by Cooperative Agreement CR820229-01 with the United States Environmental Protection Agency awarded to C.N.P. This manuscript has been reviewed by the Health Effects Research Laboratory, U.S. EPA, and approved for publication. Approval does not signify that the contents necessarily reflect the views and policies of the Agency, nor does mention of trade names or commercial products constitute endorsement or recommendations for use.

REFERENCES

Chakraborti, T., Chaudhuri, J., Harp, P., and Pope, C.N., 1994, *The Toxicologist* **14(1)**:241.
Chaudhuri, J., Chakraborti, T.K. Chanda, S.M., and Pope, C.N., 1993, *Journal of Biochemical Toxicology* **8(4)**:207-214.
Costa, L.G., Hand, H., Schwab, B.W., and Murphy, S.D., 1982, *Toxicology* **25**:79-97.
D'Agostino, G., Kilbinger, H., Chiari, M.C., and Grana, E., 1986, *Journal of Pharmacology and Experimental Therapeutics* **239(2)**:522-528.
Ecobichon, D.J., 1991, In *Casarett and Doull* (4th edition, M. O. Amdur, J. Doull and C. D. Klassen, Eds.,) pp 565-622, Pergammon Press, New York.
Feuerstein, T.J., Lehmann, J., Sauermann, W., Velthovon, V., and Jackisch, R., 1992, *Brain Research* **572**:64-71.
Finn, I.B., Iuvone, P.M., Holtzman, S.G., 1990, *Neuropharmacology* **29**:625-631.
Glowinski, J. and Iversen, L.L., 1966, *International Journal of Neurochemistry* **13**:655-669.
Huff, R.A., and Abou-Donia, M.B, 1994, *Journal of Neurochemistry* **62**:388-391.
Huff, R.A., Corcoran, J.J., Anderson, J.K. and Abou-Donia, M.B., 1994, *Journal of Pharmacology and Experimental Therapeutics* **269**:329-335.
Jett, D.A., Abdallah, E.A.M., El-Fakahany, E.E., Eldefrawi, M.E., and Eldefrawi, A.T., 1991, *Pesticide Biochemistry and Physiology* **39**:149-157.
Katz, L.S., and Marquis, J.K., 1989, *Toxicology and Applied Pharmacology* **101**:114-123.
Marchi, M, Ruelle, A., Andrioli, G.C., and Raiteri, M.,1990, *Brain Research* **520**:347-350.
Marquis, J. K. 1986. *Contemporary Issues in Pesticide Toxicology and Pharmacology* (ed. F. Homburger). Cambridge, Karger.
Moser, V.C., McCormick, J.P., Creason, J.P., and MacPhail, R.C., 1988, *Fundamental and Applied Toxicology* **11**:189-206.
Pope, C.N., Chakraborti, T.K., Chapman, M.L., Farrar, J.D., and Arthun, D, 1991,*Toxicology* **68**:51-61.
Russell, R.W., and Overstreet, D.H., 1987, *Progress in Neurobiology* **28**:97-129.
SAS. SAS/STAT User's Guide. 1988. Version 6.03 ed. Cary, North Carolina: SAS Institute.
Ward, T. R., Ferris, D.J., Tilson, H.A., and Mundy, W.R., 1993, *Toxicology and Applied Pharmacology* **122**:300-307.
Weiler, M. H., Misgeld, U., and Cheong, D.K., 1984, *Brain Research* **296**:111-120.
Weiler, M.H., 1989, *Experientia Supplement* **57**:204-211.

EFFECT OF PHYSICAL/CHEMICAL STRESSORS ON THE CHOLINERGIC SYSTEM IN RAT

Satu M. Somani

Southern Illinois University School of Medicine
Department of Pharmacology
P.O. Box 19230
Springfield, Illinois 62794-9230

EXERCISE AND CHOLINERGIC SYSTEM

Synaptic transmission is mediated by acetylcholine-neurotransmitter between pre-ganglionic and post ganglionic nerve fibre in the sympathetic pathway and those sites that use this neurotransmitter are called cholinergic. The sites that use norepinephrine as neurotransmitter are called adrenergic. This presentation primarily concerns with cholinergic system. It is comprised of neurotransmitter acetylcholine, its receptors (muscarinic and nicotinic) and enzymes involved for its synthesis - choline acetyltransferase (ChAT) and degradation - acetylcholinesterase (AChE). The level of acetylcholine in the CNS is regulated by its anabolism through ChAT and catabolism through AChE. The neurotransmitters play an important role for human beings to adjust and adapt to any changes in their environment. Alterations in the central neurotransmitter system due to physical exercise and/or chemical stressors have not received much attention. Of the few studies that have specifically examined biochemical markers of the brain cholinergic system, the two most frequently measured indices have been the biosynthetic enzyme of acetylcholine (ACh), choline acetyltransferase (ChAT), and the degradative enzyme, acetylcholinesterase (AChE).

Apparent discrepancies in the literature regarding the effects of stressors on cholinergic function have been common. For example, ChAT activity diminished in rat brain exposed to immobilization stress (Gottesfeld et al., 1978). ChAT activity increased in the rat cerebral cortex after acute and repeated electroshock (Longoni et al., 1976; Oesch, 1974). Unchanged ChAT activity after exposure to immobilization for 2 h has been reported in different brain regions: brainstem, striatum and hippocampus (Gilad et al., 1983; Gottesfeld et al., 1978; Tucek et al, 1978). None of these previous studies directly compared the effect of physical exercise and chemical stressors such as physostigmine (Phy), a cholinesterase inhibitor on both of the cholinergic enzymes. We have reviewed Phy - an overview as pretreatment drug for organo - phosphate intoxication (Somani and Dube, 1989).

Enzymes of the Cholinesterase Family, Edited by Daniel M. Quinn et al.
Plenum Press, New York, 1995

Table 1. Training protocol for exercising rats

Week	Belt Speed (m/min)	Inclination (% grade)	Duration at each speed (min)	Total time (min)
1	8.2, 15.2, 19.3	6	10	30
2	8.2, 15.2, 19.3	6	10	30

Reprinted from Somani et al., 1991, with kind permission from Elsevier Science Ltd, The Boulevard, Langford Lane, Kidlington OX5 1GB, UK.

EFFECT OF PHYSOSTIGMINE AND EXERCISE TRAINING ON ChAT And AChE IN BRAIN REGIONS

In order to more clearly delineate these effects on central cholinergic system, Somani et al., (1991) have studied the effect of exercise training the (Table 1), subacute Phy administration, or the combination of these two treatments on the changes in the biosynthetic and degradative enzymes for ACh. They reported that these changes were differentially expressed within subregions of the brain as shown in Tables 2 and 3.

The marked differences (up to 5-20-fold) have been observed in the regional distribution of ChAT and AChE (Somani et al., 1991), consistent with the known cholinergic innervation to the areas examined (Eckstein et al., 1988). Remarkably, the short- and long-term changes in ChAT and AChE activity elicited by the physical or chemical stressors also showed regional selectivity. For example, the only brain region where ChAT activity responded to exercise training was the brainstem, a region which is involved in maintaining critical autonomic functions related to the cardiopulmonary system and where ACh has potent actions (Arneric et al., 1990). However, an alternative interpretation is that the changes

Table 2. Effect of repeated dose of physostigmine (70 µg/kg,i.m.) daily for 2 wk and exercise training for 2 wk on choline acetyl transferase (ChAT) activity (in nmoles/mg of protein/hr) in different brain regions of rat. Values are mean+S.E.M. (n = 4)

Groups		Treatment	Corpus striatum	Cerebral cortex	Brainstem	Hippocampus
I		Sedentary Control	484.7 ± 7.8	128.1 ± 9.7	266.5 ± 12.4	146.1 ± 19.0
II		Exercise Training	475.4 ± 19.4	107.6 ± 2.8	194.0* ± 5.8	113.7 ± 3.2
III		Subacute Phy				
	IIIa	Sacrificed 20 min after Phy	370.1* ± 3.4	122.6 ± 11.0	215.9* ± 18.8	124.9 ± 9.1
	IIIb	Sacrificed 24 hr after Phy	430.1* ± 14.3	114.0 ± 4.4	207.9* ± 8.7	110.9 ± 6.1
IV		Subacute Phy acute exercise				
	IVa	Sacrificed 20 min after Phy	461.1* ± 5.9	122.0 ± 5.9	224.7 ± 22.0	109.5 ± 2.0
	IVb	Sacrificed 24 hr after Phy	332.2 ± 1.3	118.4 ± 3.1	218.3* ± 9.4	105.3* ± 4.6
V		Subacute Phy + exercise training				
	Va	Sacrificed 20 min after Phy	445.6* ± 15.9	105.5 ± 13.7	213.4* ± 17.2	110.1 ± 5.2
	Vb	Sacrificed 24 hr after Phy	428.8 ± 1.0	101.7* ± 3.0	216.0* ± 3.3	107.2* ± 3.1

*Significant at p < 0.05.
Reprinted from Somani et al., 1991, with kind permission from Elsevier Science Ltd, The Boulevard, Langford Lane, Kidlington OX5 1GB, UK.

Table 3. Effect of repeated dose of physostigmine (70 µg/kg, i.m.) daily for 2 weeks and trained exercise for 2 weeks on AChE activity (µmole/hr/mg of protein) in different brain regions of rats. Values are means ± S.E.M. (n = 4).

Group		Treatment	Corpus striatum	Cerebral cortex	Brainstem	Hippocampus
I		Sedentary control	54.3 ± 1.3	5.9 ± 0.3	7.7 ± 0.3	7.9 ± 0.3
II		Exercise training	51.1 ± 2.2	5.0 ± 0.1	6.2 ± 0.2*	7.5 ± 0.3
III		Subacute Phy				
	IIIa	Sacrificed 20 min after Phy	44.5 ± 1.3	5.4 ± 0.2	4.7* ± 0.2	6.0* ± 0.4
	IIIb	Sacrificed 24 hr after Phy	48.5 ± 2.9	6.2 ± 0.2	6.4* ± 0.4	7.6 ± 0.2
IV		Subacute Phy + acute exercise				
	IVa	Sacrificed 20 min after Phy	44.2* ± 1.3	4.2* ± 0.1	6.1* ± 0.1	5.5* ± 0.2
	IVb	Sacrificed 24 hrafter Phy	47.5 ± 0.6	5.2 ± 0.3	7.2 ± 0.4	7.2* ± 0.4
V		Subacute Phy + exercise training				
	Va	Sacrificed 20 min after Phy	42.4* ± 2.1	4.4 ± 0.3	5.5* ± 0.2	5.8* ± 0.2
	Vb	Sacrificed 24 hr after Phy	48.8 ± 1.7	5.6 ± 0.3	7.0 ± 0.4	7.2* ± 0.1

*Significant at p < 0.05.
Reprinted from Somani et al., 1991, with kind permission from Elsevier Science Ltd, The Boulevard, Langford Lane, Kidlington OX5 1GB, UK.

in ChAT and AChE activities may have occurred with the different motor neurons within the brainstem, since they provide a major contribution to the detected enzyme activities. ChAT activity in cerebral cortex is not affected by any individual stressor, which may suggest the relative sparing of cholinergic systems in higher association centers involved with cognitive function. The combination of these two stressors (Phy + Exercise Training) did show an interaction to reduce ChAT activity in cortex and in hippocampus, an area of brain involved in learning and memory processes. In contrast, ChAT activity in corpus striatum was depressed significantly in all groups receiving Phy and remain depressed even 24 h following withdrawal of Phy. The cholinergic system in corpus striatum, which is normally involved in motor control, is essentially unaffected by exercise but susceptible to chemical stressor such as Phy. However, cholinergic parameters in various regions of brain change differently to altered stress conditions, such as cold and swimming (Fibiger et al., 1969; Godfrey et al., 1984; Hata et al., 1980). This study supports this concept, since regional differences in cholinergic activities were also observed by exercise training and Phy.

The brainstem regulates respiratory and cardiovascular functions. Exercise training alone increases blood pressure and heart rate (Roskoski et al., 1974) and down regulates ChAT activity (Benarroch et al., 1986). Increases in medullary tissue concentrations of ACh by ChE inhibitors also elevated blood pressure and respiration (Benarroch et al., 1986). It is plausible that when tissue levels of ACh increase, ChAT activity is down regulated as a compensatory mechanism to normalize cholinergic transmission and, hence, blood pressure. Mechanistically, these changes may be initiated as follows. Under normal conditions, the tissue level of ACh is regulated by the net synthesis and degradation of ACh by ChAT and AChE, respectively. Immediately following AChE inhibition, the tissue level of ACh increases and thereby, initiates processes that decrease both ChAT and AChE activity. Chronically elevated ACh concentrations eventually down regulate ChAT and AChE activities to normalize cholinergic transmission; whereas, acute exercise initially increases tissue ACh levels by enhancing biosynthesis without affecting degradation. This phenomenon does not appear to be expressed in corpus striatum, hippocampus or cerebral cortex. However, chronically elevated levels of ACh still down regulate ChAT and AChE activities. In order to verify this hypothesis, both tissue concentrations and the turnover of ACh need to be studied. Tissue release and uptake of AChE due to exercise is not known, and nor is the effect of exercise on sensitivities to cholinergic receptors. Somani et al. (1991) have suggested that the biosynthetic and degradative enzymes for ACh in brain regions involved with control of motor, autonomic and cognitive functions are affected by Phy or exercise training or the combination of both in a regionally selective pattern. The data are consistent with the hypothesis that the responsiveness of these brain regions to the physical and/or chemical stressors is a function of the level of ongoing cholinergic transmission and that elevations in ACh levels due to AChE inhibition may have long-term effects on ChAT and AChE activities through a negative feedback mechanism.

EFFECT OF PHYSOSTIGMINE AND EXERCISE TRAINING ON FAST AND SLOW MUSCLE

ChAT and AChE activities were found predominantly in cholinergic neurons (Rossier, 1977) and specifically tend to be high at cholinergic synapses (Godfrey et al., 1984). Dettbarn has reported the distinctive difference in AChE activity in fast and slow muscles after reinnervation in rats (Dettbarn, 1979). We observed a significant decrease in ChAT activity in fast twitch EDL muscle, but not in soleus muscle (Table 4) (Babu et al., 1993). This finding may be explained by considering the animal locomotion and the type of muscles involved in locomotion. Fast twitch muscles are active primarily during locomotion whereas slow twitch muscles are active while the animals are at rest, regardless of their locomotive state. This active involvement of EDL during exercise may lead to a decrease in ChAT. But AChE was inhibited in both fast and slow muscles significantly due to exercise training (Table 4).

Exercise decreased AChE activity in EDL muscle and this muscle is very active during exercise. Our finding of reduction in ChAT activity in EDL muscle due to training is consistant with the recent finding that exercise decreases ChAT activity in adrenal gland of young rats (Tumer et al., 1992). In trained exercise and Phy administered rats, the AChE activities were significantly lower even after 24 hr in both the muscles. The inhibition was more than 50% at the 20 min period. However, subacute Phy + trained exercise affected ChAT only in EDL and not in soleus muscle. This is a significant observation since no change was expected in the passive soleus muscle. Our findings reveal that exercise not only affected the active EDL muscle in regard to ChAT activity but affected the slow soleus also by decreasing AChE activity. Exercise has prolonged the inhibition of AChE in both the muscles. Exercise and Phy not only has a cumulative effect immediately after exercise but also prolonged this effect up to 24 hr. Recovery of AChE activity was observed in trained rats but not in trained + Phy administered rats. Though further studies are needed to clearly elicit the mechanism of action of these two stressors the present study indicate that 1) physical exercise plays an imporant role in prolonging the action of the drug not only in active muscle like EDL but also in passive muscle like soleus; hence, it is important to consider another tissue along with the target tissue when evaluating for drug effects; and 2) when evaluating an anticholinesterase agent for its effects, care should be taken to avoid a cumulative effect by a physical stressor like exercise. These results, when considered with our previous studies (Somani et al., 1991), provide evidence that the exercise and physostigmine additively decrease the AChE activity by modifying the functional activity of cholinergic system.

Table 4. Choline acetyltransferase and acetylcholinesterase activities (percent of control) in fast (EDL) and slow (soleus) muscles in subacute administration of Phy (70 mg/kg, i.m., twice daily for 2 wk) and/or exercised (once daily for 2 wk) rats

Group	Treatment	Time of sacrifice after treatment	Cholin acetyltransferase EDL	Cholin acetyltransferase Soleus	Acetylcholinesterase EDL	Acetylcholinesterase Soleus
I	Sedentary control	20 min	100%	100%	100%	100%
IIa	Exercise training	20 min	67.8 ± 2.1*	123.9 ± 4.9*	57.1 ± 4.5*	54.9 ± 3.6*
IIb	Exercise training	24 hr	86.5 ± 3.2	109.1 ± 2.9	79.2 ± 2.9	73.4 ± 2.8
IIIa	Subacute physostigmine	20 min	89.6 ± 3.3	94.4 ± 4.9	55.7 ± 7.1*	57.3 ± 7.3*
IIIb	Subacute physostigmine	24 hr	110.5 ± 6.0	116.4 ± 4.8	$68.6 \pm 10.$	67.1 ± 3.6*
IVa	Subacute physostigmine + single acute exercise	20 min	96.7 ± 7.8	89.8 ± 11.1	$62.8 \pm 10.$	31.7 ± 4.9*
IVb	Subacute physostigmine + single acute exercise	24 hr	105.1 ± 3.9	111.1 ± 8.9	88.5 ± 5.7	50.6 ± 7.3*
Va	Subacute physostigmine + exercise training	20 min	65.3 ± 3.8*	98.4 ± 3.5	42.8 ± 2.8*	29.2 ± 4.8*
Vb	Subacute physostigmine + exercise training	24 hr	66.7 ± 5.6*	108.5 ± 7.2	62.8 ± 12.8	65.8 ± 4.8*

Values are means $\pm$ SEM. Control value of ChAT in EDL, 10.77 ± 0.23; Soleus, 8.81 ± 0.29 of ACh synthesized/mg protein/hr. AChE control values in EDL, 0.70 ± 0.13; Soleus, 0.82 ± 0.05 μmoles/mg protein/hr. Statistical significance at p<0.05(*).
Reprinted from Babu et al., 1993, with kind permission from Elsevier Science Ltd, The Boulevard, Langford Lane, Kidlington OX5 1GB, UK.

EFFECT OF PHYSICAL AND CHEMICAL STRESSORS ON CHOLINESTERASE ACTIVITY IN THE BRAIN

Physical exercise evokes a number of enzymatic changes in the body, especially in muscle and liver (Harri et al., 1983; Vihko et al., 1978). Exercise is one of the important factors that alter ChE activity (Pedzikiewicz et al., 1984; Ryhanen et al., 1988). The intensity of these changes depends upon the type and severity of exercise (Holloszy and Booth, 1976; Pawlowska et al., 1985). Dube et al. (1990) studied the effect of different intensities of acute exercise (physical stress) on ChE activity in RBC and tissues. Exercise at 50, 80 and 100% VO_2 max produced a significant increase in ChE (p < 0.05) activity (116% - 108% of control) in RBC of rat; whereas, different intensities of acute exercise (50, 80 and 100% VO_2 max) alone produced a slight, but statistically insignificant decrease in brain ChE activity (92-87% of control). This increase in ChE activity of RBC may be due to secondary effect of hypoxia, increased hemoconcentration, and sequestration of RBC from spleen during initial time points to cope with the increased demand of the body during exercise (Pawlowska et al., 1985). Later, there may be a cholinergic acclimatization after 10 min. Endurance training followed by an acute bout of exercise significantly decreased the ChE activity of RBC during 5-60 min. (Somani and Dube, 1992). Dube et al. (1990) showed that acute exercise alone, irrespective of the intensity, affects the ChE enzyme to a moderate degree in RBC and heart, without affecting brain, diaphragm and thigh muscle. These findings are in agreement with Ryhanen *et al.* (1988) and are contrary to the finding of Pedzikiewicz *et al.* (1984), who have shown a slight increase (3%) in brain ChE activity after a single exercise event. Holmstedt (1971) has reported that physical exercise accelerates nerve activity in the CNS, resulting in an increased amount of acetylcholine in the nerve endings, and hence increased ChE inhibition.

Somani and Dube (1992) reported that the central and peripheral responses were altered due to an interactive effect of acute exercise and endurance training in the presence of a chemical stressor such as Phy. They showed that acute exericse (AE) or endurance training (ET) for six weeks did not significantly alter ChE activity in brain. However, AE as well as training, both modify the pharmacodynamic effect of Phy on ChE activity in brain.

AE transiently increased ChE activity in RBC, which returned to normal within 10-15 min. ET decreased ChE activity of RBC without affecting other tissues. AE enhances the rate of regeneration and thus decreases time of recovery of enzyme as compared to Phy alone. However, ET potentiates ChE inhibition in RBC and various tissues due to decreased clearance of Phy, ET for 6 weeks may be beneficial to prolong and potentiate the ChE inhibition by Phy. Phy has a low margin of safety—a slight increase in dose causes toxic symptoms. ET may help reduce the required dose of Phy, thereby reducing its toxic effects. This may be advantageous if Phy is used as a pretreatment drug against organophosphate intoxication.

REFERENCES

Arneric SP, Giuliano R, Ernsberger P, Underwood MD, and Reis DJ. Synthesis, release receptor binding of acetylcholine in the C_1 area of the rostral ventrolateral medulla: contributions in regulation arterial pressure. Brain Res. 1990; 511:98-112.

Babu SR, Somani SM, and Dube SN. Effect of physostigmine and exercise on choline acetyltransferase and acetylcholinesterase activities in fast and slow muscles of rat. Pharm. Biochem. and Behav. 1993; 45:713-717.

Benarroch EE, Granata AR, Ruggiero DA, Park OH, and Reis DJ. Neurons of C_1 area mediate cardiovascular response initiated from ventral medullary surface. Am. J. Physiol. 1986; 250:R932-945.

Clark DD and Sokoloff L. Circulation and energy metabolism of the brain. In: Basic Neurochemistry. Siegel GJ, Agranoff BW, Albers RW, and Molinoff, PB (Eds.) Raven Press 1993; 645-680.

Dube, S.N., Somani SM, and Babu SR. Concurrent acute exercise alters central and peripheral responses to physostigmine. Pharm. Biochem. and Behav. 1993; 46:827-834.

Dube SN, Somani SM, and Colliver JA. Interactive effects of physostigmine and exercise on cholinesterase activity in RBC and tissues of rat. Arch. Int. Pharmacodyn. Ther. 1990; 307:71-82.

Eckstein FP, Baughman RW, and Quinn J. An anatomical study of cholinergic innervation in rat cerebral cortex. Neuroscience 1988; 25(2):457-474.

Fibiger HC. The organization and some projections of cholinergic neurons of the mammalian forebrain. Brain Res. Rev. 1982; 4:327.

Gilad GM, Rabey JM, and Shenkman L. Strain-dependent and stress-induced changes in rat hippocampal cholinergic system. Brain Res. 1983; 267:171-174.

Godfrey DA, Park JL, and Ross CD. Choline acetyltransferase and acetylcholinesterase in centrifugal labyrinthine bundles of rats. Hearing Res. 1984; 14:93.

Gottesfeld Z, Kvetnansky R, Kopin IJ, and Jacobowitz DM. Effects of repeated immobilization stress on glutamate decarboxylase and choline acetyltransferase in discrete brain regions. Brain Res 1978; 152:374-378.

Harri M, Dannenberg T, Oksanen-Rossi R, Hohtola E, and Sundin U. Related and unrelated changes in response to exercise and cold in rats: A reevaluation. J. Appl. Physiol: Respir. Environ. Exercise Physiol. 1984; 57:1489-1497.

Hata T, Kita T, Higash T, and Ichide S. Total acetylcholine content and activities of cholineacetyltransferase and acetylcholinesterase in brain and duodenum of SART stressed (repeated cold stressed) rat. J. Pharmacol 1986; 41:475-485.

Holloszy JO and Booth W. Biochemical adaptations to endurance exercise in muscle. Annu. Rev. Physiol. 1976; 38:273-291.

Holmstedt B. Distribution and determination of cholinesterase in mammals. Bull. WHO 1971; 44:99-107.

Longoni R, Mulas A, Oderfeld-Novak B, Pepu IM, and Pepeu G. Effect of single and repeated electroshock applications on brain acetylcholine levels and choline acetyltransferase activity in the rat. Neuropharmacology 1976; 15:283-286.

Oesch F. Trans-synaptic induction of choline acetyltransferase in the preganglionic neuron of the peripheral sympathetic nervous system. J. Pharmacol. Exp. Ther. 1974; 188:439-466.

Pawlowska D, Moniuszko-Jankoniuk J, and Soltys M. Parathion-methyl effect on the activity of hydrolytic enzymes after single physical exercise in rats. Pol. J. Pharmacol. Pharm. 1985; 37:629-638.

Pedzikiewicz J, Piaskowska E, and Pytasz M. Acetylcholinesterase (E.C.3.1.1.7) in the skeletal muscle and brain of rats after exercise and long term training. Acta. physiol. pol. 1984; 35:469-47.

Roskoski R Jr, Mayer HE, and Schmid PG. Choline acetyltransferase activity in guinea-pig heart in vitro. J Neurochem. 1974; 23:1197-1200.

Ross CD and Godfrey DA. Distributions of choline acetyltransferase and acetylcholinesterase activities in layers of rat superior colliculus. J. Histochem. Cytochem. 1985; 33(7):631-641.

Rossier J. Choline acetyltransferases: A review with special reference to its cellular and subcellular localization. Rev. Neurobiol. 1977; 20:284-334.

Ryhanen R, Kajovaara M, Harri M, Kaliste-Korhonen E. and Hanninen O. Physical exercise affects cholinesterases and organophosphate response. Gen. Pharmacol. 1988; 19:815-818.

Somani SM and Dube, SN. Endurance training changes central and peripheral responses to physostigmine. Pharmacol. Biochem. Behav. 1992; 41:773-781.

Somani SM, Dube SN, Garcia V, Buckenmeyer P, Mandalaywala RH, Verhulst SJ, Knowlton RG. Influence of age on caloric expenditure during exercise. Int. J. Clin. Pharmacol. Ther. Toxicol. 1992; 30:1-6.

Somani SM, Babu SR, Arneric S, and Dube SN. Effect of cholinesterase inhibitor and exercise on choline acetyltransferase and acetycholinesterase activities in rat brain regions. Pharmacol. Biochem. and Behav. 1991; 39:337-343.

Tucek S, Zelena J, Ge I, Vyskocil F. Choline acetyltransferase in transected nerves, denervated muscles and Schwann cells of the frog: Correlation of biochemical, electron microscopical and electrophysiological observations. Neurosciences 1978; 3:709-724.

Tumer N, Hale C, Lawler J, and Strong R. Modulation of tyrosin hydroxylase gene expression in the rat adrenal gland by exercise: Effects of age. Mol. Brain. Res. 1992; 14:51-56.

Vihko V, Salminen A, Rajamski J. Oxidation and lysosomal capacity in skeletal muscle of mice after endurance training of different intensities. Acta Physiol. Scand. 1978; 104:74-81.

EVALUATION OF THE DIRECT ACTIONS OF HI-6 IN REVERSING SOMAN-INDUCED TETANIC FADE

Michael Adler, Donald M. Maxwell, Richard E. Sweeney, and
Sharad S. Deshpande

U.S. Army Medical Research Institute of Chemical Defense
Aberdeen Proving Ground, Maryland 21010-5425

INTRODUCTION

The most widely accepted mechanism for the acute toxicity of organophosphorus anticholinesterase agents is irreversible inhibition of acetylcholinesterase (AChE), an enzyme present at all known cholinergic synapses (Taylor, 1990). Inhibition of AChE results in accumulation of acetylcholine (ACh), which then leads to aberrant cholinergic transmission (Katz and Miledi, 1973). The precise nature of the abnormality varies with the synapse, and can include depolarization, desensitization, repetitive firing or sustained activation (Hobbiger, 1976; Adler *et al.*, 1992).

At present, the standard therapeutic strategy in the U. S. for organophosphate toxicity is aimed at counteracting the muscarinic actions of excess ACh with atropine, reactivating inhibited AChE by using the oxime, pyridine-2-aldoxime (2-PAM) and protecting a critical pool of AChE molecules by pretreatment with the carbamate, pyridostigmine (Dunn and Sidell, 1989). For organophosphorus AChE inhibitors that undergo rapid aging (e.g., soman), 2-PAM is relatively ineffective in providing protection. This is presumably the result of the 2-PAM-mediated reactivation rate being too slow to produce significant recovery of AChE prior to completion of the aging reaction (Shih *et al.*, 1991). However, the bispyridinium oxime, HI-6 ([[[(4-aminocarbonyl)pyridino]-methoxy]methyl]-2-[(hydroxyimino)methyl]-pyridinium dichloride) is effective even against soman toxicity, especially when preceded by pyridostigmine and co-applied with atropine. The greater efficacy of HI-6 has been attributed to its faster rate of reactivation relative to 2-PAM, thereby allowing a greater fraction of AChE to be recovered prior to aging of soman-inhibited AChE (Schoene *et al.*, 1983).

It has also been suggested that differences in the reactivation rate may not entirely account for the superior efficacy of HI-6. Instead, the protective action of HI-6 was proposed to be mediated by direct actions of the oxime on the target tissues. These direct actions may be considered to complement the therapeutic benefit of AChE reactivation (Wolthuis *et al.*, 1981), or to be of primary importance (Alkondon *et al.*, 1988; Van Helden *et al.* 1991). In

Enzymes of the Cholinesterase Family, Edited by Daniel M. Quinn et al.
Plenum Press, New York, 1995

most systems, the nature of the direct action is unknown, however, in skeletal muscle, analysis of single channel recordings have revealed that HI-6 produces a rapid block of open ACh activated channels (Alkondon *et al.*, 1988). This would limit the duration of the endplate response and therefore oppose the action of AChE inhibitors.

Although evidence for the existence of direct actions of HI-6 is beyond dispute, the role of these direct actions in mediating the beneficial effect of HI-6 is not as clear. To obtain convincing evidence that the direct actions of HI-6 are important *in vivo*, it is necessary to show that these actions persist under appropriate conditions of synaptic activity. For diaphragm muscle, it should be possible to demonstrate that the direct effect of HI-6 is sustained during stimulation at frequencies consistent with the ventilatory cycle. Previous studies did not address this issue since they were performed with either low frequency stimulation (Alkondon *et al.*, 1988) or with brief trains repeated at relatively long intervals (Van Helden *et al.*, 1991). Since these stimulation patterns do not accurately model the high duty cycle of respiratory muscle in rodents, it is necessary to show that direct effects of HI-6 still persist under conditions that more realistically resemble the normal ventilatory cycle (De Candole *et al.*, 1953; Wright, 1954; Gill, 1963; Cohen, 1979). To address this issue, experiments were performed in the rat phrenic nerve-hemidiaphragm preparation using 20 or 50 Hz trains repeated at regular intervals of 1.25 sec. In addition, an alternative stimulation pattern was developed to approximate the systematic recruitment of motor units that occurs during normal respiration. This will be referred to as "contoured" pulse trains. For both stimulation paradigms, complications from reactivation were eliminated by allowing a four hour interval between soman exposure and HI-6 addition. The results suggest that HI-6, even when administered after completion of aging, was still effective in restoring tension production. However, the beneficial effects of HI-6 were highly use dependent and underwent marked attenuation with increases in stimulation frequency.

METHODS

Soman Exposure

Male Sprague-Dawley rats weighing 200-250 g were used in all experiments. Prior to soman exposure, the animals were pretreated with an i.m. dose of 17 mg/kg atropine methylnitrate for 10 min followed by a s.c. injection of a 3 LD50 dose of soman (339 µg/kg) in saline; the injection volumes were 0.5 ml/kg. Atropine methylnitrate was used to minimize peripheral autonomic symptoms such as bronchoconstriction and increases in airway secretions (Adler *et al.*, 1992). Rats exposed to 339 µg/kg soman died within 7.5-12.2 min after injection (mean ± S.E. = 10.3 ± 1.4 min). Prior to death, all soman-injected rats exhibited convulsions characterized by violent whole-body tremors. Control animals received 17 mg/kg atropine methylnitrate followed 10 min later by saline and were symptom-free.

Contractility Measurements

Soman-intoxicated rats were decapitated immediately after death; control rats were killed by exposure to excess CO_2 followed by decapitation. The left hemidiaphragm muscle was carefully removed with approximately 2 cm of phrenic nerve and mounted in temperature controlled (37° C) twitch baths. The bath contained Tyrode's solution of the following composition (mM): NaCl, 137; KCl, 2.7; $NaHCO_3$, 11.9; NaH_2PO_4, 0.3; $MgCl_2$, 0.5; $CaCl_2$, 1.8 and glucose, 5.6. The pH was maintained at 7.4 by bubbling with a gas mixture of 95% O_2 and 5% CO_2. Muscle tensions were recorded with Grass FT 0.03 isometric force-displacement transducers and displayed on a Gould Model 2800 chart recorder and a Honeywell 101 instrumentation tape recorder.

Stimulation of the phrenic nerve was achieved using bipolar platinum electrodes. Supramaximal pulses (4-6 V) were delivered from Grass S-88 stimulators at 20 or 50 Hz. The trains lasted 0.55 sec and were repeated 30 times per trial at intervals of 1.25 sec. A 5 min rest period was provided between trials to achieve optimal consistency during consecutive episodes.

Contoured Pulses

The pulses were generated by a computer program written in Fortran-77 and assembly language for a Zenith computer equipped with a Data Translation analog input/output card (DT 2801). The number of pulses in the train as well as the pulse amplitude and frequency were specified independently. To execute a trial, the program received synchronizing information in the form of an analog signal from the Clampex subroutine of the acquisition and analysis software, pClamp (Axon Instruments, Foster City, CA). This signal was sampled continuously until a negative-to-positive excursion greater than a threshold level was found. When the triggering pulse was detected, the program generated the appropriate sequence of analog pulses. The signals were amplified using an operational amplifier in a voltage follower configuration to provide greater current capability, and the output was applied to the nerve stimulating electrode. The pulse amplitudes increased from just above threshold to that producing maximal twitch tension during the course of the train. As with the previous stimulations, a total of 30 trains were elicited per trial.

Data Analysis

Contractility data was analyzed using the "Clampan" subroutine of pClamp. Since soman-treated muscles were generally unable to sustain tetanic tension, the records were quantified by measuring the area under each contraction record (Van Helden *et al.,* 1991). Unless stated otherwise, values are expressed as the mean ± S.E. of 8 muscles, one from each animal.

RESULTS

Supramaximal Stimulation

To probe the direct actions of HI-6 under realistic conditions, rats were pretreated with atropine methylnitrate and exposed to 339 µg/kg soman (3 LD50). Following convulsion and death, the diaphragm was removed, mounted in a twitch bath and tested for its ability to maintain tension during repetitive stimulation. The data with 20 Hz trains are shown in Fig. 1. Under control conditions, muscle tensions elicited by brief, repetitive trains of 20 Hz were well-maintained and generally increased with each train (Fig. 1A). After exposure to 339 µg/kg soman, the tensions were reduced to 24% of control and showed a progressive decrement especially during the last 5-10 responses (Fig. 1B). The addition of HI-6 to these muscles 4 hr after soman exposure, and therefore after completion of aging, led to a marked recovery in muscle tension. Considering the entire trial of 30 trains, tensions were restored to 87% of control in the presence of 0.3 mM HI-6 and to 91% of control following increases in the HI-6 concentration to 1 mM (Fig. 1D). Thus, the beneficial effects of HI-6 persisted when trains were evoked repetitively at 20 Hz.

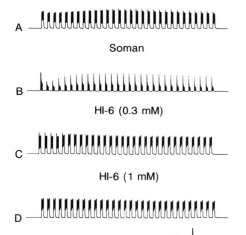

Control

A

Soman

B

HI-6 (0.3 mM)

C

HI-6 (1 mM)

D

20 g

10 sec

Figure 1. Contractions elicited by 30 brief trains at 20 Hz. A 4 hr period was allowed before addition of HI-6 to permit completion of the aging reaction. Aging was determined to be complete since no significant reactivation was observed after a 30 min incubation with 1 mM HI-6. Record A was obtained from a control rat; records B-D were obtained from a rat exposed to 339 µg/kg soman. Reproduced by permission from Adler *et al.*, 1994, European Journal of Pharmacology, Environmental Toxicology and Pharmacology Section, 270:9-16.

The efficacy of HI-6 was considerably reduced, however, when stimulation frequencies were raised to 50 Hz (Fig. 2). After exposure to 339 µg/kg soman, pronounced tetanic fade was evident even during the first train (Fig. 2B). The second 50 Hz train was further attenuated, followed by a more gradual reduction of tension thereafter. The decrease in area between the second and first train was 53% (Fig. 2B) and that between the last and first train was 81%. Addition of 0.3 mM HI-6 led to a partial antagonism of soman-induced muscle weakness (Fig.

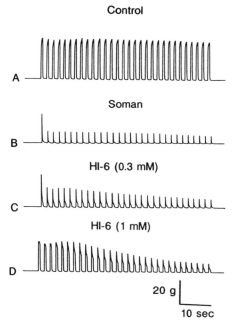

Control

A

Soman

B

HI-6 (0.3 mM)

C

HI-6 (1 mM)

D

20 g

10 sec

Figure 2. Contractions elicited by 30 brief trains at 50 Hz. Record A was obtained from a control rat; records B-D were obtained from a rat exposed to 339 µg /kg soman. HI-6 was added 4 hr after soman exposure to allow for completion of aging. Reproduced by permission from Adler *et al.*, 1994, European Journal of Pharmacology, Environmental Toxicology and Pharmacology Section, 1994, 270:9-16.

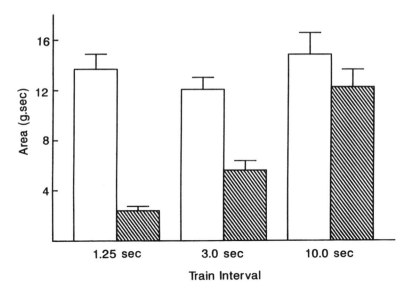

Figure 3. Histograms comparing three different train intervals (1.25, 3 and 10 sec) on the efficacy of HI-6; The open and hatched bars represent areas of the 1st and 30th train. Muscles were obtained from rats exposed to 339 μg/kg soman. HI-6 (1 mM) was added 4 hr later. Stimulation parameters were: pulse duration, 0.1 msec; train duration, 0.55 sec; voltage, supramaximal. The data represent the mean ± S.E. obtained from 4-6 muscles.

2C), and a further antagonism was observed following incubation with 1 mM HI-6 (Fig. 2D). However, restoration of the indirectly-elicited muscle tension by HI-6 was not sustained, and successive responses exhibited progressive attenuation with continued stimulation. These results suggest that the direct action of HI-6 is insufficient to counteract tetanic fade resulting from soman exposure when muscles are stimulated at a relatively high duty cycle.

The effect of different train intervals on the direct effect of HI-6 at 50 Hz is shown in Fig. 3. The histograms represent areas of the first and last trains from muscles stimulated at intervals of 1.25, 3 and 10 sec in the presence of 1 mM HI-6. As is clear from Fig. 3, muscle tensions fell markedly during stimulation at both 1.25 and 3 sec intervals. However, the tensions in the diaphragm muscle were better sustained when trains were elicited at 10 sec intervals, undergoing less than a 20% reduction in area at the end of 30 trains. The latter is in accord with the results of Van Helden *et al.* (1991).

Contoured Pulse Stimulation

From the data obtained with supramaximal fixed frequency stimulations, the importance of the direct action of HI-6 would appear to depend on how closely the stimulation pattern corresponds to the *in vivo* respiratory rhythm. The diaphragmatic contraction *in vivo* has an approximately triangular shape in which motor units are recruited progressively during the course of inspiration, accompanied by increases in the phrenic motor neuron firing frequency (Gill, 1963; Cohen, 1979). To approximate the normal inspiratory activity of the diaphragm muscle, we developed a computer program to generate pulses that would permit simultaneous increases in both the amplitude and frequency of stimuli that are applied to the phrenic nerve. This method has been designated as contoured pulses.

Under control conditions, tensions underwent a small but consistent increase in area during the 30 consecutive trains (Fig 4A), a pattern similar to that observed with fixed-

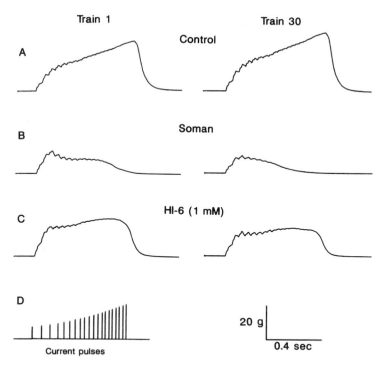

Figure 4. Records of diaphragmatic tensions obtained from the 1st and 30th train by the contoured pulse method. The individual trains lasted 0.55 sec and consisted of 20 pulses increasing in frequency from 14 Hz to 80 Hz and in amplitude from 1.1 V to (threshold) to 4.6 V (maximal) (record D). The parameters were selected to produce a control area similar to that observed during a 50 Hz supramaximal stimulation. Record A was obtained from one rat; the remaining records were acquired from a second animal that was injected with 339 µg/kg soman 4 hr earlier.

frequency trains. Data from soman-exposed animals showed the typical pattern of tetanic fade especially at frequencies above 30 Hz (Fig. 4B). Contrary to expectations, however, the fade was not strictly frequency-dependent. This presumably arose because the amplitude of the stimuli increased along with the frequency, causing recruitment of previously quiescent neuromuscular junctions. At the conclusion of the trains, a decrement in diaphragmatic tension was still observed in soman-exposed animals, although the reduction was less severe than for responses elicited by 50 Hz supramaximal trains. Addition of HI-6 caused a marked recovery of muscle tension; in the example shown, the area of the 1st train was restored to 84% of control (Fig. 4C). Moreover, with contoured pulse stimulation, a smaller decrement was detected for the remaining trains in the trial. In the presence of 1 mM HI-6, the area integrated over the entire 30 trains was 56% of control and even the last train retained a sizable area compared to that observed with supramaximal 50 Hz pulses (see Fig 2D).

DISCUSSION

The results of the present investigation demonstrate a role for the direct action of HI-6 in reversing diaphragmatic neuromuscular fatigue after exposure of rats to a 3 LD50 dose of soman. To ensure that direct actions of HI-6 were responsible for the observed improvements in muscle dynamics, the addition of HI-6 was delayed for 4 hr to permit completion of the aging reaction. Although the aging rate of soman-inhibited AChE is rapid in most tissues (half-time ≤ 6 min), slower rates (half-time ≈ 20) min have been reported in the rat diaphragm preparation (Clement, 1982). The 4 hr period was therefore chosen to eliminate possible ambiguities in interpretation arising from AChE reactivation.

The most important new finding of the current study is that the contribution of the direct action of HI-6 depends markedly on the frequency, duration and pattern of nerve stimulation. Using conventional supramaximal pulses, soman-induced decrements of tensions elicited at 20 Hz were effectively antagonized with 0.3 mM HI-6. At 50 Hz, however, 1 mM HI-6 was required to antagonize reductions following single brief tetanic pulses or trials elicited at 10 sec intervals (Fig. 3). With more frequent train rates, even 1 mM HI-6 was inadequate, and successive trains exhibited a progressive decline.

The relative contribution of the direct actions of HI-6 in antagonizing neuromuscular fatigue may be expected to vary during the course of soman exposure. The respiratory-related units which drive the diaphragm muscle *in vivo* undergo complex alterations leading ultimately to respiratory arrest (De Condole *et al.*, 1953). Prior to cessation of firing, these units increase their discharge frequency and become arrhythmic. In addition, the hypoxia, CO_2 accumulation and alterations in the airway compliance that occur during soman exposure may also act through reflex pathways to modify the respiratory rhythm and frequency (Cohen, 1979). If it is assumed that the direct action of HI-6 is exerted principally on peripheral structures, in particular, the diaphragm muscle, it is expected that the oxime would augment muscle contractility during the low frequency components of each respiratory cycle and become less effective as the discharge frequency increased. A similar loss of efficacy would be expected as the interval between cycles became shorter. If the contoured pulses more accurately reflect neuromuscular activity than fixed frequency paradigms, a more complete recovery of neuromuscular activity may be expected by HI-6 than would be concluded from data obtained by conventional tetanic stimulation. However, even under this condition, the restoration of muscle tensions was incomplete and repetitive stimulation still caused tensions to decline over time.

It is concluded that the direct action of HI-6 is of value in the therapy of organophosphate toxicity, especially for agents that undergo rapid aging. However, the direct action may be of a more limited nature than previously considered since the beneficial effects on muscle tension are not maintained during sustained stimulation at frequencies above 20 Hz. Moreover, as shown by previous investigators (Alkondon *et al.*, 1988; Van Helden *et al.*, 1991), the direct effects of HI-6 are readily reversible and necessitate the continuous presence of the oxime.

REFERENCES

Adler, M., Moore, D.H. and Filbert, M.G., 1992, Mechanism of soman-induced contractions in canine tracheal smooth muscle, *Arch. Toxicol.* 66:204-210.

Alkondon, M., Rao, K.S. and Albuquerque, E.X., 1988, Acetylcholinesterase reactivators modify the functional properties of the nicotinic acetylcholine receptor ion channel, *J. Pharmacol. Exp. Thera.* 245:543-555.

Clement, J.G., 1982, HI-6: Reactivation of central and peripheral acetylcholinesterase following inhibition by soman, sarin and tabun *in vivo* in the rat, Biochem. Pharmacol. 31:1283-1287.

Cohen, M.I., 1979, Neurogenesis of respiratory rhythm in the mammal, *Physiol. Rev.* 59:1105-1173.

De Candole, C.A., Douglass, W.W., Lovatt-Evans, C., Holmes, R., Spencer, K.E.V., Torrance, R.W. and Wilson, K.M., 1953, The failure of respiration in death by anticholinesterase poisoning, *Brit. J. Pharmacol. Chemother.* 8:466-475.

Dunn, M. and Sidell, F.R., 1989, Progress in medical defense against nerve agents. *JAMA* 1262:649-652.

Gill, P.K., 1963, The effects of end-tidal CO_2 on the discharge of individual phrenic motoneurons. *J. Physiol, (Lond.)* 168:239-257.

Hobbiger, F., 1976, Pharmacology of anticholinesterase drugs, in Handbook of Experimental Pharmacology, ed. E. Zaimis, (Springer-Verlag, Berlin) Vol 42, pp. 487-581.

Katz, B. and Miledi, R., 1973, The binding of acetylcholine to receptors and its removal from the synaptic cleft, *J. Physiol. (Lond.)* 231:549-574.

Shih, T-M., Whalley, C.E. and Valdes, J.J., 1991, A comparison of cholinergic effects of HI-6 and pralidoxime-2-chloride (2-PAM) in soman poisoning, neuromuscular actions of nerve agents, *Toxicol. Lettrs.* 55:131-147.

Schoene, K., Steinhanses, J. and Oldiges, H., 1983, Reactivation of soman inhibited acetylcholinesterase *in vitro* and protection against soman *in vivo* by bispyridinium-2-aldoximes, *Biochem. Pharmacol.* 32:1649-1651.

Taylor, P., 1990, Anticholinesterase agents, in Goodman and Gilman's the Pharmacological Basis of Therapeutics, eds A.G. Gilman *et al.* (Pergamon Press, NY) pp. 131-149.

Van Helden, H.P.M., de Lange, J., Busker, R.W. and Melchers, B.P.C., 1991, Therapy of organophosphate poisoning in the rat by direct effects of oximes unrelated to ChE reactivation, *Arch. Toxicol.* 65:586-593.

Wolthuis, O.L., Berends, F. and Meeter, E., 1981, Problems in the therapy of soman poisoning, *Fund. Appl. Toxicol.* 1:183-192.

Wright, P.G., 1954, An analysis of the central and peripheral components of respiratory failure produced by anticholinesterase poisoning in the rabbit, *J. Physiol. (Lond.)* 126:52-70.

CLINICAL BLOOD CHOLINESTERASE MEASUREMENTS FOR MONITORING PESTICIDE EXPOSURES

B. W. Wilson,[1] S. Padilla,[2] J. R. Sanborn,[3] J. D. Henderson,[1] and
J. E. Billitti[1]

[1] University of California
Davis, California 95616
[2] US Environmental Protection Agency
Research Triangle Park, North Carolina 27711
[3] California Environmental Protection Agency
Sacramento, California 95815

INTRODUCTION

Measurement of blood cholinesterase (ChE) activity, especially acetylcholinesterases (AChE, EC 3.1.1.7) and butyrylcholinesterases (BChE, EC 3.1.1.8), is of worldwide interest and importance. They are one of the few biomarkers that function both as an indicator of exposure and also as evidence of an adverse effect. Determining blood ChEs is required for monitoring farmworkers in the state of California in the United States. In addition, ChE determinations are used in emergency wards to diagnose exposure to anti-ChEs, as part of the submissions to regulatory agencies to set safe levels of pesticides, to monitor wildlife and their exposure to dangerous anti-ChE chemicals, and to provide evidence of the use or escape of nerve gases.

Recently the reliability of ChE data submitted for regulatory and diagnostic purposes has been examined by the US and California Environmental Protection Agencies. Issues raised included variability of clinical laboratory results (Christenson et al., 1994; Wilson et al., 1992) and lack of validated and interlaboratory standard operating procedures.

This paper reviews our recent experiences in examining the reliability of thiocholine ester colorimetric assays, modified from the assay of Ellman et al. (1961), in the monitoring of blood ChEs from humans, domestic animals and wildlife.

ASSAYS AND CONDITIONS

Three common assays for ChE activities are: the Michel test, based on pH changes following hydrolysis of choline esters (Michel, 1949); the Johnson and Russell assay, in

Enzymes of the Cholinesterase Family, Edited by Daniel M. Quinn et al.
Plenum Press, New York, 1995

Table 1. Selected assays for ChE activity

Test	Basis	Type	Substrate	Plus/minus
Michel, 1949	Change in pH	Kinetic	ACh	Natural Substrate, Cheap / Slow
Johnson & Russell, 1976	Radiometric	End Point	ACh	Micro / Costly, Disposal
Ellman et al., 1961	Colorimetric	Kinetic	AcTh	Cheap / Unnatural Substrate

ACh = acetylcholine; AcTh = acetylthiocholine.

which radioactive acetate is separated into the toluene phase following hydrolysis of tritiated acetylcholine (Johnson and Russell, 1975); and the Ellman assay, based on the reaction of dithiobisnitrobenzoate (DTNB) with the thiocholine released by hydrolysis of thiocholine esters (Ellman et al., 1961). Some strengths and weaknesses of the assays are indicated in Table 1.

REGULATORY BACKGROUND

EPA Workshop

When an internal study by the Office of Pesticide Programs of the Environmental Protection Agency (EPA) revealed problems in their large data base of reports submitted to set safety levels for pesticides in food, a workshop on ChE methodologies was held (Wilson et al.,1992) bringing together government agency, industry and university scientists. The presentations documented the lack of standards and guidelines for the submission of data to the agency. Following the meeting a round robin test was conducted to see how well clinical and research laboratories could determine ChE activity in blood and brain by using the hydrolysis of thiocholine esters. Each laboratory used its own version of the basic assay with samples of tissues inhibited with organophosphate (OP) and carbamate (CB) pesticides. Several of the clinical laboratories used an automated Hitachi spectrophotometer with a reagent kit and instructions from Boehringer-Mannheim. Research laboratories used microassays run on multiple well plate readers (Doctor et al., 1988). The results (Wilson et al.,1993; Wilson et al., Submitted) revealed a lack of reproducibility of results from laboratory to laboratory, suitability of the instructions for reagent kits and the programming of the automated instrument, especially when measuring the AChE activity of the rat erythrocyte.

ChE Monitoring in California

At the same time, the Department of Pesticide Regulation of the California EPA surveyed clinical laboratories approved to monitor blood ChE of farmworkers. They also found several methods in use and a lack of guidelines and standard operating procedures (Choi et al., 1993). Approximately half of the laboratories used the Michel pH method, others used manual thiocholine/colorimetric and automated thiocholine-based assays. As a result of the survey, a project was started to develop a "gold standard" guideline to help laboratories demonstrate the reliability of their assays.

Testmate Kit

A third project began when US National Institute of Occupational Safety and Health (NIOSH) epidemiologist Kyle Steenland called our attention to the Test-Mate, a portable

Table 2. Thiocholine assay conditions

	Assay				
	Ellman	B/M manual	B/M automatic	Sigma	Testmate
Wavelength	412 nm	405 nm	480 nm	405 nm	470 nm
Substrate/conc	AcTh/0.5 mM*	AcTh/5.4 mM	AcTh/5.4 mM	PropTh/4 mM	AcTh/1 mM
Buffer pH	8.0	7.2	7.2	7.2	7.4
DTNB conc	0.32 mM	0.24 mM	0.24 mM	0.25 mM	0.3 mM

AcTh = Acetylthiocholine (*1 mM in Round Robin); PropTh = Propionylthiocholine; DTNB = Dithionitrobenzoate

colorimeter developed by Magnotti and colleagues (Magnotti et al., 1988). The instrument is hard-wired for a thiocholine-based assay and programmed to monitor human blood ChE levels from a drop of blood. The instrument first was used by occupational medicine physicians to examine exposure to pesticides in Nicaragua (Magnotti et al., 1987). We compared the performance of the instrument with that of a portable multiwell plate reader, determining enzyme activity with both instruments. Differences in the conditions of the several thiocholine assays discussed here are listed in Table 2.

BIOCHEMICAL BACKGROUND

Although all ChEs are equal in the sense that they rely upon serine-catalytic sites and hydrolyze choline esters, some are more equal than others with regard to their substrate specificities and the extent to which they have been studied. Different ChEs have distinctive substrate preferences and other biochemical specificities. For example, AChEs are inhibited by excess acetylthiocholine (AcTh), above approximately 1-2 mM, but BChEs continue to increase their rate of substrate hydrolysis with increasing substrate concentration. Birds and some other vertebrates (unlike mammals) have no AChE activity on their red blood cells. The plasma ChEs are another difference between species. Rat plasma ChE activity is 40-50 % AChE whereas human plasma ChE is virtually all BChE.

Although there have been a number of excellent comparative studies, few have involved the number of animals, including the human, necessary to establish "normal" ranges. Normal ranges of values for the human published in the instructions that accompany commercial kits (see Boehringer-Mannheim Catalog #450035 and #124117 and Sigma Diagnostics Procedure No. 422) do not address differences in optimum conditions for the determination of AChE and BChE.

EPA ROUND ROBIN

The EPA Round Robin test was designed to compare results between clinical and research laboratories. Rats were dosed with organophosphate (chlorpyrifos) and carbamate (carbaryl) pesticides. Blood and brain samples were taken: RBCs and plasma were separated, and the RBCs were treated with Triton X-100; brains were homogenized; and frozen samples were shipped to participating laboratories. The clinical laboratories used an assay kit and an automated, programmable multisample spectrophotometer of Boehringer-Mannheim/Hitachi. In the case of the RBC, the more inhibited was the sample, the more in error were the

Table 3. Round robin enzyme activities

Assay conditions	Plasma activity		RBC activity	
	Low Dose	High Dose	Low Dose	High Dose
410 nm, 25 C	67.6 ± 2.8	34.8 ± 3.5	51.7 ± 21.2	1.8 ± 2.0
410 nm, 37 C	60.5 ± 7.3	27.7 ± 5.4	46.8 ± 11.6	18.7 ± 17.8
480 nm, 37 C	54.0 ± 5.6	24.0 ± 1.7	71.0 ± 12.5	43.3 ± 7.5
Grand Mean	60.7 ± 6.8	28.8 ± 5.5	56.5 ± 12.8	21.3 ± 20.9

Mean ± Standard Deviation.

results from the clinical laboratories. Inhibitions were often much less than those found by the plate reader assays.

A major reason for the discrepancy was that the assays conducted according to the instructions and programming of Boehringer-Mannheim did not take into account oxidation of DTNB by the rat RBCs in the absence of AcTh. Tissue blanks were not recommended by the manufacturer, nor were they possible without reprogramming the machines. Even after correcting for blanks (Table 3), RBC values were more variable than were those from plasma and brain.

The high blank of rat RBCs is discussed in some detail by Loof at the ChE Methodology Workshop (Wilson et al., 1992) in which he concludes "Depending on the time set for reading, the background reaction, which cannot be inhibited by any phosphorus ester or carbamate, becomes the dominant part of the overall reaction. *If a sample blank is not introduced, false high results ...are found.* (Italics added.)

Other major differences between the assays of the clinical laboratories and the research laboratories were the pH of the assay buffers, substrate concentrations and the wavelength of the readings (Table 2). The peak absorbance of DTNB is between 405-415 nm, wavelengths that correspond to the Soret band of hemoglobin absorption. To avoid such interference, the automated assay of Boehringer-Mannheim/Hitachi determines DTNB absorption at 480 nm, significantly reducing the sensitivity of the assay (Wilson and Henderson, 1992).

OPTIMUM CONDITIONS FOR HUMAN ChEs

Conditions appropriate for assaying human ChEs were investigated in collaboration with the California EPA. Experiments were performed to examine the conditions of the Ellman et al. (1961) assay for the determination of AChE and BChE of humans and to compare the optimum conditions for the microplate reader with those recommended in the instructions for the kits of Boehringer-Mannheim Corporation and Sigma Chemical Company. Assays were performed with whole blood, washed and hemolyzed RBCs and plasma. Experiments such as the one shown in Figure 1 comparing the effect of acetylthiocholine concentration and pH on total AChE activity established that the assay conditions of Ellman et al. (1961) were optimal for the human enzymes.

Problems with the two commercial kits tested included choice of substrate, concentration of substrate and pH. For example, the Boehringer-Mannheim kit does not employ optimal conditions for AChE activity by using pH 7.2 buffer and a substrate concentration of 5.4 mM; both conditions reduced the activity of the enzyme. Assays lost BChE activity by running at rate-limiting substrate concentrations. It is not possible to recommend a single optimal AcTh concentration for both RBC AChE and plasma BChE; AChE enzymes are

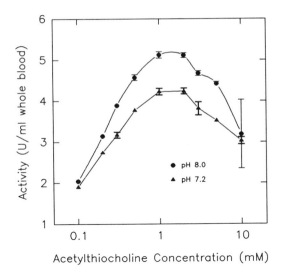

Figure 1. Effect of substrate concentration on AChE of human blood.

inhibited by AcTh in excess of 1-2 mM and BChE enzymes are not. Choosing another substrate does not help. The Sigma kit used propionylthiocholine, a substrate that is not optimal for the RBC enzyme (instructions with the Sigma kit indicate that the manufacturer focuses on using it to screen for plasma BChE variants to prevent adverse responses to succinylcholine and similar drugs during surgery, rather than emphasizing its use to monitor human blood enzymes for pesticide exposures). The need for a "tissue blank" to correct for the high endogenous DTNB oxidation by the rat RBC was not necessary with human RBC samples because normal activity in the human is 10-fold higher than in rats, minimizing the effect of background DTNB oxidation.

TESTMATE

The performance of the Testmate kit was examined in collaboration with the State of California. Workers harvesting peaches in an orchard previously sprayed with azinphos-methyl (Guthion) were examined with the Testmate procedure and with the microplate assay using blood drawn by venous puncture. The blood was iced and brought to the laboratory where it was assayed for AChE activity. The data showed a better than 90 percent correlation between assay methods (Figure 2), even though, like the automated Boehringer-Mannheim assay, absorption was determined at a wavelength removed from the optimum. In addition to validating the assays, the data provided evidence for reduced ChE activity in the blood of some of the workers.

The Testmate measures enzyme activity one sample at a time, a major inconvenience for large scale sample processing. One alternative is to adapt a multiwell plate reader to field studies by using a generator or a voltage converter and an automobile 12 volt battery, taking the instrument to the samples, rather than taking the samples to the laboratory. To accomplish this in a way useful to field biologists we lyophilized reagents (with sucrose as a filler) in 96 well plates. In this way, approximately thirty samples can be run in triplicate in less time than it would take to run a dozen samples with the Testmate.

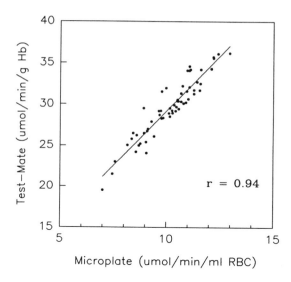

Figure 2. Comparison of AChE assay methods.

CONCLUSION: RESPONSIBILITIES

The work reported here supports several recommendations:

- *Pesticide levels.* The procedures used to set safe levels of pesticides need to be modified in so·far as ChE measurements are concerned as was recommended by Wilson et al.(1993) in their report to the EPA Scientific Advisory Panel. Standard Operating Procedures and Good Laboratory Practices should be implemented for the participating laboratories.
- *Optimization of assay conditions.* Assays should be optimized for the pesticides and test animals under study including the human.
- *Manufacturer's instructions.* Companies that manufacture the kits for ChE assays used in clinical laboratories by personnel not trained in enzyme kinetics or biochemical toxicology should write instructions that spell out the limitations of their assays. By the same token, clinical laboratories undertaking such work should employ staff with sufficient biochemical expertise to understand the limitations of the assays and to be able to modify the recommendations of the manufacturer.
- *Conversion factors.* Baseline data obtained by assays that have been optimized for the samples concerned and run by documented standard operating procedures are needed to determine conversion factors between assays. This would permit transfer of information from one study to another, assisting in the detection of potentially harmful exposures. A farmworker in California may move from a region where the Michel pH test is performed to one where the clinical laboratory uses the Boehringer-Mannheim automated instrument to one using a manual Sigma propionylthiocholine procedure. Because the values derived from different methods cannot be compared directly, a way is needed to convert from one assay condition to another. Conversion factors can readily be derived so long as the dose/activity curves are linear. Preliminary conversion factors for some of the

Table 4. Conversion factors between assays

Sample	Boehringer/Mannheim	Sigma Diagnostic
RBC	1.43	3.04
Plasma	0.59	0.38

Factor converts measured activity to Ellman assay activity.

conditions discussed here (established by diluting rather than inhibiting the enzyme activity) are given in Table 4.

- *Testmate.* The results of these and other studies (Magnotti et al., 1988, Magnotti et al., 1987, Wilson and Henderson, 1992) attest to the reliability of the Testmate instrument. Our experience suggests that it, and similar instruments, are useful for field studies where more expensive and elaborate instruments are unavailable. Examples are state agency personnel immediately responding to a report of workers in a field suspected of being exposed to harmful levels of a pesticide, and wildlife care center personnel examining the blood of animals brought into the center with muscle weakness and convulsions. Because of the "hard-wired" program of the Testmate, it does not replace instruments that can be set to different wavelengths, sampling times and calculations of activities. Nevertheless, its relatively low cost makes it a cholinesterase "Best Buy" when used as directed for what it was designed to do. If funds are available, and multiple sample runs are necessary, 96 well plate readers can be hooked up to voltage converters or generators and moved to the field.

CLOSING

Bench scientists know that enzyme assays must be adjusted to the jobs they are called upon to perform, but we found that clinical laboratories submitting data for diagnostic and regulatory purposes relied upon generic instructions when assaying ChE activity for government agencies. Specific standard operating procedures for ChE assays applicable to the conditions and to the needs of the agencies are needed. Perhaps the lesson to be learned from the widespread use of assays performed under less than optimal conditions for diagnostic and regulatory purposes is the recognition of what may happen when a gap opens between bench laboratory research and those that apply it.

ACKNOWLEDGMENTS

The authors acknowledge the assistance of W.S. Brimijoin (Mayo Clinic), P.D. Dass (Miles, Inc.), B.P. Doctor (Walter Reed Army Institute of Research), G. Elliot (Dupont), D. Lenz (USAMRICD), R. Pearson (Hazleton), R. Spies (Pharmaco-LSR), and the collaboration of members of the Department of Occupational Medicine, UC Davis and the Department of Pesticide Regulation, California EPA. Supported in part by the UC Davis Agricultural Health and Safety Center, NIOSH Cooperative Agreement #U07/CCU906162-02; the UC Davis Center of Environmental Health Sciences, NIEHS ESO5707; US Air Force, AFOSR-91-0226; and the UC Toxic Substances Program.

REFERENCES

Choi, R., O,Malley, M., Sanborn, J.R., 1993, Cholinesterase Laboratory Survey Report, June 13, 1993, HS-1668, CA Dept Pesticide Regulation, 1220 N Street, Sacramento CA 95814.

Christenson, W.R., Van Goetham, D.L., Scroeder, R.S., Wahle, B.S., Dass, P.D., Sangha, G.K. and Thyssen, J.H., 1994, *Toxicol. Letters* 71:139-150.

Doctor, B.P., Toker L., Roth E., and Silman, I., 1987, *Anal. Biochem.* 166:399-403.

Ellman, G.L., Courtney, K.D., Andres, V. Jr., and Featherstone, R.M., 1961, *Biochem. Pharmacol.* 7:88-95.

Johnson C.D. and Russell, R.L., 1975, *Anal. Biochem.* 64:229-238.

Magnotti, R.A. Jr., Dowling, K, Eberly, J.P., and McConnell, R.S., 1988, *Clinica Chemica Acta.* 315:315-332.

Magnotti, R.A. Jr., Eberly, J.P., Quarm, D.E.A., and McConnell, R.S., 1987, *Clin. Chem.* 33:1731-1735.

Michel, H.O., 1949, *J. Lab. Clin. Med.* 34:1564-1568.

Wilson, B.W., Padilla, S., Henderson, J.D., 1994, *J. Toxicology and Environ. Health.* Submitted.

Wilson, B.W., Padilla, S., and Henderson, J.D., 1993, *Progress Report on Round Robin Cholinesterase Tests,* US EPA Scientific Advisory Panel.

Wilson, B.W. and Henderson, J.D., 1992, *Reviews of Environ. and Contam. and Toxicol.* 128:55-69.

Wilson, B.W., Jaeger, B., and Baetcke, K. (Eds)., 1992, *Proceedings of the EPA Workshop on Cholinesterase Methodologies,* Arlington, VA, December 4-5, 1991, Washington, D.C.: Office of Pesticide Programs, US EPA.

REPEATED DOSING WITH CHLORPYRIFOS INCREASES ACETYLCHOLINESTERASE IMMUNOREACTIVITY IN RAT BRAIN

S. Padilla,[1]* S. Chiappa,[2] C. Koenigsberger,[2] V. Moser,[1] and S. Brimijoin[2]

[1] Cellular and Molecular Toxicology Branch, Neurotoxicology Division
U.S. Environmental Protection Agency
Research Triangle Park, North Carolina 27711
[2] Department of Pharmacology
Mayo Clinic
Rochester, Minnesota 55905

INTRODUCTION

From a practical standpoint the toxicological significance of acetylcholinesterase (AChE) outweighs all other rationales for continued study of this enzyme. For many reasons, we need to understand how environmental chemicals affect AChE in the tissues of insects, wildlife, and mammals. New knowledge of this sort could improve our ability to determine the hazards of exposure, define safe levels of exposure, monitor the consequences of exposure, and devise effective treatments for overexposure. With this objective, we have been concerned with characterizing the in vivo effects of anticholinesterase insecticides in mammals.

It is commonly assumed that the primary mechanism of action of organophosphate insecticides is the inhibition of AChE. This enzyme maintains the proper synaptic level of the neurotransmitter acetylcholine in the central and peripheral nervous systems. When AChE activity is depressed due to exposure to cholinesterase-inhibiting insecticides, the ordinarily rapid breakdown of acetylcholine is retarded. This causes overstimulation of target cells, under extreme conditions leading to a "cholinergic crisis", which can be rapidly fatal.

Chlorpyrifos (Dursban® ; O,O'-diethyl-O-(3,5,6-trichloro-2-pyridyl)-phospho-rothionate) is the most widely used organophosphate insecticide in the United States with a 1993 agricultural and non-agricultural usage estimate of 19-27 million pounds (Aspelin, 1994). Normally, after a single exposure to cholinesterase-inhibiting insecticides, AChE activity is only depressed for a few days and begins to recover soon after. Recovery of AChE activity after a single dose of chlorpyrifos is, by contrast, exceptionally retarded, much

---

*Address correspondence to Dr. Padilla. Tel: (919)541-3956; Fax (919) 541-4849.

Enzymes of the Cholinesterase Family, Edited by Daniel M. Quinn et al.
Plenum Press, New York, 1995

slower than from other phosphorothionate insecticides such as parathion and methyl para-thion (Pope et al., 1991). Animals exposed to chlorpyrifos exhibit a dose-dependent fall in brain AChE activity lasting for weeks (Pope et al., 1991; Bushnell et al., 1993; Hooser et al., 1988; Padilla et al., 1994). This prolonged depression of AChE activity could be explained in one of two ways: (1) by accumulation of chlorpyrifos or its active metabolites in tissue depots with subsequent slow release, or (2) by impairment of AChE resynthesis. The present experiments were undertaken to evaluate these two possibilities. The data is this report were derived from a larger paper which is being published concurrently (Chiappa et al., 1995).

MATERIALS AND METHODS

All animal procedures were in strict conformance with the NIH Guide for the Care and Use of Laboratory Animals and were approved by the Animal Care Committee of the Health Effects Laboratory of the U.S. Environmental Protection Agency. Male, Long-Evans rats (Charles River Breeding Laboratories, Raleigh, NC), 60-70 days old, were housed individually in suspended plastic cages and allowed free access to water. Rats were dosed with chlorpyrifos according to one of two regimens.

Weekly, sc Dosing Regimen: One set of animals received subcutaneous injections of chlorpyrifos in peanut oil. These animals were weighed daily and each was maintained at 350 ± 5 g body weight. Dosages were 60 ("high dose"), 30, or 15 mg/kg, or vehicle only (control). Injections were given at weekly intervals for four weeks (i.e., up to five doses). Animals were sacrificed, with no further injection, at 1, 3, and 5 weeks after starting treatment (dosing phase) and also at 7 and 9 weeks (recovery phase). The experiment was repeated with a second set of controls and "high-dose" rats sacrificed at 3 and 5 weeks.

Daily, Oral Dosing Regimen: In a separate experiment intended to produce maximal AChE inhibition, rats with free access to food received chlorpyrifos in corn oil by oral gavage. During week one, toxicant was given at 100 mg/kg for 3 days. This level of dosage was not well tolerated, so after allowing a day for recovery, a reduced dose of 75 mg/kg was given. For weeks two through four, the rats received 75 mg/kg each day, Monday through Friday. Only 4 out of the original 10 rats survived until tissues were collected at 4 weeks, approximately 24 hrs after the last dosing.

At varying times after dosage, rats were euthanized by carbon dioxide anesthesia followed by decapitation. Brains were quickly removed and dissected. In selected cases, cerebellum, hippocampus and frontal regions forward of the optic chiasm were sampled; in general, however, these areas were reserved for unrelated studies. Remaining parts of the telencephalon, diencephalon and brainstem were collected from all rats. Care was taken to ensure that these "global brain samples" had a consistent composition with closely compa-rable weights and AChE activities in all samples from a given experimental group. Dia-phragm, femoral biceps muscle, and liver were collected from a few rats.

Immediately after dissection, tissue pieces (divided at the midline for separate analyses of protein and mRNA) were frozen on dry ice for storage at -80 C. Tissue extraction for ELISA and AChE assay were performed on ice. Frozen tissue samples were thawed and homogenized in ten volumes of 10 mM Tris buffer at pH 7.4 and containing 0.9% (w/v) NaCl, 1 mM EDTA, 0.05% (v/v) Triton X-100 and 1% (w/v) bovine serum albumin (BSA; Sigma Chemical). Extracts were centrifuged at 10,000 x g for 15 min at 4 C and the supernatants were retained for assay.

AChE immunoreactivity (AChE-IR) was determined by a two-site enzyme-linked immunoadsorbent assay (ELISA) using an adaptation of published procedures (Brimijoin et al., 1987; Griesmann et al., 1991). Intra-assay coefficients of variation ranged from 7 to 9%.

To facilitate comparison between runs, all ELISA data are expressed as percentages of the mean value from the concurrently assayed controls.

AChE enzyme activity was determined in triplicate by a spectrophotometric method (Ellman et al., 1961) with 1 mM acetylthiocholine as substrate. Butyrylcholinesterase (E.C. 3.1.1.8) was inhibited by ethopropazine at a final concentration of 10^{-4} M.

Data were examined for significant differences using analysis of variance (ANOVA) and the unpaired Student's t-Test. Individual data were normalized by expression as percentages of the mean control value in a given assay with a given tissue. The criterion for significance was adjusted to allow for effects of multiple comparisons and to keep the overall probability of type I error below 5%. Regression analysis was used to ascertain whether a dose response relation existed. More detailed descriptions of the methods may be found elsewhere (Chiappa et al., 1995).

RESULTS

Weekly, sc, Dosing Regimen

As would be expected, the degree of AChE inhibition by chlorpyrifos in vivo was the least in the samples from the low dosage group (15 mg/kg per week); the progressively higher doses (i.e., 30 and 60 mg/kg) caused more rapid and greater inhibition. At each of the three time points during dosing, there was a significant correlation between dose of chlorpyrifos and degree of AChE inhibition ($p < 0.001$). Inhibition of AChE activity in the global brain samples never exceeded 80%, even after 4 weeks of treatment at the 60 mg/kg level (Fig. 1, upper panel). During the recovery phase (7-9 weeks into dosing), brain AChE activity increased slowly (Fig. 1, upper panel). At 9 weeks, the lowest dosage group was no longer significantly affected (data not shown), but the rats in the higher dosage groups still exhibited activities about 30% below control (Fig. 1, upper panel).

The effect of chlorpyrifos dosing on the amounts of AChE protein was determined by ELISA for AChE-IR in aliquots of the same global brain extracts used for enzyme assay. AChE-IR was not decreased by any dosage at any time. Instead, a modest increase (15%) was apparent after 3 or 5 weeks of dosing in the 60 mg/kg group (Fig. 1, lower panel). No change in the amount of AChE-IR occurred after treatment with 15 mg/kg, but, compared with vehicle-treated controls, AChE-IR in global brain samples increased after chlorpyrifos at a dose of 30 mg/kg after 5 weeks of dosing (data not shown). At 3 and 5 weeks, regression analyses showed a significant relationship between dose of chlorpyrifos and level of AChE-IR in the global brain samples ($p < 0.01$). On the other hand, no significant changes in AChE-IR were seen elsewhere in the brain (forebrain, hippocampus, cerebellum) or in diaphragm or liver (data not shown).

Daily, Oral Dosing Regimen

This dosing regimen not only resulted in a much higher total dose of chlorpyrifos but also was more toxic, as 60% of the animals died. The AChE activity fell sharply throughout the brain after oral chlorpyrifos for 4 weeks. Final activities ranged between 9% to 27% of control depending on the brain area analyzed (Fig. 2). Again, there were no decreases in AChE-IR. Instead, as with the weekly, sc dosing, this dosing regimen increased brain AChE-IR, but more markedly. Thus, AChE-IR increased up to 56% above control levels in global brain samples, and there were significant increases in the other brain regions examined as well (Fig. 2).

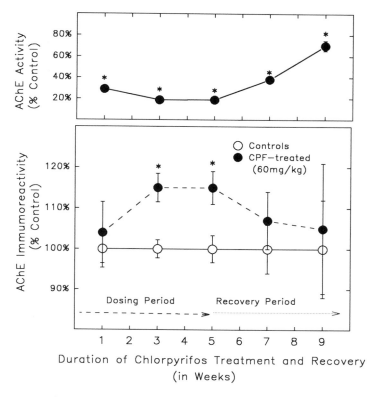

Figure 1. Time course of AChE activity (upper panel) and immunoreactivity (lower panel) in the global brain samples of animals treated weekly with 60 mg/kg chlorpyrifos (sc) for 5 doses and then allowed to recover for 4 more weeks. Samples were obtained one week after injection during the dosing phase. Mean values (±sem) of AChE inhibition are represented as percentages of the concurrently assayed control means (4.9 umol/min/g). Immunoreactivity is expressed as percentages (±sem) of mean control value (average across all experiments, 11.7 ± 2.1 mOD/min/mg wet weight). The number of rats per group ranged from 6 to 18. *p≤0.001.

AChE mRNA Northern blotting with a ^{32}P-labeled cDNA probe for murine AChE served to quantitate AChE message levels in contralateral samples for comparison with enzyme activity and immunoreactivity. In global brain extracts from rats given weekly injections of chlorpyrifos at a dosage of 60 mg/kg, s.c., there were no statistically significant changes in AChE message levels, either during dosing or during recovery (data not shown). Message levels did appear to increase by 87% in the brains of animals dosed using the daily oral dosing regimen, but because of the need to combine samples for adequate sensitivity, this determination was based on a single pool of 4 brains. Overall, therefore, the data fail to prove that AChE mRNA increased in response to treatment with chlorpyrifos, though they clearly show that message levels did not decrease.

DISCUSSION

It is apparent from the above results that the unusually prolonged depression of AChE activity seen after dosing with the commonly used insecticide, chlorpyrifos, is not due to inhibition of resynthesis of AChE. In fact, chlorpyrifos (or its metabolites) appears to

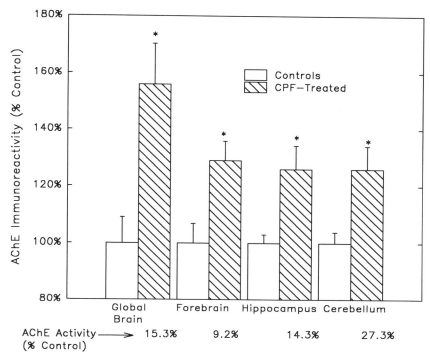

Figure 2. AChE immunoreactivity and AChE activity after daily, oral chlorpyrifos dosing (4 rats per group). Tissues were harvested at 4 weeks, approximately 24 hours after the last dose. Results are given as means ± sem. AChE-IR is expressed as percentages of control. *p≤0.05. The amount of AChE residual activity (as a percentage of control mean) is given below each tissue.

enhance the levels of AChE protein in the brain in a time- and dose-dependent manner. We cannot say at present if this might be a general response to cholinesterase inhibitors or if it is compound-specific.

What could have caused the increased AChE levels in the chlorpyrifos-treated rats? It is possible that a prolonged or severe depression of AChE activity precipitates an up-regulation in the synthesis of AChE. The idea that organophosphate treatment might up-regulate the production of AChE is not altogether new. Some earlier observations on the recovery of enzyme activity in cultured muscle cells suggested that AChE synthesis is transiently accelerated after enzyme inhibition (Wilson & Nieberg, 1983; Cisson & Wilson, 1977). Although we could not demonstrate significant elevation of AChE mRNA after chlorpyrifos administration, the data do not rule out an increase in the message level, or a postranslational effect leading to greater production of brain AChE. Another possible mechanism for AChE accumulation would be a reduced rate of degradation of AChE due to inhibition of protease activity in the brain. Many organophosphate compounds inhibit serine proteases (O'Brien, 1963; Cohen & Oosterbaan, 1963; Casale et al., 1989) and therefore could prolong the life span of intracellular and extracellular proteins. Moreover, there is new evidence that occupation of the hydrolytic site of neural AChE by an inhibitor may actually stabilize the enzyme in vivo (Harvey Alan Berman, personal communication). Experiments on AChE turnover in brain would help discriminate between these possibilities.

It is interesting to note that no matter how long or how intensively cholinesterase-inhibiting compounds are administered in vivo, there is always some residual AChE activity in brain tissue. In other words, it is virtually impossible to inhibit all the brain AChE activity in a rat which survives the dosing (Lancaster, 1972; Miyamoto, 1969; Wecker et al., 1977).

Why AChE activity should persist at low levels in the brain is unknown at present. We can, however, rule out two explanations for this residual activity: (1) that the residual activity was associated with a physically protected compartment which never "saw" the inhibitor or (2) that there was a subset of AChE activity which was kinetically refractory to inhibition (Milatovic & Dettbarn, 1994). The relevant findings are described elsewhere (Chiappa et al., 1995). First, we observed that chorpyrifos oxon (the active metabolite of chlorpyrifos) was not excluded from any compartment of brain slices in vitro. Second, the "residual" AChE activity present in the tissue from highly dosed animals has the same IC_{50} to chlorpyrifos oxon as does AChE from brains of control animals.

Taking into consideration the presence of residual AChE activity and the increased amounts of AChE-IR, we speculate that these two facts may be related. That is, that the residual AChE activity may represent newly synthesized enzyme which has not yet encountered the organophosphate. We have not proved that AChE synthesis is accelerated after exposure to chlorpyrifos. Nevertheless, it remains possible that chlorpyrifos, and related organophosphorus compounds, could promote residual AChE activity by enhancing enzyme biosynthesis. In any case, because chlorpyrifos treatment does lead to increases in the amount of immunoreactive AChE protein, explanations for the slow recovery of enzyme activity after dosing with chlorpyrifos should focus on the propensity of this agent for depot storage and gradual bioactivation (Chambers & Carr, 1993).

In conclusion, we suggest that the present study represents one example of a productive approach toward elucidating the multifaceted toxicity of anticholinesterases. By combining the methodologies of enzyme assay, immunoassay, and specific mRNA determination, it becomes possible to identify factors that define rates of enzyme inhibition and recovery after exposure to cholinesterase-inhibiting compounds. In addition, this approach has the potential to reveal direct adverse effects of specific compounds on cholinergic neurons in the brain, spinal cord, and peripheral nervous system. Finally, it appears that toxicological experiments of this sort may provide new insights into the mechanisms that normally regulate cholinesterases in vivo.

ACKNOWLEDGMENTS

The authors would like to thank Phillip Bushnell for supplying the rats used in the first study; Kristin L. Kelly, Thomas H. Delay, LaShawn Poinsette and Sue Willig for dosing many of the animals; and Valerie Z. Wilson for performing some of the dissections. We also wish to thank Kaye Riggsbee, James Allen and Mike McFarland for weighing and weight-maintenance of the rats. This work was supported in part by NIH Grant NS 18170 from the National Institute of Neurological Diseases and Stroke. The research described in this article has been reviewed by the Health Effects Research Laboratory, U.S. Environmental Protection Agency, and approved for publication. Approval does not signify that the contents necessarily reflect the views and policies of the Agency nor does mention of trade names and commercial products constitute endorsement or recommendation for use.

REFERENCES

Aspelin, A.L. (1994) Office of Pesticide Programs, U.S. Environmental Protection Agency, Washington, DC
Brimijoin, S., Hammond, P., & Rakonczay, Z. (1987) J. Neurochem. 49, 555-562
Bushnell, P.J., Pope, C.N. & Padilla, S. (1993) J. Pharmacol. Exp. Ther. 266, 1007-1017

Casale, G.P., Bavari, S. & Connolly, J.J. (1989) Fundam. Appl. Toxicol. **12**, 460-468

Chambers, J.E. & Carr, R.L. (1993) Fundam. Appl. Toxicol. **21**, 111-119

Chiappa, S., Padilla, S., Koenigsberger, C., Moser, V.C. & Brimijoin, S. (1995). Biochem. Pharmacol., In press.

Cisson, C.M. & Wilson, B.W. (1977) Biochem. Pharmacol. **26**, 1955-1960

Cohen, J.A. & Oosterbaan, R.A. (1963) in Cholinesterases and Anticholinesterase Agents G. B. Koelle, ed. (Springer, Berlin,), vol. XV, pp. 299-373

Ellman, G.L., Courtney, K.D., Andres, V.J. & Featherstone, R.M. (1961) Biochem. Pharmacol. **7**, 88-95

Griesmann, G.E., McCormick, D.J. & Lennon, V.A. (1991) J. Immunol. Methods **138**, 25-29

Hooser, S.B., Beasley, V.R., Sundberg, J.P. & Harlin, K. (1988) Am. J. Vet. Res. **49**, 1371-1375

Lancaster, R. (1972) J. Neurochem. **19**, 2587-2597

Milatovic, D. & Dettbarn, W.-D. (1994) The Toxicologist **14**, 243

Miyamoto, J. (1969) Residue Reviews **25**, 251-264

O'Brien, R.D. (1963) in Metabolic Inhibitors R.M. Hochster, J.H. Quastel, eds. (Academic Press, New York,), vol. 2, pp. 205-241

Padilla, S., Wilson, V.Z., & Bushnell, P.J. (1994) Toxicology, **92**, 11-25

Pope, C.N., Chakraborti, T.K., Chapman, M.L., Farrar, J.D. & Arthun, D. (1991) Toxicol. **68**, 51-61

Wecker, L., Mobley, P.L., & Dettbarn, W-D. (1977) Biochem. Pharmacol. **26**, 633-637

Wilson, B.W. & Nieberg, P.S. (1983) Biochem. Pharmacol. **32**, 911-918

BISQUATERNARY OXIMES AS ANTIDOTES AGAINST TABUN AND SOMAN POISONING

Antidotal Efficacy in Relation to Cholinesterase Reactivation

G. Amitai,[1,*] I. Rabinovitz,[1] G. Zomber,[1] R. Chen,[2] G. Cohen,[1] R. Adani,[1]
and L. Raveh[1]

[1] Division of Chemistry
[2] Division of Environmental Sciences
IIBR
P.O. Box 19
Ness Ziona 74100, Israel

INTRODUCTION

Quaternary pyridinium aldoximes such as 2-PAM were first introduced by Ginsburg and Wilson (1955) as nucleophilic reactivators of organophosphoryl (OP)-inhibited AChE. During the last forty years there were several attempts to improve the reactivation and antidotal efficacy of oximes against OP poisoning, e.g., by introducing a second pyridinium oxime group as in Toxogonin (TOX) or TMB-4. These oximes displayed remarkable efficacy against poisoning by certain OP nerve agents such as sarin and tabun (Heilborn and Tolagen, 1963) . However, rapid aging of soman-inhibited AChE caused considerable difficulties in the antidotal treatment of soman poisoned animals (Coult and Marsh, 1960). The Hagedorn oximes HI-6 and HLo-7 which were introduced in the seventies were the first oximes which also displayed antidotal efficacy against soman in mice and guinea pigs (Loeffer, 1976 and Schoene et al., 1973). Due to its relatively low toxicity and remarkable reactivation potency HI-6 was extensively studied and was proposed to replace the first generation oximes in the military antidotal kit. However, some contradictory data were obtained for the antidotal efficacy of HI-6 against tabun poisoning. Boskovic et al.(1984) have demonstrated limited antidotal action of HI-6 (combined with atropine and diazepam) against tabun in mice, rats and dogs. Schoene et al.(1973) and Clement (1983, 1987) showed that HI-6 is practically devoid of any antidotal efficacy against tabun in mice. In contrast, Hamilton and Lundy (1989) have demonstrated that HI-6 is an effective antidote against tabun when administered in conjunction with atropine and diazepam in rhesus monkeys. It was further noted that bispyridinium salts such as SAD-128 and HH-54 which do not have an oxime moiety in their

---

[*] Address correspondence to Dr. Amitai.

Enzymes of the Cholinesterase Family, Edited by Daniel M. Quinn et al.
Plenum Press, New York, 1995

molecular structure could protect against soman poisoning in mice (1976). Moreover, it was demonstrated that restoration of diaphragm activity by HI-6 in soman poisoned rats could not be attributed to AChE reactivation (Wolthuis et al., 1978). Therefore, alternative mechanisms which are unrelated to AChE reactivation were proposed for the antidotal action of oximes. Indeed, moderate antimuscarinic (Amitai et al., 1980) and antinicotinic activity (Kuhnen-Clausen et al., 1983) was demonstrated for certain bispyridinium oximes and salts. Based on the observed anticholinergic activity of existing oximes we have designed bis-quaternary oximes, designated AB-oximes, in which one pyridinium ring was substituted by a quinuclidinium moiety. Some of these AB-oximes are relatively non-toxic in rodents and displayed high efficacy against certain OP agents in mice and guinea pigs (Amitai et al., 1985 and 1993). The study reported here describes the antidotal efficacy of AB-8, AB-13, TOX, HI-6 and HLo-7 against tabun and soman in beagle dogs and baboon monkeys. Prior to the antidotal evaluation we have determined the therapeutic doses of these oximes by monitoring their dose-dependent toxic signs in dogs and monkeys. Using TOX as a reference oxime, a calculated unit of equivalent dose (CED) was defined so that this unit is equivalent in terms of toxic effect for all oximes. In addition, cholinesterase (ChE) activity was measured both *in vitro* and *in vivo* in order to ascertain the relationship between the antidotal efficacy and ChE reactivation during recovery from OP intoxication.

MATERIALS AND METHODS

Materials

The OP agents and the AB oximes were prepared by Dr. H. Leader and B. Manistersky of the Department of Organic Chemistry, IIBR. Tabun and soman were synthesized according to standard procedures. AB-8 and AB-13 were prepared as described previously (Amitai et al., 1987). HI-6, HLo-7 and toxogonin were kindly provided by Dr. Kullmann, In San I 3 MOD, Bonn, Germany.

Animals

Mice, male and female, of CD-1 strain (Charles River, UK), Dogs, male Beagles, 9-12 months old (Harlan Olac CPB, Netherlands), Monkeys, male and female papio anubis baboons of African origin.

Methods

Determination of Calculated Equivalent Dose (CED) of Oximes in Dogs and Monkeys. CED was defined as a dose unit of an oxime which is equivalent in its toxic effect to TOX. This unit is the ratio between the oxime's MTD (minimal toxic dose) and TR (therapeutic ratio) of TOX. The TR is the ratio between the therapeutic dose and the MTD of TOX. Assuming that for each oxime, $MTD=ED_{10}/k$, where k is a constant equal to all oximes, it can be shown that MTD may be replaced by ED_{10} of the tested oxime (t), i.e.,

$$CEDt = ED_{10,t}*(TDr/ED_{10,r}).$$

Where TDr and $ED_{10,r}$ are the therapeutic dose and ED_{10} of the reference (r) oxime TOX. Since rigorous estimation of MTD requires a huge number of animals the advantage of its replacement by ED_{10} is obvious. The ED_{10} of each oxime was estimated from the experimental data, assuming the log-logistic function for the toxicity dose response curve.

According to these assumptions and definitions, the CED of oximes are equivalent in terms of their toxic effect. Thus, if the ED_{10} of a given oxime equals that of TOX , then in terms of weight , one CED unit of that oxime equals the therapeutic dose of TOX. Similarly, when one CED of an oxime exhibits a better therapeutic effect than that of another oxime it is considered to have a higher efficacy, since the two oximes are given in doses which are equivalent in terms of their toxic effect.

In Vivo Antidotal Efficacy of Oximes in Dogs and Monkeys. Tabun or soman solutions were injected s.c. by the scruff of the neck in dogs and above the gluteal muscle in monkeys. The subcutaneous LD_{50} values for tabun and soman were previously determined in dogs and monkeys using the up and down method (Dixon and Massey, 1969). $5xLD_{50}$ of tabun were administered to dogs (1357 ug/kg) and monkeys (474 ug/kg). $5xLD_{50}$ soman were injected to dogs (38 ug/kg) and $3xLD_{50}$ soman (23.7 ug/kg) were injected to monkeys. Antidotal treatment included atropine (2 mg/kg, in dogs and for soman poisoning in monkeys, 0.5 mg/kg in monkeys for tabun poisoning). Oximes were administered i.m. together with atropine 5 minutes post OP challenge in dogs and 1 min after challenge in monkeys. Each treatment group consists of six animals treated with the same oxime dose. Three doses of oximes were used: 0.3x, 1x, and 3xCED. Two control groups consisted of animals which did not receive any treatment or only atropine treatment (n = 2). Animals were observed during 7 days for survival rate (SR), cumulative mean clinical score (MCS), dynamics of recovery which was based on the change in clinical score during recovery and blood ChE activity.

Blood ChE Activity in Vivo. Blood samples were drawn from the jugular and forelimb vein in dogs and from the femoral vein or artery and forelimb vein in monkeys. ChE activity in whole blood and plasma was measured at 1, 4, 8, 24, 72 hours and 7 days after challenge, using the radiometric method (Johnson and Russel, 1975).

AChE Activity in Vitro. Purified fetal bovine serum (FBS) AChE was kindly provided by B.P. Doctor of WRAIR, Washington DC. Approximately 1-4 units of FBS-AChE were used for the inhibition and reactivation kinetic experiments. AChE activity was determined according to Ellman procedure (1961) in 50 mM Tris pH 7.4, 25°C. Reactivation of somanyl-AChE was measured at pH 9.0 in order to slow down the aging process (Coult and Marsh, 1960).

RESULTS AND DISCUSSION

CED of Oximes in Dogs and Monkeys

The following CED values were obtained for AB-8, AB-13, TOX, HI-6 and HLo-7: 35.5, 26.7, 10.8, 6.3, 5.2 mg/kg in dogs and 101.4, 83.5, 15, 69.4 and 48.4 mg/kg, in monkeys, respectively. All CED values in monkeys were significantly larger than those obtained in dogs for each oxime indicating higher tolerance for cholinergic effects in baboons. It is pertinent to note that the CED values obtained for HI-6 in dogs and baboon monkeys (6.3 and 69.4 mg/kg, respectively) are consistent with HI-6 doses used previously in dogs (10.8 mg/kg, Weger and Szinicz, 1981) and in either rhesus or marmoset monkeys (50 mg/kg, Hamilton and Lundy, 1989, Van Helden et al., 1992). In order to cover a wide range of therapeutic doses, antidotal evaluation of oximes was performed by using three different doses for each oxime: 0.3x,1x and 3xCED.

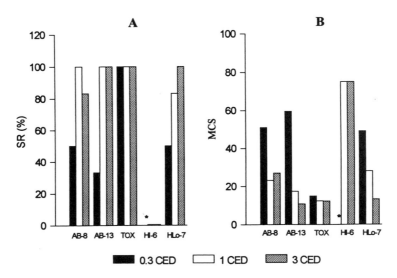

Figure 1. Survival rate, (SR, %) **(A)** and mean clinical score (MCS) **(B)** obtained for antidotal treatment following tabun poisoning in dogs. (* not determined values).

Evaluation of Antidotal Efficacy in Dogs

Tabun Poisoning. The SR and cumulative MCS values which were monitored during 7 days following $5 \times LD_{50}$ of tabun and antidotal treatment are described in figure 1.

Except for HI-6, which did not provide any protection, all other oximes exhibited good antidotal efficacy against tabun in dogs. Antidotal efficacy of AB-13, AB-8 and TOX (1xCED) provided 100% survival against tabun poisoning in dogs but was markedly decreased at 0.3xCED to 33.3, 50 and 50%, respectively, except for TOX (SR = 100%). *In vivo* reactivation of blood ChE in dogs and monkeys during the first 8 hours after tabun challenge (see figure 5) was consistent with the recovery process following tabun poisoning and treatment with oximes and atropine. The lowest MCS values were obtained for TOX at all administered doses (figure 1B). The ranking order for oximes as antidotes against tabun in dogs is: TOX>AB-13>HLo-7=AB-8.

Soman Poisoning. The SR and MCS values obtained for oximes following poisoning by $5 \times LD_{50}$ soman are presented in figure 2. The most efficacious oxime is AB-8 which provides the highest SR (83.3, 100 and 100%) and lowest MCS values (24.3, 16.7 and 8.5) at all three doses. HI-6 and HLo-7 also exhibited good antidotal efficacy at 1x and 3xCED. However, following treatment with 1xCED of either AB-13, HI-6 or HLo-7 a relapse in the animals clinical status occurred after 24 hours and recovery was incomplete even after 7 days (not shown). Blood ChE levels did not increase during the first 8 hours (figure 5) following treatment with all oximes. *In vitro* reactivation of somanyl-AChE exhibited marked potency of HI-6 and HLo-7 for the inhibited non-aged enzyme as compared to low reactivation potency of AB-8 and AB-13 (figure 6). These results indicate other mechanisms unrelated to AChE reactivation for the antidotal efficacy of oximes against soman poisoning.

Evaluation of Antidotal Efficacy in Monkeys

Tabun Poisoning. The SR and MCS values obtained from the treatment of tabun poisoned monkeys are presented in figure 3. The highest survival rate and lowest clinical

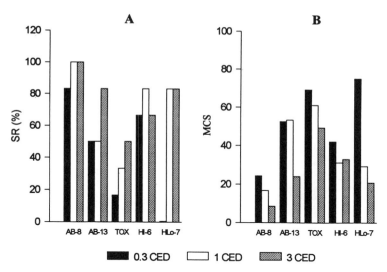

Figure 2. Survival rate, (SR, %) **(A)** and mean clinical score (MCS) **(B)** obtained for antidotal treatment following soman poisoning in dogs.

score were obtained for HLo-7, AB-13 and TOX (SR = 100, 100 and 75%, MCS = 28.5, 30.3 and 39.8 at 1xCED, respectively). HI-6 at 3xCED provided 66.7% survival with relatively low MCS (32.3). These results are consistent with those of Hamilton and Lundy in rhesus monkeys (1989). However. these authors have added diazepam as an anticonvulsant.and animals were anesthetized during treatment. The ranking order of oximes for the treatment of tabun poisoning in monkeys is: AB-13 = HLo-7 > TOX > HI-6 >> AB-8.

Soman Poisoning Figure 4 depicts the SR and MCS values for the antidotal efficacy of oximes against soman poisoning in monkeys.

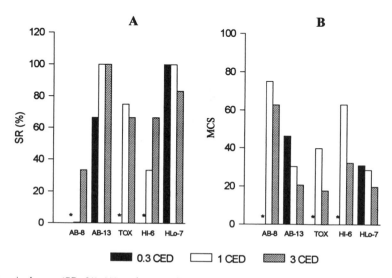

Figure 3. Survival rate, (SR, %) **(A)** and mean clinical score (MCS) **(B)** obtained for antidotal treatment following tabun poisoning in monkeys. (* denotes not determined values).

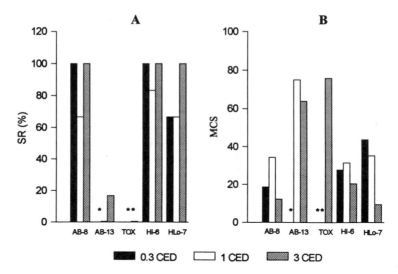

Figure 4. Survival rate, (SR, %) **(A)** and mean clinical score (MCS) **(B)** obtained for antidotal treatment following soman poisoning in monkeys. (* denotes not determined values).

The highest SR values were obtained for AB-8 and HI-6 (100% at 0.3 CED). An unexpected decrease in SR occurred with these oximes administered at 1xCED (66.7 and 83.3%). All animals survived $3xLD_{50}$ of soman poisoning when treated with 3xCED of AB-8, HI-6 and HLo-7 (figure 4A) and displayed only few clinical symptoms within the first 24 hours (MCS < 20, figure 4B). *In vivo* blood ChE activity measured during recovery showed

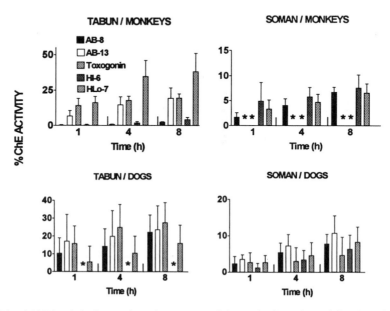

Figure 5. Blood ChE levels in dogs and monkeys measured during the first 8 hours following tabun or soman challenge and antidotal treatment.

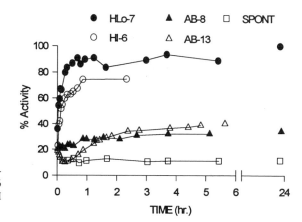

Figure 6. Time-course of *in vitro* reactivation of somanyl-AChE by HLo-7, HI-6 (0.5mM), AB-13 and AB-8 (1mM), in 50 mM Tris pH 9.0 at 25°C.

no significant reactivation within the first 8 hours (figure 5). Recovery of blood ChE occurred only at 24-72 hours post challenge and its rate was independent of the oxime dose (not shown). The late ChE recovery in blood may thus arise from *de novo* synthesis of the enzyme. Therefore, reactivation of non-aged somanyl-AChE by HI-6 and HLo-7 as shown in figure 6 may explain only partially the efficacy of these oximes against soman in monkeys. Additional mechanisms which are unrelated to ChE reactivation should be provided for explaining the outstanding antidotal efficacy of these oximes and particularly of AB-8 against soman poisoning in both dogs and monkeys.

REFERENCES

Amitai, G., Balderman, D., Bruckstein-Davidovici, R. and Spiegelstein, M., (1987), Novel antidotes for organophosphorus poisoning, US Patent 4,675,326.

Amitai, G., Bruckstein-Davidovici, R. Balderman, D. and Ashani, Y., (1985), Novel bisquaternary oximes for the treatment of organophosphorus poisoning, Fed. Proc. 44, 896.

Amitai, G., Rabinovitz, I., Chen R., Cohen, G., Zomber, G., Adani, R., Manistersky, B., Leader, H., and Raveh, L. (1993), Antidotal efficacy of biquaternary oximes against organophosphorus poisoning in various species. Proceedings of the 1993 Medical Defense Bioscience Review, Organized by US Army Medical Research and Development Command, pp. 527-533.

Amitai, G., Kloog, Y., Balderman, D. and Sokolovsky, M., The interaction of bis-pyridinium oximes with mouse brain muscarinic receptor, (1980), Biochem. Pharmacol., 29, 83-88.

Boskovic, B., Kovacevic, V. and Jovanovic, V., (1984), PAM-2 Cl, HI-6, and HGG-12 in soman and tabun poisoning. Fund. Appl. Toxicol. 4, S106-S115.

Clement, J.G., (1983), Efficacy of mono- and bis-pyridinium oximes versus soman, sarin and tabun poisoning in mice. Fund. Appl. Toxicol. 3, 533-535.

Clement, J.G., Shiloff, J. and Gennings, C., (1987), Efficacy of a combination of acetylcholinesterase reactivators, HI-6 and obidoxime, against tabun. Arch. Toxicol. 61, 70-75.

Coult, D.B. and Marsh, D.J., (1960), Dealkylation studies on inhibited acetylcholinesterase. Biochem. J. 98, 869-874.

Dixon, W.J. and Massey, F.J., (1969), Introduction to Statistical Analysis, McGraw Hill Book Company, 3rd edition, p. 380.

Ellman, G.L., Courtney, K.D., Andres V., Jr., Featherstone, R.M., (1961), A new and rapid colorimetric determination of acetylcholinesterase activity. Biochem. Pharmacol. 7, 88-95.

Hamilton, M.G., and Lundy, P.M., (1989), HI-6 therapy of soman and tabun poisoning in primates and rodents. Arch. Toxicol., 63, 144-149.

Heilbronn, E. and Tolagen B., (1965), Toxogonin in sarin, soman and tabun poisoning. Biochem. Pharmacol. 14, 73-77.

Johnson, C.D. and Russel, R.L., (1975), A rapid, simple radiometric assay for cholinesterase, suitable for multiple determinations. Anal. Biochem. 64, 229-238.

Kuhnen-Clausen, D., Hagedorn, I., Gross, G., Bayer, H., and Hucho, F., Interactions of bisquaternary pyridine salts (H-oximes) with cholinergic receptors, (1983), Arch. Toxicol. 54, 171-179.

Loefler, M., (1976), Quartare salze von pyridin 2,4-dialdoxim als gegenmittel fur organophosphatvergiftungen, Thesis, Freiburg, Germany.

Schoene, K. and Oldiges, H. (1973), Efficacy of pyridinium salts in tabun and sarin poisoning in vivo and in vitro. Arch. Int. Pharmacodyn., 24, 110-123.

Schoene, K., Steinhause, J., and Oldiges, H., (1976), Protective activity of pyridinium salts against soman poisoning in vivo and in vitro. Biochem. Pharmacol. 25, 1955-1958.

Van Helden, H.P.M,Van der Wiel, H.J., De Lange, J., Busker, R.W., Melchers, B.P.C., and Wolthuis, O.L. (1992), Therapeutic efficacy of HI-6 in soman-poisoned marmoset monkeys, Toxicol. Appl. Pharmacol., 115, 50-56.

Weger, N. and Scinicz, L., (1981), Therapeutic effects of new oximes, benactyzine and atropine in soman poisoning: Part I. Fund. Appl. Toxicol. 1, 161-163.

Wilson, I.B. and Ginsburg, S., (1955), A powerful reactivator of alkyl phosphate-inhibited acetylcholinesterase. Biochim. Biophys. Acta 18, 168-172.

Wolthuis, G.L. and Kepner, L.A., (1978), Successful oxime therapy one hour after soman intoxication in the rat, Eur. J. Pharmacol. 49, 415-425.

COMPARISON OF ACETYLCHOLINESTERASE, PYRIDOSTIGMINE, AND HI-6 AS ANTIDOTES AGAINST ORGANOPHOSPHORUS COMPOUNDS

Donald M. Maxwell,[1] Karen M. Brecht,[1] Ashima Saxena,[2] Palmer Taylor,[3] and Bhupendra P. Doctor[2]

[1] U.S. Army Medical Research Institute of Chemical Defense
Aberdeen Proving Ground, Maryland 21010-5425
[2] Walter Reed Army Institute of Research
Washington, D.C. 20307-5100
[3] University of California
San Diego, La Jolla, California 92093

INTRODUCTION

Conventional medical treatment against the toxicity of organophosphorus (OP) compounds consists of a regimen of anticholinergic drugs to counteract the accumulation of acetylcholine and oximes to reactivate OP-inhibited acetylcholinesterase (AChE) (Taylor, 1985). Reactivation of OP- inhibited AChE by oximes can generate enough active AChE in the peripheral nervous system, especially in the diaphragm, to restore normal cholinergic neurotransmission after exposure to many OP compounds. However, some OP compounds, such as soman (pinacolylmethylphosphonofluoridate), inhibit AChE and rapidly "age" into a form that cannot be reactivated by oximes (De Jong and Wolring, 1984), thereby reducing the ability of oximes to provide protection (Maxwell and Brecht, 1991). The inability of oximes to provide adequate protection against the toxicity of rapidly aging OP compounds stimulated the development of carbamate pretreatment in which carbamylation of AChE effectively protects it against inhibition by OP compounds (Leadbeater *et al.,* 1985). Spontaneous decarbamylation of AChE after the OP compound has been detoxified then generates enough active AChE to allow normal cholinergic neurotransmission. Behavioral side effects from carbamate pretreatment in the absence of exposure to OP compounds have been avoided by the use of cationic pretreatment carbamates, such as pyridostigmine, which do not enter the central nervous system (Maxwell *et al.,* 1988).

Enzymes of the Cholinesterase Family, Edited by Daniel M. Quinn et al.
Plenum Press, New York, 1995

Although both oxime and carbamate antidotal regimens are effective in preventing lethality from OP compounds, their restriction to the periphery because of their cationic charges results in postexposure incapacitation and toxic effects in experimental animals after exposure to OP compounds (Dunn and Sidell, 1989). The failure of these pharmacological approaches to provide complete protection against OP compounds led to the development of enzyme scavengers to remove OP compounds from circulation before they can inhibit neural AChE (Wolfe *et al.*, 1987). The enzyme scavenger approach was recently reviewed (Maxwell and Doctor, 1992) and found to be effective against a variety of OP compounds in rodents as well as nonhuman primates. However, several aspects of the feasibility of enzyme scavenger protection appeared to require further examination. Specifically, the number of subjects in nonhuman primate studies was small due to the extensive training that is required for sophisticated tests of behavioral incapacitation, maximal protection by enzyme scavengers in many cases was inferred from *in vitro* experiments instead of the occurrence of *in vivo* lethality, no systematic direct comparison of enzyme scavenger protection to carbamate or oxime protection had been performed, and the 1:1 stoichiometry of detoxification of OP compounds by enzyme scavengers needed improvement.

In this paper we compare oxime, carbamate and enzyme scavenger protection by using the antidotes that have been generally reported to be the most efficacious for each type of protection. Accordingly, HI-6 [1-(2-(hydroxyimino)methyl)) pyridinium-2-(4-(aminocarbonyl)pyridinium) dimethylether], pyridostigmine and AChE were chosen as the most efficacious examples of oxime, carbamate and enzyme scavenger antidotes, respectively. Antidotes were compared in mice with regard to maximal protection, relative effective dose, incidence of postexposure incapacitation and biological half-life. We also provide evidence that the combination of wild type or mutant AChE with oximes may improve the stoichiometry of AChE detoxification of OP compounds.

METHODS

Animals

Male ICR mice weighing 22 to 26 g, were obtained from Charles River Laboratories (Wilmington, MA). Mice were allowed free access to food and water before and after drug administration and were maintained on a 12-hr light/dark full spectrum lighting cycle with no twilight.

Materials

Soman and sarin (isopropylmethylphosphonofluoridate) were obtained from the Chemical Research, Development and Engineering Center (Aberdeen Proving Ground, MD) and were >98.6% pure. HI-6 was obtained from the Defense Research Establishment (Suffield, Canada). Pyridostigmine hydrobromide was purchased from Hoffmann-La Roche (Nutley, NJ). 2-(O-cresyl-4H-1:3:2-benzodioxaphosphorin-2-oxide (CBDP) was synthesized by Starks Associates (Buffalo, NY) and was 99.5% pure. Atropine sulfate and acetylthiocholine were purchased from Sigma Chemical Co. (St. Louis, MO). The chemical purities of soman, sarin and CBDP were determined by ^{31}P-nuclear magnetic resonance spectroscopy. All other drugs or chemicals were USP or reagent grade, respectively. AChE was purified from fetal bovine serum (FBS) by the procedure of De La Hoz *et al.* (1986) to electrophoretic homogeneity (>98% purity). Wild type *Torpedo* AChE and its $E_{199}Q$ mutant were expressed and purified by affinity chromatography by the procedure of Radic *et al.* (1992).

Animal Experiments

To generate results that were more comparable across species, mice were pretreated with CBDP, an inhibitor of carboxylesterase (CaE), a soman-detoxifying enzyme that is an important determinant of species variation in carbamate and oxime protection (Maxwell *et al.*, 1988; Maxwell and Brecht, 1991). Soman, HI-6 dichloride, pyridostigmine hydrobromide, atropine sulfate and FBS-AChE were administered as solutions in isotonic saline. CBDP was administered as a solution in propylene glycol containing 5% ethanol. HI-6, pyridostigmine and atropine were administered i.m., whereas soman and CBDP were administered s.c. FBS-AChE was injected i.v. in the tail vein. Injection volumes were < 5 ml/kg and animals receiving multiple injections were injected in alternate hind limbs or in the tail vein.

AChE Determinations

AChE activity was determined by the method of Ellman *et al.* (1961) by using acetylthiocholine as substrate.

Biological Half-life Determination

The pharmacokinetics of FBS-AChE, HI-6 and pyridostigmine in mice were examined by obtaining heparinized blood samples (50 µl) from the retro-orbital sinus. Biological half-lives were estimated from linear regression of a single-compartment model of elimination. The biological half-life of pyridostigmine was determined from the time course of its inhibition of endogenous AChE activity in blood after pyridostigmine administration. Blood levels of HI-6 were determined by the spectrophometric method of Kepner and Wolthuis (1978).

Toxicity Studies

All LD_{50} values for soman were determined in CaE-inhibited mice that had received 2 mg/kg of CBDP. This dosage of CBDP has been reported previously to inhibit plasma CaE without inhibiting brain, diaphragm or blood AChE (Maxwell *et al.*, 1987). The protective effect of either HI-6 in combination with 16.1 µmol/kg of atropine administered i.m. 30 sec after soman, pyridostigmine in combination with 16.1 µmol/kg of atropine administered i.m. 30 min before soman or FBS-AChE administered i.v. 60 min before soman was estimated by its alteration of the soman LD_{50} in mice. The time and route of antidote administration and atropine dosage for these antidote regimens were adopted from previous reports that maximized protection against soman by HI-6-atropine therapy (Maxwell and Brecht, 1991), pyridostigmine-atropine pretreatment (Maxwell *et al.*, 1988) and FBS-AChE pretreatment (Ashani *et al.*, 1991). LD_{50} values were calculated by probit analysis of deaths occurring within 24 hr after administration of soman at five different doses with six mice per dose (Finney, 1971).

Behavior and Motor Function

The occurrence of incapacitation after soman exposure was examined by using a modification of the behavioral scoring system of McDonough *et al.* (1989) and the inverted screen test of Koplovitz *et al.* (1989). The scores for incapacitation were assigned by using the following scale: lacrimation-normal (0), tears (1), bloody tears (2); activity-normal (0), hypoactive but responsive (1), prostrate or unresponsive (2); motor-normal (0), tremors (1), whole body convulsions (2); inverted screen test-climbed to top (0), failed to reach top (1), and fell from screen (2).

AChE Inhibition and Detoxification

AChE inhibition with soman or sarin was performed at 25°C in 50 mM sodium phosphate buffer (pH 8.0) containing 0.01% BSA; For titration studies AChE (0.14 nmol) was incubated with serial dilutions of OP compounds for 15 min and then assayed for AChE activity. Bimolecular rate constants for OP compounds were estimated from their median inhibitory concentrations (IC_{50}) by the method of Aldridge (1950). For detoxification studies AChE (0.14 nmol) was inhibited under the same conditions in the presence of 2 mM HI-6 with repeated doses of OP compounds (3 nmol of sarin or 0.2 nmol of soman) at 30 min intervals until the AChE activity decreased significantly.

Aging Rate Determinations

AChE was inhibited at 25 °C in 50 mM sodium phosphate buffer (pH 8.0) containing 0.01% BSA. At sequential time intervals after inhibition, aliquots of the inhibited AChE reaction mixture were transferred to an equal volume of 5 mM HI-6 in 50 mM sodium phosphate buffer (pH 8.0) for overnight reactivation at 25°C. Aging rate constants were calculated by the method of Hovanec and Lieske (1972).

RESULTS

The dose-response of the protection of HI-6 in combination with atropine, pyridostigmine in combination with atropine, and FBS-AChE without atropine against the lethal effects of soman are shown in Figure 1. The protective effect of each antidote regimen increased with the dose of protective compound until maximal protection was achieved. The maximal protective effects that were achieved for HI-6-atropine, pyridostigmine-atropine and FBS-AChE were virtually identical (Table 1). The maximal soman LD_{50} in antidote-protected mice varied in a narrow range from 952 nmol/kg for FBS-AChE to 1169 nmol/kg for pyridostigmine. Inasmuch as the soman LD_{50} in unprotected CaE-inhibited mice was 121 nmol/kg (See Table 1), all three antidotes protected against 8 to 10 LD_{50} of soman. However, the doses of each antidote to produce the same degree of protection were quite different. In

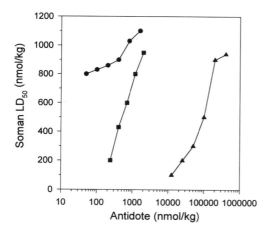

Figure 1. Dose-response of antidotes against Soman LD_{50}. Pyridostigmine (●) with 16.1 μmol/kg atropine was administered i.m. 30 min before soman, AChE (■) was administered i.v. 60 min before soman, and HI-6 (▲) with 16.1 μmol/kg atropine was administered i.m. 30 sec after soman.

Table 1. Antidote protection against soman

Antidote	Dose of Antidote to Protect against 968 nmol/kg (8 x LD$_{50}$)[a] of Soman (nmol/kg)	Maximal Protective Effect against Soman[b] (nmol/kg)
Pyridostigmine[c]	566	1169 (959-1395)
FBS-AChE	1150	952 (780-1159)
HI-6[c]	200,000	968 (755-1201)

[a]LD$_{50}$ of Soman unprotected CaE-inhibited mice was 121 nmol/kg.
[b]LD$_{50}$ of Soman with 95% confidence limits in parentheses.
[c]Atropine (16.1 umol/kg) was administered in conjunction with these compounds.

Table 2. Duration of antidote in blood

Antidote	Dose of antidote (nmol/kg)	Biological half-life (min)
HI-6	200,000	11 ± 3[a]
Pyridostigmine	750	48 ± 12[b]
FBS-AChE	1000	1550 ± 320[a]

[a]Elimination half-life determined from pharmacokinetics.
[b]Pharmacodynamic half-life determined from AChE inhibition.

comparison to pyridostigmine, twice as much FBS-AChE or 400 times as much HI-6 was required to produce protection against the lethal effects of 968 nmol/kg (8 x LD$_{50}$) of soman in CaE-inhibited mice.

The biological half-life for each of the antidotes is shown in Table 2. The half-life for HI-6 was only 11 min, whereas the half-lives for pyridostigmine and FBS-AChE were 4-fold and 140-fold greater, respectively, which may contribute to their superior *in vivo* potency.

The ability of each antidote to prevent postexposure incapacitation from soman is shown in Figure 2. Although four manifestations of soman incapacitation were examined, antidote regimens produced similar incapacitation scores for all types of incapacitation. HI-6-atropine-treated and pyridostigmine-atropine-treated mice exhibited mean incapacitation scores from 1.3 to 1.7, whereas AChE-treated groups exhibited incapacitation scores from 0.3 to 0.5. Therefore, FBS-AChE demonstrated an ability to protect against postexpo-

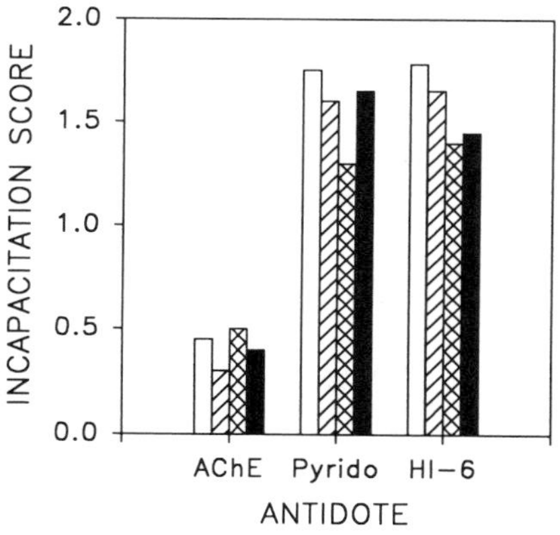

Figure 2. Effect of antidotes on incapacitation 24 hr after 8 x LD$_{50}$ of soman. Antidotes were administered as described in Fig. 1. Incapacitation is described for lacrimation (□), motor function (///), activity level (⊠), and inverted screen test (■).

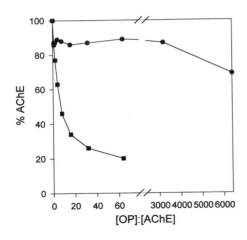

Figure 3. *In vitro* detoxification of OP compounds by AChE in the presence of 2 mM HI-6. The OP: AChE ratio describes the cumulative molar amount of sarin (●) or soman (■) detoxified per mole AChE.

sure incapacitation from soman that was far superior to either HI-6 or pyridostigmine antidotal regimens, regardless of the type of incapacitation that was examined.

The *in vitro* detoxification capability of FBS-AChE in the presence of oximes is shown in Figure 3. In comparison to a 1:1 stoichiometric removal of sarin in the absence of oxime, the detoxification of 3,200 sarin molecules per AChE active site could be achieved in the presence of 2 mM HI-6. The detoxification of soman in the presence of 2 mM HI-6 was much less than sarin, presumably because of its rapid aging to a nonreactivable OP-inhibited enzyme. The difference in aging rates between sarin and soman are quantitated in Table 3. At pH 7.5 soman-inhibited AChE aged at a rate 150 times faster than sarin-inhibited AChE. However, this problem may be addressed by site-directed mutagenesis of the active site of AChE. After inhibition by soman the $E_{199}Q$ mutant of AChE, which substitutes glutamine in place of the glutamate 199 adjacent to the active site serine, ages at a rate 600 times slower than wild type AChE (See Table 3). Since maintaining good reactivity with OP compounds is critical for their detoxification, the ability of this mutant AChE to react with soman in comparison to wild type AChE was evaluated in titration studies (data not shown) from which bimolecular inhibition rate constants were estimated. Wild type and mutant *Torpedo* AChE had bimolecular rate constants for inhibition by soman of 1.4×10^7 M^{-1} min^{-1} and 1.2×10^6 M^{-1} min^{-1}, respectively. Although the mutant enzyme aged at a rate 600 times slower than wild type, its inhibition rate constant was only about 12 times slower, which suggests that it retains a high reactivity for organophosphorus compounds.

DISCUSSION

Pyridostigmine, HI-6 and FBS-AChE all exhibited excellent protection against the lethal effects of soman, demonstrating that a variety of mechanisms (*i.e.,* AChE carbamylation, AChE reactivation and soman sequestration) can produce nearly equivalent levels of

Table 3. Aging rate constant for wild type and mutant AChE

Organophosphorus compound	Wild type (min^{-1})	$E_{199}Q$ mutant (min^{-1})
Soman	0.18 ± 0.03	0.0003 ± 0.0001
Sarin	0.0012 ± 0.0002	ND[a]

[a]Aging not detectable over a 24-hr period.

protection. Although oxime, carbamate and enzyme protection exhibited similar levels of maximal protection against the lethality of soman, FBS-AChE exhibited a superiority over HI-6 or pyridostigmine in other aspects of protection. For example, the stoichiometry between the dose of protective compound and the level of soman protection varied dramatically between compounds. The superior potency of FBS-AChE and pyridostigmine resulted from their longer half-lives (Table 2) and their mechanisms of protection. For each nanomole per kilogram of AChE that was administered, the LD_{50} of soman was increased about 0.82 nmol/kg (See Table 1), which is consistent with its mechanism of protection as a stoichiometric scavenger, inasmuch as the *in vitro* stoichiometry for the reaction between the toxic stereoisomers of soman and FBS-AChE is 1:1 (Ashani *et al.* 1991). Carbamates are more efficient than scavengers in providing protection, because it is only necessary for carbamates to carbamylate a small amount of synaptic AChE (<5 nmol/kg of AChE is present in rodents) to provide protection, whereas scavengers must sequester virtually all of a lethal dose of an organophosphorus compound (*i.e.,* 121 nmol/kg per LD_{50} for soman in mice) to provide protection. Thus, each nanomole per kilogram of pyridostigmine increased the LD_{50} of soman 1.7 nmol/kg, which was greater than twice the protective efficiency of FBS-AChE. In contrast, a large excess of HI-6 (*i.e.,* 200 nmol/kg of HI-6 per nmol/kg of soman) was required to ensure rapid enough reactivation of soman-inhibited AChE to provide protection, because the rate of AChE reactivation is directly dependent on the concentration of oxime.

The most compelling advantage of FBS-AChE over either pyridostigmine or HI-6 antidotal regimens was the difference in the levels of incapacitation observed in the survivors of soman exposure that were produced by each protective regimen. The survivors of soman exposure that were treated with FBS-AChE had much less incapacitation as shown by lacrimation, motor dysfunction, activity level and the inverted screen test than did the survivors treated with HI-6-atropine or pyridostigmine-atropine. Protection against postsoman incapacitation comparable to that observed with FBS-AChE has only been reported with HI-6 and pyridostigmine when an anticonvulsant was added to the protective regimen (McDonough *et al.*, 1989).

Although protection against soman by pyridostigmine (Maxwell *et al.*, 1988), HI-6 (Maxwell and Brecht, 1991) and FBS-AChE (Ashani *et al.*, 1991) have been demonstrated in normal mice, we compared these antidotes in CaE-inhibited mice in order to facilitate the extrapolation of our observations to other species, particularly primates. The soman LD_{50} in CaE-inhibited mice is very similar to the soman LD_{50} in nonhuman primates (Dirnhuber *et al.*, 1979), species devoid of endogenous plasma CaE. In addition, the level of maximal protection of HI-6-atropine and pyridostigmine-atropine regimens against soman in CaE-inhibited mice (Table 1) is similar to the protection achieved by these compounds in primates (Dirnhuber *et al.*, 1979; Hamilton and Lundy, 1989). Therefore, the maximal protection achieved by AChE in this study may provide an estimate of the maximal level of protection against soman that may be achieved in primates. In fact, recent studies of FBS-AChE protection against soman in rhesus monkeys suggest that protection against multiple LD_{50} doses of soman may be achieved with no postexposure incapacitation (Doctor *et al.*, 1993).

The major disadvantage of enzyme scavengers for OP detoxification is their large size and 1:1 stoichiometry for removal of OP compounds. These disadvantages may be overcome by combining AChE with oxime treatment. The ability of AChE to remove sarin, for example, increased 3,200-fold in the presence of HI-6. Oxime treatment does not significantly increase AChE's ability to detoxify OP compounds, such as soman, that inhibit AChE and age to a form that is no longer reactivated by oximes. However, site-directed mutagenesis was selectively used to modify AChE so that the inhibition and aging rate constants for the reaction of soman with mutant AChE resembled the corresponding rate constants for the reaction of sarin with wild type AChE. Therefore, the combination of tailor-made mutant enzymes with oxime-induced OP turnover suggests a promising future for detoxification of OP compounds by enzyme scavengers.

REFERENCES

Aldridge, W.M. (1950) *Biochem. J.* 46, 451-460.
Ashani, Y., Shapira, S., Levy, D., Wolfe, A.D., Doctor, B.P., & Raveh, L. (1991) *Biochem. Pharmacol.* 41, 37-41.
De Jong, L.P.A., & Wolring, G.Z. (1984) *Biochem. Pharmacol.* 33, 1119-1125.
De La Hoz, D., Doctor, B.P., Ralston, J.S., Rush, R.S., & Wolfe, A.D. (1986) *Life Sci.* 39, 195-199.
Dirnhuber, P., French, M.C., Green, D.M., Leadbeater, L., & Stratton, J.A. (1979) *J. Pharm. Pharmacol.* 31, 295-299.
Doctor, B.P., Blick, D.W., Caranto, G., Castro, C.A., Gentry, M.K., Larrison, R., Maxwell, D.M., Murphy, M.R., Schutz, M., Waibel, K., & Wolfe, A.D. (1993) *Chem. Biol. Interactions* 87, 285-293.
Dunn, M.A., & Sidell, F.R. (1989) J. Am. Med. Assoc. 262, 649-652.
Ellman, G.L., Courtney, K.D., Andres, V, Jr., & Featherstone, R.M. (1961) *Biochem. Pharmacol.* 7, 88-95.
Finney, D.J. (1971) *Probit Analysis*, 3rd ed., pp 50-124, Cambridge University Press, Cambridge.
Hamilton, M.G., & Lundy, P.M. (1989) *Arch. Toxicol.* 63, 144-149.
Hovanec, J.W., & Lieske, C.N. (1972) *Biochemistry* 11, 1051-1056.
Kepner, L.A., & Wolthuis, O.L. (1978) *Eur. J. Pharmacol.* 48, 377-382.
Koplovitz, I. Romano, J.A., & Stewart, J.R. (1989) *Drug Chem. Toxicol.* 12, 221-235.
Leadbeater, L., Inns, R.H., & Rylands, J.M. (1985) *Fundam. Appl. Toxicol.* 5, S225-S231.
Maxwell, D.M., Brecht, K.M., & O'Neill, B.L. (1987) *Toxicol. Lett.* 39, 35-47.
Maxwell, D.M., Brecht, K.M., Lenz, D.L., & O'Neill, B.L. (1988) *J. Pharmacol. Exp. Ther.* 246, 986-991.
Maxwell, D.M., & Brecht, K.M. (1991) *Neurosci. Biobehav. Rev.* 15, 135-139.
Maxwell, D.M., & Doctor, B.P. (1992) in *Chemical Warfare Agents* (Somani, S.M., Ed.) pp. 195-207, Academic Press, San Diego.
McDonough, J.H., Jaax, N.K., Crowley, R.A., Mays, M.Z., & Modrow, H.E. (1989) *Fundam. Appl. Toxicol.* 13, 256-276.
Radic, Z., Gibney, G., Kawamoto, S., MacPhee-Quigley, K., Bongiorrno, C., & Taylor, P. (1992) *Biochemistry* 31, 9760-9767.
Taylor, P. (1985) in *The Pharmacological Basis of Therapeutics* (Gilman, A.G., Goodman, L.S., Rall, T.W., & Murad, F., Eds.) pp 110-129, Macmillan, New York.
Wolfe, A.D., Rush, R.S., Doctor, B.P., Koplovitz, I. & Jones, D. (1987) *Fundam. Appl. Toxicol.* 9, 266-270.

INHALATION TOXICOKINETICS OF C(±)P(±)-SOMAN AND (±)-SARIN IN THE GUINEA PIG

H. P. Benschop, L. P. A. de Jong, and J. P. Langenberg

TNO Prins Maurits Laboratory
Postbox 45
2280AA Rijswijk
The Netherlands

INTRODUCTION

For several years we have been involved in research on the toxicokinetics of nerve agents, especially C(±)P(±)-soman (Due et al., 1994 and previous papers). We have performed these investigations in order to provide a quantitative basis for the toxicology of nerve agents, which was nonexistent before we started our work. Such a quantitative basis is needed to develop therapy and prophylaxis of nerve agent intoxication on a rational basis (Langenberg et al., 1991). We have used rats, guinea pigs and marmoset primates. This selection of species was made in order to extrapolate results, at a later stage, to humans via physiologically based modelling (Langenberg et al., 1993).

In order to measure the levels of soman isomers in blood and in tissues, two specific problems have to be dealt with. First of all, it should be taken into account that synthetic soman consists of four stereoisomers. These stereoisomers are described as C(+)P(+), C(+)P(-), C(-)P(+), and C(-)P(-), in which C stands for asymmetry in the pinacolyl moiety and P for asymmetry at phosphorus. Sarin is only asymmetric at phosphorus, which means that we can describe these stereoisomers as P(+)- and P(-)-sarin. The P(-)-isomers of nerve agents inhibit AChE with rate constants that are ca. four orders of magnitude higher than those of the P(+)-isomers. Accordingly, the P(-)-isomers are extremely toxic whereas the P(+)-isomers are hardly toxic (Benschop & De Jong, 1988). It follows that the toxicokinetics of P(+)-isomers should be strictly differentiated from those of the P(-)-isomers.

The second problem that has to be dealt with is the high reactivity of nerve agent stereoisomers in blood and tissues, which requires immediate stabilization after sampling. As an internal standard in our analysis, we use C(±)P(+)-soman perdeuterated in the pinacolyl moiety or (-)-sarin perdeuterated in the isopropyl moiety. The isomers and the added internal standard are extracted from the stabilization mixture. The subsequent gas chromatographic analysis involves thermodesorption/cold trap (TCT) injection and two-di-

Enzymes of the Cholinesterase Family, Edited by Daniel M. Quinn et al.
Plenum Press, New York, 1995

mensional chromatography, in which the second, the analytical column is coated with an optically active phase in order to separate the stereoisomers and the internal standard. In this way, nerve agent stereoisomers can be analyzed at minimum detectable concentrations of ca. 1-5 pg.ml^{-1} blood, which is well below the level of toxicological relevance (Due et al., 1993; Benschop & Van Helden, 1993).

EXPERIMENTAL

Animals

Male albino outbred guinea pigs of the Dunkin-Hartley type were purchased from Charles River (Sulzfeld, Germany). The animals were allowed to eat and drink *ad libitum*. Anesthesia was performed with racemic ketamine hydrochloride (40 mg/kg, im). The animals were atropinized by ip administration of atropine sulfate (17.4 mg/kg).

Analysis of Soman and Sarin Stereoisomers in Blood

For work-up of blood samples, the soman isomers were stabilized by addition of acetate buffer containing aluminum sulfate, (±)-neopentyl sarin, and the internal standard D$_{13}$-C(±)P(+)-soman, before extraction over Sep-Pak C$_{18}$ columns (Millipore, Bedford, Ma, USA) as described elsewhere. For (±)-sarin the same procedure was used, with D$_7$-(-)-sarin as the internal standard. The concentrations of soman and sarin stereoisomers in the extracts were determined by two dimensional gas chromatography with TCT injection and selective detection (NP), finally on a chiral stationary phase, i.e., L-Chirasil-Val for C(±)P(±)-soman and Cyclodex B (Chrompack, Middelburg, The Netherlands) for (±)-sarin (Benschop & Van Helden,1993).

Apparatus for Generation of Nerve Agent Vapor and Exposure of Animals

The apparatus is described in detail elsewhere (Benschop and Van Helden, 1993).

Toxicokinetics of C(±)P(±)-soman and (±)-sarin Via Nose-only Exposure

Toxicokinetic experiments were performed by nose-only exposure of anesthetized, atropinized and restrained guinea pigs to a concentration of C(±)P(±)-soman vapor in air of 48 ± 2 and 24 ± 5 mg.m^{-3} for 8 min, yielding 0.8 and 0.4 LCt50, respectively. Furthermore, animals were exposed for 5 h to a C(±)P(±)-soman vapor concentration of 160 ± 16 μg.m^{-3} (0.02 ppm). Toxicokinetic studies of 0.8 and 0.4 LCt50 of (±)-sarin were performed by 8 min nose-only exposure of animals to (±)-sarin vapor concentrations of 38 ± 4 and 19 ± 2 mg.m^{-3}, respectively.

Blood samples were taken just before exposure, during the exposure and up to ca. 1 h after the start of the exposure. A small portion of the blood sample was used for the determination of AChE activity; the larger part was used for gas chromatographic analysis. Throughout the exposures, the respiration of the animals was monitored.

RESULTS AND DISCUSSION

Intravenous Toxicokinetics

First we have investigated the toxicokinetics of C(±)P(±)-soman after intravenous administration of the agent. Doses of C(±)P(±)-soman varying between 6 and 1 LD50 were administered intravenously to artificially ventilated and atropinized animals. Blood samples were taken from a carotid cannula.

Some general conclusions from our studies are (Benschop & De Jong, 1991):

1. The P(+)-isomers of soman and sarin are degraded within a few seconds, mostly by enzymatic hydrolysis, but also by some covalent binding.
2. The toxic P(-)-isomers are relatively stable and circulate at toxicologically relevant concentrations for at least one hour. Degradation is mostly by way of covalent binding.
3. The curves for the blood levels versus time of the C(±)P(-)-isomers of soman are best described with three-exponential functions at high doses and with two-exponential functions at low doses.
4. The toxicokinetics become increasingly nonlinear with decreasing dose. This phenomenon is explained by assuming that an increasing fraction of a decreasing dose is eliminated by binding to highly reactive but nonspecific binding sites.
5. Interspecies variations in the toxicokinetics are considerable. Mostly because rats have a large amount of aspecific binding sites for soman in blood, guinea pigs are better models for primates than rats.

Inhalation Toxicokinetics of C(±)P(±)-soman

In case of chemical warfare, exposure to volatile agents like (±)-sarin and C(±)P(±)-soman is by way of inhalation. The agents are absorbed in the respiratory tract, mostly in the upper part of it (Ainsworth & Shephard, 1961). The exposure may last from a few seconds up to several hours, depending on the situation in the field. Therefore, the concentration curves of the soman isomers in blood will have shapes which are rather different from those after iv bolus administration. This can have important consequences for treatment of intoxications. Within this context, we have now investigated the toxicokinetics of C(±)P(±)-soman and (±)-sarin after nose-only exposure of atropinized guinea pigs. Two exposure scenarios have been adopted in our experiments. The first scenario pertains to acute exposure to just sublethal Ct values of nerve agent with exposure times of 4-8 min. The second scenario involves long term-low level exposure to 0.02 ppm of C(±)P(±)-soman with an exposure time of 5 h.

For nose-only exposure, nerve agent vapor is generated in air, which is then humidified. The concentration of agent in the air is measured gas chromatographically. This airstream is led to exposure modules, which consist of modified Battelle tubes each holding one guinea pig. The airstream enters a 25-ml front chamber of the tube through a wire mesh resistance, with the nose of the animal protruding into the chamber. The air is drawn from the front chamber at a bias flow of 1 l/min via a critical orifice. Respiration of the animal is superimposed on the bias flow. The differential pressure over the resistance is measured and transformed in a data system to give the respiratory minute volume (RMV) and respiratory frequency (RF). Blood samples are drawn from a carotid artery cannula, through a small hole in the Battelle tube.

We started our nose-only toxicokinetics by exposing guinea pigs to 0.8 LCt50 of C(±)P(±)-soman in 8 min. A change in the RMV and RF due to the exposure was not

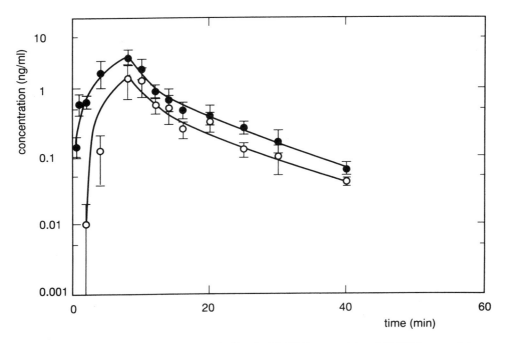

Figure 1. Mean concentrations (± sem, n=6) in blood of C(-)P(-)-soman (•) and C(+)P(-)-soman (o) versus time after nose-only exposure of anesthetized and atropinized guinea pigs to 0.8 LCt50 of soman in the course of 8 min.

observed. Figure 1 shows the measured blood levels of the C(±)P(-)-isomers of soman. The data were fitted discontinuously, with a mono-exponential absorption phase and a two-exponential phase for distribution and elimination.

It is evident that the absorption of C(-)P(-)-soman is very rapid. Already after 30 sec of exposure, this isomer is detected in blood. In contrast herewith, appearance of the C(+)P(-)-isomer has a lag time of several minutes, which is presumably due to preferential covalent binding of this isomer at the absorption site in the respiratory tract rather than to slower absorption. We found very recently that C(+)P(-)-soman binds approximately 36-fold faster to nonspecific binding sites (carboxylesterases) than C(-)P(-)-soman. Preferential binding of the C(+)P(-)-isomer is also indicated by its lower area under the curve (AUC), even though this isomer is present in soman in 22% excess over the C(-)P(-)-isomer. The time lag for penetration and lower bioavailability of the C(+)P(-)-isomer is observed in all experiments with respiratory exposure, as well as after subcutaneous administration (Due et al., 1994). The lower bioavailability of C(+)P(-)-soman is also observed after iv administration of the agent.

The terminal half-lives of the C(±)P(-)-isomers after respiratory exposure to 0.8 LCt50 are 1.5-2 times longer than those after iv bolus administration of a dose corresponding with 0.8 LD50, which suggests that some depot formation at the absorption site may occur, although the blood levels start to decrease immediately after stopping exposure. In accordance with the depot assumption, the terminal half-life is even 5 times longer when the exposure time at the same Ct is reduced twofold (4 min exposure).We also measured the AChE activity in the blood samples. It appeared that the AChE activity decreased almost linearly with time to 88% inhibition at the end of the 8 min exposure period.

In addition to nose-only exposure to 0.8 LCt50 of C(±)P(±)-soman in 8 min, we performed a similar 8 min exposure to 0.4 LCt50 in order to investigate the influence of exposure "dose" on the inhalation toxicokinetics. The curves for the blood levels of the C(±)P(-)-isomers are similar in shape for the two exposure doses. However, a comparison of the areas under the curve [e.g.,C(+)P(-)-isomer:16.0 and 4.3 ng.min.ml $^{-1}$ for 0.8 and 0.4 LCt50 , respectively] shows that the toxicokinetics are clearly nonlinear with dose. This result is in accordance with the earlier mentioned nonlinearity, presumably due to proportionally increased binding at lower dose.

Inhalation Toxicokinetics of (±)-Sarin

We also investigated the inhalation toxicokinetics of sarin in an acute exposure scenario. As a reference for the inhalation experiments, the toxicokinetics of (-)-sarin after iv bolus administration of 0.8 LD50 was first investigated. A comparison with the iv bolus toxicokinetics of C(-)P(-)-soman at the same dose (Figure 2) shows that the distribution of (-)-sarin is almost tenfold faster than that of C(-)P(-)-soman, whereas the elimination phase is much slower for (-)-sarin. The slow elimination is unexpected. We explain it tentatively as being due to a less effective enzymatic hydrolysis and covalent binding of (-)-sarin. Figure 3 shows a comparison of the blood levels of (-)-sarin and C(-)P(-)-soman both for 8 min nose-only exposure to 0.8 LCt50. Evidently, the absorption phase of (-)-sarin is similar to that of C(-)P(-)-soman. A lag time for absorption was not observed. The terminal half-lives of elimination of (-)-sarin are 4 times longer than for the C(±)P(-)-isomers of soman, in accordance with the results after iv bolus administration. However, the terminal half life is somewhat shorter than after iv administration of (±)-sarin which suggests that depot formation at the absorption site may not occur for sarin.

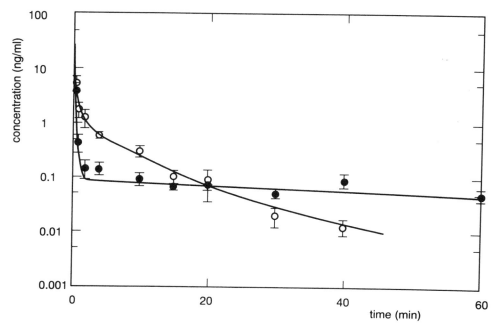

Figure 2. Mean concentrations in blood (± sem, n = 6) of (-)-sarin (•) and of C(-)P(-)-soman (o) versus time after intravenous administration of 0.8 LD50 of sarin or soman to anesthetized and atropinized guinea pigs.

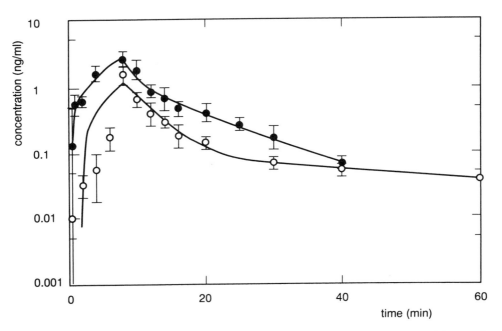

Figure 3. Mean concentrations in blood (± sem, n=6) of (-)-sarin (o) and of C(-)P(-)-soman (•)versus time after nose only exposure of anesthetized and atropinized guinea pigs to 0.8 LCt50 of (±)-sarin and C(±)P(±)-soman, respectively, in the course of 8 min.

Toxicokinetics of Long Term - Low Level Exposure to C(±)P(±)-soman

A second scenario of respiratory exposure pertains to long term-low level exposure. Such a scenario is relevant for unprotected military personnel performing duty in "toxic agent free areas", for example in command and control units and in medical units where CW casualties are treated. We studied the toxicokinetics of C(±)P(±)-soman in guinea pigs exposed nose-only to 0.02 ppm of agent for 5 h (total Ct = 48 mg.min.m^{-3}). The concentration-time courses of the C(±)(P(-)-isomers are shown Figure 4. Much to our surprise, the

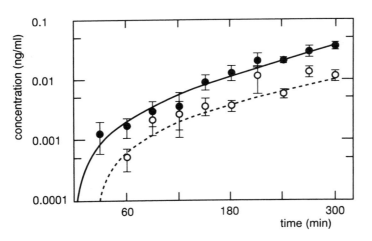

Figure 4. Mean concentrations in blood (± sem, n=6) of C(+)P(-)-soman (o) and C(-)P(-)-soman (•) versus time after nose only exposure of anesthetized and atropinized guinea pigs to 0.02 ppm of soman for 5 h.

C(-)P(-)-isomer could already be detected after 30 min of exposure (i.e., a Ct of 5 mg.min.m^{-3}) and the blood level of the C(-)P(-)-isomer increased very gradually to almost 40 pg.ml^{-1}. It took approximately half an hour longer before the C(+)P(-)-isomer appeared above the detection limits. The increase of both blood levels can be described very adequately with a mono-exponential function, with a time lag of 30 min for the C(+)P(-)-isomer.

We found that AChE activity in blood is gradually inhibited to 11% residual activity after 5 h of exposure. Brain AChE was not inhibited at all at the end of the exposure period, whereas simultaneous central and peripheral inhibition is usually observed upon acute intoxication with C(±)P(±)-soman. The progressive inhibition of AChE activity in blood should correlate with the concentrations of the independently measured blood levels of the C(±)P(-)-isomers. Therefore, we also measured the second order rate constants for inhibition of AChE on erythrocyte membranes in guinea pig blood at 38.5 °C by these two isomers (4.5 x 10^7 M.$^{-1}$min^{-1} and 1.2 x 10^8 M.$^{-1}$min^{-1} for the C(-)P(-)- and C(+)P(-)-isomers, respectively). Equation 1 was used to calculate the progression of AChE inhibition

$$-(d[E]/dt) = k_{C(-)P(-)}*[C(-P(-)-soman]*[E] + k_{C(+)P(-)}*[C(+)P(-)-soman]*[E] \qquad (1)$$

$$[C(-)P(-)-soman] = A*e^{-at} + B \qquad (2)$$

$$[C(+)P(-)-soman] = A'*e^{-a'(t-30)} + B' \text{ (lag time 30 min)} \qquad (3)$$

based on the mono-exponential equations for the blood levels of the isomers (Eqs 2-3).

As shown in Figure 5, the measured progression of inhibition is only very slightly faster than that calculated from the measured blood levels of the C(±)P(-)-isomers and their inhibition rate constants. We conclude that there is almost complete agreement between the two independently measured sets of data. To the best of our knowledge, this is the first example of an in vivo measured progression of AChE inhibition corresponding with the independently measured time course of concentrations of the anticholinesterase.

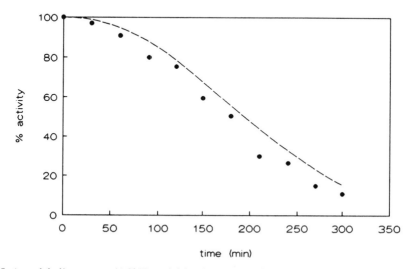

Figure 5. Acetylcholinesterase (AChE) activities (•) measured in blood samples of anesthetized and atropinized guinea pigs nose-only exposed to 0.02 ppm of C(±)P(±)-soman for 300 min. The interrupted line depicts the profile of AChE inhibition calculated from the measured concentrations of C(±)P(-)-soman in blood, based on actually measured rate constants of inhibition of AChE in guinea pig blood.

Further studies should establish the effects of long term-low level exposure. On the basis of our previous investigations, we assume that long term-low level exposure of primates instead of guinea pigs to nerve agents will result in similar blood levels of P(-)-isomers and AChE-inhibition at much lower Ct values.

ACKNOWLEDGMENTS

This research was supported in part by the US Army Medical Research, Development, Acquisition and Logistics Command under Grant No. DAMD17-90-Z-0034 and in part by the Directorate of Military Medical Services of the Ministry of Defence, The Netherlands.

REFERENCES

Ainsworth, M. & Shephard, R.J. (1961). *Inhaled Particles and Vapours* (C.N.Davies, ed.), pp 233-247, Pergamon Press, Oxford.

Benschop, H.P. & De Jong, L.P.A. (1988). *Acc. Chem Res. 21*, 368-374.

Benschop, H.P. & De Jong, L.P.A. (1991). *Neurosci. Behav. Rev. 15*, 73-77.

Benschop, H.P. & Van Helden, H.P.M. (1993). *Toxicokinetics of inhaled soman and sarin in guinea pigs.* Final Report for Grant DAMD17-90-Z-0034, U.S.Army Medical Research and Development Command.

Due, A.H., Trap, H.C., Van Der Wiel, H.J., & Benschop, H.P. (1993). *Arch.Toxicol. 67*, 706-711.

Due, A.H., Trap, H.C., Langenberg, J.P., & Benschop, H.P. (1994). *Arch. Toxicol. 68*, 60-63.

Langenberg, J.P., De Jong, L.P.A., & Benschop, H.P. (1991). Book of Abstracts, EUROTOX Congress, Maastricht, p. 244.

Langenberg, J.P., Van Dijk, C., De Jong, L.P.A., Sweeney, R.E., & Maxwell, D.M. (1993). *Proc. 1993 Med. Def. Biosc. Rev.*, Baltimore MD, 10-13 May 1993, pp. 675-684.

INHIBITION OF FISH BRAIN ACETYLCHOLINESTERASE BY CADMIUM AND MERCURY

Interaction with Selenium

S. Sen, S. Mondal, J. Adhikari, D. Sarkar, S. Bose, B. Mukhopadhyay, and Shelley Bhattacharya

Environmental Toxicology Laboratory
Department of Zoology
Visva Bharati University
Santiniketan 731235, India

ABSTRACT

Industrial chemicals are known to affect the activity of cholinesterases where Se was found to have a negligible effect.It is also reported that Se reduces human cancer death rates. Such contradictory reports prompted us to study the role of Se in the inhibition of fish brain AChE caused by Cd and Hg. I 50 concentrations of Cd and Hg were found to be 20 uM and 62 nM respectively.Since Se showed no I 50, a lower (3.1 uM) and a higher (57 uM) concentration was used. Positive interaction of Se was clear when Hg was at I 50 added either before or after Se or together, and 50% inhibition was completely abolished and 100% activity was recorded.This effect of Se was observed at both the test concentrations. Interaction of Se and Cd was remarkably different recording significant inhibition at all test systems.The activation of Hg-inhibited AChE by Se may be due to acceleration of enzyme deacetylation. In the presence of Cd Se probably binds to the esteratic site more strongly to effect a conformational change thereby decreasing the AChE activity.

INTRODUCTION

Various levels of Selenium (Se) toxicosis are on record in humans.The common symptoms of Se poisoning include muscle weakness, eye and nasal irritations, dizziness, gastrointestinal distress, respiratory failure and convulsions. All these signs indicate CNS defects to be the regulatory factor in the manifestation of Se toxcity. However, there is no report alluding a direct effect of Se on Acetylcholinesterase (AChE) in any animal system. Interestingly some ameliorative effects of Se have been discussed by various authors in decreasing cancer death (Shamberger

Enzymes of the Cholinesterase Family, Edited by Daniel M. Quinn et al.
Plenum Press, New York, 1995

et al, 1976) or antineoplastic effects with regards to benzo(a)pyrene and benzanthracene induced skin tumors in mice (Hammond and Beliles, 1980).Se is known to inhibit the formation of malonaldehyde from peroxidative tissue damage which is carcinogenic and also to act as an antidote to the toxic effects of various metals (Frost & Lish, 1975). Thus it is quite clear that the mode of action of Se is not very well understood.Considering the inhibitory role of heavy metals such as Cd and Hg in the brain of animals an attempt was made to address the role of Se in inhibition of AChE in fish brain caused by Cd and Hg.

MATERIALS AND METHODS

Chemicals

Acetylthiocholine iodide and 5,5'-dithiobis-2-nitro benzoic acid were purchased from Sigma Chemical Company, St. Louis, USA. Cadmium chloride, magnesium chloride, mercuric chloride and sodium chloride were obtained from E. Merck, India. All other reagents used were of analytical grade and purchased from either E. Merck, India or BDH, England.

Animals

A fresh water teleost,*Channa punctatus*(30-35g body weight) collected from a local source was used for the present study. They were acclimatized under laboratory conditions in glass aquaria (120cm x 90cm x90cm) for 15 days prior to toxicant exposure.

Preparation of Brain Homogenate

The fish were decapitated to dissect out the brain carefully and rapidly. The adherent blood vessels were removed immediately and a 5% homogenate was prepared with double glass distilled water at 0°C in a glass homogenizer having a teflon coated pestle. This homogenate was used as the primary source of AChE.

Estimation of AChE Activity

AChE activity was measured according to the method of Ellman et al (1961). Briefly, an aliquot of the brain homogenate was added to the reaction mixture containing Na-K phosphate buffer (0.05 M, pH 7.6), activators (NaCl-30 mM,MgCl$_2$-4 mM) and thiol indicator (0.25 mM). The substrate was added last to start the reaction in a final volume of 3.0 ml. The AChE activity was followed in Beckman Spectrophotometer (DU 640).

In vitro Inhibition Kinetics

In vitro inhibition kinetics of AChE was done according to Aldridge (1959) as modified by Jash and Bhattacharya (1983). AChE preparation was exposed to varying concentrations of the metals for different periods of preincubation and the respective reaction rates were noted. The I 50 concentration and t 0.5 were calculated from the reaction rates.

Interaction of Se was studied at two doses of the metal with the I 50 concentrations of Hg and Cd. For each set of experiment incubation periods for Hg and Cd were set at their respective t 0.5 which were also predetermined. Se was incubated for a 10 min either before or after Hg/Cd exposure. Interaction of Se was further analysed by adding both Se and Hg/Cd together.

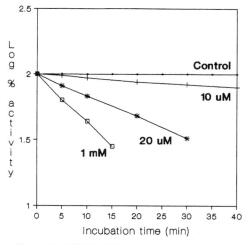

Figure 1. AChE inhibition by cadmium chloride.

RESULTS

In the present study single metal inhibition kinetics revealed that Hg is the most toxic and Se non toxic among the three metals tested to fish brain AChE. The I 50 concentrations of Cd (Fig. 1) and Hg (Fig. 2) effecting 50 % inhibition were calculated to be 20 uM and 62 nM respectively. Although both Cd and Hg are group II B metals, Hg is more toxic than Cd not only in terms of concentration but also t 0.5. Se, on the other hand, at none of the doses tested had any inhibitory effect on AChE.

Interaction of Se with Cd (Fig. 3) and Hg (Fig. 4) revealed that Se has some role in the reversal of Cd/Hg inhibited AChE. Both the higher and the lower doses of Se could effectively waive the inhibitory effect of Hg irrespective of the mode of incubation of Se, either before/after or together with Hg. However, Se incubation with Cd shows an altogether different trend. The lower dose of Se did not reduce the inhibitory effect of Cd on AChE. The

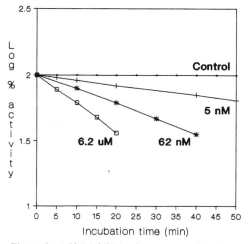

Figure 2. AChE inhibition by mercuric chloride.

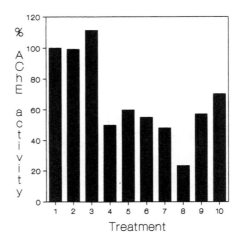

Figure 3. Interaction of Se and Cd with AChE [1=Control; 2=Se(low); 3=Se(high); 4=Cd; 5=Cd,Se(l); 6=Se(l),Cd; 7=Se(l)+Cd; 8=Cd,Se(h); 9=Se(h),Cd; 10=Se(h)+Cd].

higher dose of Se, on the contrary, exerted a greater inhibition when incubated after 20 min of preincubation of AChE with Cd.Interestingly, Se upon coincubation with Cd could remove the inhibition by Cd significantly.

DISCUSSION

Fish brain ChE was found to be inhibited by various industrial pollutants (Mukherjee and Bhattacharya,1974; Olson and Christensen, 1980). No in vitro inhibition kinetics of metals has been done with fish brain AChE, presumably considering the non specific effects of the metals.However, there are reports which suggest the neurotoxic property of Cd and

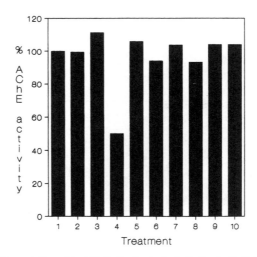

Figure 4. Interaction of Se and Hg with AChE [1=Control; 2=Se(low); 3=Se(high); 4=Hg; 5=Hg,Se(l); 6=Se(l),Hg; 7=Se(l)+Hg; 8=Hg,Se(h); 9=Se(h),Hg; 10=Se(h)+Hg].

Hg. The transport of neurotransmitter amines in synaptosomes is inhibited by Cd (Lai et al, 1981) besides damaging the cells of the cerebellar cortex (Gabbiani et al, 1976).The mechanism by which these metals act on the CNS is poorly understood (Cohen, 1980;Vallee and Ulmer, 1972). It was also opined that polyvalent cations may form complexes with molecules present in the cell membrane leading to disturbed synaptic functions (Bulman and Griffin, 1980; Maas and Colburn, 1965).

The pattern of AChE inhibiton recorded in the present study indicates a very different role of Se in presence of the two different group II B metals. Since Se is non toxic any inhibition shown by the enzyme must be due to Cd or Hg or any spontaneous reactivation manifested by the inhibited enzyme must be due to Se.Since there is no report on the mechanism of action of Se,Cd or Hg in respect of AChE activity we can only suggest a probable mode of action.Metals seem often to act as a bridge between the substrate and the protein which is thought to activate the metal by withdrawing electrons from it. The excess positive charge, thus created on the metal, is thought to withdraw electrons from the substrate (Albert, 1973). When Se is present in the medium along with Cd or Hg the competition between the two metals in the binding of the active site results in higher rate of binding of Se owing to its smaller ionic radius than either Cd or Hg. Since Se has no inhibitory effect on AChE, the rate of binding of Se appears to be the critical factor in reversing the Cd/ Hg inhibited AChE.It may be concluded that i) the activation of the Hg-inhibited AChE by Se is due to faster deacetylation of the enzyme and ii) in the presence of Cd Se probably binds to the esteratc site more strongly to effect a conformational alteration thereby decreasing the AChE activity.

ACKNOWLEDGEMENTS

The authors are grateful to DBT, DST, ICAR for research grants, to UGC for the DSA support to the department and to Dr B Bhattacharyya, Dept of Metallurgy, Jadavpur University, Calcutta for providing sodium selenite.

REFERENCES

Albert, A., 1973, Selective toxicity, Chapman and Hall, London.

Aldridge, W.N., 1959, The mechanism of inhibition of esterases by organophosphorus compounds, *Arhiv za Higijenu Rada I Toksikologiju 10* : 143-154.

Bulman, R.A., and Griffin, R.J., 1980, Actinide transport across cell membranes,*J. Inorg. Biochem. 12* : 89-92.

Cohen, M.H., 1980, Neurotoxic effects of heavy metals and metalloids, In *Biochemistry of Brain*, S Kumar,Ed. Pergamon Press, New York, pp. 453-468.

Ellman, G.L., Courtney, K.D., Andres, Jr.V., and Featherstone, R.M., 1961, A new and rapid colorimetric determination of acetylcholinesterase activity, *Biochem. Pharmacol. 7* : 88-95.

Frost,D.V., and Lish,P.M.,1975, Selenium in biology, *Ann. Rev. Pharmacol. 15* : 259-284.

Gabbiani,G., Baic,D., and Deziel,C, 1976, Toxicity of cadmium for the central nervous system, *Exp.Neurol. 18* : 154-160.

Hammond, P.B., and Beliles, R.P., 1980, Metals, In Casarett and Doull's Toxicology The Basic Science of Poisons, J.Doull, C.D.Klaassen and M.O.Amdur, Eds., Macmillan Pub.Co.Inc., New York, pp. 409-467.

Jash, N.B., and Bhattacharya, Shelley, 1983, Phenthoate-induced changes in the profiles of acetylcholinesterase and acetylcholine in the brain of *Anabas testudineus* (Bloch): Acute and delayed effect, *Toxicol. Letters 15* : 349-356.

Lai, J.C.K., Guest, J.F., Leung,T.K.C., Lim,L., and Davison, A.N., 1980, The effect of cadmium, manganese and aluminium on sodium - potassium - activated and magnesium - activated adenosine triphosphatase activity and choline uptake in rat brain synaptosomes, *Biochem.Pharmacol. 29* : 141-146.

Maas, J.W., and Colburn, R.W., 1965, Co-ordination chemistry and membrane function with particular reference to the synapse and catecholamine transport, *Nature 208* : 41-46.

Mukherjee, S., and Bhattacharya, Shelley, 1974, Effects of some industrial pollutants on fish brain cholinesterase activity, *Environ. Physiol. Biochem. 4* : 226-231.

Olson, D.L., and Christensen, G.M., 1980, Effect of water pollutants and other chemical on fish acetylcholinesterase (*in vitro*), *Environ. Res. 21* : 327-335.

Shamberger, R.J., Tytko,S.A., and Willis,C.E., 1976, Antioxidants and cancer. Part VI. Selenium and age adjusted human cancer mortality, *Arch. Environ. Health, 31* : 231-235.

Vallee, B.L., and Ulmer, D.D., 1972, Biochemical effects of mercury, cadmium and lead, *Annu.Rev. Biochem. 41 : 91-128.*

PHARMACOLOGICAL EVALUATION OF ANTIDOTES AGAINST ORGANOPHOSPHORUS INTOXICATION

S. N. Dube,* R. Bhattacharya, K. Husain and A. K. Sikder

Defence Research and Development Establishment
Gwalior-474002, India

INTRODUCTION

Organophosphorus (OP) anticholinesterase agents are highly toxic ester derivatives of phosphoric and phosphonic acids. This class of compounds has wide application in agriculture and medicine. From the pharmacological and toxicological points of view, OP agents have many implications as chemical warfare agents (Somani *et al.*, 1992).

The therapy against OP poisoning is based on co-administration of anticholinergics like atropine and the oximes. Atropine antagonises effects of endogeneously accumulated acetylcholine and oximes reactivate inhibited acetylcholinesterase (AChE) by displacing the phorphoryl residue from the active site and restore enzymatic activity (Gray, 1984). Among the oximes, pyridine aldoximes, such as pralidoxime (2-PAM), obidoxime (LuH6), and trimedoxime (TMB-4) are known to be therapeutically most effective. However, being quaternary salts, pyridine aldoximes do not readily penetrate into the central nervous system, which results in limited tissue distribution and hence poor reactivation at the central sites. This communication addresses research carried out in our Establishment to evaluate the antidotal efficacy of various oximes against OP intoxication.

MATERIALS AND METHODS

Animals

Wistar rats (100-200 g), albino swiss mice (20-25 g) and guinea pigs (300-400 g), bread and maintained under standard conditions in the Animal Facility of the Defence Research and Development Establishment, Gwalior, were used.

---

* To whom correspondence should be addressed.

Enzymes of the Cholinesterase Family, Edited by Daniel M. Quinn et al.
Plenum Press, New York, 1995

Chemicals and Drugs

The organophosphorus compounds (DFP, soman and sarin) and oximes were synthesized in the Synthetic Chemistry Division of this Establishement. Other chemicals were purchased from usual commercial sources.

Physiological Parameters

The experiments were carried out on anaesthetized rats/mice maintained on positive pressure ventilation to record various physiological parameters on the Grass Polygraph (Model 7-16P-35), as described earlier (Dube *et al.*, 1993).

Antidotal Efficacy

The protection index (PI) was carried out in mice and rats as described by Sikder *et al.* (1992). *In vitro* and *in vivo* cholinesterase reactivations in blood and various regions of brain against sarin and DFP inhibited AChE using test oximes were carried out following the method of Ellman *et al.* (1961) and the percent ChE reactivation was calculated as described earlier (Katrolia *et al.*, 1994).

Neuromuscular (NM) Function Studies

In vivo NM function on gastrocnemius muscle of rats was carried out by stimulating the sciatic nerve at 25, 50, 100 and 200 Hz, each lasting 3 sec, using a Grass Stimulator Model S88. The *in vitro* NM function was tested using the phrenic nerve diaphragm preparation of rats. The phrenic nerve was stimulated at 25, 50, 100 and 200 Hz for 3 sec each to record the tetanic response on the Grass Polygraph. The NM recovery was determined by employing the OP and oximes as per the following method described earlier (van Helden *et al.*, 1983).

```
              OP                 Oxime                              OP
              |                    |                                 |
   Test A  ----- 5' ---- B ---- 10' ---- C ---- 10' ---- D ---- 5' ---- E
              |                    |                                 |
            Wash                 Wash                              Wash
```

The overall percentage recovery from OP poisoning was analyzed on three aspects: recovery due to direct oxime action (C%-D%), recovery attributed to oxime induced reactivation of AChE (D%-E%), and persistance of NM recovery notwithstanding full AChE inhibition also called as adaptation (E%).

RESULTS AND DISCUSSION

Synthesis of OP Compounds and Oximes

Various OP compounds (DDVP, DFP, sarin, soman) and oximes (diethyxime, DEV, DA, SPK) were synthesized by standard procedures mentioned in the literature (Sikder *et al.*, 1993). Their purity was checked by various analytical procedures and were found to be >98% pure.

Effects of OP Compounds on Physiological Parameters in Rats

Intravenous administration of DFP (0.3-3.3 mg/kg) or sarin (20.3-203.4 µg/kg) in anaesthetized and positive pressure ventilated rats produced a dose dependent sustained hypertension, a bradycardia and muscle twitch potentiation of gastrocnemius muscle at 2 min. While, only higher doses of DFP (1.7 and 3.65 mg/kg) produced a significant tracheo-bronchial constriction at 30 min. Administration of tolazoline and atropine completely antogonized the hypertensive response induced by OP compounds suggesting the involvement of central muscarinic as well as adrenergic receptors.

Subcutaneous administration of various doses of sarin (20.3-101.7 µg/kg) produced a dose dependent transient hypotension and bradycardia probably attributed to the sympathetic vasodilaor fibres present in the muscle which were affected by atropine.

Skeletomotor Effect of 2-Pyridine Aldoxime (2-PAM) on Innervated and Denervated Rat Diaphragm Preparation

Skeletal myoneural junction is one of the primary targets of acute and chronic actions of OP toxicity and the oximes are powerful reactivators of ChE at the motor end-plates. It has been demonstrated that 2-PAM possesses a dual action at the skeletomotor junction, i.e. facilitation of twitch response at lower concentrations mediated through its weak anticholinesterase action and inhibition at higher concentrations mediated through a curare like action (Dube *et al.*, 1986a) suggesting its benefical role in the treatment of anticholinesterase poisoning.

Effect of Newly Synthesized Oximes on NM Function Against OP Poisoning

Various oximes were synthesized and evaluated for their antidotal efficacy on NM blockade induced by OP poisoning. A nonquaternary oxime, diethyxime, along with atropine, produced a marked antidotal effect against DDVP (dimethyl dichlorovinyl phosphate) poisoning on NM function without affecting DFP poisoning in rats (Dube *et al.*, 1986b), suggesting that only entry of oximes in the central nervous system may not increase the antidotal efficacy.

Several 3,3'-bispyridinium mono-oximes (DEV and DA series) produced recovery of NM blockade induced by sarin and soman poisoning in rat diaphragm preparation (Dube *et al.*, 1992). DA series oximes were found to be comparable with 2-PAM as effective antidotes against sarin poisoning (Sikder *et al.*, 1992). Alkylpyridinium oximes (SPK series) produced significant reversal of NM blockade and had a higher therapeutic index than 2-PAM against sarin poisoning, probably due to reactivation of ChE in cerebral cortex and corpus striatum (Table 1). These oximes show recovery of NM blockade due to reactivation of AChE as well as adaptation (Table 2).

Use of Different Adjuncts

Various drugs (anticonvulsants, bronchodilators, calcium channel blockers, carbamates) have been evaluated as an adjunct to the conventional therapy of atropine and oxime to control tremor and counteract respiratory failure, which is the primary cause of death in OP poisoning. The work carried out in our Establishment demonstrated an enhanced protection against DDVP and DFP poisoning in mice (Srivastava *et al.*, 1984, 1987)

Table 1. Effect of oximes on cholinesterase activity of blood and certain brain regions of mice exposed to DFP/sarin

	% ChE activity		
Groups	Blood	Cerebral cortex	Corpus striatum
DFP	16.4 ± 3.5^a	49.2 ± 8.7^a	38.1 ± 4.3^a
DA-9	87.4 ± 9.2^b	65.8 ± 9.1^b	38.8 ± 4.7
	(85.7)	(32.0)	(−)
DA-10	89.5 ± 9.5^b	67.8 ± 9.6^b	39.8 ± 4.3
	(88.3)	(36.1)	(−)
DA-11	86.7 ± 9.9^b	64.9 ± 9.4^b	40.3 ± 5.1
	(84.0)	(31.0)	(3.7)
Sarin	37.8 ± 8.6^a	47.3 ± 3.5^a	38.3 ± 2.6^a
DA-9	77.3 ± 9.8^c	55.4 ± 8.1^c	36.0 ± 3.2
	(63.0)	(15.3)	(−)
DA-10	79.7 ± 9.5^c	51.4 ± 6.9^c	38.6 ± 3.8
	(67.0)	(8.0)	(−)
DA-11	92.3 ± 9.2^c	50.7 ± 9.2	36.6 ± 4.2
	(87.7)	(6.4)	(−)
SPK-3	89.2 ± 9.8^c	62.6 ± 6.1^c	49.8 ± 3.2^c
	(32.6)	(31.2)	(15.3)
SPK-4	94.4 ± 9.6^c	51.7 ± 4.2^c	40.9 ± 3.5
	(91.0)	(11.1)	(0.3)
2-PAM	96.2 ± 9.7^c	46.8 ± 5.2	37.9 ± 3.3
	(92.8)	(−)	(−)

Figures in parentheses are % reactivation.
[a] $P < 0.01$ compared to control group.
[b] $P < 0.01$ compared to DFP group.
[c] $P < 0.01$ compared to sarin group.

Table 2. A comparison of the recovery of neuromuscular transmission in organophosphorus poisoned and oxime treated diaphragm muscles of rat

	Recovery of NM transmission due to		
Groups	Direct oxime action $C\% - D\%$	Oxime induced reactivation of AChE $D\% - E\%$	Adaptation $E\%$
Sarin poisoning			
DA-9	—	20.0	20.3
DA-10	—	21.8	22.6
DA-11	—	29.7	20.7
SPK-3	—	63.2	2.0
SPK-4	—	62.8	18.8
2-PAM	—	84.5	5.0
Soman poisoning			
DEV-2	—	37.9	—
DEV-3	13.3	29.5	—
DEV-9	—	75.6	—
DEV-10	2.2	43.4	—
HI-6	—	80.9	—

following therapy with various bronchodilators (salbutamol, terbutaline and metaproterenol) without affecting sarin poisoning (Bhattacharya *et al.*, 1991).

Carbamates as Pretreatment Agents

The conventional therapy with atropine, oxime and other adjuncts may not be very useful due to fast aging of soman-inhibited ChE. Various reports are available regarding use of carbamates as pretreatment agents against OP poisoning (Somani & Dube, 1989). The pretreatment (20 min) with physostigmine gives better protection than pyridostigmine, probably due to protection of its central cholinergic receptors against lethality of inhaled sarin (Vijayaraghavan *et al.*, 1992). A detailed study has shown a dose dependent efficacy of pyridostigmine pretreatment against soman poisoning on the recovery of NM blockade and on the *in vivo* protection in rats and guinea pigs (Jeevaratnam *et al.*, 1990). These results indicate that several bispyridinium mono and alkylpyridinium oximes may produce better antidotal efficacy against DDVP, DFP, sarin and soman poisoning in rodents. However, there is a need for detailed evaluation of these oximes including pharmacokinetics before being used in the clinical practice.

ACKNOWLEDGMENT

The authors thank Dr. R. V. Swamy, Director, Defence Research & Development Establishment, Gwalior, for keen interest in the study, and all the scientiest of the Synthetic Chemistry Division for providing the OP compounds and oximes.

REFERENCES

Bhattacharya, R., Kumar, P., Jeevaratnam, K., Pandey, K.S. & Das Gupta, S. (1991) Asia Pacific J. Pharmacol. **6**, 75-80.

Dube, S.N., Das Gupta, S., Vedasiromani, J.R. & Ganguly, D.K. (1986a) Ind. J. Med. Res. **83**, 314-317.

Dube, S.N., Ghosh, A.K., Jeevaratnam, K., Kumar, D., Das Gupta, S., Pant, B.P., Batra, B.S. & Jaiswal, D.K. (1986b) Jap. J. Pharmacol. **14**, 267-271.

Dube, S.N., Kumar, D., Sikder, A.K., Sikder, N., Jaiswal, D.K. & Das Gupta, S. (1992) Pharmazie **47**, 68-69.

Dube, S.N., Kumar, P., Kumar, D. & Das Gupta, S. (1993) Arch. Int. Pharmacodyn. Ther. **321**, 112-122.

Ellman, G.L., Courtney, K.D., Andres, V., Jr. & Featherstone, R.M. (1961) Biochem. Pharmacol. **7**, 88-95.

Gray, A.P. (1984) Drug Metabol. Rev. **15**, 557-589.

Jeevaratnam, K., Dube, S.N. & Das Gupta, S. (1990) Asia Pacific J. Pharmacol. **5**, 145-149.

Katrolia, S.P., Sikder, A.K., Acharya, J., Sikder, N., Jaiswal, D.K., Bhattacharya, R., Husain, K., Dube, S.N., Kumar, D. & Das Gupta, S. (1994) Pharmacol. Commun., in press.

Sikder, A.K., Pandey, K.S., Jaiswal, D.K., Dube, S.N., Kumar, D., Husain, K., Bhattacharya, R. & Das Gupta, S. (1992) J. Pharm. Pharmacol. **44**, 1038-1040.

Sikder, A.K., Ghosh, A.K. & Jaiswal, D.K. (1993) J. Pharm. Sci. **82**, 258-261.

Somani, S.M. & Dube, S.N. (1989) Int. J. Clin. Pharmacol. Ther. Toxicol. **27**, 367-387.

Somani, S.M., Solana, R.P. & Dube, S.N. (1992) In *Chemical Warfare Agents* (Somani, S.M., Ed.), Academic Press, New York, pp 67-123.

Srivastava, R.K., Ghosh, A.K., Prakash, S., Shukla, R., Srimal, R.C. & Dhawan, B.N. (1984) Eur. J. Pharmacol. **97**, 339-340.

Srivastava, R.K., Ghosh, A.K., Agarwal, M., Sachan, A.S., Kumar, P. & Sharma, S.K. (1987) Ind. J. Pharmacol. **19**, 230-233.

Van Helden, H.P.M., VanderViel, H.J. & Wolthuis, O.L. (1983) Br. J. Pharmacol. **78**, 579-589.

Vijayaraghavan, R., Husain, K., Kumar, P., Pandey, K.S. & Das Gupta, S. (1992) Asia Pacific J. Pharmacol. **7**, 257-262.

MEDICAL PROTECTION AGAINST ORGANOPHOSPHORUS TOXICITY

S. Das Gupta

Defence Research and Development Establishment
Jhansi Road, Gwalior 474002, India

INTRODUCTION

The organophosphorus (OP) anticholinesterase compounds are widely used in agriculture as pesticides and some selected compounds like diisopropylphosphorofluoridate (DFP) have been utilized in veterinary and human medicine. Besides these the highly toxic ester derivatives of phosphoric and phosphonic acid, viz. sarin, soman, VX and tabun (known as nerve agents) have a potential application as chemical warfare agents (CWA). All these compounds elicit their primary effects by phosphorylating or phosphonylating the serine hydroxyl at the active site of the enzyme acetylcholinesterase (AChE), causing essentially irreversible inhibition of the enzyme resulting in accumulation of excess acetylcholine (ACh) at various cholinergic sites causing toxic manifestations. These effects have been reviewed by Gray (1984), Lotti (1991) and Somani *et al.* (1992). The inhibited enzyme can be reactivated by certain drugs (cholinesterase reactivators) such as hydroxylamine, hydroximic acids and oximes (Wilson *et al.*, 1955). Among the oximes, pralidoxime (PAM), obidoxime or toxogonin (LuH6) and trimedoxime (TMB-4) are well known cholinesterase reactivators (Vojvodic & Boskovic, 1976). But till today there is no single oxime or an universal antidote which could protect against the lethality of CWA, so that controversy on the cholinesterase reactivating drug which will be preferred is still open. Moreover, different countries prefer different oximes for reasons still unknown. In the process of selecting essential drugs as laid down by WHO (1977) a founded decision on the drug of choice is only possible on the basis of experimental evaluation under strictly comparable conditions and for a tropical country like India this is of special importance.

COMPARATIVE EVALUATION OF KNOWN OXIMES

Initial work in the Defence R&D Establishment at Gwalior started on the comparative evaluation of the three known oximes, viz. PAM, LuH6, TMB-4, along with diethyxime (S-[2-(diethylamino)ethyl]-4-bromobenzothiohydroximate hydrochloride), a nonquaternary compound which was claimed as an universal oxime (Kokshareva *et al.*,

Enzymes of the Cholinesterase Family, Edited by Daniel M. Quinn et al.
Plenum Press, New York, 1995

1977). All these oximes were synthesized by the Chemistry Division. Biological evaluation was carried out by determining the Protective Index (PI) which is a ratio of the LD50 in animals treated with the antidotes (usually oxime along with atropine) by the LD50 of the OP compound used. The 24 hour LD50 was determined by Dixon's up and down method (Dixon, 1965).

Biochemical estimations included determination of inhibited cholinesterase at various time intervals by DFP and its reactivation by the oximes in serum and certain target organs like liver, heart and brain. Results from this study revealed that atropine along with obidoxime offered maximum protection with 33% of reactivation in serum after 1 hour of treatment although it failed to reactivate the brain cholinesterase. However, a marginal reactivation (10%) of brain cholinesterase was observed with PAM after 2 hours of treatment. It was expected that diethyxime would be more lipophilic due to the absence of quaternary nitrogen atom and hence better distribution into the brain giving more protection but that was not observed. Earlier observation by Heffron & Hobbiger (1980) indicated that reactivation in the brain fails to raise the antidotal action which is in agreement with our findings. It can be emphasized that penetration of a reactivating compound into the central nervous system (CNS) may not increase the protective index (Das Gupta et al., 1982).

ENTRY OF OXMES INTO RED BLOOD CELLS

A comparative study was also carried out on the entry of cholinesterase reactivators into red blood cells (RBC). It was observed that PAM diffuses into erythrocytes up to 50% whereas the bisquaternary oximes obidoxime and trimedoxime failed to enter the RBC due to their higher molecular weights (Ghosh et al., 1983).

PHARMACOKINETICS OF OXIMES

The biological half life ($t\frac{1}{2}$) of various cholinesterase reactivators is usually of short duration and they have been reported to be excreted rapidly from the body. Various drugs were tried earlier to prolong the retention of oximes but the results were inconclusive. Swartz & Sidell (1974) reported prolongation of $t\frac{1}{2}$ of PAM by pretreating with thiamine hydrochloride in male human volunteers. We in our laboratory while working on the pharmacokinetics of PAM reinvestigated the pretreatment studies using male and female rats to establish the influence of thiamine hydrochloride on the $t\frac{1}{2}$ of PAM and observed that retention of PAM was highly significant at 150 and 180 min in female rats while the male rats did not respond to this effect. The $t\frac{1}{2}$ of PAM increased significantly from 88 min to 212 min in female rats after pretreatment with thiamine hydrochloride (10 mg/kg, i.m.) followed by PAM (30 mg/kg, i.m.). Thus a significant sex difference on the retention of PAM was observed (Das Gupta et al., 1979). Next, we wanted to see whether this pretreatment was effective in increasing the therapeutic efficacy against DFP intoxication. Using a two compartment system we studied the therapeutic efficacy of PAM along with suitable pretreatment against DFP. Though thiamine hydrochloride pretreatment prolonged the $t\frac{1}{2}$ of PAM it failed to increase the protective efficacy. On the other hand sodium hydrogen carbonate pretreatment augmented the protective action of PAM without altering its $t\frac{1}{2}$ appreciably through increased distribution of PAM into tissue compartment (Jeevarathinam et al., 1988).

THERAPEUTIC STUDIES IN ORGANOPHOSPHORUS POISONING

THe choice of appropriate treatment for OP intoxication including nerve agent depends on the chemical agent, the severity of exposure and route(s) of exposure. Decontamination is effective in mild exposure whereas severe exposure to vapour or liquid requires immediate medical intervention. Usually it consists of administration of antidotes in the form of oxime and a cholinolytic together with artifical respiration and sometimes uses of different adjuncts. Monitoring of the enzyme AChE level is useful although controversy exists whether AChE can be regarded as a "true marker" in the poisonings. Sometimes nerve agent AChE complex becomes resistant to conventional oxime therapy and for this reason a search is on to find an universal antidote comprising of a cholinesterase reactivator and a cholinolytic.

In our laboratory various oximes have been synthesized and biological evaluation carried out against various OP compounds. Among the cholinesterase reactivators, PAM, LuH6 and TMB-4 are known to be therapeutically effective against most of the nerve agents except soman due to the rapid aging of AChE. The bisquaternary compounds HS-6 and HI-6, on the other hand, in adjunction with atropine, are somehow effective against soman intoxication. While evaluating the efficacy of these oximes we found that atropine in combination with HI-6 afforded a PI of 5.2 against soman, while LuH6 gave the highest PI of 32 against VX and 3.5 against sarin. Atropine alone failed to offer any significant protection against these nerve agents (Das Gupta et al., 1990).

Two new oximes DEV-9 (1-[3-hydroxyiminomethyl-1-pyridino]-3-[4-carbamoyl-1-pyridino]propane dibromide) and DEV-10 (1-[3-hydroxyiminomethyl-1-pyridino]-3-[carboxymethyl-1-pyridino]propane dibromide) were evaluated in DFP treated mice along with estimation of AChE in blood, cerebral cortex and corpus striatum by the Ellman method (Ellman et al., 1961). It was observed that DFP inhibited AChE in blood to 84% while in cortex and striatum it was 66% and 76%, respectively. DEV-10 along with atropine offered a protective index of 4.5 with 77% reactivation in blood while 10% reactivation was observed in cerebral cortex. None of the oximes could reactivate the deeper structure of the brain (corpus striatum). It was concluded that the protection afforded by these oximes was mainly due to peripheral reactivation (Husain et al., 1990). The continuation of these investigations resulted in synthesis of newer and effective 3,3'-bispyridinium mono oximes containing a 2-oxopropane bridge. These oximes were evaluated against DFP and sarin for in vivo protection and cholinesterase reactivation in blood and various brain regions. Sarin produced a significant inhibition of AChE activity in blood (62%), medulla (62%) and cerebellum (60%). The compounds could cross the blood brain barrier although they are quaternary salts and could reactivate the cortical AChE (15%) without affecting the deeper areas of the brain. Based on these findings it can be concluded that these newer 3,3'-bispyridinium mono oximes are effective antidotes in OP intoxication (Sikder et al., 1992).

In a recent study we have synthesized a series of long chain alkyl pyridinium oximes varying in length of the alkyl chain between eight and sixteen carbon atoms and evaluated the efficacy of these compounds against sarin intoxication. Results show that 1-cetyl-4-hydroxyiminomethylpyridinium bromide along with atropine was effective against sarin intoxication in mice as evidenced by cholinesterase reactivation in blood and cerebral cortex (Katrolia et al., 1994).

USE OF DIFFERENT CHOLINOLYTICS

Atropine is the conventional cholinolytic which is administered along with a cholinesterase reactivator, although many have tried other cholinolytics also to counteract OP poisoning. In our laboratory we evaluated the antidotal efficacy of isopropamide iodide in acute DFP poisoning in mice and found that it augmented the protection index considerably (27.2) in combination with LuH6. This drug appears to have an advantage of a long duration of action (Das Gupta et al., 1985). In a separate study, a search was pursued for an effective combination of cholinolytic and a cholinesterase reactivator against the nerve agent soman where it was observed that benactyzine in combination with HS-6 offered a significant protection (PI = 3.2; P<0.05 with respect to soman alone). This beneficial effect of benactyzine is possible due to its greater antimuscarinic effect in the CNS than atropine or dexetimide (Das Gupta et al., 1991).

USE OF ADJUNCTS IN OP THERAPY

The anticonvulsant drug benzodiazepam and congeners have been tried for treatment of convulsions caused by cholinergic crisis (Grudzinska et al., 1979). We also observed an increase in PI when diazepam was used as an adjunct in OP intoxication (Das Gupta et al., 1982; Das Gupta and Ghosh, 1985). Furthermore, use of the bronchodilator (β-2 adrenoceptor agonist) salbutamol, given along with atropine and LuH6, significantly enhanced the PI against DFP in mice but failed to offer significant protection against sarin poisoning. The reasons could be that, unlike DFP, sarin caused minimal tracheobronchial response and less inhibition of tracheal AChE at equitoxic doses (Bhattacharya et al., 1991).

Organophosphates are reported to stimulate release of intracellular calcium (Kauffman et al., 1987). The efficacy of the calcium channel blocker nifedipine as an adjunct to atropine and obidoxime therapy against DFP and sarin was evaluated. Results indicate that adjunction of nifedipine potentiated the PI significantly against DFP while it was marginal against sarin (Rohatgi et al., 1993).

PRETREATMENT STUDIES IN RODENTS

Reports on the protective efficacy against OP toxicity by carbamates have been conflicting because of species variation, different dose schedules and pretreatment time. Pretreatment studies were carried out in rats with reversible cholinesterase inhibitors in combination with antimuscarinic (atropine) and antinicotinic (mecamylamine) agents, against the three nerve agents soman, sarin and VX. It was observed that physostigmine (Phy) was more effective than pyridostigmine (Pyr) against all the nerve agents (Das Gupta et al., 1990). This corroborates our earlier finding with DFP (Das Gupta et al., 1987). A study was undertaken to evaluate the efficacy of Phy alone in guinea pigs against multiple doses of sarin. It was observed that the mean time of mortality after 3 LD50 of sarin (LD50 of sarin 0.8 μg/kg) injected subcutaneously was 16.3±2.4 min, whereas pretreating the animals with a sign free dose of Phy (0.16 mg/kg) for 10 min prolonged the survival time to 33.7±6.6 min (P<0.01) after administration of 3 LD50 of sarin. Plasma cholinesterase activity varied over the period of death 4-10% of the initial activity with 3 LD50 sarin, while the erythrocyte AChE activity varied from 10-20% of the initial value. With the Phy pretreatment a marginal protection was seen only in erythrocyte AChE, 82±3.2%, and not in plasma, 58±2.7 (Das Gupta et al., 1993).

PRETREATMENT STUDIES IN PRIMATES

Objectives of the studies were to determine a safe and optimal dose of Phy given intramuscularly (i.m.) to monkeys and assessing the time activity profile of cholinesterase enzyme assay with measurement of various physiological variables. The doses of Phy were 100 µg/kg and 50 µg/kg given i.m., while the control group received saline only. The cholinesterase activity both in plasma and erythrocytes decreased significantly reaching the nadir by 60 min after administration, the degree of inhibition being greater in animals receiving the higher dose. The pattern of recovery from its peak value was different in erythrocytes as compared to that in plasma and was also faster. Results from this study indicate that a dose of 50 µg/kg given i.m. to monkeys appears to be a safe pretreatment against nerve agent intoxication (Das Gupta *et al.*, 1992).

PRETREATMENT STUDIES AGAINST INHALED SARIN AEROSOLS

A comparative efficacy of the carbamates Phy and Pyr pretreated at various intervals prior to exposure to sarin aerosols in rats was carried out. It was observed that a symptom-free dose of Phy (0.1 mg/kg, i.m.) given 5 minutes prior to sarin aerosols significantly protected the lung ChE and increase the survival time, whereas Pyr in symptom-free dose (0.075 mg/kg, i.m.) did not offer protection. Pretreatment with Phy increased the survival time significantly when it was given 35, 20, or 5 min prior to sarin inhalation, but with Pyr an increase in survival time was observed when it was given 20 min prior to exposure. This study demonstrates that Phy gives better protection than Pyr which may be due to the protection of the central cholinergic receptors (Vijayaraghavan *et al.*, 1992). In a separate study Husain *et al.* (1993) have shown that pretreatment with these carbamates 20 min prior to sarin exposure in rats protected the animals by its action on dynamic pulmonary mechanics.

DEVELOPMENT OF INDIGENOUS AUTOINJECTOR

Nowadays emphasis to combat medical care is shifting from trained medical personnel to first aid by non-medical personnel, especially in field areas. A need was felt to develop an autoinjector indigenously. The Defence R&D Establishment successfully produced a disposable type of autoinjector and a reversible type of autoinjector. These autoinjectors have been successfully tested in field conditions (Rohatgi *et al.*, 1992).

CONCLUSION

Treatment of intoxication by an OP anticholinesterase agent is effective with combination therapy. Still, there is no single cholinesterase reactivator or drug which can counteract the intoxications caused by various CWA. Though cholinesterase reactivators will remain an important aspect in therapeutic regimen, care must be taken that these reactivators are not toxic when given in high doses, it should be stable and have a considerable retention time to couteract the toxicity. Pretreatment with the reversible carbamate Phy is promising provided a careful dose schedule is worked out for use as a prophylactic drug. Adjuncts like diazepam and nifedipine are effective in mild toxicity.

ACKNOWLEDGMENTS

The author thanks Dr. R. V. Swamy, Director, Defence R&D Establishment, Gwalior, for his encouragement in writing this article. The author places on record his sincere thanks to all the scientists of the Division of Pharmacology & Toxicology of DRDE for their dedicated work in carrying out the various experiments and providing the author all the support needed for completion of the projects. The author also thanks al the synthetic chemists of DRDE for providing the chemicals, and Mr. G. R. Khanwilkar for typing of the manuscript.

REFERENCES

Bhattacharya, R., Kumar, P., Jeevaratnam, K., Pandey, K.S. & Das Gupta, S. (1991) Asia Pacific J. Pharmacol. **6**, 75-80.

Das Gupta, S., Moorthy, M.V., Chowdhri, B.L. & Ghosh, A.K. (1979) Experientia **35**, 249.

Das Gupta, S., Ghosh, A.K., Moorthy, M.V., Jaiswal, D.K., Chowdhri, B.L., Purnanand & Pant, B.P. (1982) Pharmazie **37**, 605.

Das Gupta, S. & Ghosh, A.K. (1985) Pharmazie **40**, 499-500.

Das Gupta, S., Ghosh, A.K. & Jeevaratnam, K. (1987) Pharmazie **42**, 206-207.

Das Gupta, S. Bhattacharya, R., Prunanand & Pant, B.P. (1990) Pharmazie **45**, 801-802.

Das Gupta, S., Ghosh, A.K., Chowdhuri, B.L., Asthana, S.N. & Batra, B.S. (1991) Drug & Chem. Toxicol. **14**, 283-291.

Das Gupta, S. & Selvamurthy, W. (1992) Abstract XI Annual Conference of Indian Academy of Neurosciences & Colloquium on Cellular and Molecular Advances in Neuropharmacology, Lucknow, Oct. 13-15, p 83.

Das Gupta, S. (1993) DRDE Project Closure Report No. RD-P1-90/DRDE-134, pp 1-27.

Dixon, W.J. (1965) J. Am. Statist. Assoc. **60**, 967-978.

Ellman, G.L., Courtney, K.D., Andres, V., Jr. & Featherstone, R.M. (1961) Biochem. Pharmacol. **7**, 88-95.

Ghosh, A.K., Moorthy, M.V., Chowdhri, B.L. & Das Gupta, S. (1983) Pharmazie **38**, 790-791.

Gray, A.P. (1984) Drug Met. Reviews **15**, 557-589.

Grudzinska, E., Gidynaska, T. & Rump, S. (1979) Arch. Intern. Pharmacodyn. **238**, 344-350.

Hefferon, P.P. & Hobbiger, F. (1980) J. Pharmacol. **68**, 313-314.

Husain, K., Sikder, A.K., Raza, S.K., Das Gupta, S. & Jaiswal, D.K. (1990) Current Sci. **59**, 1338-1339.

Jeevaratnam, K., Ghosh, A.K., Srinivasan, A. & Das Gupta, S. (1987) Pharmazie **43**, 114-115.

Katrolia, S.P., Sikder, A.K., Acharya, J., Sikder, N., Jaiswal, D.K., Bhattacharya, R., Husain, K., Dube, S.N., Deo Kumar & Das Gupta, S. (1994) Pharmacol. Commun., in press.

Kauffman, F.C., Davis, L.H. & Whittaker, M. (1987) Sixth Med. Chem. Def. Biosc. Rev., 115-122.

Kokshareva, N.V., Kovtum, S.D., Kagan, Y.S., Mizyukova, I.G. & Medvedev, B.M. (1977) Byull. Eksp. Biol. Med. **83**, 29-32.

Lotti, M. (1991) Med. J. Australia **154**, 51-55.

Rohatgi, S. Das Gupta, S., Dube, S.N. & Rao, N.B.S.N. (1992) Armed Forces Med. J. **48**, 223-227.

Sikder, A.K., Pandey, K.S., Jaiswal, D.K., Dube, S.N., Deo Kumar, Husain, K., Battacharya, R. & Das Gupta, S. (1992) J. Pharm. & Pharmacol. **44**, 1038-1048.

Somani, S.M., Solana, R.P. & Dube, S.N. (1992) In *Chemical Warfare Agents* (Somani, S.M., Ed.), Academic Press, pp 67-123.

Swartz, R.D. & Sidell, F.R. (1974) Proc. Soc. Exp. Biol. Med. **146**, 419-424.

Vijvodic, V. & Boskovic, B. (1976) In *Medical Protection Against CWA*, SIPRI Publ., Almquist & Wiskell, pp 65-73.

Wilson, I.B. & Ginsberg, S. (1955) Biochim. Biophys. Acta **18**, 168-170.

WHO: a) 1st Report Techn. Rep. Ser., No. 615, Geneva, 1977; 2nd Report Techn. Rep. Ser., No. 641, Geneva, 1979.

IDENTIFICATION OF A 155 kDa FRACTION THAT POSSESSES NEUROPATHY TARGET ESTERASE ACTIVITY

C. E. Mackay, B. D. Hammock, and B. W. Wilson

Departments of Avian Science, Environmental Toxicology and Entomology
University of California
Davis, California 95616

Neuropathy target esterase (NTE) is a protein that hydrolyses phenyl valerate in the presence of paraoxon, but is inhibited by mipafox. Many consider it the putative site of action for organophosphorus compounds that cause organophosphate induced delayed neuropathy (OPIDN). Williams and Johnson (1981) suggested that NTE was a 155 kDa protein. Thomas et al. (1993) used 3-(9'-mercaptononylthio)-1,1,1-trifluoropropan-2-one (MNTFP) bound to sepharose to obtain evidence that this protein was responsible for NTE activity. Recently, Glyn et al. (1994) isolated a 155 kDa peptide with a biotinylated saligenin phosphate analog. To date, no one has isolated the active enzyme.

The biological source for NTE was whole brains of 18-day old chick embryos. They were homogenized, centrifuged at 10,000xg for 20 minutes and the microsomes separated by centrifugation of the supernatant at 100,000xg for 60 min. NTE activity was solubilized from the resultant pellet by phospholipase A_2 treatment (Seifert and Wilson 1994).

Initial purification of NTE was based on gel filtration chromatography using a 5x100 cm S-400 HR column (Pharmacia). Solubilized chick brain extracts of 1 ml were combined 1:1 with a tris buffer containing 500 mM NaCl and 0.3% Triton X-100 or 0.1% W1 and applied to the gel filtration column. Samples were eluted with the same tris buffer used to dilute the extract. Comparing this to known commercial standards, it was possible to determine the mass of the activity to be about 200 ± 30 kDa. The Mipafox inhibition curves suggest it is the same enzyme responsible for the NTE activity in the initial extract. This is the first report of NTE partitioning in gel filtration with a mass less that about 1,000 kDa.

Subsequent purification involved preparative isoelectric focusing (IEF) and agarose native gel electrophoresis (AGE). IEF was performed on the 200 kDa eluates using a Rotofor (Biorad) containing 40 ml water and 5% 3/10 Biolyte. This permitted the isolation of the NTE activity within the precipitate at a pI of 4.5±0.6 and showed a 5-fold concentration of NTE over gel exclusion and a 4-fold increase over the initial extract.

Subsequent AGE separation on a 0.75% agarose gel containing 0.1% W1 and 250 mM sucrose demonstrated two distinct NTE bands: one that remained within 1 cm of the top of the gel with a PV hydrolase activity that was exclusively NTE, and another that ran with the solvent front in which PV hydrolase activity accounted for only 12% of the total present. Analysis of the mipafox inhibition curves for NTE in these two fractions showed that the smaller one possessed a greater sensitivity to the inhibitor with an I50 of 85 ± 10 μM compared to 280 ± 24 μM for the larger.

The identity of the NTE activity was confirmed using immobilized MNTFP bound to sepharose. There was a time-dependant loss of NTE activity from the solution when the solubilized protein from the larger fraction was combined with the ligand that was not seen with the smaller NTE fraction. Elution of the ligand with Laemmli buffer and subsequent SDS-PAGE demonstrated a single protein band with an apparent mass of 138-157 kDa. This is in agreement with the observations of Williams and Johnson (1981), Thomas et al. (1993) and Glyn et al. (1994) suggesting that the activity in the larger band was exclusively NTE.

These results suggest that the earlier reports of solubilized NTE possessing a mass of greater than 1,000 kDa were the result of a hydrophobic aggregation and that the NTE activity observed was due to a 200 kDa subunit of that mass. Interestingly, although the fractions of NTE activity in our study possess the same approximate pIs, they exhibit different mipafox inhibition profiles and ligand affinities. The approach presented here will permit kinetic analysis on the natural enzyme to further characterize its activity and seek to identify its endogenous substrate.

REFERENCES

Glynn, P., Read, D., Guo, R., Wylie, S., and Johnson, M.K., 1994, *Biochem. J.* 301, 551-556
Seifert, J., and Wilson, B.W., 1994, *Comp Biochem Physiol C.* 108:337-341.
Thomas, T.C., Szekacs, A., Hammock, B.D., Wilson, B.W., and McNamee, M.G., 1993, , *Chem-Biol Interact.* 87, 347-360.
Williams, D.G., and Johnson, M.K., 1981, *Biochem J.* 199, 323-33.

INTERLEUKIN 6 MODULATES AChE ACTIVITY IN CNS NEURONS

D. Clarençon, E. Multon, M. Galonnier, M. Estrade, C. Fournier,
J. Mathieu, J. C. Mestries, G. Testylier, P. Gourmelon, and A. Fatôme

CRSSA Emile Pardé
BP 87
38702, La Tronche, France

ABSTRACT

Whole-body radiation injury induces a series of pathological effects depending on the dose of radiation. Amongst these, the life-threatening pathologies include hematological, digestive and neurological syndromes, which predominate respectively for increasing doses. The neurological syndrome is classically described for doses exceeding 30 Gy for humans. However, some alterations in nervous system function occur for doses as low as a few grays. These include both EEG abnormalities and significant changes in brain neurochemistry, or nonspecific disorders like fever, hypothalamic-pituitary-adrenal axis activation, altered cardiovascular function and behavioral changes. The mechanisms of these responses is poorly known yet, but all these troubles are reminiscent of the central actions of cytokines. Indeed, during the early phase of the acute radiation syndrome, an inflammatory process with systemic release of cytokine has been described (Hérodin et al., 1992). Previously, using the microspectrophotometric method to measure in vivo AChE rat brain activity (Testylier & Gourmelon, 1987), we observed an early decrease in enzymatic activity, following a 10 or 15 GY whole-body gamma exposure. For this dose range, this effect is not due to a direct alteration of the protein (Nayer & Srinivasan, 1975) but represents a functional response of the CNS. The object of the present study was to find possible IL-6 systemic effects on the central nervous system.

Two groups of rats, anesthetized with chloral hydrate (80 $mg \cdot kg^{-1} \cdot hr^{-1}$ IV), received either saline or IL-6 (870 $\mu g \cdot kg^{-1}$) IV injections. AChE activity was recorded every 30 min. We observed a significant decrease of AChE activity (15 to 20%) 105 min after IL-6 injection when compared to saline. No recovery of AChE activity was observed, up to two hours after administration. Then, we studied the in vitro effect of IL-6 on isolated mouse neurons. The activities of both membrane-bound and released enzyme were not modified up to four hours after exposure to IL-6 (5 to 500 $ng \cdot ml^{-1}$). These results suggest that IL-6 induces AChE decrease by an indirect mechanism. Very few reports refer to a relationship between cytokines and cholinergic systems. If IL-1 is known to decrease extracellular ACh in the hippocampus, no report mentions IL-6 effects on AChE activity in live animals. Our *in vivo* data suggest that circulating IL-6 could act, through a mechanism still to be found, by modulating neuronal AChE activity after radiation exposure.

REFERENCES

Hérodin, F. et al. (1992) *Blood* **80**, 688-695.
Nayar, G.N.A. & Srinivasan, S. (1975) *Radiation Research* **64**, 657-661.
Testylier, G. & Gourmelon, P. (1987) *Proc. Natl. Acad. Sci. USA* **84**, 8145-8149.

HI-6 IS INCAPABLE OF REACTIVATING TABUN-PHOSPHONYLATED HUMAN ACETYLCHOLINESTERASE

Chunyuan Luo, Jinsheng Yang, and Manji Sun

Institute of Pharmacology and Toxicology
Beijing 100850, China

A 1:10 (w/v) ratio of human brain homogenate in phosphate buffer (67 mmol·L^{-1}, pH 7.2) was respectively added to sarin or VX (50 nmol·L^{-1}), tabun or soman (100 nmol·L^{-1}). A 1:60 (v/v) ratio of human RBC suspension in isotonic phosphate buffer (pH 7.2) was added to nerve agents (all 87 nmol·L^{-1}). Inhibition was carried out at 0°C for 10 min in the case of soman, or at 37°C for 10 min for other nerve agents. Reactivation was performed at 4°C for 6 h for soman-AChE, or at 37°C for 30 min for other nerve agents, with 1 mmol·L^{-1} of oximes in the case of soman-RBC AChE or 0.1 mmol·L^{-1} for the other conjugates. Results showed that mono-oximes (2-PAM, HI-6) and di-oximes (TMB_4, LuH_6) were powerful reactivators of sarin- or VX-phosphonylated human acetylcholinesterases (RBC, brain). However, HI-6 was incapable of reactivating the tabun-phosphonylated human acetylcholinesterases. The most effective reactivators were TMB_4 and LuH_6.

Table 1. Oxime reactivation of phosphonylated human AChE

AChE	Oxime	Reactivation (%)[a]			
		Sarin	VX	Soman	Tabun
Human RBC	HI-6	97±5	98±10	35±3	1±1
	2-PAM	99±8	97±4	2±2	8±3
	TMB_4	93±2	96±2	10±3	80±12
	LuH_6	95±3	96±3	7±3	48±8
Human brain	HI-6	88±10	87±6	0	1±1
	2-PAM	83±5	60±15	2±1	1±1
	TMB_4	69±8	93±4	0	42±8
	LuH_6	81±4	88±5	0	22±6

[a]Values are means ± standard errors of 3 to 4 observations.

PATTERN OF INHIBITION OF ACETYLCHOLINESTERASES IN DIFFERENT REGIONS OF THE BRAIN BY TWO ORGANOPHOSPHORUS HOMOLOGUES, IN RELATION TO THEIR DIFFERENTIAL NEUROTOXICITY

P. Santhoshkumar, Subramanya Karanth, and T. Shivanandappa

Toxicology Unit
ICP Department
Central Food Technological Research Institute
Mysore-570 013
India

INTRODUCTION

Although OP compounds are anticholinesterases their differential *in vivo* inhibition of acetylcholinesterase (AChE) in the nervous system is poorly understood (Tripathi and Dewey, 1989). Bromophos (BR) and Ethylbromophos (EB) are the two structurally homologous OP insecticides which show distinct differences in their toxicity as well as the time-course of neurotoxic symptoms. EB is 24 fold more toxic (LD_{50}(rat) -125 mg/kg b.w) than BR and produces characteristic neurotoxic symptoms such as tremors and lacrimation which are absent in the case of latter. The purpose of this study was to investigate the biochemical basis of the differential toxicity of the OP homologues in the rat with reference to the AChE inhibition in different regions of the brain.

METHODS

Albino rats (CFT-wistar) were intubated with single equitoxic doses ($1/5LD_{50}$) of BR (600 mg/kg b.w) and EB (20 mg/kg b.w) and the time course of AChE ihibition in the whole brain and in different regions of the brain, viz., cortex, cerebellum, striatum, hippocampus, thalamus, brain stem, pons and optic chiasma was measured.

RESULTS AND CONCLUSIONS

The degree of AChE inhibition in the whole brain varied greatly. Maximum inhibition of AChE was seen at 12 h and 24 h time points in the case of EB and BR respectively. The response of different areas of the brain to the two OP compounds varied significantly. Maximum inhibition of AChE by both the OP compounds was seen in the hippocampus whereas the cortex was least affected. The brain stem was least sensitive to BR but most sensitive to EB. BR produced maximum inhibition in the pons of all the brain regions. The recovery from *in vivo* AChE inhibition in the brain was faster in the case of EB (12 days) and some what slow (> 16 days) in the case of BR. The effect of single equimolar (0.15 m mole/kg b.w) doses of BR and EB on the pattern of AChE inhibition in the brain clearly shows that EB is more potent AChE inhibitor than BR. Among the different regions of the brain, the cortex was least affected by both the OPs while the inhibition in the striatum was somewhat equal. EB produced maximum inhibition in the brain stem.

Since both OP compounds are highly hydrophobic their relative rate of metabolism may be important in the differential inhibition of AChE in the brain regions. It is possible that besides differential AChE inhibition in the brain regions, interactions with the noncholinesterase targets (receptors) may be an important causative factor in the pattern of cholinergic symptoms (Eldefrawi *et al.*, 1988).

ACKNOWLEDGMENTS

The first two authors are grateful to CSIR (New Delhi) for the award Fellowships.

REFERENCES

Eldefrawi, M.E.; Schweizer, G; Bakry, N.M.; Valdes, J.J., (1988) Densensitization of the nicotinic acetylcholine receptor by diisopropylfluorophosphate. J.Biochem. Toxicol. 3:21-32.
Tripathi, H.; Dewey, W.L., (1989) Comparison of the effects of diisopropylfluorophosphate, sarin, soman and tabun on toxicity and brain acetylcholinesterase activity in mice. J. of Tox. and Env. Health, 437-446.

ORGANOPHOSPHATE-SENSITIVE CARBOXYLESTERASE ISOZYME(S) IN THE RAT LIVER

Subramanya Karanth and T. Shivanandappa

Toxicology Unit
Infestation Control and Protectants Department
Central Food Technological Research Institute
Mysore-570 013, India

INTRODUCTION

Carboxylesterases (CEs) (EC 3.1.1.1) are known to play an important role in the detoxication of organophosphorus (OP) compounds either by hydrolysing them (A-esterases) or by binding them (B-esteraes) according to the classification of Aldridge (1953). CE isozymes have been purified from rat liver and have also been shown to bind preferentially to certain compounds like disulfiram (Mentlein *et al.*, 1980; Zemaitis *et al.*, 1976). Tissue CE are important targets for OP compounds in vivo and their inhibition does not have toxic physiological consequence but may afford protection by binding the OPs (Nousiainen, 1984). It is not clearly known if there is any specific isozyme sensitive to OPs or is there any differential susceptibility of the isozymes to structurally different OP compounds. The present study is aimed at identifying the differential sensitivity of CEs in the rat liver to the two OP compounds, Bromophos and Monocrotophos.

RESULTS AND CONCLUSIONS

Female albino rats were orally intubated with single 1/5 LD_{50} doses of Bromophos (LD_{50} = 3000 mg/kg b.w) and Monocrotophos (LD_{50} = 20 mg/kg b.w). Animals were sacrificed after 24 hrs and 12,000 g supernatant of the liver was used as the enzyme source which was subjected to native polyacrylamide gel electrophoresis (PAGE). The enzymes were located on the gel using a-naphthyl acetate and fast blue B salt.

The results indicated that Bromophos is a potent inhibitor of CE than Monocrotophos. The PAGE pattern from *in vivo* experiments showed the selective inhibition of the first major band (from +ve end) by Bromophos whereas the last band was more sensitive to inhibition by Monocrotophos. The CE isozymes not only exhibited a differential sensitivity among the bands but also between the OPs. From the results it appears that the band-I could be important in the binding (sequestering) of Bromophos whereas the last band may be important for

Monocrotophos binding. Therefore band-I could be viewed as 'B' esterase for Bromophos and not for Monocrotophos. This clearly presents the inadequacy of the earlier classification of CEs based on their interaction with OPs. In view of this there appears to be a need for a re-look into the accepted classification of CEs which is not universal. On the other hand, it may be possible to categorize OPs based on their interaction with the CEs.

ACKNOWLEDGMENTS

The first author is grateful to CSIR (New Delhi-India) for the award of Fellowship.

REFERENCES

Aldridge, W.N. *Biochem. J.*, 1953, 53, 110-117.
Mentlein, R.; Heiland, S.; Heymam, E. *Arch. Biochem. Biophys.* 1980, 200, 547-559.
Nousiainen, U. "Inducibility of carboxylesterases by xenobiotics in the rats", Academic dissertation, University of Kuopio, Kuopio, Finland. 1984.
Zemaitis, M.A.; Greene, G.E. *Biochem. Pharmacol.*, 1976, 25, 453-459.

FENTHION TREATMENT PRODUCES TISSUE-, DOSE-, AND TIME-DEPENDENT DECREASES IN MUSCARINIC SECOND MESSENGER RESPONSE IN THE ADULT RAT CNS

P. Tandon,[1] C. N. Pope,[2] S. Barone, Jr.,[3] W. Boyes,[3] H. A. Tilson,[3] and S. Padilla[3]

[1] Department Biology, Boston College
  Boston, Massachusetts
[2] Toxicol. Program, NLU
  Monroe, Louisiana
[3] Neurotoxicology Division, US Environmental Protection Agency
  Research Triangle Park, North Carolina

POSTER SUMMARY

Fenthion (dimethyl 3-methyl-4-methylthiophenyl phosphorothionate) belongs to a class of organophosphate compounds used as insecticides throughout the world. It is commonly assumed that cholinesterase-inhibiting insecticides do not cause long-term toxic sequelae; however, some studies indicate that there may be persistent effects after the depressed cholinesterase activity has returned to normal (Karczmar, 1984). Exposure to fenthion has been reported to produce permanent ocular degeneration in laboratory animals and visual dysfunction in humans (Imai et al., 1983). We have reported that a single, moderate (100 mg/kg) dose of fenthion produces long-term changes in muscarinic receptor density and function in the rat retina in the absence of overt pathology (Tandon et al., 1994).

The present series of experiments investigates the effects of lower fenthion dosages on muscarinic receptor function in the retina and the cortex either using a single or repeated dosing regimen. Fenthion was administered as a single dose in corn oil at 0, 10, 50 or 100 mg/kg (sc) at day 0 to adult, male, Long-Evans rats. The animals were killed 4, 14 or 56 days after dosing. For the repeated dosing regimen, animals received 0, 10, 25, or 50 mg/kg fenthion (sc) 2 days/wk for 13 weeks. Cholinesterase (ChE) activity, muscarinic receptor (mChR) density and carbachol-stimulated release of inositolphosphates (IP) were measured in the retina and frontal cortex (Johnson & Russell, 1975; Yamamura & Snyder, 1974; Berridge et al., 1982). In the acute dosing experiment, although both ChE activity and mChR density decreased with increasing dose, there appeared to be no correlation among the depression in ChE activity, down-regulation of mChR and depression of IP release. Moreover, carbachol-stimulated IP release was depressed only in the retina; fenthion (50 or 100 mg/kg) produced a decrease in retinal IP release which was still evident 56 days after the single fenthion dose. In the repeated dosing regimen, muscarinic receptor response was depressed at all doses in both tissues by 45 days into the dosing period; however, by 104 days after the last fenthion administration, the cortical carbachol-stimulated IP release was normal, whereas the retinal IP release remained depressed in the animals which received the 25 or 50 mg/kg dosage. We conclude that (i) a single exposure to fenthion produces a dose-dependent decrease in muscarinic receptor function specifically in the rat retina; whereas (ii) repeated dosing with fenthion depressed muscarinic receptor response both in the retina and cortex; (iii) only the retina showed persistent muscarinic receptor dysfunction.

REFERENCES

Berridge, M. J., Downes, C. P., & Hanley, M. R. (1982) Biochem. J. **206**, 587-595.
Imai, H., Miyata, M., Uga, S., & Ishikawa, S. (1983) Environ. Res. **30**, 453-465.
Johnson, C. D., & Russell, R. L. (1975) Anal. Biochem. **64**, 229-238.
Karczmar, A.G. (1984) Fund. Appl. Toxicol. **4**, S1-S17.
Tandon, P., Padilla, S., Barone, Jr., S., Pope, C.N., & Tilson, H.A. (1994) Toxicol. Appl. Pharmacol. **125**, 271-280.
Yamamura, H. I., & Snyder, S. H. (1974) Brain Res. **78**, 320-326.

PROTECTION OF GUINEA PIGS AGAINST SOMAN INHALATION BY PRETREATMENT ALONE WITH HUMAN BUTYRYLCHOLINESTERASE

N. Allon,[1] L. Raveh,[1] E. Gilat,[1] J. Grunwald,[1] E. Manistersky,[1] E. Cohen,[2] and Y. Ashani[1]

[1] Israel Institute for Biological Research
Ness Ziona, Isreal
[2] The Hebrew University
Faculty of Agriculture
Rehovot, Israel

Multi-drug therapy (i.e. pyridostigmine, atropine, oximes, diazepam) is generally considered as effective against organophosphorus (OP) toxicity. However, this treatment raises several problems: 1. It is limited in its protection range. 2. It does not eliminate several post exposure symptoms. and 3. For fast acting OPs self injection of atropine and oximes may not be feasible. Prophylactic antidotes that will scavenge the OP in the blood may afford reasonable solutions for these limitations. Human butyrylcholinesterase (HuBChE) was previously evaluated by us both *in vitro* and *in vivo* as a single prophylactic antidote against the lethal effects of nerve agents. Remarkable protection has been demonstrated in HuBChE-treated mice (Raveh et al., 1993), rats (Raveh et al.,1993; Brandeis et al., 1993) and monkeys (Ashani et al., 1993) following an iv exposure to lethal doses of sarin, soman, VX and tabun. Since inhalation challenge is the most realistic simulation of exposure to nerve agents, the protection afforded by HuBChE was tested against inhaled soman. Awake animals caged in a whole body plethysmograph designed and built in our laboratory, were exposed for 45 to 75 sec to 417-450 µg/L soman. Five out of 8 animals with 28 to 45 nmol circulating HuBChE/animal were completely protected against $1.3\text{-}2.4\text{x}LD_{50}$ doses of inhaled soman. Two animals displayed slight tremors and ataxia while one guinea pig exposed to $2.6\text{x}LD_{50}$ died. A linear correlation (r = .946) was established between nmols of inhaled soman and the reduction in the levels of circulating HuBChE with a slope of 0.13. Since the soman to HuBChE ratio required to inhibit *in vivo* the exogenous enzyme is approximately 1.2, it is suggested that only 16% of inhaled soman reached the circulation.

A group of 3 guinea pigs pretreated with 110 nmol (10mg/animal) HuBChE was exposed to soman vapor in four successive sessions at 1h intervals. Animals were protected against a single dose of as high as $2.0\text{x}LD_{50}$ soman and accumulated dose of up to $5.5\text{x}LD_{50}$.

A plot of residual activity of HuBChE vs. the dose of inhaled soman gave a nonlinear titration curve of HuBChE in the blood. This nonlinearity is consistent with the involvement of other detoxification processes that compete with HuBChE on the sequestration of soman. The development of physiological tolerance to successive exposure to soman vapor may explain, in part, the observation that the overall efficacy of pretreatment with HuBChE was higher than the predicted value based on blood levels of exogenously administered HuBChE.

Compared with prophylaxis with HuBChE alone, the protective ratio in control animals pretreated with pyridostigmine and administered conventional post exposure therapy was approximately 2.3; however, in contrast to HuBChE-treated guinea pigs, all surviving animals showed severe toxic symptoms. Of all antidote regimens tested in our laboratory, prophylaxis alone with HuBChE afforded an almost perfect protection against inhalation of soman.

ACKNOWLEDGMENT

This work was supported by the U.S. Army Medical Research and Development Command under Contract DAMD17-90-C-0033.

REFERENCES

Ashani, Y., Grunwald, J., Grauer, E., Brandeis, R., Cohen, E., and Raveh, L., 1993, *Proc. of the 1993 Medical Defense Bioscience Review*, USAMRDC, Baltimore, MD, pp. 1025-1034.

Brandeis, R., Raveh, L., Grunwald, E., Cohen, E., and Ashani, Y., 1993, *Pharmacol. Biochem. Behav.*, 46:889-896.

Raveh, L., Grunwald, J., Marcus, D., Papier, Y., Cohen, E., and Ashani, Y., 1993, *Biochem. Pharmacol.* 45:2465-2474.

HUMAN BUTYRYLCHOLINESTERASE AS PROPHYLAXIS TREATMENT AGAINST SOMAN

Behavioral Test in Rhesus Monkeys

E. Grauer,[1] L. Raveh,[1] J. Kapon,[1] J. Grunwald,[1] E. Cohen,[2] and Y. Ashani[1]

[1] Israel Institute for Biological Research
Ness-Ziona, Israel
[2] The Hebrew University
Faculty of Agriculture
Rehovot, Israel

Pretreatment with butyrylcholinesterase from human origin (HuBChE) was shown by us to prevent lethality in mice and rats (Raveh et al., 1993) and monkeys (Ashani et al., 1993) and to prevent soman-induced cognitive deficits in rats (Brandeis et al., 1993), without the need for post exposure therapy. Here, HuBChE was tested for its ability to prevent, when administered alone, the behavioral toxicity induced by soman exposure in rhesus monkeys. In addition, HuBChE itself was tested for possible behavioral side effects that will prevent its use in prophylactic therapy. To this end, animals were assigned to one of three treatments: 1. HuBChE alone (13 (n = 3) and 34 (n = 1) mg, iv). 2. Pretreatment with pyridostigmine followed by exposure to soman (0.9 and $3.3xLD_{50}$, iv, n = 4) and post treatment with TMB4/atropine/benactyzine (TAB). 3. Prophylactic administration of HuBChE (21-26 mg/monkey) followed by exposure to soman (3.2-$3.6xLD_{50}$, iv, n = 4). The behavioral test was a spatial discrimination paradigm. In the HuBChE alone group, three monkeys showed no behavioral alterations at all following this single administration, while one monkey treated with the low dose, showed a transient increase in some parameters immediately following the injection and a day later. In group 2, treated with the currently available treatment, severe toxic signs were seen in all animals immediately following exposure and monkeys exposed to the higher soman dose did not perform the task for the next two days. In addition, during performance of the behavioral task, some of these animals showed a continuous impairment in working memory. In group 3, four monkeys were pretreated with HuBChE and exposed to the higher dose range of soman (approx. 3.2 to $3.6XLD_{50}$ of soman). One animal was completely protected by the enzyme and showed no toxicity sign following exposure and no behavioral alterations afterward. In two others animals a slow, unharacteristic development of mild toxicity signs developed immediately following soman, and their behavioral performance was impaired similarly to that seen after treatment with TAB. In the fourth monkey typical toxic signs developed following exposure to the poison.

Approximately 1:1 nmol ratio of soman to HuBChE was found to provide a reasonable protection against multiple LD_{50} of soman with minimum manifestation of behavioral deterioration. If the enzyme does not afford full protection, no prediction can be made about the severity of the ensuing behavioral deficits. The assessment of the antidotal efficacy of the enzyme in non-human primates increases the reliability of the animal-to-human extrapolation of the prophylaxis use of HuBChE.

ACKNOWLEDGMENT

This work was supported by the U.S. Army Medical Research and Development Command under contract DAMD17-90-C-0033.

REFERENCES

Ashani, Y., Grunwald, J., Grauer, E., Brandeis, R., Cohen, E., and Raveh, L., 1993, *Proc. of the 1993 Medical Defense Bioscience Review*, USAMRDC, Baltimore, MD, pp. 1025-1034.

Brandeis, R., Raveh, L., Grunwald, E., Cohen, E., and Ashani, Y., 1993, *Pharmacol. Biochem. Behav.*, 46:889-896.

Raveh, L., Grunwald, J., Marcus, D., Papier, Y., Cohen, E., and Ashani, Y., 1993, *Biochem. Pharmacol.* 45:2465-2474.

EFFICACY OF PROPHYLAXIS WITH HUMAN BUTYRYLCHOLINESTERASE AGAINST SOMAN AND VX POISONING

A Comparative Analysis

L. Raveh,[1] J. Grunwald,[1] E. Cohen,[2] and Y. Ashani[1]

[1] Israel Institute for Biological Research
Ness-Ziona, Israel
[2] The Hebrew University
Faculty of Agriculture
Rehovot, Israel

Prophylactic treatment with human butyrylcholinesterase (HuBChE) as a single pretreatment against organophosphorus (OP) anti-ChE compounds confers excellent protection against soman and VX poisoning in mice, rats (Raveh et al., 1993; Brandeis et al., 1993) and monkeys (Ashani et al., 1993). *In vivo* sequestration of soman and VX by exogenous HuBChE was consistent with the stoichiometry determined *in vitro*. However, when VX or soman were sequentially administered (i.e., *in vivo* titration) manifestation of poisoning symptoms were observed earlier with VX compared with soman-exposed animals. The goals of this study were: a) To assess the HuBChE/OP molar ratios required to completely prevent manifestation of toxic symptoms following poisoning with soman and VX. b) To compare the *in vivo* rate of sequestration of both OPs in the circulation, and c) To provide explanation for the apparent differences in scavenging efficacy of HuBChE towards VX and soman.

The molar ratio HuBChE/OP required to completely prevent post exposure toxic signs were 0.5 and 1.1 for soman and VX respectively, in rats. These values increased to >0.8 and 1.25 in monkeys. In both species a linear relationship was observed between the concentration of HuBChE in blood and the amount of sequestered OP. Thus, one equivalent of the enzyme detoxified in rats approximately 0.5 mol of either soman or VX. In monkeys, one equivalent of HuBChE neutralized 0.9 and 0.5 mol of soman and VX, respectively. The reaction between the OPs and HuBChE in blood, either *in vivo* or *in vitro*, was significantly faster with soman compared with VX. In addition, the rate of the reaction of VX with HuBChE in blood of monkeys was reduced compared with the rate of its sequestration in HuBChE-buffered solution. The minor but consistent differences observed between the rate of sequestration of soman and VX in blood of animals are likely to result from either the differences observed in the bimolecular rate constants, or to differences in the pharmacokinetics properties of the two OPs or to a combination of both factors. Comparative studies utilizing side-by-side fetal bovine serum acetylcholinesterase and HuBChE may shed more light on the mechanisms underlying the observed differences. It is clear, however, that HuBChE provided remarkable protection against both soman and VX without the need for post-exposure treatment.

ACKNOWLEDGEMENT

This work was supported by the U.S. Army Medical Research and Development Command under Contract DAMD17-90-C-0033.

REFERENCES

Ashani, Y., Grunwald, J., Grauer, E., Brandeis, R., Cohen, E., and Raveh, L., 1993, Human butyrylcholi-nesterase as a general scavenging antidote for nerve agents toxicity, *Proc. of the 1993 Medical Defense Bioscience Review*, USAMRDC, Baltimore, MD, pp. 1025-1034.

Brandeis, R., Raveh, L., Grunwald, E., Cohen, E., and Ashani, Y., 1993, Prevention of soman-induced cognitive deficits by pretreatment with human butyrylcholinesterase in rats, *Pharmacol. Biochem. Behav.*, 46:889-896.

Raveh, L., Grunwald, J., Marcus, D., Papier, Y., Cohen, E., and Ashani, Y., 1993, Human butyrylcholinesterase as a general prophylactic antidote for nerve agent toxicity: In vitro and in vivo characterization, *Biochem. Pharmacol.* 45:2465-2474.

PREVENTION OF BRAIN DAMAGE AND BEHAVIORAL PERFORMANCE CHANGES FOLLOWING AN IV INJECTION OF SOMAN AND VX IN RATS PRETREATED WITH HUMAN BUTYRYLCHOLINESTERASE

T. Kadar,[1] L. Raveh,[1] R. Brandeis,[1] J. Grunwald,[1] E. Cohen,[2] and
Y. Ashani[1]

[1] Israel Institute for Biological Research
Ness Ziona, Israel
[2] The Hebrew University
Faculty of Agriculture
Rehovot, Israel

Soman and VX are potent irreversible cholinesterase inhibitors. At near LD_{50} doses, both agents cause severe toxic signs related to excessive cholinergic activity at the peripheral and central nervous system (CNS). Survivors of organophosphorus (OP) poisoning suffer irreversible brain lesions which deteriorate with time and affect their behavior and performance. Therefore, the prevention of these lesions is of utmost importance. The aim of the present study was to evaluate the histopathological effects of soman and VX on the CNS of rats following a single iv injection, to correlate the severity of brain lesions with the behavioral changes and to assess the efficacy of human butyrylcholinesterase (HuBChE) pretreatment in preventing soman and VX-induced brain lesions. Using the Morris Water Maze we have demonstrated that administration of soman (Brandeis et al., 1993) and VX induced significant long-term cognitive deficits in rats that were prevented by a pretreatment with HuBChE. Following the behavioral tests, at approximately 2 weeks and one month post exposure, animals were sacrificed and their brains were processed for histological evaluation. A severe brain damage was observed in most of the non protected animals following administration of 0.9-$1.1 LD_{50}$ of either soman (~65nmol/rat) or VX (~12nmol/rat). Lesions were most prominent in the hippocampal CA1 layer, in the piriform cortex and in the thalamus, and varied in their distribution and severity among animals. It should be pointed out that in several cases animals that showed typical toxic signs did not display brain damage and their performance was normal and similar to that of the control group. This observation is consistent with reports on the variability among animals in their response to OP poisoning.

Pretreatment with 42-44 nmol HuBChE/rat prevented the histopathological damage observed following administration of $1.5xLD_{50}$ soman and VX , even in areas known to be sensitive to OP intoxication such as the hippocampal CA1 layer. Morphobehavioral correlation revealed a close relationship between the cognitive impairment and the pathological findings. Animals showing remarkable cognitive deficits exhibited also severe pathological lesions. Results clearly demonstrate that exogenously administered HuBChE confers excellent prophylactic protection in rats from the toxic effects of both soman and VX, with no need for post exposure therapy.

ACKNOWLEDGMENT

This work was supported by the U.S. Army Medical Research and Development Command under Contract DAMD17-90-C-0033.

REFERENCES

Brandeis, R., Raveh, L., Grunwald, E., Cohen, E., and Ashani, Y., 1993, Prevention of soman-induced cognitive deficits by pretreatment with human butyrylcholinesterase in rats, *Pharmacol. Biochem. Behav.*, 46:889-896.

INACTIVATION OF THE CATALYTIC ACTIVITIES IN HUMAN SERUM BUTYRYLCHOLINESTERASE BY METAL CHELATORS

C. D. Bhanumathy and A. S. Balasubramanian

Neurochemistry Laboratory
Christian Medical College and Hospital
Vellore 632 004
India

Human serum butyrylcholinesterase (BChE) which shows 53% of sequence identity to acetylcholinesterase (AChE) exhibits an arylacylamidase (AAA) and a peptidase activity. Earlier studies in our laboratory have shown that the peptidase activity in BChE was inhibited by metal chelators and metal ions can stimulate the peptidase activity. (Rao and Balasubramanian, 1993). In pursuance of those studies we describe now that metal chelators can inhibit not only the peptidase but also the cholinesterase and AAA activities.

EDTA inhibited the peptidase activity exclusively without affecting the cholinesterase and AAA activities.1,10-phenanthroline and N,N,N',N'-tetrakis (2-pyridyl-methyl) ethylene diamine (TPEN) inhibited all the three activities (cholinesterase, AAA and peptidase). The metal chelators are dissociable by prolonged dialysis resulting in full recovery of activity. Transition metal ions like Zn^{2+}, Co^{2+}, Mn^{2+}, Ni^{2+}, but not Ca^{2+} or Mg^{2+} could partially reverse the cholinesterase, AAA and peptidase activities that was inactivated by TPEN.

EDTA pretreated BChE could bind to a Zinc -chelate Sepharose column and be eluted from the column by EDTA or imidazole. When histidine residues in the enzyme were modified by diethylpyrocarbonate, binding to the column was abolished suggesting that histidine residues are involved in the metal binding. All the three activities (cholinesterase, AAA and peptidase) were co-eluted from the column under all conditions.

The sequence $HGYE(X)_{107}H$ (residues 438-548) is a possible metal binding site in human BChE. Similar sequences are also found in the zinc exopeptidases carboxypeptidase A and B as well as AChE (Balasubramanian and Bhanumathy, 1993). The sequence is conserved in BChE from several sources (Cygler et al., 1993).

The fact that EDTA inhibited only the peptidase but not the cholinesterase and AAA activities suggests that more than one metal binding site is present in BChE. The identity of these metal binding sites in BChE remains to be established.

ACKNOWLEDGMENT

This work was supported by a grant from the Department of Science and Technology, New Delhi.

REFERENCES

Cygler, M., Schrag, J.D., Sussman, J.L., Harel,M., Silman, I.,Gentry, M.K.,and Doctor, B.P.,1993, Relationship between sequence conservation and three-dimensional structure in a large family of esterases, lipases and related proteins, Protein Science 2,366-382.

Rao, R.V., and Balasubramanian, A.S.,1993, The peptidase activity of human serum butyrylcholinesterase : Studies using monoclonal antibodies and characterization of the peptidase, J. Protein. Chem.12,103-110.

Balasubramanian, A.S.,and Bhanumathy, C.D., 1993, Noncholinergic functions of cholinesterases, FASEB J. 7,1354-1358.

PROTECTIVE EFFICACY OF PHYSOSTIGMINE AND PYRIDOSTIGMINE AT VARIOUS PRETREATMENT TIMES AGAINST INHALED SARIN IN RATS

R. Vijayaraghavan, K. Husain, P. Kumar, K. S. Pandey, and S. Das Gupta

Defence Research and Development Establishment
Gwalior-474002, India

Physostigmine (Phy) and pyridostigmine (Pyr) are tertiary and quaternary carbamates, respectively. They reversibly inhibit acetylcholinesterase. These carbamates have been found to be good prophylactic agents for organophosphorus (OP) intoxication in various animal species. In most of the cases the effectiveness of the carbamates has been established against parenterally administered OP compounds. In the present study the effectiveness of a symptom-free dose (a dose that failed to produce any obvious effects) of Phy and Pyr at varous pretreatment intervals against sarin aerosols in rats is examined.

Wistar male rats were exposed to sarin aerosols to a concentration of 51.2 mg/m^3 in a dynamically operated whole body exposure chamber. They were pretreated either with saline (control) or Phy (0.1 mg/kg) or Pyr (0.075 mg/kg). Various pretreatment times were used for Phy and Pyr (50, 35, 20 and 5 min) prior to sarin exposure. The rats were either (i) sacrificed immediately after 15 min of exposure to sarin for the estimation of cholinesterase (ChE) in cerebral cortex, spinal cord, lung and blood, or (ii) allowed to die in the chamber while being continuously exposed to sarin for the study of survival time. A group of rats, treated with Phy or Pyr at various pretreatment times without exposure to sarin, was kept for the estimation of ChE in different tissues for comparison.

Following exposure to sarin aerosols, the ChE activity was significantly decreased in all the tissues, with lung showing maximum inhibition of 53.8% and spinal cord showing minimum inhibition of 36.9%. Phy given alone significantly inhibited ChE activity in cerebral and peripheral tissues 20 min after treatment, whereas Pyr inhibited ChE activity only in peripheral tissues. Phy given 20 min prior to sarin exposure reduced the inhibition of ChE by sarin in cerebral as well as peripheral tissues, and significantly increased the survival time. Pyr given 20 min prior, reduced inhibition of ChE only in peripheral tissues and the increase in the survival time was less than with Phy. Phy given 35 min or 5 min prior to sarin increased the survival time whereas Pyr did not. Pretreatment with Phy or Pyr 50 min prior to sarin exposure did not protect the inhibition of ChE by sarin in all the tissues and there was no significant change in the survival time.

The present study demonstrates that pretreatment with Phy gives better protection than Pyr and the pretreatment tim 20 min prior to exposure is optimal for protection against sarin administered by inhalation.

INDUCTION OF CARBOXYLESTERASE ISOENZYMES AND ALTERED ACETYLCHOLINESTERASE DURING OP RESISTANCE IN MOSQUITO *CULEX QUINQUEFASCIATUS*

N. Gopalan, S. Prakash, K. M. Rao and B. K. Bhattacharya

Defence Research and Development Establishment
Gwalior-474002, India

Insecticide resistance development is an ability in a strain of insects to tolerate doses of toxicants which could otherwise prove to be lethal to the majority of individuals in a normal population of the same species. So far 700 insect species have been reported to have developed insecticide resistance (Hollomon, 1993; WHO, 1957). Malathion, an organophosphorus ester (OP) insecticide, is widely used for control of mosquitoes. The biochemical mechanism of development of malathion resistance has been attributed to either change in susceptibility of the target enzyme acetylcholinesterase (AChE) to the insecticide or enhanced insecticide detoxifying enzyme carboxylesterase (CaE) activity (Rajagopal, 1977).

In the present study, we investigated the development of malathion resistance in laboratory culture as well as field collected filariasis vector *Culex quinquefasciatus*. In cultured mosquito samples, the larvae were given a constant insecticide pressure of LC_{90} at every generation up to the 25th generation. There was many-fold increase in LC_{50} and LC_{90} values. Though total AChE activity did not change significantly, IC_{50} values of paraoxon, diisopropylfluorophosphate, N,N'-diisopropylphosphorodiamidic fluoride and propoxur increased nearly 1.2 to 10-fold, indicating lesser sensitivity of AChE from resistant larvae towards these inhibitors.

Total CaE activity in larvae from resistant generations increased significantly at the end of the 25th generation using α-naphthyl acetate as substrate. Determination of the Michaelis constant (K_m) of partially purified CaE obtained from resistant larvae showed differential affinities for different substrates. Maximum affinity was observed for α-naphthyl propionate with two distinct K_m values (0.058 mM and 1.80 mM) followed by α-naphthyl butyrate (0.457 mM) and α-naphthyl acetate (0.601 mM). Two K_m values for α-naphthyl propionate is probably due to differential affinities of CaE isoenzymes for the substrate. To study the different isoenzymes, partially purified CaE was separated on 7.5% acrylamide gels and enzyme was stained by incubating the gels in buffered substrates. Naphthol produced was stained using fast blue dye. With α and β-naphthyl acetate intense A1 and A2 isoenzymes were dominant CaE bands. With α and β-naphthyl propionate, an additional B2 band became distinct, while with α and β-naphthyl butyrate only the B2 band was prominent. In heterogeneous resistant mosquito populations in the field, in which total CaE activity at times does not increase significantly, a differential isoenzyme profile would give a better marker for detection of malathion resistance. Arrhenius temperature plots showed a lowered transition temperature for CaE activities with all the above three substrates in resistant samples, showing a metabolic adaptation process.

REFERENCES

Holloman, D.W. (1993) Chemistry & Industry **22**, 892-895.

Rajagopal, R. (1977) Ind. J. Med. Res. **66**, 27-28.

WHO (1957) Technical Report Series **125**.

STRUCTURE-ACTIVITY OF PYRIDINIUM OXIMES AS ANTIDOTES TO ORGANOPHOSPHORUS ANTICHOLINESTERASE AGENTS

A. K. Sikder, S. N. Dube, and D. K. Jaiswal

Defence Research and Development Establishment
Gwalior-474002, India

The mechanism of toxicity of organophosphorus (OP) nerve agents is by irreversible inhibition of acetylcholinesterase (AChE). OP esters first phosphorylate the serine hydroxyl moiety at the enzyme active site. Subsequent O-dealkylation of one of the ester side chains of the OP ester leads to irreversible enzyme inhibition by the phenomenon called aging. Therapy in OP toxicity is based either on alleviating the effects of excess acetylcholine (ACh) on receptors by shielding the receptors, viz. atropine (muscarinic antagonist), or use of pyridinium oximes, which reactivate the inhibited enzyme. A disadvantage in using quaternary pyridinium aldoxime is its impermeability through the blood brain barrier and hence nonavailability for brain AChE reactivation.

In this study, keeping the above in view, we synthesized lipophilic long chain N-alkyl 2- and 4-(hydroxyiminomethyl)pyridinium bromides with alkyl chain lengths ranging between 8 and 16 carbon atoms. Determination of acid dissociation constants showed lower pK_a values for 2-substituted derivatives (7.78-7.87) showing their dissociability at physiological pH, as well as reactivating potential, while 4-substituted derivatives showed higher pK_a values (8.3-8.5). Increase in lipophilicity decreased the dissociation constant with significant effect on reactivation of sarin (isopropyl methylphosphonofluoridate) inhibited AChE. *In vivo* experiments showed greater reactivation of brain AChE, inhibited after administration of sarin, and the protection index (increase in LD_{50}) increased compared to a conventional oxime like pralidoxime.

Another series of compounds was synthesized containing asymmetrically substituted 3,3′-bispyridinium mono oximes bridged by 2-oxopropane groups, in which the pK_a of the oximino group ranged between 8.3 and 9.24. Though these compounds were not good *in vitro* reactivators of diisopropylfluorophosphate (DFP) inhibited AChE, they are effective in protecting against DFP and sarin poisoning in the mouse. Acetylcholine receptor mediated beneficial effects at neuromuscular junctions have been envisaged as the probable reason for such protective actions.

A NON-CHOLINERGIC FUNCTION FOR ACETYLCHOLINESTERASE

S. A. Greenfield

University Department of Pharamcology
Mansfield Road, Oxford
OX1 3QT, United Kingdom

BACKGROUND

The idea that acetylcholinesterase (AChE) may have non-cholinergic functions is not new. For twenty years it has been known that certain brain regions, such as the substantia nigra, contain large amounts of AChE but disproportionately low levels of its conventional substrate acetylcholine and its synthesising enzyme choline acetyltransferase (Silver, 1974). Moreover it has also been established for almost twenty years that AChE exists in soluble forms (Chubb and Smith, 1975a) one of which is released (Chubb and Smith, 1975b). The substantia nigra is one particular brain area where there is a disproportionately large amount of AChE relative to only modest levels of choline acetyltransferase: moreover a form of AChE is released in a fashion that is insensitive to agents either stimulating or blocking the cholinergic receptor (Greenfield, 1991).

It seems likely therefore that in the substantia nigra, AChE has a novel, non-cholinergic role. This role appears more related to the functioning of dopaminergic neurons since AChE is released almost exclusively from the somata/dendrites of these cells (Greenfield, 1991). Moreover, within the substantia nigra, AChE and dopamine appear to be intimately related: AChE secretion is evoked by direct application of amphetamine (Greenfield and Shaw, 1982) and reduced by alpha-methyl-para-tyrosine, which blocks *de novo* dopamine synthesis (Dally and Greenfield, 1994). Once in the extracellular space, secreted AChE may underlie an unusual form of neuronal signalling, involving an action independent of hydrolysis of the substrate acetylcholine. Non-cholinergic actions of AChE have been observed in both *in vivo* (Last and Greenfield, 1987) and *in vitro* (Greenfield et al., 1989) recordings from nigral neurons themselves. In addition raised levels of both endogenous (Jones and Greenfield, 1991) and exogenous AChE (Greenfield et al., 1984; Weston and Greenfield, 1986) within the substantia nigra have been associated with increases in movement as a result of enhanced activity of the nigrostriatal pathway (see Greenfield, 1991).

'Non-cholinergic' AChE may play an important role not just in the normal, mature substantia nigra, but in pathological conditions as well. It is well known that degeneration of the DA-containing neurons of the substantia nigra are pivotal to the pathology of

Enzymes of the Cholinesterase Family, Edited by Daniel M. Quinn et al.
Plenum Press, New York, 1995

Parkinson's disease. However it is only now being realised, through circumstantial evidence, that AChE, acting in a non-cholinergic capacity, might be fundamentally linked to both neuronal homeostasis and regeneration. First, it has long been established that AChE has a non-classical role in development (see Silver, 1974): indeed certain regions of brain, such as the auditory cortex, display transient AChE staining which vanishes by maturity (Kutscher, 1991): moreover antibodies to AChE can produce functional deficits (Bean et al., 1991). Secondly, Layer and co-workers (1993) have shown that AChE can affect neurite outgrowth in the retina independent of its conventional catalytic action, whilst Arendt et al. (1992) have discovered that during neuronal degeneration, the distribution of different forms of AChE in the adult reverts to that seen in the immature brain. Thirdly, an abnormal form of AChE is released during degeneration in Alzheimer's disease (Navartnam et al., 1991). In this disorder there is also selective loss of secretable (compared to membrane bound) AChE from the adrenal medulla, an organ obviously not causally implicated in CNS degeneration (Appleyard and McDonald, 1991). Not only does this peripheral organ actually share a common feature with brain tissue, namely the secretion of AChE, but in addition the adrenal medulla and brain both show a decreased catecholamine content in Parkinson's disease (Stoddard et al, 1989). These findings suggest that secretion of AChE may be a pivotal factor in neurodegeneration. Fourthly, AChE levels are significantly depleted in the Parkinsonian substantia nigra (McGeer and McGeer, 1971), and indeed release of AChE from the rat substantia nigra is virtually abolished following extensive destruction of dopaminergic neurons by the selective toxin 6-hydroxydopamine (Greenfield et al., 1983). Moreover, ultrastructural studies have confirmed that within the substantia nigra, only dopaminergic nigrostriatal neurons are likely to release AChE (Henderson and Greenfield, 1984). Fifthly, introduction of exogenous AChE into the cisterna magna of the rat model of hemi-parkinsonsism can have an ameliorative effect, lasting for several days, on the motor disorder (Greenfield et al., 1986). Hence, the processes underlying development and degeneration may be in some way linked, with 'non-cholinergic' AChE as the common factor. Finally, in a cell-free system, AChE has actually been shown to form a bond with dopamine,

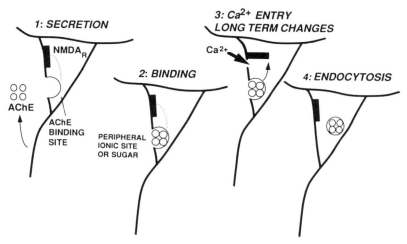

Figure 1. Scheme showing the possible four steps in the non-classical signalling of AChE in a nigral neuron, depeicted with only proximal sections of dendrites. The AChE molecule is shown as a tetramer, the binding site by the invagination and the NMDA receptor by the black rectangle.

the transmitter deficient in Parkinson's disease, and thereby to retard its autoxidation (Klegeris et al., 1994).

However if 'non-cholinergic' AChE really represents a novel form of neuronal signalling, in both health and disease, then we would expect to be able to describe the phenomenon in terms of four criterion: (i) calcium-dependent release contingent upon a specific pre-synaptic signal, (ii) binding to a specific target site, (iii) an action demonstrably distinct from the hydrolysis of the conventional substrate acetylcholine, (iv) removal from the extracellular target.

(i) EVOKED RELEASE OF AChE

A picture is gradually emerging (Greenfield, 1985, 1991) of some of the factors that influence the functional state of nigral cells and concomitant secretion of AChE. At the ultimate ionic level, secretion is independent of overall neuronal excitability and indeed is resistant to blockade of fast sodium channels by tetrodotoxin (TTX). Rather, AChE is released from the dendrites of nigrostriatal cells in a calcium-dependent manner: more specifically, an L-type calcium channel, exclusive to the dendrites, has been implicated at least in part as fulfilling the requirements for dendritic secretion of AChE (Llinas et al., 1984; Nedergaard et al., 1988, 1989). This conductance, which would operate independent of events at the soma, appears to be triggered more by incoming afferent inputs. Hence secretion of AChE in the substantia nigra may well differ from release of substances at the classic axonal synapse, in that it does not reflect the activity of the cell from which it is released, but rather the activity of the incoming inputs.

More recently, we have thus focused attention on the major inputs to the substantia nigra. Application of substances known to be released by these afferents, i.e. glutamate (NMDA) (Jones et al., 1994a) and serotonin (Dickie and Greenfield, 1994) does indeed evoke secretion of AChE. Moreover, stimulation of major inputs, namely striatum (Taylor and Greenfield, 1989), dorsal raphe nucleus (Dickie and Greenfield, 1994) and subthalamus (Jones et al., 1994) all result in enhanced secretion of dendritic AChE. The critical question is then, how afferent-evoked secretion of AChE from dopaminergic dendrites in the substantia nigra is subsequently relevant to movement.

Studies on the functional relevance of this evoked secretion of AChE in the substantia nigra have been hampered by the poor time resolution of established assays. However we have adapted a chemiluminscent assay for acetylcholine (Israel and Lesbatts, 1981) for detecting secretion of AChE 'on-line' *in vitro* (Llinas and Greenfield, 1987) and subsequently *in vivo* (Taylor et al., 1989, 1990; Taylor and Greenfield, 1989). Using this system it has been possible to demonstrate that AChE secretion is sensitive to ongoing movements (Jones and Greenfield, 1991; Jones et al., 1994a) and to the sensory stimulation of a flashing light (Jones et al., 1991).

(ii) A POSSIBLE BINDING SITE FOR 'NON-CHOLINERGIC' AChE

Any binding site for AChE has yet to be demonstrated. However indirect evidence suggests that AChE acting in a non-cholinergic capacity could indeed bind to the external surface of target neurons and glia. When exogenous, purified AChE that has been catalytically inactivated, is introduced into the substantia nigra, there is a significant displacement of endogenous AChE from the substantia nigra specifically, but not from neighbouring brain areas serving as controls (Dickie et al., 1994). In addition, it is worth noting that the idea of a specific binding site for AChE on the plasmamembrane has a precedent, in the light of the recent identification and cloning of a receptor for a protein (phospholipase A2) which, like AChE, is also a secreted esterase (Lambeau et al., 1994).

(iii) NON-CHOLINERGIC ACTION OF AChE

A non-cholinergic action of AChE has already been demonstrated upon neurite outgrowth in the chick retina, (Layer et al., 1993). We have now shown a similar trophic action of AChE on neurite outgrowth in the developing substantia nigra maintained in 'organotypic' cell cultures: not only does additional AChE promote neuron growth but AChE, independent of its catalytic site, actually appears essential for nigral cell survival. Incubation of the nigral cell cultures with an inhibitor of the catalytic site (ecothiophate) has no effect on the apparent health of dopaminergic cells. However, application of BW 284C51, which blocks in addition a peripheral anionic site on the AChE molecule, results in a significant reduction in surviving neurons (Jones et al., 1994b).

Since neuronal plasticity and development can be associated with influxes of calcium ions acting as intracellular triggers, it was tempting to explore whether the non-cholinergic action of AChE might involve calcium entry into neurons. We are currently using intracellular recording of nigral neurons *in vitro* to test this hypothesis. A certain rostral population of neurons in the *in vitro* substantia nigra displays, in the presence of particular sodium and potassium ion channel blockers (tetrodotoxin and tetraethylammonium), a calcium-mediated autorhythmicity. These calcium-dependent potentials are markedly enhanced following addition of AChE. Moreover, recombinant AChE raised from either human or murine sources has similar effects to material purified from commercial preparations from electric eel. Hence any actions of contaminants can be eliminated. In addition, we can also eliminate sugar residues on the AChE molecule itself as playing any important role in this action on neurons.[*] The human AChE, which was raised in E.Coli and was thus non-glycosylated, had actions on nigral neurons that were indistinguishable from the glycosylated recombinant mouse AChE.

Preliminary experiments with selective inhibitors of different parts of the AChE molecule, as performed already in the tissue cultures of substantia nigra described above, are suggesting that again, it is the peripheral anionic site that may participate in the non-cholinergic actions of AChE. It is interesting to note that this peripheral site (Hucho et al., 1991), which is not actually needed in the hydrolysis of acetylcholine, would nonetheless contribute to the characteristic phenomenon of substrate inhibition of AChE: in neurodegeneration, this substrate inhibition virtually vanishes (Arendt et al., 1992), suggesting that modifications to the peripheral site may well play a part in the involvement of 'non-cholinergic' AChE with neurodegeneration, as suggested in section (i).

'Non-cholinergic' AChE might thus induce calcium entry into neurons in ways that are important for the developing/degenerating substantia nigra, as well as for the normal functioning of dopaminergic neurons in the adult. Indeed the entry of calcium in this way could in turn trigger intracellular changes which would account for the long-term effects on movement observed following AChE administration (Greenfield et al., 1984; Hawkins and Greenfield, 1992).

However, calcium influx into neurons is not always beneficial, but in excess can result in 'excitotoxicity' where the mitochondria become swollen with calcium and oxidative phosphorylation ceases (Greenfield, 1992; Turski and Turski, 1993). The nigral neurons used in our studies are particularly sensitive to such reductions in intracellular ATP since they

---

[*] On the other hand, AChE may bind to glia using sugar residues, see Klegeris et al., this volume.

possess particular potassium channels ('K-ATP' channels) that are activated when cell metabolism starts to become compromised (Murphy and Greenfield, 1992). Hence, in these neurons, any excessive entry of calcium ions could result in a hyperpolarization. Preliminary data do indeed show that high doses of the excitatory amino acid can result in a hyperpolarization that is reversed by the K-ATP channel blocker tolbutamide. A similar effect can also be seen with prolonged applications of AChE: moreover NMDA applied in the presence of AChE can induce a hyperpolarization in a neuron where either substance on its own has no marked effect. These current observations thus suggest that AChE can enhance calcium entry into nigral neurons, possibly by modification to the NMDA channel. This hypothesis is consistent with recording from neurons in the cerebellum where AChE, acting in a non-cholinergic capacity, can enhance the effects of excitatory amino acids (Appleyard and Jahnsen, 1992).

(iv) TERMINATION OF 'NON-CHOLINERGIC' AChE

Clues from earlier work suggest that the action of 'non-cholinergic' AChE could be terminated by uptake into nigral cells. Studies on the behavioural actions of AChE infused into the substantia nigra have shown that the action can be long-term (a matter of days), and thus not readily attributable to the continued presence of the protein molecule in the extracellular space (Hawkins and Greenfield, 1992). Furthermore, the uptake of AChE by endothelial cells of capillaries in the superior cervical ganglion, a region also secreting the protein, has already been demonstrated in ultrastructural studies showing the presence of AChE-reactive coated pits (Jessen et al., 1978). Ultrastructural studies in the substantia nigra itself (Henderson and Greenfield, 1984) have revealed the presence of AChE in a class of neuron (non-dopaminergic pars reticulata cells) where the protein is nonetheless undetected in the Golgi apparatus. If these cells are synthesising AChE, either it is arrested in the rough endoplasmic reticulum (RER), or is in very rapid transit through the Golgi apparatus. Alternatively, the AChE detected in these cells may be extracellular in origin. Hence a possible interpretation of AChE deposits seen outside the RER of these non-dopaminergic cells is that they have been endocytosed from elsewhere.

More recently we have obtained very suggestive evidence that nigral neurons can indeed take up AChE. In brief, exogenous AChE is selectively retained in the substantia nigra *in vivo* in an energy-dependent manner: selective retention of the protein does not occur at 4°C nor *post mortem*; nor is any retention observed when exogenous AChE is perfused into regions outside of the substantia nigra. Moreover, this phenomenon is not attributable to the action of particular nigral proteases, nor to a facilitated site-selective trapping or conversely diffusion of AChE. The most obvious conclusion from this data then is that AChE is selectively endocytosed by nigral cells (Dickie et al., 1994).

In a pilot collaboration with Dr David Vaux in the Sir William Dunn School of Pathology, Oxford, we have demonstrated *directly* with the use of fluorescent imaging techniques that immunocytochemically identified dopaminergic neurons can become double labelled after local infusion with tagged AChE. We have observed colocalization of deposits of labelled, exogenous AChE with tyrosine hydroxylase containing cells: this effect is not seen in brain regions outside of the substantia nigra, nor when vehicle lacking labelled AChE is perfused instead (see Dickie et al., 1994). Hence AChE might have a non-cholinergic action at the level of the cell surface, analogous for example to insulin, in that in both cases modulatory actions occur at the plasmamembrane yet are terminated by admission into the interior of the cell.

CONCLUSIONS

The substantia nigra serves as a good example of a brain region where AChE may function in a non-cholinergic capacity to underscore a novel form of neuronal signalling. There is some evidence that the protein is released from neurons in response to a specific incoming signal such as activation of the NMDA receptor by glutamate, that it binds competitively to a specific extracellular site by means of the peripheral anionic site and thereby enhances calcium entry through the NMDA receptor complex, whereupon the AChE molecule is endocytosed into the neuron. However, we still need to know a great deal more about these events. Perhaps one of the most intriguing questions concerns the possible interaction of the AChE molecule with part of the NMDA receptor complex. One attractive candidate target for the AChE acting in this way is the modulatory subunit of the NMDA receptor, activation of which causes a tenfold enhancement in calcium influx, and which presents a 550 amino acid sequence into the extracellular space (Seeburg, 1993).

One final consideration is whether 'non-cholinergic' AChE might ever be exploited in developing a novel therapy for Parkinson's disease and other degenerative disorders. We have seen that, by promoting calcium influx, the nonclassical action of AChE would be beneficial in a physiological situation, but harmful in pathological situations entailing excessive calcium entry and /or when cell metabolism was already marginalised. Hence in treating already existing degenerative conditions, blocking the non-cholinergic action of AChE might reduce an important contributing factor, that of excitotoxicity. This would be an appealing strategy compared with more direct attempts to block all calcium entry into neurons, which would also block any basis for neuronal adaptation. Instead, by blocking only the enhancing modulation effected by 'non-cholinergic' AChE, calcium entry via NMDA receptors would still be possible, and a potentially pathological situation diverted into a physiological one. In this way a fuller understanding of novel neuronal mechanisms such as 'non-cholinergic' AChE may open up a truly new and productive approach to combating neurodegeneration.

REFERENCES

Appleyard, M.E., & McDonald, B. (1991) Lancet 338, 1085-1086.
Appleyard, M., & Jahnsen, H. (1992) Neuroscience. 47, 291-301.
Arendt, T., Bruckner, M.K., Lange, M., & Bigl, V. (1992) Neurochem. Int. 21, 381-396.
Bean, A. Xu-Z., Chai, S.Y., Brimijoin, S., & Hokfelt, T. (1991) Neurosci. Lett. 133, 145-149.
Chubb, I.W., & Smith, A.D. (1975a) Proc. Roy. Soc. Lond. B 191, 245-261.
Chubb, I.W., Smith, A.D. (1975b) Proc. Roy. Soc. Lond. B 191, 263-269.
Dally, J. J., & Greenfield, S. A. (1994) Neurochem. Int. 25, 339-344.
Dickie, B., & Greenfield, S.A. (1994) NeuroReport 5, 769-772.
Dickie, B., Budd, T.J., Vaux, D.J., Greenfield, S.A. (1994) Eur. J. Neurosci. (In Press).
Greenfield, S.A. (1985) Neurochem. Int. 11, 55-77.
Greenfield, S.A. (1991) Cell. Mol. Neurobiol. 11, 55-77.
Greenfield, S.A. (1992) Essays in Biochemistry 27, 103-118.
Greenfield, S.A., & Shaw, S.G. (1982) Neuroscience 7, 2883-2893.
Greenfield, S.A., Grunewald, R.A., Foley, P., & Shaw, S.G. (1983) J. Comp. Neurol. 214, 87-92.
Greenfield, S.A., Jack, J.J.B., Last, A.T.J., & French, M. (1988) Exp. Brain Res. 70, 441-444.
Greenfield, S.A., Nedergaard, S., & French, M. (1989) Neuroscience 29, 21-25.
Greenfield, S.A., Chubb, I.W., Grunewald, R.A., Henderson, Z., May, J., Protnoy, S., Weston, J., & Wright, M.C. (1984) Exp. Brain Res. 54, 513-520.
Greenfield, S.A., Appleyard, M.E., & Bloomfield, M.R. (1986) Behav. Brain Res. 21, 47-54.
Hawkins, C.A., & Greenfield, S.A. (1992) Behav. Brain Res. 159-163.
Henderson, Z., & Greenfield, S.A. (1984) J. Comp. Neurol. 230, 278-286.
Hucho, F., Jarv, J., & Weise, C. (1993) Pharmacol. Sci. 12, 422-426.

Israel, M., & Lesbats, B. (1981) Neurochem. Int., 3, 81-90.

Jessen, K.R., Chubb, I.W., & Smith, A.D. (1978) J. Neurocytol. 7, 145-154.

Jones, S.A., & Greenfield , S.A. (1991) Eur. J. Neurosci. 3, 292-295.

Jones, S. A., Dickie, B. G. M., Klegeris, A., & Greenfield, S. A. (1994) J. Neural Transm. (in press).

Jones, S. A., Holmes, C., T.C., B., & Greenfield, S. A. (1994) Cell Tissue Res. (in press).

Klegeris, A., Korkina, L. G., & Greenfield, S. A. (1994) Free Radic. Biol. Med. (in press).

Llinas, R. R., & Greenfield, S. A. (1987) Proc. Natl. Acad. Sci. USA. 84, 3047-3050.

Kutscher, C.L. (1991) Brain Res. Bull. 27, 641-649.

Last, A. T. J., & Greenfield, S.A. (1987) Exp. Brain Res. 66, 394-400.

Lambeau, G., Anciaqn, P., Barhanin, J., & Lazdunski, M. (1994) J. Biol. Chem. 269, 1575-1578.

Layer, P., Weikart, T., & Alber, R. (1993) Cell Tissue Res. 273, 219-226.

McGeer, P.L., & McGeer, E.G. (1971) Arch. Neurol. 25, 265-268.

Murphy, K.P. J.S., & Greenfield, S.A. (1992) J. Physiol. 453, 167-183.

Navaratnam, D.S., Priddle, J.D., McDonald, B. Esiri, M.M., Robinson, J.R., & Smith, A.D. (1991) Lancet, 337, 447-450.

Nedergaard, S., Bolam, J.P., & Greenfield, S.A. (1988) Nature 333, 174-177.

Nedergaard, S., Webb, C., & Greenfield, S.A. (1989) Acta Physiol. Scand. 135, 67-68.

Seeburg, P.H. (1993) TiPS 14, 297-303.

Silver, A. (1974) The Biology of the Cholinesterases, North Holland, Amsterdam.

Stoddard, S.L., Tyce, G.M., Ahlskog, J.E., Zinsmeister, A.R., & Carmichael, S.W. (1989) Exp. Neurol. 104, 22-27.

Taylor, S.J., & Greenfield, S.A. (1989) Brain Res. 505, 153-156.

Taylor, S. J., Haggblad, J., & Greenfield, S. A. (1989) Neurochem. Int. 15, 199-205.

Taylor, S.J., Jones, S.A., Haggblad, J., & Greenfield, S.A. (1990) Neuroscience 37, 71-76.

Turski, L., & Turski, W.A. (1993) Towards an understanding of the role of glutamate in neurodegenerative disorders: energy metabolism and neuropathology.

Weston, J., & Greenfield, S.A. (1986) Neuroscience 17, 1079-1088.

NON-CHOLINERGIC FUNCTION OF CHOLINESTERASES

K. M. Kutty, V. Prabhakaran, and A. R. Cooper

Janeway Child Health Centre
Department of Pathology and Pediatrics
Janeway Place, St. John's, NF A1A 1R8, Canada
Faculty of Medicine
Memorial University of Newfoundland
St. John's, Newfoundland, Canada

INTRODUCTION

Alles and Hawes in 1940, suggested that there was a difference between "serum cholinesterase" and "erythrocyte cholinesterase" (Alles & Hawes, 1940). The erythrocyte cholinesterase (ChE) was found to be similar to the ChE associated with the cholinergic nervous system. This enzyme is involved in the hydrolyses of acetylcholine at the neuromuscular junction and therefore called acetylcholinesterase (AChE; EC 3.1.1.7). On the other hand, serum ChE has no role in cholinergic function, hence called pseudo-cholinesterase (PChE; EC 3.1.1.8) or non-specific ChE. AChE, because of its known function, was designated as true cholinesterase. Serum PChE is derived from the liver and is present in the serum of horse, man, rat, mouse, etc. It is known that PChE is found not only in the serum but also in adipose tissue, intestine, smooth muscle cells, white matter of the brain and many other tissues (Silver, 1974). However, the tissue specific enzymes, other than of the liver origin, are not secreted into the blood. There is also no evidence to indicate that tissue enzymes are derived from the plasma. In spite of such an ubiquitous distribution and survival during the evolution of this emzyme, we are still uncertain as to its biological role. To justify the existence of PChE, various proposals have been made describing its biological role, which has no connection with the cholinergic functions. Our major interest is to find out whether PChE has any role in lipid metabolism.

CHOLINESTERASE AND LIPID METABOLISM

One of the earlier proposals (Clitherow et al., 1963) implicated a role for PChE in lipid metabolism. The primary contention is that butyrylcholine, which is toxic and is formed as an intermediate during lipid metabolism, has to be removed from the system and PChE

Enzymes of the Cholinesterase Family, Edited by Daniel M. Quinn et al.
Plenum Press, New York, 1995

plays a role in such a detoxification process. Another suggestion was that serum PChE participates in choline homeostasis and thus controls the synthesis of acetylcholine (Funnel & Oliver, 1965). These proposals can not be substaniated in view of the following findings. When we inhibited PChE in rats and mice, by treating them with iso-OMPA, over a period of 7 to 31 days, we failed to discover either any accumulation or even the presence of cholinesters in serum, liver or urine of these animals (Kutty et al., 1989). This is in spite of the fact that about 70-80% of inhibition of PChE was maintained in these animals.

Furthermore, these animals did not show any toxicological effects of the known anticholinesterase agents belonging to the organophosphate and carbamate compounds. This leads to two conclusions: a) PChE has no role in cholinergic functions and b) there is no known biological substrate for this enzyme to act on. This leaves the question as to what other role PChE might have?

To begin with, it was initially thought that AChE in the red cell membrane had a structural function. Inhibition of AChE in the red cell was reported to increase the ionic permeability and fragility of these cells (Kutty et al., 1976). Therefore, it was proposed that AChE could increase the structural stability of the red cell membrane. The assumption was that the active site of the enzyme could interact with the phosphorylcholine of lecithin associated with the lipoproteins in the membrane. A similar assumption was also made in relation to PChE and low-density-lipoprotein (LDL). Many patients with hyperlipoprote-inemia, including nephrotic syndrome, have both elevated serum PChE activity and LDL (Way et al., 1975). Furthermore, a patient who was poisoned by an organophosphate insecticide had extremely low levels of serum PChE and LDL (Kutty et al., 1975). However, when this patient was treated with atropine and other drugs, the patient not only recovered but also increased both LDL and PChE concurrently with the recovery. Similarly, both rats (Kutty et al., 1977) and guinea pigs (Kutty et al., 1975), when treated with antiChE agents, also showed a significant reduction in serum LDL with the inhibition of the enzyme. In spite of these observations, a complete explanation cannot be provided regarding the biological role of PChE. This is because of the fact that only about 10% of serum PChE activity can be recovered in the LDL fraction (Chu et al., 1978). One consistent observation is that many patients who are either hyperlipoproteinemic, diabetic or obese show increased serum PChE activity (Chu et al., 1978; Kutty et al., 1981a). The common characteristics of these patients are elevated serum triglycerides (TG) and very-low-density-lipoprotein (VLDL).

CHOLINESTERASE IN EXPERIMENTAL ANIMAL MODELS

Because of these observations in patients, we conducted studies in experimental animal models of obesity and diabetes. Both genetically obese (ob/ob) and diabetic (db/db) mice showed markedly increased PChE activity in the serum and liver (Kutty et al., 1983; Kutty et al., 1981b). Restriction of the diet decreased serum PChE activity and body weight in these animals. To find out whether PChE increase is genetically determined in obesity, we artificially created obesity in mice with gold-thioglucose (Kutty et al., 1981b). These obese animals again showed increased serum and liver PChE activity. To find out if the enzyme increase is species specific, we studied the changes in Zucker fat rats (Kutty et al., 1984). These fat rats showed increased serum and liver PChE activity as well as TG. The common event in all these animals is that they have increased lipogenesis and are hyperphagic. Hence, it appeared to us that excessive caloric intake leads to increased lipogenesis and TG synthesis which might be the reason for the increase in serum PChE activity. Studies in obese mice also indicated that there is increased synthesis of PChE in the liver (Kean et al., 1986).

To study the relationship between serum lipids, especially TG, we investigated the lipid characteristics and clinical characteristics in patients who showed serum ChE activity

Table 1. Lipid characteristics of those with serum PChE > 5000 units

	Chol	HDL chol	LDL chol	Trig	ChE
I Normal N=7	5.1 ± 0.94	1.6 ± 0.32	2.9 ± 0.53	1.3 ± 0.66	5194 ± 201
IIa N=22	7.3 ± 1.5	1.4 ± 0.32	5.3 ± 1.4	1.3 ± 0.41	5293 ± 166
IIb N=69	7.5 ± 1.3	1.1 ± 0.25	4.8 ± 1.1	3.6 ± 1.3	5328 ± 776
IV N=31	7.3 ± 1.9	0.94 ± 0.22	3.2 ± 1.4	7 ± 2.7	5183 ± 971
V N=11	15.4 ± 5.4	0.59 ± 0.15	3.3 ± 3.5*	24.9 ± 10.3	6157 ± 638

*N=8

Total N=140

of greater than 5000 units (See Table 1). The upper limit of the normal range is less than 3500 units. A significant majority of these patients showed increased serum TG and/or cholesterol. Only a small percentage of patients with normal ChE activity showed high PChE activity. However, clinical features of those patients with high PChE activity are obesity, diabetes, family history of heart disease, alone or in combination (See Table 2). Furthermore, patients with both increased serum lipids and PChE activity showed a decrease in these variables after treatment for hyperlipoproteinemia. All these studies show a strong association between PChE and abnormal TG lipid metabolism. To obtain a better understanding of these relationships, we used streptozotoxin induced diabetic rats and observed the following findings (Annapurna et al., 1991).

Table 2. Clinical characteristics of patients with cholinesterase activity > 5000

Initials, age and sex	Clinical status
1. GW - 7 year old male	- Overweight and family history of heart disease and diabetes.
2. SS - 53 year old female	- Overweight.
3. HB - 54 year old male	- Family history of heart disease and diabetes.
4. MB - 38 year old male	- Strong family history of heart disease.
5. AA - 67 year old male	- Hypertension.
6. JP - 63 year old male	- Alcoholism.
7. JA - 30 year old female	- Obesity.
8. JR - 12 year old male	- Family history of heart disease and hyper-cholesterolemia
9. EP - 67 year old female	- Family history of heart disease.
10. BB - 28 year old male	- Hypertension.
11. RC - 36 year old male	- Obesity.
12. GH - 39 year old male	- Family history of heart disease and diabetes.
13. HB - 54 year old male	- Heart disease.
14. TM - 22 year old female	- Obesity.
15. MB - 13 year old male	- Family history of hyper-cholesterolemia
16. DS - 34 year old male	- Obesity.
17. SS - 35 year old female	- Abdominal pain.
18. BH - 70 year old female	- Diabetes.
19. SK - 24 year old male	- Overweight and family history of heart disease.

CHOLINESTERASE IN DIABETIC RATS

With the induction of diabetes in rats, along with the increase in serum glucose, both TG and PChE concentrations elevated. When the diabetes was controlled with insulin, TG and PChE decreased concomitantly. Since lipoprotein lipase is decreased in these diabetic animals, the increase in serum TG could occur both as a result of increased synthesis and delayed clearance. Heparin treatment of these diabetic animals decreased serum TG without decreasing PChE. This indicates the increase in serum PChE activity in diabetes is due to the over-production and secretion of TG and VLDL from the liver. In order to verify this assumption, we treated the diabetic rats with iso-OMPA, which is a known inhibitor of only PChE. Treatment with this drug decreased not only the serum TG and VLDL but also the liver TG. All these studies lead to this question, what dictates the induction of PChE and how is it involved in the VLDL synthesis and secretion?

CONCLUSION

In conclusion, what might be the role of PChE not only serum or liver but also in various other tissues and cells?

ACKNOWLEDGMENTS

We thank Miss Laurie Powell in assisting with the preparation of the manuscript.

REFERENCES

Alles, G.A. & Hawes, R.C. (1940) J. Biol. Chem. 133, 375-390.
Annapurna, V., Senciall, I., Davis, A.J. & Kutty, K.M. (1991) Diabetologia 34, 320-324.
Chu, M.I., Fontaine, P., Kutty, K.M., Murphy, D. & Redheendran, R. (1978) Clin. Chim. Acta. 85, 55-59.
Clitherow, J.W., Mitchard, M. & Harper, N.J. (1963) Nature 199, 1000-1001.
Funnel, H.S. & Oliver, W.T. (1965) Nature 208, 689-690.
Kean, K.T., Kutty, K.M., Huang, S.N. & Jain, R. (1986) Jour. Amer. Coll. of Nutr. 5, 253-261.
Kutty, K.M., Jacob, J.C., Hutton, P.J. & Peterson, S.C. (1975) Clin. Biochem. 8, 379-383.
Kutty, K.M., Chandra, R.K. & Chandra, S. (1976) Exp. 32, 289-290.
Kutty, K.M., Redheendran, R. & Murphy, D. (1977) Experientia 33, 420-421.
Kutty, K.M., Jain, R., Huang, S.N. & Kean, K.T. (1981a) Clin. Chim. Acta. 115, 55-61.
Kutty, K.M., Huang, S.N. & Kean, K.T. (1981b) Experientia 37, 1141-1142.
Kutty, K.M., Kean, K.T., Jain, R. & Huang, S.N. (1983) Nutrition Research 3, 211-216.
Kutty, K.M., Jain, R., Kean, K.T. & Peper, C. (1984) Nutrition Research 4, 99-104.
Kutty, K.M., Annapurna, V. & Prabhakaran, V. (1989) Biochem. Soc. Trans. 17, 555-556.
Silver, A. (1974) In The Biology of Cholinesterase. (Ed), North-Holland Publishing Co., Amsterdam, Oxford. 443-449.
Way, R.C., Hutton, C.J. & Kutty, K.M. (1975) Clin. Biochem. 8, 103-107.

GLYCOSYLATED INACTIVE FORMS OF CHICKEN BUTYRYLCHOLINESTERASES AND THEIR POSSIBLE FUNCTIONS

Paul G. Layer, Christoph Ebert, Sven Treskatis, Thomas Weikert, and Elmar Willbold

Technische Hochschule Darmstadt
Institut für Zoologie
Schnittspahnstraße 3, 64287 Darmstadt, Germany
Max-Planck-Institut für Entwicklungsbiologie
Spemannstraße 35/IV, 72076 Tübingen, Germany

CHOLINESTERASE EXPRESSION AT NONSYNAPTIC SITES OF THE EARLY CHICK EMBRYO

The spatiotemporal expression of cholinesterases during early chick neurogenesis suggests that both cholinesterases may be involved in neuronal target finding and specification (for review Layer, 1990, Layer and Willbold 1994). Histologically, expression of cholinesterases during establishment of the segmented nerve-muscle connectivities in the trunk of the chick embryo is most instructive (Layer et al., 1988). Before motoraxons leave, AChE is expressed in both their original cell bodies and their myotomal target cells, while BChE is elevated in cells along the pathways of growing axons. In particular, BChE can be detected in the rostral sclerotome earlier than any other reported molecule (see Keynes et al., 1990). Thus the expression of BChE may convey some information, e.g. for cell growth regulation or/and for supporting adhesive functions. Thus both cholinesterases are expressed shortly before neurites extend. Do cholinesterases somehow affect neurite growth? Could they represent a novel class of cell adhesion molecules, adding to the so far known three major groups of cell adhesion molecules, the cadherins, the integrins and the members of the immunoglobulin superfamily (see Rathjen, 1991)?

In contrast, binding of the peanut agglutinin (PNA; it binds to a Gal-β1-3-NAcGal disaccharide of glycoproteins) is found in the caudal part of the sclerotome (Stern et al., 1986). We have described similar histological situations with cholinesterases being associated with neuronal differentiation and neurite growth, while PNA binding seems to delineate noninvaded areas (Liu and Layer, 1984), e.g. during cranial nerve growth (Layer and Kaulich, 1991), rhombomere formation (Layer and Alber, 1990), or during axon navigation into the limb buds (Alber et al., 1994). In this system, motor axons navigate from AChE-

Enzymes of the Cholinesterase Family, Edited by Daniel M. Quinn et al.
Plenum Press, New York, 1995

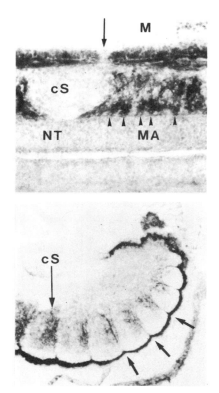

Figure 1. (upper) BChE is expressed in rostral half of sclerotome fostering navigating motoraxons (MA, marked by arrowheads), while the caudal half of the sclerotome (cS) is BChE-negative. The arrow marks the border between two adjacent somites (M, myotome; NT, neural tube). (lower) PNA binding is found in the BChE-negative and axonfree caudal half of the sclerotome (cS). The thick arrows mark borders between somites. Does BChE support and PNA binding proteins inhibit neurite growth?

positive motoneurons to their destination sites near the mesenchymal chondrogenic core (which is transiently AChE-positive). During growth, the axons are surrounded by BCHE-producing cells (presumably early Schwann cells). The chondrogenic core itself will not be invaded by the axons. It binds strongly the PNA lectin.

The asymmetric distribution of molecules mediating cell-cell interactions seems to be required for directing growing motoraxons and for a proper establishment of synaptic contacts in the target regions. Accordingly, molecules found in the rostral sclerotome (or comparable locations) could support neurite outgrowth, while molecules found in parts that form the boundaries (e.g. the caudal sclerotome, chondrogenic core of limb bud) could inhibit neurite growth.

In the first part of this article we are going to summarize our previous experiments that have demonstrated that cholinesterases indeed can influence cell growth and neurite differentiation (Layer, 1990; Layer et al., 1993; Willbold and Layer, 1994). In the second part, more recent work will be presented that is aimed at characterizing the specific forms of butyrylcholinesterases and their particular glyco-decoration as found in the chick, that possibly could contribute to these functions (Treskatis et al., 1992; Weikert and Layer, 1994; Weikert et al., 1994a, b). In particular, cholinesterases expressing the HNK-1 and the peanut lectin (PNA) binding carbohydrate epitopes will be dealt with, since they are candidates for positive and negative neurite growth regulation, respectively.

IN VITRO EVIDENCE FOR ROLES OF CHOLINESTERASES IN CELL GROWTH AND NEURITE GROWTH REGULATION

Laminar Histogenesis Including Cell and Neuropil Growth Is Regulated by BChE in Retinospheroids

Functional implications of cholinesterases for the laminar histogenesis of coherent neural networks have been demonstrated during in vitro regeneration of retina-like structures of the chick embryo. To this end, the effects of inhibition of BChE on laminar retinogenesis in so-called retinospheroids have been investigated (Willbold and Layer, 1994). Retinospheroids arise from dissociated embryonic chicken retinal cells in rotation culture. In the presence of the BChE inhibitor iso-OMPA, the number of spheroids/dish is increased, and their diameter is decreased by about 20%, corresponding to about 50% volume size. As a corollary, the course of histotypical differentiation is dramatically accelerated. Thus as a consequence of BChE inhibition, both organisation of nuclear cell layers and of plexiform-like (neuropil) areas, as detected by an antibody to the fasciculation protein F11 (Rathjen et al., 1987), is temporally advanced by at least two days. Moreover, AChE is almost fully diminished in these areas.

Neurite Growth Is Altered by Anticholinesterases under Various In Vitro Assay Conditions, Indicative of an Adhesive Role of Cholinesterases

We have provided direct evidence showing that neurite growth from various neuronal tissues in the chick embryo can be modified by some, but not all anticholinesterases (Layer et al., 1993). Distinct morphological changes, such as defasciculation, and disorientation of neurite bundles are induced by iso-OMPA, BW284C51 and Bambuterol in retinal explants grown on striped laminin carpets (Walter et al., 1987). Moreover, after culturing dissociated tectal cells on microplates, we have determined the neuritic material produced in individual microwells with an ELISA assay, that detects the neurite-specific G4 antigen in picogram amounts. Thereby, the quantitative effect of anticholinesterases on neurite outgrowth is directly compared with their cholinesterase inhibitory action. BW284C51 and ethopropazine, inhibiting acetylcholinesterase (AChE) and butyrylcholinesterase (BChE), respectively, strongly decrease neurite growth in a dose-dependent manner. However, echothiophate that inhibits both cholinesterases, does not change neuritic growth.

These studies allowed us to conclude that a) cholinesterases can regulate axonal growth, that b) both AChE and BChE can perform such a secondary function, and c) that this function is not due to the enzyme activity *per se*, but rather, due to a secondary site(s), most likely by adhesive mechanisms.

Cholinesterases Belong to a New Protein Family

Is there further evidence that cholinesterases may be involved in cell-cell or cell-matrix communication? Despite a close similarity to the active center of serine proteases (Gly-Asp-Ser-Gly-Gly), cholinesterases present no further similarities with this group of enzymes. Instead, it was found by sequence data comparisons that cholinesterase-like proteins form a separate molecular family (Gentry and Doctor, 1990; this issue), including the *Drosophila* cell adhesion molecules *neurotactin* and *glutactin*, as well as other nonrelated proteins such as thyroglobuline, rabbit liver esterase and esterase-6 from *Drosophila* (Chatonnet and Lockridge, 1989; Krejci et al., 1991).

These data show that cholinesterases possess cell adhesion molecule-like properties. Therefore, it is not surprising that after blocking cholinesterases we find effects on neurite growth. Yet, it is not clear whether this secondary or "nonclassical" function is mediated by the esteratic function, i.e. by the active site of the molecule (thereby ACh being involved), by other sites on the protein, by their carbohydrate calyx, or a mixture of these.

CARBOHYDRATE EPITOPES ON CHOLINESTERASES FOR ADHESIVE FUNCTIONS?

The major objective of this study therefore is to consider the modification of cholinesterases by covalently linked carbohydrates. Carbohydrates are powerful mediators of cell recognition. Membrane bound carbohydrates are the linkage targets for many hormones, toxins, viruses, bacteria or cells (review see Sharon and Lis, 1993). Different monoclonal antibodies recognizing specific dN-linked carbohydrates originally raised against AChE from *Torpedo* (Liao et al., 1991, 1992) bind to muscle and brain AChE from different fishes and mammals, however, not to erythrocyte AChE. Thus, the recognized carbohydrate epitopes are highly conserved during vertebrate evolution. After removal of sialic acid residues from neuronal and glial cells in culture, an increased activity both of AChE and BChE was observed (Stefanovic et al., 1975). Although it is not clear whether the activity of cholinesterases can be trapped or regulated by a direct linkage of sialic acid to specific parts of the AChE molecule, this system resembles the modification of NCAM by sialic acid during embryogenesis (Rutishauser et al., 1985). Accordingly, growth cone behavior of motor axons in the plexus region has been shown to depend on polysialic acid (Tang et al., 1994).

Inactive Butyrylcholinesterases from Chick Serum Have Different Glycosylation Patterns

Recently, we have isolated active BChE from serum (s-BChE$_{81}$) and brain (b-BChE$_{74}$) of adult chicken. Moreover, besides an active form we have found three inactive forms of BChE in adult chicken serum, called is-BChE$_{75}$, is-BChE$_{62}$, is-BChE$_{54}$ (Weikert et al., 1994b). The numbers indicate their different subunit sizes due to different degrees of glycosylation (Treskatis et al., 1992). Following sucrose gradient centrifugation, the active s-BChE$_{81}$ is found in a peak corresponding to a tetramer (Fig. 2), while is-BChE$_{75}$ found at the dimer position has very low or even zero enzyme activity. Consistent with this, binding of tritiated DFP is also very low.

Lectin binding and deglycosylation experiments on these various types of BChE's revealed their distinct patterns of glycosylation. The active serum BChE turned out to be of the triantennary type. The brain-derived b-BChE$_{74}$ is a tetrameric form with high specific activity, most of which is membrane-bound; it is not PIPLC- but protease-releasable. It binds the lectins GNA, DSA and ConA and is sensitive to N-glycosidase F, typical of a glycoprotein of the hybride type. Moreover, two inactive BChE's from serum of 62 and 54 kD molecular weight are also N-glycosylated proteins, while the is-BChE$_{75}$ is an O-glycosylated protein.

The HNK-1 Carbohydrate Epitope Is Found on Most, But Not All Chicken Butyrylcholinesterases

The socalled L2/HNK-1 sugar epitope on cell adhesion molecules of the immuno-globulin superfamily (Kruse et al., 1984; Tucker et al., 1988) itself has been found to

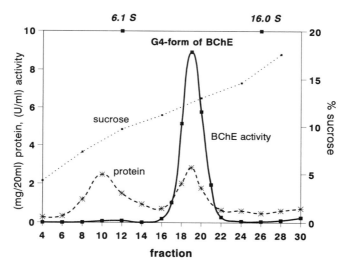

Figure 2. Separation of nonactive is-BChE$_{75}$ from active S-BChE$_{81}$ by centrifugation on a 5-20% sucrose gradient. BChE from whole serum of adult chicken was purified by ammonium sulfate precipitation, anion exchange and procainamide affinity chromatography. Note that protein is divided into two distinct peaks at 4.9 S and 10.3 S; BChE activity is contained almost exclusively in the 10.3 S peak.

contribute to the adhesive function (Riopelle et al., 1986; Cole and Schachner, 1987). Therefore it has been suggested that the mere presence of this epitope on a glycoprotein may indicate its adhesive nature. AChE from *Torpedo* has been reported to bear this epitope (Bon et al., 1987).

In a histological study, the expression of AChE, BChE and HNK-1 binding have been compared during cranial nerve growth (Layer and Kaulich, 1991). A meshwork of HNK-1 was noted that spatio-temporally closely matched with BChE expression. In addition, migrating AChE-positive cells become decorated by HNK-1 when approaching their target. This indicated that the HNK-1 epitope on cholinesterases is highly regulated, and that HNK-1 matrices may help directing cells and growing neurites to their targets.

The major inactive form of BChE from adult chicken serum, is-BChE$_{75}$, is a molecule with a very different glycosylation pattern. Both its protein bands near 75 kD bind strongly the HNK-1 IgM (Fig. 3). This molecule belongs to the rare type of O-glycosylated proteins, since it is sensitive to O-glycosidase but insensitive to N-glycosidase F treatment and, after its desialisation by neuraminidase, efficiently binds the peanut lectin. The brain-derived b-BChE$_{74}$ is a second representative of a HNK-1-positive BChE. It is noteworthy that only the two bands at 74 kD bind the IgM, while the copurified higher bands are HNK-1-negative. In addition, the BChE's of 62 and 54 kD molecular mass both strongly express the HNK-1 epitope. Thus, of five different types of butyrylcholinesterases investigated, four express the HNK-1 epitope while one is devoid of it.

These data show that both active and inactive forms of BChE from different tissues of the chick express the HNK-1 epitope. These include secreted BChE molecules from serum, and a membrane-associated BChE from brain. Only one type, namely the active BChE$_{81}$ from serum has been found so far that is HNK-1-negative. Moreover, the HNK-1 epitope can be expressed on three different types of glycoproteins, the chemistry and the significance of which is still unknown.

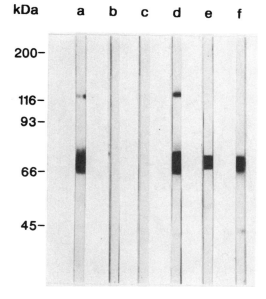

Figure 3. The is-BChE$_{75}$ protein is detected by BChE mAbs. Immunoreactivity of is-BChE$_{75}$ on PVDF membranes with the BChE-specific mAbs 1, 2, 3, 5, 6 (lanes a - e), after separation on 12% PAGE and electroblotting of the protein. Note that mAbs 1, 5 and 6 bind strongly, mAbs 2 and 3 do not bind. mAbs 1 and 5 also detect dimers. Lane f shows immunoreactivity with HNK-1 IgM.

The Paradox of the Inactive 75 Kd BChE from Chick Serum

The inactive molecules are fascinating molecules in the context of investigation of nonclassical functions of cholinesterases, since they are void of a catalytic activity but still may have retained the postulated secondary function. In particular, the case of the inactive 75 kD BChE from chick serum has become interesting for the following reasons. It is not labelled by tritiated DFP and thus is an inactive BChE. It expresses the HNK-1 epitope, but most surprisingly, turned out to be an O-glycosylated protein (Weikert and Layer, 1994). If that is so, how then could the HNK-1 epitope be associated with the protein?

The HNK-1 epitope as being a 3-sulfated glucuronic acid lactoseries (Chou et al., 1986) is generally accepted to be bound to one of the three antennaes of a triantennary complex N-glycosylated protein; alternatively, it can exist as part of certain glycolipids (Ariga et al., 1987). Our findings indicate that the 75 kD protein is not N-glycosylated, since there is no shift by N-glycosidase treatment. Either the shift is obscured for some reason, so that this molecule could still be both N- and O-glycosylated; or, alternatively, we should consider whether an HNK-1 containing glycolipid could be associated with the 75 kD protein. However, this seems unlikely under our conditions of protein denaturation. Clearly, more work is needed to finally clarify the glyco-chemistry of the 75 kD BChE protein.

Growth-Supportive and Growth-Suppressive Glycoepitopes on One and the Same Cholinesterase Molecule?

Thus the 75 kD BChE protein is O-glycosylated and PNA-positive, but at the same time shows the HNK-1 epitope. Remarkably, this chemical paradox presents an exciting biological face: with the HNK-1 and the PNA receptor on it, this molecule could have both a growth supportive and a growth-suppressive function. Histologically, PNA has been suggested to indicate the presence of barriers or boundaries to neurite growth (see above). We have started to study whether the growth adversive function of PNA on the 75 kD BChE can overcome the growth-supportive function mediated by the HNK-1 epitope. Preliminary in vitro data are consistent with this hypothesis. Obviously, we are left with the questions: how are the two epitopes regulated, and does such a molecule have any physiological meaning?

In summary, this study has drawn attention to the following points, all supporting and further clarifying nonclassical functions of cholinesterases:

1. further evidence for growth and differentiation regulation by cholinesterases is accumulating;
2. besides the enzyme activity, the particular glyco-decoration of cholinesterases may be relevant for their role as a new class of adhesive molecules;
3. the adhesion-related HNK-1 glycoepitope is highly regulated on cholinesterases;
4. when present on cholinesterases, the HNK-1 epitope may support neurite growth;
5. an inactive BChE from serum bears both the HNK-1 and the PNA epitope; it is an O-glycosylated cholinesterase;
6. we postulate that the PNA epitope may counteract the HNK-1 function on one and the same BChE molecule.

ACKNOWLEDGMENTS

We wish to thank our collegues R. Alber, T. Brümmendorf, S. Kaulich, S. Kotz, L. Liu, F.G. Rathjen, M. Reinicke, S. Rommel, O. Sporns, and Bettina Weiß, who have contributed to the progress of this project. We thank A. Chatonnet, A. Gierer, S. Greenfield, F. Hucho, J. Massoulié, H. Meinhardt, and L. Puelles for helpful discussions on different aspects of this work.

REFERENCES

Alber, R., Sporns, O., Weikert, T., Willbold, E., and Layer, P.G. 1994, *Anat. Embryol.* 190: 429-438.
Ariga, T., Kohriyama, T., Freddo, L., Latov, N., Saito, M., Kon, K., Ando, S., and Suzuki, M. 1987, *J. Biol. Chem.* 262: 848-853.
Bon, S., Méflah, K., Musset, F., Grassi, J., and Massoulié, J. 1987, *J. Neurochem.* 49: 1720-1731.
Chatonnet, A., and Lockridge, O. 1989, *Biochem. J.* 260: 625-634.
Chou, D.K.H., Ilyas, A.A., Evans, J.E., Costello, C., Quarles, R.H., and Jungalwala, F.B. 1986, *J. Biol. Chem.* 261: 11717-11725.
Cole, G.J., and Schachner, M. 1987, *Neurosci. Lett.* 78: 227-232.
Hoffman, S., Crossin, K.L., and Edelman, G.M. 1988, *J. Cell. Biol.* 106: 519-532.
Keynes, R., Cook, G., Davies, J., Lumsden, A., Norris, W., and Stern, C. 1990, *J. Physiol. Paris* 84: 27-32.
Krejci, E., Duval, N., Chatonnet, A., Vincens, P., and Massoulié, J. 1991, *Proc. Natl. Acad Sci. (USA)* 88: 6647-6651.
Kruse, J., Mailhammer, R., Wernecke, H., Faissner, A., Sommer, I., Goridis, C., and Schachner, M. 1984, *Nature* 311: 153-155.
Layer, P.G., Alber, R., and Rathjen, F.G. 1988, *Development* 102: 387-396.
Layer, P.G. 1990, *BioEssays* 12: 415-420.
Layer, P.G., and Alber, R. 1990, *Development* 109: 613-624.
Layer, P.G., and Kaulich, S. 1991, *Cell Tissue Res.* 265: 393-407.
Layer, P.G., Weikert, T., and Alber, R. 1993, *Cell Tissue Res.* 273: 219-226.
Layer, P.G., and Willbold, E. 1994, *Int. Rev. Cytol.* 151: 139-181.
Liao, J., Heider, H., Sun, M.C., Stieger, S., and Brodbeck, U. 1991, *Eur. J. Biochem.* 198: 59-65.
Liao, J., Heider, H., Sun, M.C., and Brodbeck, U. 1992, *J. Neurochem.* 58: 1230-1238.
Liu, L., and Layer, P.G. 1984, *Dev. Brain Res.* 12: 173-182.
Rathjen, F.G., Wolff, J.M., Frank, R., Bonhoeffer, F., and Rutishauser, U. 1987, *J. Cell Biol.* 104: 343-353.
Rathjen, F.G. 1991, *Curr. Opin. Cell Biol.* 3: 992-1000.
Riopelle, R.J., McGarry, R.C., and Roder, J.C. 1986, *Brain Res.* 367: 20-25.

Rutishauser, U., Watanabe, M., Silver, J., Troy, F.A., and Vimr, E.R. 1985, *J. Cell Biol.* 101: 1842-1849.

Sharon, N., and Lis, H. 1993, *Sci. Amer.* 268: 74-81.

Stefanovic, V., Mandel, P., and Rosenberg, A. 1975, *Biochemistry* 14: 5257-5260.

Stern, C.D., Sisodiya, S.M., and Keynes, R.J. 1986, *J. Embryol. Exp. Morph.* 91: 209-226.

Tang, J., Rutishauser, U., and Landmesser, L. 1994, *Neuron* 13: 405-414.

Treskatis, S., Ebert, C., and Layer, P.G. 1992, *J. Neurochem.* 58: 2236-2247.

Tucker, G.C., Delarue, M., Zada, S., Boucaut, J.C., and Thiery, J.P. 1988, *Cell Tissue Res.* 251: 457-465.

Walter, J., Kern-Veits, B., Huf, J., Stolze, B., and Bonhoeffer, F. 1987, *Development* 101: 685-696.

Weikert, T., and Layer, P.G. 1994, *Neurosci. Lett.* 176: 9-12.

Weikert, T., Rathjen, F.G., and Layer, P.G. 1994a, *J. Neurochem.* 62: 1570-1577.

Weikert, T., Ebert, C., Rasched, I., and Layer, P.G. 1994b, *J. Neurochem.* 63: 318-325.

Willbold, E., and Layer, P.G. 1994, *Eur. J. Cell Biol. 64: 192-199.*

POSSIBLE CHOLINERGIC AND NON-CHOLINERGIC ACTIONS OF TRANSIENTLY EXPRESSED ACETYLCHOLINESTERASE IN THALAMOCORTICAL DEVELOPMENT PROJECTIONS

Richard T. Robertson, Ron S. Broide, Jen Yu, and Frances L. Leslie

College of Medicine
University of California, Irvine
Irvine, California 92717

INTRODUCTION

Recent studies from this laboratory and others have demonstrated that acetylcholinesterase (AChE) activity is expressed transiently in developing cerebral cortex of mammals (Kostovic and Goldman-Rakic, 1983; Kostovic and Rakic, 1984; Kristt and Waldman, 1981; Robertson, 1987; Robertson et al., 1988; Robertson and Yu, 1994; Schlaggar and O'Leary, 1994). Studies in rodents have demonstrated that the transiently expressed AChE is produced by the protein synthetic apparatus in the perikaryal cytoplasm of thalamic neurons and is transported along thalamocortical axons to their terminals in primary sensory cortex (Robertson, 1991; Robertson et al., 1988; Robertson and Yu, 1994). The AChE activity can first be detected in rats around the time of birth, reaches peak intensity during the second postnatal week, and then subsides to adult levels during the third postnatal week (Robertson, 1987). Interestingly, this temporal pattern of AChE expression coincides closely with the timing of thalamocortical axonal ingrowth into cerebral cortex (Catalano et al., 1991; Kageyama and Robertson, 1993; Schlaggar and O'Leary, 1994).

The function of the transiently expressed AChE activity in developing thalamocortical projections remains unknown. AChE is of course well known for its catabolic function for the neurotransmitter acetylcholine (Ach) in cholinergic transmission and AChE typically is found both in cholinergic and in cholinoceptive neurons. Curiously, however, AChE is also found in some systems that are believed to be neither cholinergic nor cholinoceptive, raising the possibility that AChE may have functions other than the hydrolysis of acetylcholine. It remains unclear whether transiently expressed AChE may have a cholinergic, or a non-cholinergic function. On the one hand, the transiently expressed AChE could be

Enzymes of the Cholinesterase Family, Edited by Daniel M. Quinn et al.
Plenum Press, New York, 1995

serving a cholinergic role in developing cortex, although thalamocortical neurons are believed not to be cholinergic. On the other hand is the suggestion that transiently expressed AChE may have a non-cholinergic and morphogenic role in neuronal development (Greenfield, 1984; Layer et al., 1987; Robertson, 1987; Robertson and Yu, 1994; Small, 1990).

This chapter summarizes some results from recent work in our laboratories that has investigated possible functions of the transiently expressed AChE in cortical development. One set of studies indicates that chronic inhibition of AChE catalytic activity does not produce detectable disturbances in normal morphological development of thalamocortical projections, arguing against a direct role for the esterase activity of AChE in morphological development. A second set of studies suggests a possible cholinergic role by demonstrating remarkable regional and temporal relationships between transient AChE expression and transient expression of nicotinic receptors in developing cerebral cortex.

CHRONIC INHIBITION OF AChE ESTERASE ACTIVITY AND THALAMOCORTICAL DEVELOPMENT

The first set of studies was based on the hypothesis that the AChE activity expressed transiently in developing thalamocortical projections may serve to aid the thalamocortical axons in growing into their terminal fields in cortex. That is, the transient AChE may be involved in a proteolytic process that helps thalamocortical axons penetrate through the cortical neuropil (Robertson and Yu, 1994). We tested this hypothesis by examining the effects of chronic AChE inhibition on thalamocortical development (Ling et al., 1995). Our previous studies (Catalano et al., 1991; 1995; Kageyama and Robertson, 1993) provided data on normal patterns of development and demonstrated that growing thalamocortical axons had reached their mature laminae of termination by postnatal day 5.

On the day of birth (postnatal day 0; P0) newborn rat pups received surgical implants of Elvax sheets (Silberstein and Daniel, 1982) infiltrated with the irreversible, non-competitive AChE inhibitor phospholine iodide (PI). Elvax sheets were placed unilaterally over sensory cortex and provided a slow release of PI onto the cerebral cortex. Previous in vitro and in vivo tests had demonstrated that the release of PI from the implanted Elvax sheets extended over 4-5 days effectively inhibited AChE activity during that period. Because the PI-impregnated Elvax sheets were placed unilaterally, we could compare the morphological development of thalamocortical projections in the experimental (PI treated) hemisphere and the control untreated hemisphere within the same animal. Morphological development of thalamocortical projections was assessed by two techniques. In some experimental animals, the laminar patterns of thalamocortical projections to visual cortex were studied using anterograde transneuronal transport of horseradish peroxidase conjugated to wheat germ agglutinin, following its injection into both eyes (Kageyama and Robertson, 1993). This compound is actively taken up by retinal ganglion cells and is transported transneuronally to label the entire retino-geniculo-cortical pathway. Other animals were perfused with aldehydes and the fixed brains received placements of the carbocyanine dye DiI (1,1'-dioctadecyl-3,3,3',3'-tetramethylindocarbocyanine perchlorate) to label thalamocortical axons in visual and somatosensory regions of cerebral cortex (Catalano et al., 1991; 1995; Godement et al., 1987; Kageyama and Robertson, 1993). Tissue sections were processed for AChE histochemistry to determine the effectiveness of the Elvax-PI treatment.

The photomicrograph shown in Figure 1A demonstrates that AChE activity throughout the cortex is dramatically suppressed in the hemisphere receiving an Elvax implant. The implant did not disrupt the normal structure of the cortex; Nissl stains provided no evidence of necrosis or unusual atrophy.

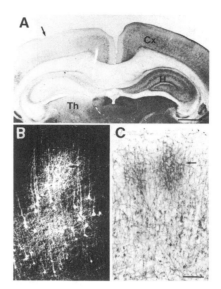

Figure 1. Chronic AChE inhibition and thalamocortical in-growth. A: AChE histochemically reacted section through the forebrain showing markedly decreased AChE activity in the Elvax-PI treated left hemisphere. Arrow indicates placement of Elvax. Cx: cerebral cortex; H: hippocampus; Th: thalamus. Calibration bar = 1mm. B: Fluorescence photomicrograph showing normal appearing DiI labeling of thalamocortical axons in the "barrel" region of somatosensory cortex in an Elvax-PI treated case. C: Histochemically reacted section from a normal control animal, showing AChE staining of two "barrels". Calibration bar in 'C' = 100µm for 'B' and 'C'.

Figure 1B presents a confocal fluorescence photomicrograph showing patterns of DiI labeling in an Elvax-PI treated somatosensory cortex. Although AChE histochemistry revealed marked reduction of AChE activity, the pattern of thalamocortical labeling in the "barrels" of somatosensory cortex appears normal. This was a reliable finding in both somatosensory and in visual cortices; indeed both visual subjective analyses and quantitative computer densitometric analyses detected no difference between experimental and control hemispheres (Ling et al., 1995).

These results demonstrate that sustained inhibition of AChE histochemical activity through slow release of PI from implanted Elvax sheets does not appear to disrupt the process of ingrowth of thalamocortical axons into layer IV of sensory cortex. Thus, normal thalamo-cortical development at this early stage, at least at this level of analysis, appears not to require AChE activity. The present results, however, do not necessarily indicate that AChE has no role in thalamocortical development. The present data indicate only that the *esterase* activity of AChE is not necessary for normal thalamocortical ingrowth. Recent results from other investigations have demonstrated that AChE, independent of its esterase activity, may interact with extracellular matrix molecules to affect cell activity, including neurite out-growth (Small et al., 1995).

TRANSIENTLY EXPRESSED AChE AND NICOTINIC RECEPTORS

A second line of research is suggestive of a possible cholinergic role for the transiently expressed AChE (Broide et al., 1995a; 1995b). This suggestion is based upon studies demonstrating a close correspondence, both regionally and temporally, between the transient expression of AChE and the transient expression of nicotinic receptors in developing cortex. These results come out of a larger study of patterns of cholinergic receptors in developing rodent cerebral cortex.

A developmental series of brains was studied from animals aged P0 to young adults. The distribution of nicotinic receptors was examined by autoradiographic localization of $[^{125}I]$ α-bungarotoxin, which binds to the α_7 subunit of the nicotinic receptor, and by

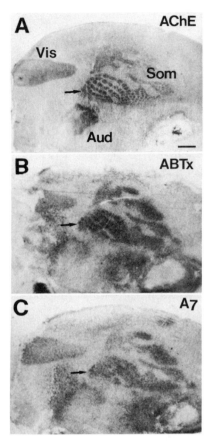

Figure 2. Distributions of cholinergic markers in tangential sections through developing rat somatosensory cortex. A: AChE histochemical reaction product reveals visual (Vis), somatosensory (Som),and auditory (Aud) cortical areas. Calibration bar = 1 mm for all three photomicrographs. B: Distribution of nicotinic receptors demonstrated by [^{125}I] α-bungarotoxin (ABTx) binding. C: Patterns of the α$_7$ subunit of the nicotinic receptor demonstrated by *in situ* hybridization. Arrows indicate the "barrel" field of somatosensory cortex.

autoradiographic studies of *in situ* hybridization to localize messenger RNA (mRNA) for the α$_7$ subunit (Seguela et al., 1993). The distributions of these markers were compared to the distribution of AChE histochemical activity. Series of adjacent tissue sections from each brain were processed by each of these techniques, allowing the close comparison of the patterns of each marker.

The photomicrographs in Figure 2 illustrate our results in tissue taken from a seven day old rat pup. These photomicrographs present a series of three tangential sections, which are cut in a plane parallel to the pial surface, showing the cerebral cortex in general and the somatosensory region of cortex in particular. These markers reveal the "ratunculus" representation of the body surface in somatosensory cortex.

The pattern of AChE histochemical staining is shown in Figure 2A. As described previously (Robertson, 1987; Robertson and Yu, 1994; Schlaggar and O'Leary, 1994), AChE is expressed by developing thalamocortical axons and so the AChE histochemical reaction product indicates the pattern of thalamocortical terminations in somatosensory cortex. The barrel cortex is particularly noticeable. Figure 2B presents an adjacent section, processed on autoradiographic film for [^{125}I] α-bungarotoxin labeling, and demonstrates a corresponding pattern for nicotinic binding sites. Additionally, Figure 2C presents the results of *in situ* hybridization analysis of the next section in this series, which reveals the pattern of expression of mRNA for the α$_7$ subunit of the nicotinic receptor. These data reveal a remarkably close correspondence between AChE histochemical activity and nicotinic receptors in developing rat somatosensory cortex.

In addition to the regional correspondence of the three markers, all three markers are transient and show a remarkable temporal correspondence. For a given cortical layer, times of onset and peak activity are similar. It is of particular interest that the expression of these three markers appears on P1 in the deepest layer of the developing cortex, layer VI, when thalamocortical axons are first entering this layer. A few days later, all three markers appear in layer IV, and show remarkably similar distributions. However, AChE histochemical activity shows earlier onset in each cortical layer and also shows a considerably more abrupt decline of activity than that of the α_7 subunit message.

We have demonstrated that the transiently expressed AChE is expressed by thalamocortical axons, including their terminals in somatosensory cortex. The locus of the nicotinic binding sites is less clear. Evidence for expression of mRNA for the α_7 subunit in cortex almost undoubtedly indicates its presence in cortical neurons, and thus at least a portion of the nicotinic binding sites is likely to be on cortical neurons. However, our studies (Broide et al., 1995a) also reveal expression of mRNA for the α_7 subunit in neurons of the ventral posterior nucleus of the thalamus. These are the neurons that send the AChE positive thalamocortical axons to somatosensory cortex and the receptor protein may be transported to the axon terminals. Thus, nicotinic binding in developing somatosensory cortex of the rat may be a property both of cortical neurons and of thalamocortical axons. The observation that AChE expression in thalamocortical axons slightly precedes expression of nicotinic receptors in cortical neurons suggests that the growing AChE positive thalamocortical axons may induce expression of the pattern of nicotinic receptors in cortical neurons.

The function(s) of the transient nicotinic binding sites in developing cortex remains unknown. These nicotinic receptors would be expected to be activated by acetylcholine (Ach) but the source of the endogenous Ach is not clear. Thalamocortical projections are believed not to be cholinergic, but three possible sources of acetylcholine do exist. In rodents, a few cholinergic local circuit neurons are present in cortex (Levey et al., 1984), but their time of appearance and regional distribution show no obvious relationship to the pattern of nicotinic receptors. Perhaps the most likely source of Ach would be provided by axons from the basal forebrain cholinergic system. Although this projection system develops slightly later than the thalamocortical system, these cholinergic afferents are reaching somatosensory cortex during the time of intense nicotinic receptor expression (Calarco and Robertson, 1995; Robertson et al., 1990). Another possible source of Ach that should be considered is the circulating blood (Small, 1995), especially so considering the weak blood-brain barrier during early development.

Nicotinic receptors comprised of α_7 subunits have been demonstrated to gate Ca^{2+}, and a vast literature has demonstrated the importance of intracellular Ca^{2+} levels in neural development and plasticity. It is likely that the effects that nicotinic activation may have on neural development and plasticity (Lauder, 1993; Leslie, 1993; Lipton et al., 1988; Zheng et al., 1994) are mediated through altered Ca^{2+} fluxes.

CONCLUSIONS

The function of transiently expressed AChE in developing thalamocortical systems remains a mystery. While the esterase activity of AChE appears to be unnecessary for at least gross aspects of morphological development of thalamocortical axons, it may be that some more subtle aspect of thalamocortical development is affected but remains undetected. The possibility of a non-cholinergic role for AChE, unrelated to its esterase function, needs to be explored, particularly with regard to possible interactions with extracellular matrix molecules. On the other hand, the close correspondence between AChE expression and nicotinic receptor expression is certainly intriguing, and continues to suggest the presence of cholinergic mechanisms in thalamocortical development.

ACKNOWLEDGMENTS

Supported by NIH grant NS 30109 and the Roosevelt Warm Springs Foundation.

REFERENCES

Broide, R.S., O'Connor, L.T., Smith, M.A., Smith, J.A., and Leslie, F.M., 1995a, Developmental expression of α_7 neuronal nicotinic receptor mRNA in rat sensory cortex and thalamus, *Neurosci.* in press.

Broide, R.S., Robertson, R.T., and Leslie, F.M., 1995b, Modulation of α_7 nicotinic receptors in the developing rat somatosensory cortex, submitted for publication.

Calarco, C.A., and Robertson, R.T., 1995, Development of basal forebrain projections to visual cortex: DiI studies in rats, *J. Comp. Neurol.*, 335: in press.

Catalano, S.M., Robertson, R.T., and Killackey, H.P., 1991, Early ingrowth of thalamocortical afferents to the neocortex of the prenatal rat, *Proc. Nat'l. Acad. Sci., USA* 88:2999-3003.

Catalano, S.M., Robertson, R.T., and Killackey, H.P., 1995, Rapid alteration of thalamocortical axon morphology follows peripheral damage in the neonatal rat, *Proc. Nat'l. Acad. Sci., USA* 92:in press.

Godement, P., Vanselow, J., Thanos, S., and Bonhoeffer, F., 1987, A study in developing visual systems with a new method of staining neurones and their processes in fixed tissue, *Development*, 101: 697-713.

Greenfield, S., 1984, Acetylcholinesterase may have novel functions in the brain. *Trends Neurosci.* 8:1-26.

Kageyama, G.H. and Robertson, R.T., 1993, Development of geniculocortical projections to visual cortex in rat: evidence for early ingrowth and synaptogenesis. *J. Comp. Neurol.* 335:123-148.

Kostovic, I. and Goldman-Rakic, P. S., 1983, Transient cholinesterase staining in the mediodorsal nucleus of the thalamus and its connections in the developing human and monkey brain, *J. Comp. Neurol.* 219:431-447.

Kostovic, I. and Rakic, P., 1984, Development of prestriate visual projections in the monkey and human fetal cerebrum revealed by transient cholinesterase staining, *J. Neurosci.* 4:25-42.

Kristt, D. A. and Waldman, J. V., 1981, The origin of the acetylcholinesterase rich afferents to layer IV of infant somatosensory cortex: a histochemical analysis following lesions, *Anat. Embryol.* 163:31-41.

Lauder, J.M., 1993, Neurotransmitters as growth regulatory signals: role of receptors and second messengers, *Trends Neurosci.*, 16: 233-240.

Layer, P., Alber, R., and Sporns, O., 1987, Quantitative development and molecular forms of acetyl- and butyrylcholinesterase during morphogenesis and synaptogenesis of chick brain and retina, *J. Neurochem.*, 49: 175-182.

Leslie, F.M., 1993, Neurotransmitters as neurotrophic factors, in Loughlin, S.E., and Fallon, J.H., (Eds.) *Neurotrophic Factors*, Academic Press, New York, pp. 565-598.

Levey, A.I., Wainer, B.H., Rye, D.B., Mufson, E.J., and Mesulam,M.-M., 1984, Choline acetyltransferase-immunoreactive neurons intrinsic to rodent cortex and distinction from acetylcholinesterase-positive neurons, *Neurosci.*, 13: 341-353.

Ling, J.J., Yu, J., and Robertson, R.T., 1995, Sustained inhibition of esterase activity of transiently expressed acetylcholinesterase does not disrupt early geniculoocortical ingrowth to developing rat visual cortex, submitted for publication.

Lipton, S.A., Frosch, M.P., Philips, M.D., Tauck, D.L. and Aizenman, E., 1988, Nicotinic antagonists enhance process outgrowth by rat retinal ganglion cells in culture, *Science*, 239: 293-1296.

Robertson, R.T., 1987, A morphogenic role for transiently expressed acetylcholinesterase in developing thalamocortical systems? *Neurosci. Lett.* 75:259-264.

Robertson, R.T., 1991, Transiently expressed acetylcholinesterase activity in developing thalamocortical projection neurons, in Massoulie, J., Bacou, F., Barnard, E., Chatonnet, A., Doctor, B.P., and Quinn, D.M., (Eds.), *Cholinesterases; Structure, Function, Mechanism, Genetics, and Cell Biology*; American Chem. Soc., Washington, D.C., pp. 358-365.

Robertson, R.T., Kageyama, G.H., Gallardo, K.A., and Yu, J., 1990, Development of basal forebrain cholinergic projections to cerebral cortex: AChE histochemical studies in rats and hamsters, *Neurosci. Abstr.*, 16: 1151.

Robertson, R.T., Hanes, M.A. and Yu, J., 1988 Investigations of the origins of transient acetylcholinesterase activity in developing rat visual cortex. *Dev. Brain Res.* 41:1-23.

Robertson, R.T. and Yu, J., 1993 Acetylcholinesterase and neural development: new tricks for an old dog? *News Physiol. Sci.* 8:266-272.

Schlaggar, B.L., and O'Leary, D.D.M., 1994, Early development of the somatotopic map and barrel patterning in rat somatosensory cortex, *J. Comp. Neurol.*, 346:80-96.

Seguela, P., Wadiche, J., Dineley-Miller, K., Dani, J.A., and Patrick, J.W., 1993, Molecular cloning, functional properties, and distribution of rat brain α_7 : A nicotinic cation channel highly permeable to calcium, *J. Neurosci.*, 13: 596-604.

Silberstein, G.B., and Daniel, C.W., 1982, Elvax 40P implants: sustained, local release of bioactive molecules influencing mammary ductal development, *Dev. Biol.*, 93:272-278.

Small, D.H., 1990, Non-cholinergic actions of acetylcholinesterases: proteases regulating cell growth and development? *Trends Biochem. Sci.* 15:213-216.

Small, D.H., 1995, A function for butyrlcholinesterase? *J. Neurochem.* 64:466-467.

Small, D.H., Reed, G., Whitefield, B., and Nurcombe, V., 1995, Cholinergic regulation of neurite outgrowth from isolated chick sympathetic neurons in culture, *J. Neurosci.*, 14: in press.

Zheng, J.Q., Felder, M., Connor, J.A., and Poo, M.-M., 1994, Turning of nerve growth cones induced by neurotransmitters, *Nature*, 368: 140-144.

EVIDENCE FOR A PUTATIVE ACETYLCHOLINESTERASE UPTAKE MECHANISM WITHIN THE SUBSTANTIA NIGRA

T. C. Budd,[1] B. G. M. Dickie,[1] D. Vaux,[2] and S. A. Greenfield[1]

[1] University Department of Pharmacology
Mansfield Road, Oxford, OX1 3QT
England, United Kingdom
[2] Sir William Dunn School of Pathology
South Parks Road, Oxford, OX1 3RE
England, United Kingdom

INTRODUCTION

There is a paucity of cholinergic innervation within the substantia nigra. However, a soluble form of acetylcholinesterase (AChE) is secreted from the dendrites of nigral dopaminergic neurones in a K^+ evoked and Ca^{2+} dependent manner that is insensitive to cholinergic agonists and antagonists (Greenfield, 1991). These observations suggest that secreted AChE may serve another function; indeed, *in vivo* studies have demonstrated that AChE has long term functional consequences (Hawkins and Greenfield, 1992). This long term action may possibly be due to released AChE being incorporated back into neurones as already shown in the periphery (Jessen *et al.*, 1978). To test this hypothesis, AChE (2-20 pM) was perfused via push-pull cannulae into the substantia nigra of conscious and anaesthetised guinea pigs. The amount of AChE recovered in the effusate from the substantia nigra was significantly reduced in both conscious and anaesthetised animals, but not in animals where the cannulae were deliberately placed in extra-nigral sites. This reduction in AChE was not apparent when the experiments were repeated immediately *post mortem* or when the medium perfused into the substantia nigra was cooled to approximately 4°C . This suggests that AChE retention within the substantia nigra involves an active uptake mechanism which is absent, or non functional, in the extra-nigral sites. Moreover, perfusion of enzymatically denatured (boiled) AChE only elevates basal levels of endogenous AChE detected in effusate from the substantia nigra. This may be due to competition between the boiled exogenous and secreted endogenous AChE for a putative membrane binding site.

Substitution of biotinylated AChE (b-AChE) for AChE, enabled uptake of AChE into nigral cells to be visualised directly by fluorescence microscopy. Deposits of b-AChE were observed (with an avidin fluorochrome conjugate) around the tips of cannulae placed in the substantia nigra but not at the tips of cannulae located extra-nigrally. B-AChE immunofluorescence was observed in both tyrosine hydroxylase -immunopositive (dopaminergic) and -negative cells within the substantia nigra. These results suggest that within the substantia nigra, secreted AChE may be subject to a temperature and energy dependent uptake mechanism which may underlie the long term functional effects of AChE.

REFERENCES

Greenfield, S.A. (1991) *Cell Mol. Neurobiol.* 11:55-77.
Hawkins, C.A., & Greenfield, S.A. (1992a) *Behav. Brain Res.* 48:153-157.
Jessen, K.R., Chubb, I.W., & Smith, A.D. (1978) *J. Neurocytol.* 7:145-154.

ACTIVATION OF PERITONEAL MACROPHAGES BY ACETYLCHOLINESTERASE IS INDEPENDENT OF CATALYTIC ACTIVITY

A. Klegeris, T. C. Budd, and S. A. Greenfield

University Department of Pharmacology
Mansfield Road
Oxford, OX1 3QT
England, United Kingdom

INTRODUCTION

To date the non-cholinergic effects of acetylcholinesterase (AChE) in the central nervous system have been associated with its action on neuronal cells. Two distinct modes of non-classical activity of AChE have been suggested: first, that AChE is a neuromodulatory agent modifying neuronal activity (Greenfield, 1991); secondly, that AChE is playing the role of a cell adhesion molecule during the development and differentiation of the central nervous system (Layer et al., 1993). Although these hypotheses imply involvement of different cellular mechanisms, a common assumption for both is an interaction between AChE and the neuronal cell membrane in a fashion that is currently unknown.

In a previous study (Klegeris et al., 1994), we used rat peritoneal macrophages as an accessible model system for central neurons to investigate interactions of AChE with the cell membrane. We found that purified electric eel AChE was capable of inducing a 'respiratory burst' in macrophages which did not involve the catalytic activity of AChE and which was mannose and divalent cation dependent. This effect suggested that there may be two possible mechanisms involved: either a mannose-fucose receptor (MFR) on the macrophage membrane, which would bind the surface sugars of AChE, or a lectin-like sugar binding site embedded in the AChE molecule itself.

RESULTS

Here we report a series of experiments which attempt to differentiate these potential mechanisms. The possibility of a lectin-like activity for AChE was tested by affinity chromatography on immobilized monosaccharides, gel-diffusion and erythrocyte agglutination. To test the involvement of the MFR, inhibition studies with specific ligands of this receptor (neoglycoproteins) were performed. We found no evidence for a lectin-like activity of AChE. On the other hand, the activation of macrophages by AChE was specifically inhibited in the presence of mannosylated, but not glucosylated or galactosylated, BSA. Furthermore human recombinant AChE, which is unglycosylated, failed to activate peritoneal macrophages. We conclude that the interaction between electric eel AChE and peritoneal cells is mediated by a macrophage MFR.

We also showed that the specific inhibitor of AChE, BW284C51, can interfere with the macrophage response to low-potency stimuli (e.g. AChE and tuftsin), without affecting the high-amplitude response to serum-opsonized zymosan, and without having any apparent cytotoxic effects. Our studies also indicate that several other proteins are capable of activating macrophages (horseradish peroxidase, mannosylated BSA). Therefore AChE activation of these cells could be an example of a more general phenomenon of signalling between neurons and brain macrophages.

REFERENCES

Greenfield, S.A. (1991) *Cell. Mol. Neurobiol.* 11:55-77.
Klegeris, A., Budd, T.C., & Greenfield, S.A. (1994) *Cell. Mol. Neurobiol.* 14:89-98.
Layer, P.G., Weikert, T., & Alber, R. (1993) *Cell Tissue Res.* 273:219-226.

FUNCTIONAL SIGNIFICANCE OF BUTYRYLCHOLINESTERASE FOR FORMATION OF NEURAL NETWORKS AS TESTED BY CHICKEN RETINOSPHEROIDS

Elmar Willbold and Paul G. Layer

Technische Hochschule Darmstadt
Institut für Zoologie
Schnittspahnstraße 3
64287 Darmstadt, Germany

During chicken neurogenesis, butyrylcholinesterase (BChE) and acetylcholinesterase (AChE) are expressed in specific patterns between final cell proliferation and differentiation. BChE activity is transiently produced by neuroblasts in the ventricular layer during their last cell cycle. Soon after they have finished their last mitosis, they migrate to the mantle layer. They begin to differentiate and rapidly accumulate AChE (Layer, 1983). This recurrent pattern hints to a possible mutual interdependance of the two enzymes. Additionally we could show that both cholinesterases are also very suitable substrates for outgrowing axons and that they are able to regulate neurite growth in vitro, however not via their enzymatic activity (Layer et al., 1993). Since cholinesterases share significant sequence homologies with *Drosophila* adhesion molecules, and since they can bear the HNK-1 epitope (Weikert and Layer, 1994), a more general and active regulatory role of the enzymes during embryogenesis can be assumed.

To further follow up these issues, we investigated the effects of inhibition of BChE and AChE, respectively, on laminar histogenesis in *retinospheroids*. Retinospheroids are histotypically organized regenerates of the embryonic chicken retina. They arise by aggregation and proliferation of dissociated embryonic retinal cells in a rotation culture system. This system is suitable to study processes of retinogenesis under defined conditions.

The addition of the irreversible BChE inhibitor tetraisopropylpyrophosphoramide (iso-OMPA) to developing retinospheroids fully inhibits BChE expression. Unexpectedly, iso-OMPA also suppresses the expression of AChE to 35 - 60%. Long time incubation experiments showed that AChE suppression by the BChE inhibitor cannot be explained by a direct cross inhibition of iso-OMPA on AChE. Morphologically, the number of spheroids/dish is increased, and their diameter is decreased by about 20%, corresponding to about 50% volume size. The release of AChE into the culture medium is also inhibited by iso-OMPA to more than 85% (Layer et al., 1992). Histochemically we could show that the diminished AChE-activity is localized in the fibre rich areas of the spheroids. Surprisingly, the inhibition of BChE also leads to a dramatic acceleration of the temporal course of histotypical differentiation. Using the fibre-specific antibody F11 (Rathjen et al., 1987) we could demonstrate that the development of the fibre areas is at least advanced by two days. Concomitantly, the columnar organization of cellular areas is also significantly accelerated (Willbold and Layer, 1994). In contrast, treatment of retinospheroids with the reversible AChE inhibitor BW284C51 initially increases the expression of AChE before then leading to a rapid decay of the spheroids.

We conclude, that the cellular expression of AChE is regulated by both the activities of BChE and of AChE within neuronal tissues. Moreover, the amount of BChE must be somehow involved in cell cycle control, in regulating neurite outgrowth patterns, and in the developmental switch from proliferation to differentiation.

REFERENCES

Layer, P. G. (1983) *Proc. Natl. Acad. Sci. USA* 80: 6413-6417.
Layer, P. G., Weikert, T., & Willbold, E. (1992) *Cell Tissue Res.* 268: 409-418.
Layer, P. G., Weikert, T., & Alber, R. (1993) *Cell Tissue Res.* 273: 219-226.
Rathjen, F. G., Wolff, J. M., Frank, R., Bonhoeffer, F., & Rutishauser, U. (1987) *J. Cell Biol.* 104 343-353.
Weikert, T., & Layer, P. G. (1994) *Neurosci. Lett.* 176: 9-12.
Willbold, E., & Layer, P. G. (1994) *Eur. J. Cell Biol.* 64: 192-199.

NEUROTROPHIC FUNCTION OF CIRCULATING ACETYLCHOLINESTERASE IN *APLYSIA*

M. Srivatsan and B. Peretz

Department of Physiology
University of Kentucky Medical Center
Lexington, Kentucky 40536

INTRODUCTION

The catalytic function of acetylcholinesterase (AChE) is well documented (Massoulie & Bon, 1982). Yet, the significance of AChE's presence in non-cholinergic sites is poorly understood. In *Aplysia*, a marine gastropod, AChE is present in the gill (Peretz & Estes, 1974), hemolymph and cholinergic as well as non-cholinergic neurons (Giller & Schwartz, 1971; Srivatsan & Peretz, 1993). Hemolymph from mature *Aplysia* exhibits neurotrophic property (Schachar & Proshansky, 1983). The elevated AChE activity in hemolymph of mature *Aplysia* (Srivatsan et al., 1992) and its link to neuronal function (Peretz et al., 1992; Srivatsan & Peretz, 1993) prompted us to investigate if AChE has a neurotrophic function.

METHODS

Young adult Aplysia, avg. age of 74 ± 15 days, from mariculture facility at Miami Univ., Florida were used. Leibowitz culture medium (L-15, GIBCO) was made isotonic to *Aplysia* serum by the addition of salts and served as control. Dissection, dissociation of neurons and maintenance of the cultures were carried out as per earlier report (Schachar & Proshansky, 1983). Neurons from the same animal were maintained in the following culture conditions: 1) L-15 by itself (control); 2) equal proportions of L-15 and hemolymph; 3) as in 2, with hemolymph AChE inhibited by BW284c51, a membrane impermeable inhibitor blocking both the catalytic and peripheral sites and 4) as in 2, with hemolymph AChE inhibited by edrophonium chloride which blocks the catalytic site. From photomicrographs taken at regular intervals, neurite length was measured using the NIH Image 1.41 software.

RESULTS AND CONCLUSIONS

The results showed that 1) A significant increase in the number (>400%) and length (>65%) of neurites was observed in cultures containing hemolymph (Fig.1b). 2) AChE in

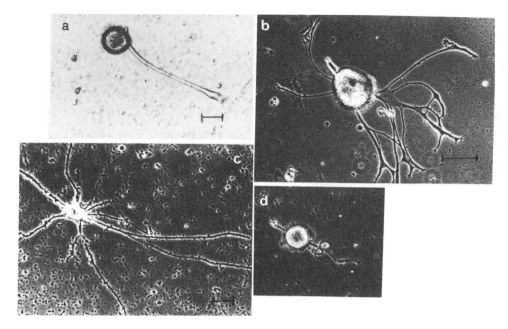

Figure 1. Neurite growth from neurons cultured in different media for 36 hours: a) L-15 alone, b) L-15 + hemolymph (10 units of AChE activity/ml), c) L-15 + hemolymph + BW284c51 and d) L-15 + hemolymph + edrophonium chloride.

hemolymph apparently enhanced neurite growth since hemolymph AChE inhibited by BW284c51 did not result in increased neurite growth (Fig.1c). 3) This neurotrophic effect of AChE was not due to its catalytic function since inhibiting only the catalytic site with edrophonium chloride did not affect the increase in number (>300%) or length (>60%) of growing neurites (Fig.1d). 4) Addition of carbachol (analogue of ACh,1,10 and 100μM) to the culture did not inhibit neurite growth (not shown here) indicating that AChE's neurotrophic function is non-cholinergic. Neuronal AChE presumably aids in target recognition (Robertson,1987) and neurite growth (Layer et al., 1993) during development. Whereas our results demonstrate that circulating AChE has a neurotrophic effect on the fully differentiated neurons regenerating in culture.

REFERENCES

Giller, E. Jr. & Schwartz, J.H.(1971)*J.Neurophysiol.*, 34:108-115.
Layer, P.G., Weikert, T. & Alber, R.(1993)*Cell Tissue Res.*, 273:219-226.
Peretz, B. & Estes, J.(1974)*J. Neurobiol.*, 5:3-19.
Peretz. B., Srivatsan, M. & Hallahan, B.(1992)*Soc. Neurosci. Abst.*,18: 583.
Robertson, R.(1987).*Neurosci. Lett.* 75:259-264.
Massoulie, J. & Bon, S.(1982)*Ann. Rev. Neurosci.* 5:57-106.
Schachar, S. & Proshanski, E.(1983)*J. Neurosci.* 3:2403-2413.
Srivatsan, M. & Peretz, B.(1993)*Soc.Neurosci. Abst.*, 19: 569.
Srivatsan, M., Peretz, B., Hallahan, B. & Talwalker, R.(1992)*J. Comp. Physiol.B.*, 162:29-37.

CHOLINESTERASES IN ALZHEIMER'S DISEASE

M.-Marsel Mesulam

Center for Behavioral and Cognitive Neurology and the Alzheimer Program
Departments of Neurology and Psychiatry
Northwestern University Medical School
320 E Superior Street, Searle Research Building Rm. 11-450
Chicago, Illinois, 60611

Alzheimer's disease (AD) is a relatively common, relentlessly progressive and currently untreatable condition of unknown cause. Amyloid plaques and neurofibrillary tangles are the two major neuropathological lesions in AD. Histochemical experiments with acetylthiocholine or butyrylthiocholine as the substrate indicate that plaques as well as tangles contain acetylcholinesterase (AChE) and butyrylcholinesterase (BChE) activity (Friede, 1965; Mesulam, et al., 1987; Mesulam, et al., 1987; Perry, et al., 1978). As expected, the AChE activity in these lesions is inhibited by BW284C51 whereas the BChE activity is inhibited by Iso-OMPA. Electronmicroscopic studies indicate that this cholinesterase activity is associated with the paired helical filaments of neurofibrillary tangles and with the neuritic as well as the amyloid component of plaques (Carson, et al., 1991).

The source of the cholinesterases in plaques and tangles has not yet been identified definitively. Axons and neuronal cell bodies of the human cerebral cortex contain intense AChE activity which could provide a potential source for the plaque and tangle-bound enzyme. Although the AChE activities in neurons, axons, plaques and tangles share the common property of selective inhibition by BW284C51, there are also enzymatic properties and inhibitor selectivities that differentiate the AChE of plaques and tangles from the AChE of cortical neurons and axons. For example, the optimal pH for obtaining the AChE reaction-product in plaques and tangles is considerably lower than the optimal pH for obtaining the AChE reaction-product in neurons and axons. The plaque and tangle-bound enzyme is also much more susceptible to inhibition by carboxypeptidase inhibitor, by the non-specific protease inhibitor bacitracin and by the aryl-acylamidase inhibitor 5-hydroxytryptophan (Wright, et al., 1993). Furthermore, the neurons and axons of the normal cerebral cortex contain very little BChE whereas large numbers of plaques and tangles in AD display intense BChE activity.

These observations suggest that the AChE as well as the BChE of plaques and tangles may have an extra-neuronal origin. Recent experiments have indicated that AChE and BChE activities with pH preferences and inhibitor selectivities identical to those of the plaque and tangle-bound enzymes are found in the astrocytes and oligodendrocytes of control and AD

Enzymes of the Cholinesterase Family, Edited by Daniel M. Quinn et al.
Plenum Press, New York, 1995

brains (Wright, et al., 1993) . Neuroglia can therefore be considered as a potential source of the plaque and tangle-bound cholinesterases. In the normal brain, this enzyme may be confined to the glial cell body whereas in AD it may be secreted into plaques and perhaps tangles.

Some of our observations suggest that the distribution of glial BChE may be of particular relevance to AD. In non-AD control brains, the density of BChE-positive glia (especially in the deep layers of cortex and the subjacent white matter) is higher in entorhinal and inferotemporal cortex, two regions with a high susceptibility to the pathology of AD, than in primary somatosensory and visual cortex, two areas with a relatively lower suscep-tibility to the disease process. In comparison to age-matched control specimens, AD brains also have a significantly higher density of BChE glia in entorhinal and inferotemporal regions but not in the primary somatosensory or visual areas (Wright, et al., 1993). These results raise the possibility that a high density of BChE glia may act as a marker for increased susceptibility to AD or that the density of BChE glia may increase as part of the overall pathological process in AD. It is interesting to note that a significant increase in the biochemically determined total BChE activity has been reported in AD (Perry, et al., 1978).

The suggestion that cholinesterases may play a role in the pathogenesis of AD may initially appear unlikely, especially in view of the pivotal role which has been attributed to amyloid in the causation of this disease. Despite the numerous pathological changes that are associated with AD, only the extracellular accumulation of Aß amyloid plaques is specific and differentiates this condition from other dementing diseases. The Aß peptide is a 4 kd fragment of a membrane spanning, naturally occuring Amyloid Precursor Protein (APP). Increased production and point mutations of APP or deviant sequestration of the Aß fragment by apolipoprotein E have been invoked as causative mechanisms for plaque formation and AD (Citron, et al., 1992; Mullan, et al., 1993; Saunders, et al., 1993; Yankner, et al., 1991).

Despite this universally acknowledged pivotal role of amyloid in the pathogenesis of AD, numerous observations are also indicating that the amyloid deposits are not by themselves sufficient for triggering the neuritic degeneration and dementia characteristic of this disease. The deposition of amyloid does not initially cause tissue pathology (Yamaguchi, et al., 1989) and may need to go on for years before the emergence of neuritic degeneration and clinically recognizable dementia (Rumble, et al., 1989). There is also very little correlation between dementia severity and plaque formation (Arriagada, et al., 1992; Berg, et al., 1993), and many non-demented elderly individuals have plaque densities in the range seen in AD (Dayan, 1970; Delaère, et al., 1993; Dickson, et al., 1991; Katzman, et al., 1988; Mizutani, et al., 1992) .

It appears that amyloid plaques may need to undergo a lengthy process of transfor-mation at the site of initial deposition before they become pathogenic. A recent observation raises the possibility that BChE reactivity may participate in the "ripening" and eventual pathogenicity of Aß deposits in AD. In an unselected sample of 4 demented and 4 age-matched non-demented brains, the total cortical area covered by plaque-like Aß amyloid and BChE deposits was measured at two regions of the temporal cortex with the help of computerized densitometry. Demented as well as age-matched non-demented brains con-tained Aß and BChE positive plaques. The total cortical area covered by the Aß precipitates was higher in demented individuals but there was overlap with the values seen in the specimens from non-demented individuals. On the other hand, the proportional amyloid plaque area displaying BChE reactivity was very significantly and 5-6 fold higher in the demented than in the non-demented group and there was no overlap between the two populations (Mesulam and Geula, in press) . The BChE reactivity of amyloid plaques is therefore a more powerful differentiating factor of AD from age-matched control brains than the amount of amyloid deposition. These results have led us to speculate that a progressively more extensive BChE reactivity of plaques could participate in their transformation from a relatively benign pre-clinical form to pathogenic structures associated with neuritic degeneration and dementia.

The physiological role served by BChE remains mysterious. BChE is the product of a different gene than AChE (Taylor, et al., 1994). Its distribution in the brain is much more restricted than that of AChE and it does not seem to have a role in conventional cholinergic neurotransmission. Detoxifying, peptidasic, growth promoting and morphogenetic functions have also been attributed to BChE (Massoulié, et al., 1993; Rao, et al., 1993). Conceivably, BChE may join a number of other substances such as a_1-antichymotrypsin (Abraham, et al., 1990), protease nexin 1 (Rosenblatt, et al., 1989), and a_1-antitrypsin (Gollin, et al., 1992) which are also found in association with the amyloid plaque deposits and which may modify the microenvironment of the plaque as well as its influence upon local neuritic reactions. For example, BChE could induce neurite extension (and, eventually, degeneration) through a growth promoting effect which has been identified in developing neural systems and which may be independent from its enzymatic activity (see Layer in this volume).

The excessive deposition of insoluble Aß amyloid in the cerebral cortex is a necessary and specific factor in the pathogenesis of AD (Citron, et al., 1992; Yankner, et al., 1991). However, this conclusion needs to be reconciled with the inconsistent temporal and quantitative relationship between Aß deposits and either tissue injury or dementia (Arriagada, et al., 1992; Berg, et al., 1993; Dickson, et al., 1991; Rumble, et al., 1989). One explanation is that the deposition of Aß is a necessary but not sufficient upstream factor and that additional downstream events are required for the development of AD. The characterization of these downstream events is crucial for understanding how AD evolves and also for finding ways in which the disease process can be altered.

Our findings raise the possibility that the cholinesterases associated with the histopathological lesions of AD, especially the BChE reactivity of amyloid plaques, may constitute one of the important downstream events in the complex life cycle of Aß plaques. If this speculation were to find further support, it would raise the interesting possibility that BChE-inhibitors may play an important and hitherto unexpected role in the prevention or treatment of AD. Since human BChE displays numerous allelic polymorphic variants (Soreq, et al., 1992) it would also be of considerable interest to determine if some of these are more closely associated with a susceptibility to AD than others.

REFERENCES

Abraham, C., Shirahama, T., & Potter, H. 1990. a1-Antichymotrypsin is associated solely with amyloid deposits containing the ß-protein: Amyloid and cell localization of a1-antichymotrypsin. *Neurobiol Age*, 11: 123-129.

Arriagada, P. V., Growdon, J. H., Hedley-Whyte, E. T., & Hyman, B. T. 1992. Neurofibrillary tangles but not senile plaques parallel duration and severity of Alzheimer's disease. *Neurol*, 42: 631-639.

Berg, L., McKeel, D. W., Miller, J. P., Baty, J., & Morris, J. 1993. Neuropathological indexes of Alzheimer's disease in demented and nondemented persons aged 80 years and older. *Arch Neurol*, 50: 349-358.

Carson, K. A., Geula, C., & Mesulam, M. M. 1991. Electron microscopic localization of cholinesterase activity in Alzheimer brain tissue. *Brain Res*, 540: 204-8.

Citron, M., Oltersdorf, T., Haass, C., McConlogue, L., Hung, A. Y., Seubert, P., Vigo-Pelfrey, C., Lieberburg, I., & Selkoe, D. J. 1992. Mutation of the ß-amyloid precursor protein in familial Alzheimer's disease increases ß-protein production. *Nature*, 360: 672-674.

Dayan, A. D. 1970. Quantitative histological studies on the aged human brain. II. Senile plaques and neurofibrillary tangles in senile dementia (with an appendix on their occurence in cases of carcinoma). *Acta Neuropathol*, 16: 95-102.

Delaère, P., He, Y., Fayet, G., Duyckaerts, C., & Hauw, J.-J. 1993. ßA4 deposits are constant in the brain of the oldest old: An immunohistochemical study of 20 French centenarians. *Neurobiol Age*, 14: 191-194.

Dickson, D. W., Crystal, H. A., Mattiace, L. A., Masur, D. M., Blau, A. D., Davies, P., Yen, S. H., & Aronson, M. K. 1991. Identification of normal and pathological aging in prospectively studied nondemented elderly humans. *Neurobiol Age*, 13: 179-189.

Friede, R. L. 1965. Enzyme histochemical studies of senile plaques. *J Neuropath Exper Neurol*, 24: 477-491.

Gollin, P. A., Kalaria, R. N., Eikelenboom, P., Rozemuller, A., & Perry, G. 1992. Alpha-1-antitrypsin and alpha-1-antichymotrypsin are in lesions of Alzheimer's disease. *Neuroreport*, 3: 201-203.

Katzman, R., Terry, R., DeTeresa, R., Brown, T., Davies, P., Fuld, P., Renbing, X., & Peck, A. 1988. Clinical, pathological and neurochemical changes in dementia: A subgroup with preserved mental status and numerous neocortical plaques. *Ann Neurol*, 23: 138-144.

Massoulié, J., Pezzementi, L., Bon, S., Krejci, E., & Vallette, F.-M. 1993. Molecular and cellular biology of cholinesterases. *Prog Neurobiol*, 41: 31-91.

Mesulam, M.-M. & Geula, C. in press. Butyrylcholynesterase reactivity differentiates the amyloid plaques of aging from those of dementia. *Ann Neurol*.

Mesulam, M.-M., Geula, C., & Morán, A. 1987. Anatomy of cholinesterase inhibition in Alzheimer's disease: Effect of physostigmine and tetrahydroaminoacridine on plaques and tangles. *Ann Neurol*, 22: 683-691.

Mesulam, M. M., & Morán, A. 1987. Cholinesterases within neurofibrillary tangles related to age and Alzheimer's disease. *Ann Neurol*, 22: 223-8.

Mizutani, T., & Shimada, H. 1992. Neuropathological background of twenty-seven centenarian brains. *J Neurol Sci*, 108: 168-177.

Mullan, M., Tsuji, S., Miki, T., Katsuya, T., Naruse, S., Kaneko, K., Shimizu, T., Kojima, T., Nakano, I., Ogihara, T., Miyatake, T., Ovenstone, I., Crawford, F., Goate, A., Hardy, J., Roques, P., Roberts, G., Luthert, P., Lantos, P., Clark, C., Gaskell, P., Crain, B., & Roses, A. 1993. Clinical comparison of Alzheimer's disease in pedigrees with the codon 717 Val-Ile mutation in the amyloid precursor protein gene. *Neurobiol Age*, 14: 407-419.

Perry, E. K., Perry, R. H., Blessed, G., & Tomlinson, B. E. 1978. Changes in brain cholinesterases in senile dementia of Alzheimer type. *Neuropath Appl Neurobiol*, 4: 273-277.

Rao, R. V., & Balasubramanian, A. S. 1993. The peptidase activity of human serum butyrylcholinesterase: Studies using monoclonal antibodies and characterization of the peptidase. *J Prot Chem*, 12: 103-110.

Rosenblatt, D. E., Geula, C., & Mesulam, M.-M. 1989. Protease nexin I immunostaining in Alzheimer's disease. *Ann Neurol*, 26: 628-634.

Rumble, B., Retallack, R., Hilbich, C., Simms, G., Multhaup, G., Martins, R., Hockey, A., Montgomery, P., Beyreuther, K., & Masters, C. L. 1989. Amyloid A4 protein and its precursor in Down's syndrome and Alzheimer's disease. *New Eng J Med*, 320: 1446-1452.

Saunders, A. M., Strittmatter, W. J., Schmechel, D., St. George-Hyslop, P. H., Pericak-Vance, M. A., Joo, S. H., Rosi, B. L. G., J. F., Crapper-McLachlan, D. R., Alberts, M. J., Hulette, C., Crain, B., Goldgaber, D., & Roses, A. D. 1993. Association of apolipoprotein E allele e4 with late-onset familial and sporadic Alzheimer's disease. *Neurol*, 43: 1467-1472.

Soreq, H., Gnatt, A., Loewenstein, Y., & Neville, L. F. 1992. Excavations into the active-site gorge of cholinesterases. *Trends Biochem Sci*, 17: 353-358.

Taylor, P., & Brown, J. H. (1994). Acetylcholine. In G. J. Siegel et al. (Ed.), *Basic Neurochemistry: Molecular, Cellular and Medical Aspects* (pp. 231-260). New York: Raven Press.

Wright, C. I., Geula, C., & Mesulam, M. M. 1993. Neuroglial cholinesterases in the normal brain and in Alzheimer's disease: relationship to plaques, tangles, and patterns of selective vulnerability. *Ann Neurol*, 34: 373-84.

Wright, C. I., Geula, C., & Mesulam, M. M. 1993. Protease inhibitors and indoleamines selectively inhibit cholinesterases in the histopathologic structures of Alzheimer disease. *Proc Natl Acad Sci U S A*, 90: 683-6.

Yamaguchi, H., Nakazato, Y., Hirai, S., Shoji, M., & Harigaya, Y. 1989. Electron micrograph of diffuse plaques. Initial stages of senile plaque formation in the Alzheimer brain. *Amer J Path*, 135: 593-597.

Yankner, B. A., & Mesulam, M.-M. 1991. ß-amyloid and the pathogenesis of Alzheimer's disease. *New Eng J Med*, 325: 1849-1856.

A ROLE FOR ACETYLCHOLINE AND ACETYLCHOLINESTERASE IN THE REGULATION OF NEURITE OUTGROWTH

Implications for Alzheimer's Disease

D. H. Small and G. Reed

Department of Pathology
The University of Melbourne
Parkville, Victoria 3052
Australia

BIOCHEMICAL PATHOLOGY OF ALZHEIMER'S DISEASE

Alzheimer's disease is a progressive dementia that is most commonly associated with ageing. The hallmark of Alzheimer's disease pathology is the deposition of amyloid in the brain, both intracellularly in the form of neurofibrillary tangles and extracellularly in the form of amyloid plaques and congophilic angiopathy. The extracellular amyloid is composed principally of a 4 kDa polypeptide referred to as the amyloid protein or ßA4. The amyloid protein is derived by proteolysis of a larger precursor protein (APP) which is a normally constituent of many healthy cells (Kang et al., 1987). The importance of APP and amyloid in the pathogenesis of the disease is highlighted by the fact that several point mutations in the APP gene have been found to cause severe early-onset forms of familial Alzheimer's disease (Goate et al., 1991; Chartier-Harlin et al., 1991; Mullan et al., 1992).

THE CHOLINERGIC HYPOTHESIS OF ALZHEIMER'S DISEASE

The cholinergic system is more profoundly affected in Alzheimer's disease than other neurotransmitter systems. Studies by Whitehouse et al. (1982) demonstrated a decrease in the number of cholinergic neurons in the basal forebrain of patients with Alzheimer's disease. This finding and other studies (reviewed by Holttum & Gershon, 1992) have led to the "cholinergic hypothesis", which states that certain clinical features of Alzheimer's disease (e.g, amnesia) can be explained by the loss of cholinergic innervation to regions of the brain such as the hippocampus and neocortex, regions critical for the processing and storage of memories.

Enzymes of the Cholinesterase Family, Edited by Daniel M. Quinn et al.
Plenum Press, New York, 1995

Despite the controversial nature of the cholinergic hypothesis of Alzheimer's disease, AChE inhibitors have been examined for their therapeutic efficacy. The inhibitor tacrine (trade name Cognex[R]) has been reported to be of some therapeutic benefit (Summers et al., 1981; Kaye et al., 1982; Eagger et al., 1991), although it is unclear whether tacrine causes any long-term improvement in patients. It is also unclear whether the reported therapeutic benefits of tacrine are related to a boost in cholinergic neurotransmission (Holttum & Gershon, 1992).

One factor which needs to be considered when assessing the value of cholinesterase inhibitors for the treatment of Alzheimer's disease is that cholinesterases may have functions which are unrelated to neurotransmission (Small, 1990). Such non-classical functions of cholinesterases may also be inhibited by tacrine or other inhibitors. The consequences of such an inhibition need to be factored into any assessment of efficacy. This consideration is important because of the high levels of AChE and BChE found in amyloid plaques. The distribution of AChE and BChE in amyloid plaques may not be a consequence of the degeneration of cholinergic synapses (Mesulam & Geula, 1990). Instead, AChE and BChE may have functions unrelated to neurotransmission in the region of the amyloid plaque.

NON-CLASSICAL FUNCTIONS OF AChE

It has long been suspected that AChE has functions unrelated to the inactivation of cholinergic neurotransmission. Numerous studies have demonstrated that the developmental and topographic expression of AChE does not closely match the distribution of cholinergic synapses (Silver, 1974). This finding has led to the view that AChE has functions unrelated to the termination of cholinergic neurotransmission.

The presence of multiple molecular forms of AChE synthesised by one cell type also suggests that AChE has multiple functions. Our own studies have shown that chromaffin cells of the adrenal medulla, which possess characteristics similar to sympathetic neurons, synthesise three major species of AChE which are targetted extracellularly (Michaelson et al., 1994).

It is now apparent that earlier reports of protease activities associated with affinity purified preparations of AChE (Chubb et al., 1980; Chubb et al., 1983) were based upon the co-purification of distinct proteases with cholinesterases. Complete separation of protease activities from esterase activities has now been achieved for eel AChE (Checler et al., 1994) and fetal bovine AChE (Michaelson & Small, 1993), the two preparations originally reported to possess "intrinsic" protease activities. Thus, it seems likely that the primary function of AChE and BChE is to hydrolyse acetylcholine (Small, 1994).

THE ROLE OF AChE IN NEURITE OUTGROWTH

There is now evidence that AChE has important functions during early embryogenesis. These functions are in all probability unrelated to the regulation of cholinergic neurotransmission. Studies by Layer and coworkers (Layer et al., 1988, 1992; Layer, 1991; Layer & Kaulich, 1991), Roberson and coworkers (Robertson, 1987; Robertson et al., 1988; Robertson & Yu, 1993) and Small et al. (1992) have shown that AChE is expressed very early in embryogenesis, and that the expression of AChE correlates well with the period in which neurite outgrowth is maximal. Thus, AChE may function in the regulation of neurite outgrowth.

AChE could regulate neurite outgrowth through its action on acetylcholine. There is increasing evidence that neurotransmitters regulate processes of neural differentiation (Hohmann et al., 1988; Schambra et al., 1989; Lauder, 1993), including neurite outgrowth (Mattson, 1988). Studies by Lipton et al. (1988) and Lipton & Kater (1989) have demonstrated that acetylcholine can inhibit neurite outgrowth from retinal ganglion cells through a nicotinic receptor-mediated mechanism. More recently, studies by Pugh & Berg (1994) using chick ciliary ganglion have shown that neurite retraction is mediated in a calcium-dependent manner by nicotinic receptors which bind alpha-bungarotoxin. In addition to its trophic effects, acetylcholine may also have *tropic* actions on neurite outgrowth. Studies by Zheng et al. (1994) have shown that gradients of acetylcholine can induce the turning of growth cones in cultures of *Xenopus* embryonic spinal neurons.

CHOLINERGIC REGULATION OF NEURITE OUTGROWTH FROM SYMPATHETIC NEURONS

We have examined the possibility that AChE expression may be associated with the action of acetylcholine on neurite outgrowth. The presence of high levels of AChE on the surface of neurites of sympathetic neurons in culture (Rotundo and Carbonetto, 1987) prompted us to study the effect of cholinergic agonists and antagonists on neurite outgrowth from chick sympathetic neurons (Small et al., 1994b). These studies have demonstrated that a variety of cholinergic agents can influence the extent of neurite outgrowth. Cholinergic agonists, such as acetylcholine and carbamylcholine, inhibit neurite outgrowth from chick sympathetic neurons in culture (Small et al., 1994b). Carbamylcholine is more potent that acetylcholine in its ability to inhibit neurite outgrowth. This difference in potency is due to the hydrolysis of acetylcholine by endogenous AChE in the culture, as the potency of acetylcholine was increased by the presence of cholinesterase inhibitors such as tacrine, BW284c51 or edrophonium (Small et al., 1994b).

Previous studies have shown that nicotinic receptors modulate neurite outgrowth in culture (Lipton et al., 1988; Pugh & Berg, 1994; Zheng et al., 1994). However, nicotine only weakly inhibited neurite outgrowth in our studies. The lack of efficacy of nicotine may be a consequence of the long time course of incubation (24 hr), during which most nicotinic receptors would be expected to have desensitized. In contrast to nicotine, oxotremorine (a muscarinic receptor agonist) strongly inhibited neurite outgrowth, indicating that G-protein linked muscarinic receptors can influence neurite outgrowth. The involvement of a muscarinic receptor-mediated mechanism is supported by the observation that the effect of carbamylcholine can be partially blocked by 50 nM pirenzipine a selective M1 receptor antagonist (Fig. 1).

We have also examined the role of intracellular second messengers in mediating inhibition of neurite outgrowth (Small et al., submitted for publication). Several lines of evidence indicate that an increase in intracellular calcium levels mediates the effects of cholinergic agents on neurite outgrowth. Stimulation of muscarinic receptors on sympathetic

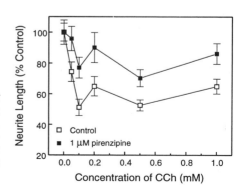

Figure 1. Effect of 50 nM pirenzipine on the ability of carbamylcholine to inhibit neurite outgrowth from chick sympathetic neurons maintained in culture for 24 hr. Figure shows quantitative computer-assisted image capture analysis of neurite length. Data are means ± SEM for four culture wells. Five fields per well, each field bearing approximately 20 neurons, were analysed.

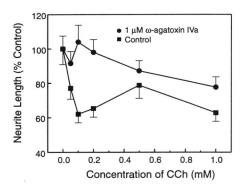

Figure 2. Effect of 1 μM omega-agatoxin IVa on the ability of carbamylcholine to inhibit neurite outgrowth from chick sympathetic neurons maintained in culture for 24 hr. Figure shows quantitative computer-assisted image capture analysis of neurite length. Data were obtained and analysed as described in Fig. 1.

neurons is known to result in the generation of a slow EPSP through the inhibition of M-type potassium channels. This slow EPSP can in turn open voltage-gated calcium channels, resulting in an influx of intracellular calcium (Constanti & Brown, 1981).

Consistent with this hypothesis, we have found that the action of carbachol on neurite outgrowth can be blocked by 1 μM omega-agatoxin IVa, a selective blocker of P-type calcium channels.

AChE MAY STIMULATE NEURITE OUTGROWTH THROUGH A NON-CATALYTIC MECHANISM

The idea that AChE could also exert an effect on neurite outgrowth through an adhesion mechanism comes from the observation that cholinesterases share amino acid sequence homology to the *Drosophila* adhesion proteins neurotactin (Barthalay et al., 1990; de la Escalera et al., 1990) and glutactin (Olson et al., 1990), and that certain forms of AChE have an HNK-1 like carbohydrate epitope (Bon et al., 1987), commonly associated with adhesion molecules. Cell adhesion to an underlying substratum is a necessary condition for neurite outgrowth in culture.

To examine the ability of AChE to stimulate neurite outgrowth through an adhesion mechanism, we have studied neurite outgrowth from sympathetic neurons grown on a substrate of AChE. Tissue culture dishes were coated with 100 μg/ml of polylysine followed by purified fetal bovine serum AChE (10 μg/ml) and/or heparan sulphate proteoglycan (HSPG)(10 μg/ml) purified from cultures of postnatal day 3 mouse brain cells. We found that when the neurons were grown on a substrate containing a mixture of AChE and HSPG, neurite outgrowth was dramatically stimulated (Small et al., 1994b). Neither AChE nor HSPG on its own was sufficient to stimulate neurite outgrowth. Pre-treatment of the AChE was diisopropyl fluorophosphate (to inactivate the esterase activity) did not abolish this effect, demonstrating that the effect of AChE on neurite outgrowth was not associated with the esterase activity. Further purification of the AChE on a antibody affinity column did not remove the neurite outgrowth promoting properties of the enzyme, indicating that the ability of AChE to promoting neurite outgrowth was an intrinsic property of the enzyme or was associated with a protein which was bound very tightly to the purified enzyme.

THE FUNCTION OF AChE IN AMYLOID PLAQUES

Overexpression of AChE and BChE occurs in the region of the neuritic amyloid plaque. This abnormal cholinesterase deposition may provide some clues which can explain the processes which lead to the formation of neuritic plaques. Studies by Mesulam & Geula (1990) have shown that the distribution of AChE and BChE staining in the region of amyloid plaques does not correlate with the density of cholinergic innervation. This suggests that the AChE present in plaques may have a function unrelated to cholinergic neurotransmission. The presence of BChE, an enzyme that is not a marker of cholinergic synapses, in amyloid plaques supports this view. Within the amyloid plaque, AChE is found in both the plaque core, and in the dystrophic neurites which commonly surround the plaque (Carson et al., 1991). The presence of AChE in the neuritic component around the plaque may be related to the role of AChE in neurite outgrowth in development. Interestingly, APP and AChE are coordinately expressed in the developing chick brain (Small et al., 1992). Furthermore, both AChE and APP can promote neurite outgrowth and show a similar dependence upon the presence of heparan sulphate proteoglycans for this effect (Small et al., 1994a). Thus AChE and APP may have related functions in the regulation of neurite outgrowth during development or during regeneration and repair.

The simplest interpretation of our data is that disturbances in normal trophic functions in the region of the amyloid plaque lead to overexpression of proteins associated with neuritic outgrowth or regeneration. The presence of cholinesterases in plaques and tangles may be a consequence of this disturbance. Overexpression of APP in the region of the plaque could result in increased amyloid protein production, which ultimately would cause increased amyloid deposition in the plaque. The identification and correction of the trophic defect could have therapeutic implications.

ACKNOWLEDGMENTS

This work was supported by grants from the National Health and Medical Research Council of Australia.

REFERENCES

Barthalay, Y., Hipeau-Jacquotte, R., de la Escalera, S., Jiminez, F., and Piovant, M., 1990, *Drosophila* neurotactin mediates heterophilic cell adhesion, EMBO J. 9:3603-3609.

Bon, S., Méflah, K., Musset, F., Grassi, J., and Massoulié, J., 1987, An immunoglobulin M monoclonal antibody, recognizing a subset of acetylcholinesterase molecules from electric organ of *Electrophorus* and *Torpedo*, belongs to the HNK-1 anti-carbohydrate family, *J. Neurochem.* 49:1720-1731.

Carson, K. A., Geula, C., and Mesulam, M.-M., 1991, Electron microscopic localization of cholinesterase activity in Alzheimer brain tissue, *Brain Res.* 540:204-208.

Chartier-Harlin, M., Crawford, F., Houlden, H., Warren, A., Hughes, D., Fidani, L., Goate, A., and Rossor, M., 1991, Early-onset Alzheimer's disease caused by mutations at codon 717 of the ß-amyloid precursor protein gene, *Nature* 353:844-846.

Checler, F., Grassi, J., and Vincent, J.-P., 1994, Cholinesterases display genuine arylacylamidase activity but are totally devoid of intrinsic peptidase activities, *J. Neurochem.* 62:756-763.

Chubb, I. W., Hodgson, A.J., and White, G. H., 1980, Acetylcholinesterase hydrolyzes substance P, *Neuroscience* 5:2065-2072.

Chubb, I. W., Ranieri, E., White, G. H., and Hodgson, A. J., 1983, The enkephalins are amongst the peptides hydrolyzed by purified acetylcholinesterase, *Neuroscience* 10:1369-1377.

Constanti, A., and Brown, D. A., 1981, M-currents in voltage-clamped mammalian sympathetic neurons, *Neurosci. Lett.* 24:289-294.

de la Escalara, S., Bockamp, E. O., Moya, F., Piovant, M., Jimenez, F., 1990, Characterization and gene cloning of neurotactin, a *Drosophila* transmembrane protein related to cholinesterases, *EMBO J* 9:3593-3601.

Eagger, S. A., Levy, R., and Sahakian, B. J., 1991, Tacrine in Alzheimer's disease, *Lancet* 337:989-992.

Goate, A. M., Chartier-Harlin, M., Mullan, M., Brown, J., Crawford, F., Fidani, L., Giuffra, L., Haynes, A., Irving, N., James, L., Mant, R., Newton, P., Rooke, K., Roques, P., Talbot, C., Pericak-Vance, M.,

Roses, A., Williamson, R., Rosor, M., Owen, M., and Hardy, J., 1991, Segregation of a missense mutation in the amyloid precursor protein gene with familial Alzheimer's disease, *Nature* 349:704-706.

Hohmann, C. F., Brooks, A. R., and Coyle, J. T., 1988, Neonatal lesions of the basal forebrain cholinergic neurons result in abnormal cortical development, *Dev. Brain Res.* 42:253-264.

Holttum, J. R., and Gershon, S., 1992, The cholinergic model of dementia, Alzheimer type: progression from the unitary transmitter concept, *Dementia* 3:174-185.

Kang, J., Lemaire, H., Unterbeck, A., Salbaum J.M., Masters, C.L., Grzeschik, K., Multhaup, G., Beyreuther, K., and Muller-Hill, B., 1987, The precursor of Alzheimer's disease amyloid A4 protein resembles a cell-surface receptor, *Nature* 325:733-736.

Kaye, W. H., Sitaram, N., Weingartner, H., Ebert, M. H., Smallberg, S., and Gillin, J. C., 1982, Modest facilitation on memory in dementia with combined lecithin and anticholinesterase treatment, *Biol. Psychiatry* 17:275-280.

Lauder, J. M., 1993, Neurotransmitters as growth regulatory signals: role of receptors and second messengers, *Trends Neurosci.* 16:233-240.

Layer, P. G., Alber, R., and Rathjen, F. G., 1988, Sequential activation of butyrylcholinesterase in rostral half somites and acetylcholinesterase in motoneurones and myotomes preceding growth of motor axons, *Development* 102:387-396.

Layer, P. G., Weikert, T., and Willbold, E., 1992, Chicken retinospheroids as developmental and pharmacological in vitro models: acetylcholinesterase is regulated by its own and by butyrylcholinesterase activity, *Cell Tissue Res.* 268:409-418.

Layer, P. G., 1991, Cholinesterases during development of the avian nervous system, *Cell. Mol. Neurobiol.* 11:7-33.

Layer, P. G., and Kaulich, S., 1991, Cranial nerve growth in birds is preceded by cholinesterase expression during neural crest cell migration and the formation of an HNK-1 scaffold, *Cell Tissue Res.* 265:393-407.

Lipton, S. A., Frosch, M. P., Phillips, M. D., Tauck, D.L., and Aizenman, E., 1988, Nicotinic antagonists enhance process outgrowth by rat retinal ganglion cells in culture, *Science* 239:1293-1296.

Lipton, S. A., and Kater, S. B., 1989, Neurotransmitter regulation of neuronal outgrowth, plasticity and survival, *Trends Neurosci.* 12:265-270.

Mattson, M. P., 1988, Neurotransmitters in the regulation of neuronal cytoarchitecture, *Brain Res. Rev.* 13:179-212.

Mesulam, M.- M., and Geula, C., 1990, Shifting patterns of cortical cholinesterases in Alzheimer's disease: implications for treatment, diagnosis, and pathogenesis, *Adv. Neurol.* 51:235-240.

Michaelson, S., and Small, D. H., 1993, A protease is recovered with a dimeric form of acetylcholinesterase in fetal bovine serum, *Brain Res.* 611:75-80.

Michaelson, S., Small, D. H., and Livett, B. G., 1994, Expression of dimeric and tetrameric acetylcholinesterase isoforms on the surface of cultured bovine adrenal chromaffin cells, *J. Cell. Biochem.* 55:398-407.

Mullan, M., Crawford, F., Axelman, K., Houlden H., Lilius, L., Winblad, B., and Lannfelt, L., 1992, A pathogenic mutation for probable Alzheimer's disease in the APP gene at the N-terminus of ß-amyloid, *Nature Genet.* 1:345-347.

Olson, P. F., Fessler, L. I., Nelson, R. E., Sterne, R. E., Campbell, A. G., and Fessler, J. H., 1990, Glutactin, a novel *Drosophila* basement membrane-related glycoprotein with sequence similarity to serine esterases, *EMBO J.* 9:1219-1227.

Pugh, P. C., and Berg, K., 1994, Neuronal acetylcholine receptors that bind α-bungarotoxin mediate neurite retraction in a calcium-dependent manner, *J. Neurosci.* 14:889-896.

Robertson, R. T., 1987, A morphogenic role for transiently expressed acetylcholinesterase in thalamocortical development?, *Neurosci. Lett.* 75:259-264.

Robertson, R. T., Hohmann, C. F., Bruce, J. L., and Coyle, J. T., 1988, Neonatal enucleations reduce specific activity of acetylcholinesterase but not choline acetyltransferase in developing rat visual cortex, *Dev. Brain Res.* 39:298-302.

Robertson, R. T., and Yu, J., 1993, Acetylcholinesterase and neural development: new tricks for an old dog?, *News Physiol. Sci.* 8:266-272.

Rotundo, R. L., and Carbonetto, S. T., 1987, Neurons segregate clusters of membrane-bound acetylcholinesterase along their neurites, *Proc. Natl. Acad. Sci. USA* 84:2063-2067.

Schambra, U. B., Sulik, K. K., Petrusz, P., and Lauder, J. M., 1989, Ontogeny of cholinergic neurons in the mouse forebrain, *J. Comp. Neurol.* 288:101-122.

Silver, A., 1974, *The Biology of Cholinesterases*, Elsevier, Amsterdam.

Small, D. H., 1990, *Trends Biochem. Sci.* 15:213-216.

Small, D. H., Nurcombe, V., Moir, R., Michaelson, S., Monard, D., Beyreuther, K., and Masters, C. L., 1992, Association and release of the amyloid protein precursor of Alzheimer's disease from chick brain extracellular matrix, *J. Neurosci.* 12:4143-4150.

Small, D. H., 1994, A function for butyrylcholinesterase?, *J. Neurochem.* (in press).

Small, D. H., Nurcombe, V., Reed, G., Clarris, H., Moir, R., Beyreuther, K., and Masters, C. L., 1994a, A heparin-binding domain in the amyloid protein precursor of Alzheimer's disease is involved in the regulation of neurite outgrowth, *J. Neurosci.* 14:2117-2127.

Small, D. H., Reed, G., Whitefield, B., and Nurcombe, V., 1994b, Cholinergic regulation of neurite outgrowth from isolated chick sympathetic neurons in culture, *J. Neurosci.* (in press).

Summers, W. K., Viesselman, J.O., Marsh, G. M., and Candelora, K., 1981, Use of THA in treatment of Alzheimer-like dementia: pilot study in twelve patients, *Biol. Psychiatry* 16:145-153.

Whitehouse, P. J., Price, D. L., Struble, R. G., Clark, A. W., and Delong, M. R., 1982, Alzheimer's disease and senile dementia: loss of neurons in the basal forebrain, *Science* 215:1237-1239.

Zheng, J. Q., Felder, M., Connor, J. A., and Poo, M. M., 1994, Turning of nerve growth cones induced by neurotransmitters, *Nature* 368:140-144.

CHOLINESTERASE INHIBITORS

From Preclinical Studies to Clinical Efficacy in Alzheimer Disease

E. Giacobini[*]

Department of Pharmacology - #1222
Southern Illinois University School of Medicine
P. O. Box 19230
Springfield, Illinois 62794-9230

PHARMACOLOGICAL BASIS FOR A CHOLINOMIMETIC THERAPY OF ALZHEIMER DISEASE

The formulation of a cholinomimetic strategy to treat Alzheimer disease calls for an improvement in function of central cholinergic synapses resulting in enhanced cognitive capacities of the patient. The pharmacological basis for such an intervention is the effect of acetylcholine (ACh)-enhancing drugs such as cholinesterase inhibitors (ChEI). One example is prostigmine's effect at the neuromuscular junction which produces an improvement of neuromuscular transmission (Thesleff, 1990) (Table I). This effect has been utilized for many decades to treat myasthenic patients' defect in nicotinic cholinergic receptors (Osterman, 1990). Similarly, the central effect of physostigmine (PHY) or diisopropylfluorophosphate (DFP) topically or systemically administered has been well documented both experimentally and in humans (Segal, 1988; Karczmar, 1974; McCormick et al., 1993; Thal et al., 1983) (Table I).

The potential of a ChEI such as PHY to improve cognitive function in AD is best exemplified by changes induced on cerebral glucose metabolism measured with fluorode-oxyglucose and PET scan before and after administration of .25 mg PHY i.v. (Tune et al., 1991). A significant increase in glucose metabolism in several brain regions affected by the disease is seen in the patient simultaneously with an improvement in the result of Necker cube task (designs a cube) a few minutes after the injection of the drug (Tune et al., 1991). This effect of a ChEI can be maintained also after prolonged (3-9 months) treatment as demonstrated by the effect of tacrine (tetrahydroaminoacridine) on cerebral glucose metabolism and nicotinic receptors (Nordberg, 1993).

---

[*] Phone: 217/785-2185; FAX: 217/524-0145; Email: SM0095@SPRINGB.

Enzymes of the Cholinesterase Family, Edited by Daniel M. Quinn et al.
Plenum Press, New York, 1995

Table I.

		Conc. or dose/kg	[ACh] in synapse	Pharmacological effect	Clinical effect
NMY*F	Topical prostigmine	6 μM	10^{-8}-10^{-7}M	incr. EPP[†]	improvement of NM transmission
CNS[‡F]	Syst. physostigmine	Change 300 μg	X 2-8	EEG spiking	improvement of cognitive function
	Syst. DFP	5 mg	X 2-8	fast-low amp EEG arousal seizures	"
	Topical physostigmine	100 μM	X 2-5	depolarization prol. discharges	"

*NMY = neuromuscular junction.
[†]EPP = endplate potential.
[‡]Thesleff, 1990; Osterman, 1990; Karczmar, 1974; Segal, 1988; McCormick et al., 1993; Thal et al., 1983.

Several factors converge to regulate and maintain steady-state levels of ACh in brain. These factors can be estimated dynamically *in vivo* by microdialysis in rat cortex. The data reported in Table II are from studies performed in our laboratory (Hallak and Giacobini, 1986; Cuadra et al., 1994). They show that the hydrolyzing capacity of rat cortex for ACh estimated as acetylcholinesterase (AChE) activity is approximately 5-fold higher than the synthesizing capacity for ACh estimated as choline acetyltransferase (ChAT) activity. The concentration of extracellular ACh measured by means of microdialysis *in vivo* without the addition of a ChEI in the probe, is in the order of 5 nM in the awake, freely moving animal. The total level of ACh measured *post mortem* following microwave irradiation of the head is of the order of 32 nmol/g w.w. For clinical purposes, it is important to obtain an estimate of the corresponding values of ACh metabolism in the human brain. The data of Table III represent the results of investigations performed in different laboratories. Due to obvious difficulties in examining human brain tissue *in vivo* the data can not be considered to be as accurate as in the experimental animal; however, they represent a useful baseline for a therapeutical strategy. In bioptic brain tissue of age-matched neurological controls (non-AD patients) the ratio between synthesis and hydrolysis of ACh is 24 (DeKosky et al., 1992). In AD cerebral tissue from biopsy the same ratio is reduced to 13, a 46% decrease (DeKosky et al., 1992).

Acetylcholine biosynthesis measured radiometrically in human brain tissue from biopsy is also decreased by 53% in AD patients (Bowen, 1983). The level of extracellular ACh in human brain can be estimated directly by microdialysis (hippocampal region,

Table II. Rat frontal cortex

ChAT*	AChE*	Ratio	ACh[†]Extracel.	ACh[‡]Total
111	500	0.2	5	32

*μmoles/ACh/g prot/hr (Hallak and Giacobini, 1986).
[†]nM ACh, microdial., awake (Cuadra et al., 1994).
[‡]nmol ACh/g, total, radiowave (Hallak and Giacobini, 1986).

Table III. Human frontal cortex

	ChAT*	AChE*	Ratio	ACh[†]Synthesis	ACh levelCSF/Extracel.
Controls	6	.25	24	6.2	33[a]-300[b]
Alzheimer	2.3	.18	13	2.9	15[c]-152[d]
% Decrease	62	28	46	53	48

*µmol/ACh/hr/g prot, biopsy (DeKosky et al., 1992).
[†]dpM/ACh/min/mg prot (Bowen, 1983) biopsy.
[a]lumbar CSF, nM ACh (Tohgi et al., 1994)
[b]hippocampal microdialysis - awake, nM ACh (Greaney et al., 1991).
[c]lumbar CSF, nM ACh (Tohgi et al., 1994)
[d]intraventricular CSF, awake, nM ACh (Giacobini et al., 1988).

Greaney et al., 1992); indirectly in intraventricular CSF sampling in the awake patient (Giacobini et al., 1988); or in the lumbar CSF (Tohgi et al., 1994). These data indicate that the difference in ACh levels between controls and AD patients is at least in the order of 50%. We can, therefore, postulate that a ChEI capable of doubling ACh levels, by increasing its concentration from approximately 150 nM to 300 nM would reestablish a close to normal level of the neurotransmitter. Cholinesterase inhibitors which are capable of penetrating into the brain have been shown to rapidly inhibit AChE in CSF (ENA 713; Enz, 1994) and to increase CSF ACh levels in humans (PHY, Giacobini et al., 1988).

THE SECOND GENERATION OF CHOLINESTERASE INHIBITORS

Development of new ChEI has strongly advanced during the last five years (Giacobini, 1991). A total of 13 ChEI (Table IV) is presently being tested clinically throughout the world. Tacrine is the compound at the most advanced stage of clinical research. At least 2,000 AD patients treated with this drug have already been studied (Table IV). Other amino acridine derivatives and active metabolites of tacrine such as velnacrine (1-0H metabolite; HP029) suronacrine (HP128) and 11-tetrahydro-2-4-methanoacridine (SM-10888; Natori et al., 1990) are also being tested in AD therapy. Multi-center trials at different stages are in progress for several ChEI to define their efficacy, safety and dose response relationships (Table IV). The diversity of chemical structures and pharmacological characteristics among these compounds is striking. Both reversible (carbamates or non-carbamate compounds) and irreversible (e.g. organophosphate) inhibitors (metrifonate) are being tested. A new derivative of PHY, heptyl-physostigmine (HEP) (eptastigmine, MF-201) is in phase II clinical trial. Together with PHY, two other ChEI natural products, the Huperzines and galanthamine are being tested. Second generation ChEI are characterized by their high penetration through the blood brain barrier, potent, long-lasting and selective inhibition of brain AChE with low peri-pheral cholinergic side effects. The appearance of hematologic complications (neutropenia or agranulocytosis) is a new finding that, together with hepatotoxicity, needs to be considered carefully in future drug development (Table IV).

EFFICACY OF CHOLINESTERASE INHIBITORS IN AD THERAPY. CHOLINESTERASE INHIBITION OR ACETYLCHOLINE LEVELS?

On the basis of preclinical and clinical studies we proposed that both cholinesterase (ChE) inhibition and increase in brain ACh would be directly related to the clinical efficacy

Table IV. Cholinesterase inhibitors: AD clinical trials (1994)

Compound	Country	Company	Clinical phase	Side effects comments
Physostigmine slow release	USA	Forest	III	N.A.
ENA 713	USA/Europe	Sandoz	II	Low side effects
Heptyl-physostigmine	USA/Italy	Mediolanum	II	Hematology[†]
E-2020	USA/Japan	Eisai	III	Low side effects
MDL 73,745	USA/Europe	Marion-Merrell Dow	II	Low side effects
Metrifonate	USA/Germany	Bayer/Miles	II	Low side effects
Tacrine (THA)*	USA/Europe	Warner-Lambert	IV	Hepatotoxicity
Velnacrine (HP029) Suronacrine (HP128)	USA/Europe	Hoechst-Roussel	II	Hematology [‡] Hepatotoxicity
Galanthamine	GermanyUSA	Shire Pharm. Ciba-Geigy	II	Low side effects
Huperzine A	China	Chinese Acad. Sci.	III	N.A.
HP290	England	Astra Arcus	I	N.A.
CP-118,954	USA	Pfizer	II	N.A.

*other indications: HIV, tardive dyskinesia.
[†]neutropenia.
[‡]agranulocytosis.
N.A. = data not available.

of the inhibitor (Becker and Giacobini, 1988a,b). Our animal experiments have shown that changes in brain ACh concentration that follow ChE inhibition are not solely the result of the magnitude of ChE inhibition but are affected by the specific pharmacological properties of individual drugs (DeSarno et al., 1989; Giacobini, 1993) (Table V). These properties include effects on the inhibition of ChE as well as on neurotransmitter release (DeSarno et al., 1989; Messamore et al., 1993). Table V presents a comparison of effects of six ChEI on cortical ChE activity as well as ACh release measured with microdialysis with no ChEI in the probe in non-anesthetized awake rats. It can be seen that following PHY administration, a maximal ChE inhibition in cortex of 60% will produce a maximal increase in ACh release

Table V. Comparison of effects of cholinesterase inhibitors on AChE activity and ACh release in rat cortex*

Drug	Dose mg/kg	% max. AChE inhibition	Peak time AChE inhibition (min)	Duration AChE inhibition (hrs)	% ACh maximum increase	Peak time ACh effect (min)	Duration ACh effect (hrs)	Cholinergic side effects[†]
Physostigmine	.3	60	30	2	4000	60	1.5	+++
Heptyl-physostigmine	5	90	60	27	3000	90	10	++
Metrifonate	80	70	60	24	1800	60	6	+
MDL 73,745	10	80	60	> 9	1100	60	7	++
THA	5	50	60	10	500	90	3	+
Huperzine A	.5	50	30	> 6	220	60	6	+

*Messamore et al., 1993; Cuadra et al., 1994; Zhu and Giacobini (submitted).
[†]fasciculations, tremor, splay.

Table VI. Next generation of drugs for AD therapy:
3 Targets

I	Less toxic than Tacrine
II	Show improvement of daily living
III	Slow down deterioration/performance decline

of 4000% while a 50% ChE inhibition produced by HUP-A (L-Huperzine-A) or THA (tacrine) will cause only a 230-500% increase in ACh. At the 50% level of ChE inhibition, cholinergic side effects will also be more pronounced with PHY than with the other two drugs. Duration of ACh increase will also vary with different drugs from a minimum of 1.5 hrs for PHY to a maximum of 10 hrs for HEP (Table V). Based on these experimental data, it seems too limited to use ChE inhibition in plasma or RBC as the primary targets of therapeutic effects. A restatement seems to be justified of the therapeutic strategy in terms of CNS ACh concentration and steady-state duration of ACh levels rather than ChE inhibition (Becker and Giacobini, 1988a,b; Giacobini and Cuadra, 1994).

NEXT GENERATION OF ChEI FOR AD THERAPY

The recent 30-week randomized controlled trial of a high dose (160 mg/kg) of THA (Knapp et al., 1994), demonstrates a mean improvement of four points on ADAS-Cog and at least 3.0 points on the MMSE after 6 months. The ability of patients to tolerate this high dose is limited, therefore; almost two-thirds of the patients were unable to complete the trial (Knapp et al., 1994). The results suggest a positive effect of the drug on patient deterioration with a significant gain of several months per year of treatment. These results need to be confirmed with longer trials up to 24-36 months in order to demonstrate a real slow-down of the disease course.

An attenuation of disease progression corresponding to a 12-month gain with a 24-month long ChEI therapy would represent a significant therapeutic effect. There are limitations with the present THA therapy. About one third of the patients entering the THA trial can not continue the treatment because of liver toxicity. Within this selected population of patients only 30-50% are improved. The development of new ChEI with a lower rate of side effects and toxicity will allow a higher level of prolonged steady-state inhibition promoting a functional level of cholinergic transmission and more significant clinical efficacy (Table VI).

ACKNOWLEDGMENTS

The author wishes to thank Diana Smith for typing and editing the manuscript. Supported by National Institute on Aging #P30 AG08014; Mediolanum Farmaceutici (Milan, Italy).

REFERENCES

Becker, R. E. and Giacobini, E., 1988a, Mechanisms of cholinesterase inhibition in senile dementia of the Alzheimer type: clinical, pharmacological and therapeutic aspects, *Drug Dev. Res.* 12:163-195.

Becker, R. E. and Giacobini, E ., 1988b, Pharmacokinetics and pharmacodynamics of acetylcholinesterase inhibition: can acetylcholine levels in the brain be improved in Alzheimer's disease? *Drug Dev. Res.* 14:235-246.

Bowen, D. M., 1983, Biochemical assessment of neurotransmitter and metabolic dysfunction and cerebral atrophy in Alzheimer's disease. In: *Banbury Report 15: Biological Aspects of Alzheimer's Disease*, Cold Spring Harbor Laboratory, pp. 219-231.

Cuadra, G., Summers, K., and Giacobini, E., 1994, Cholinesterase inhibitors effects on neurotransmitters in rat cortex in vivo, *J. Pharmacol. Exptl. Ther.* 270:277-284.

DeKosky, S. T., Harbaugh, R. E., Schmitt, F. A., Bakay, R. A. E., Chui, H. C., Knopman, D. S., Reeder, T. M., Shetter, A. G., Senter, J. H., and Markesbery, W. R., 1992, Cortical biopsy in Alzheimer's disease: diagnostic accuracy and neurochemical, neuropathological and cognitive correlations, *Ann. Neurol.* 32:625-632.

DeSarno, P., Pomponi, M., Giacobini, E., Tang, X. C., and Williams, E., 1989, The effect of heptyl-physostigmine, a new cholinesterase inhibitor, on the central cholinergic system of the rat, *Neurochemical Res.* 14(10):971-977.

Enz, A., 1994, Acetylcholinesterase inhibitors (AChEI) as a potential use for Alzheimer's disease (AD) therapy: differences in mechanisms of enzyme inhibition, *Abstr. 5th Intl. Meeting on Cholinesterases*, India: Madras, pp. 107.

Giacobini, E., 1991, The second generation of cholinesterase inhibitors: pharmacological aspects. In: *Cholinergic Basis for Alzheimer Therapy*, Becker, R. E. and Giacobini, E., eds. Cambridge: Birkhauser Boston, pp. 247-262.

Giacobini, E., 1993, Pharmacotherapy of Alzheimer's disease: new drugs and novel strategies. In: *Alzheimer's Disease: Advances in Clinical and Basic Research*, Corain, B., Iqbal, K., Nicolini, M., Winblad, B., Wisniewski, H. and Zatta P., eds. New York: John Wiley & Sons Ltd., pp. 529-538.

Giacobini, E. and Cuadra, G., 1994, Second and third generation cholinesterase inhibitors: from preclinical studies to clinical efficacy. In: *Alzheimer Disease: Therapeutic Strategies*, Giacobini, E. and Becker, R., eds. Cambridge: Birkhauser Boston, pp. 155-171.

Giacobini, E., Becker, R. E., McIlhany, M., and Kumar, V., 1988, Intracerebroventricular administration of cholinergic drugs: Preclinical trials and clinical experience in Alzheimer patients. In: *Current Research in Alzheimer Therapy*, Giacobini, E. and Becker, R., eds. New York: Taylor and Francis, pp. 113-122.

Greaney, M., Marshall, D., During, M., Bailey, B., Acworth, I., 1992, Ultrasensitive measurement of acetylcholine release in the conscious human hippocampus and anesthetized rat striatum using microdialysis. *Soc. Neurosci. Abst.*

Hallak, M. and Giacobini, E., 1986, Relation of brain regional physostigmine concentration to cholinesterase activity and acetylcholine and choline levels in rat, *Neurochemical Research* 11(7):1037-1048.

Karczmar, A. G., 1974, Brain acetylcholine and seizures. In: *Psychobiology of Convulsive Therapy*, Fink, M., Kety, S., McGaugh, J. M. and Williams, T.A., eds. Washington: Winston and Sons, pp. 251-270.

Knapp, M. J., Knopman, D. S., Solomon, P. R., Pendlebury, W. W., Davis, C. S., and Gracon, S. I., 1994, A 30-week randomized controlled trial of high-dose tacrine in patients with Alzheimer's disease, *J. Amer. Med. Assoc.* 271(13):985-991.

McCormick, D. A., Wang, Z., and Huguenard, J., 1993, Neurotransmitter control of neocortical neuronal activity and excitability, *Cerebral Cortex* 3:387-398.

Messamore, E., Warpman, U., Ogane, N., and Giacobini, E., 1993, Cholinesterase inhibitor effects on extracellular acetylcholine in rat cortex, *Neuropharmacology* 32(8):745-750.

Natori, K., Okazaki, Y., Irie, T. et al., 1990, Pharmacological and biochemical assessment of SM-10888, a novel cholinesterase inhibitor, *Jap. J. Pharmacol.* 53:145-155.

Nordberg, A., 1993, Clinical studies in Alzheimer patients with positron emission tomography, *Behavioral Brain Res.* 57:215-224.

Osterman, P. O., 1990, Current treatment of myasthenia gravis. In: *Progress in Brain Res. Vol. 84*, Aquilonius, S. M. and Gillberg, P. G., eds. Netherlands: Elsevier, pp. 151-160.

Segal, M., 1988, Synaptic activation of a cholinergic receptor in rat hippocampus, *Brain Res.* 452:79-86.

Thal, L. J., Masur, D. M., Fuld, P. A., Sharpless, N. S., and Davies, P., 1983, Memory improvement with oral physostigmine and lecithin in Alzheimer's disease. In: *Banbury Report: Biological Aspects of Alzheimer's Disease*, Vol. 15, Katzman, R., ed. Cold Spring Harbor, pp. 461-469.

Thesleff, S., 1990, Functional aspects of quantal and non-quantal release of acetylcholine at the neuromuscular junction. In: *Progress in Brain Res. Vol. 84*, Aquilonius, S. M. and Gillberg, P. G., eds. Netherlands: Elsevier, pp. 93-100.

Tohgi, H., Abe, T., Hashiguchi, K., Saheki, M., Takahashi, S., 1994, Remarkable reduction in acetylcholine concentration in the cerebrospinal fluid from patients with Alzheimer type dementia, *Neuroscience Letters* 177:139-142.

Tune, L., Brandt, J., Frost, J. J., Harris, G., Mayberg, H., Steele, C., Burns, A., Sapp, J., Folstein, M. F., Wagner, H. N., and Pearlson, G. D., 1991, Physostigmine in Alzheimer's disease: effects on cognitive functioning, cerebral glucose metabolism analyzed by positron emission tomography and cerebral blood flow analyzed by single photon emission tomography, *Acta Psychiatrica Scandinavica (Suppl.)* 366:61-65.

GENETIC PREDISPOSITION FOR VARIABLE RESPONSE TO ANTICHOLINESTERASE THERAPY ANTICIPATED IN CARRIERS OF THE BUTYRYLCHOLINESTERASE "ATYPICAL" MUTATION

Y. Loewenstein,[1] M. Schwarz,[1] D. Glick,[1] B. Norgaard-Pedersen,[2]
H. Zakut,[3] and H. Soreq[1]

[1] Department of Biological Chemistry
The Hebrew University of Jerusalem, Israel
[2] Department of Clinical Biochemistry, Statens Seruminstitut
Division of Biotechnology, Copenhagen, Denmark
[3] Department of Obstetrics and Gynecology
The Edith Wolfson Medical Center
The Sackler Faculty of Medicine
Tel Aviv University, Israel

ABSTRACT

Anticholinesterases were recently approved for treating patients suffering from Alzheimer's disease (AD) in an attempt to balance their cholinergic system. These drugs are targeted at acetylcholinesterase (AChE) but also inhibit butyrylcholinesterase (BuChE), known for its numerous genetic variants. The most common of these is the "atypical" phenotype created through a replacement of Asp70 by Gly (D70G) due to a point mutation. The "atypical" enzyme causes prolonged postanesthesia apnea following succinylcholine administration for muscle relaxation and displays a considerably reduced sensitivity to various other inhibitors. The allelic frequency of "atypical" BuChE was studied in different populations and revealed distinct patterns particular to various ethnic groups. Recently, a relatively high allelic frequency of 0.06 was found in a population of Georgian Jews, differing by up to 4-fold from the incidence in other populations (Ehrlich et al., Genomics, in press). This implies that in groups of AD patients from diverse ethnic origins, a significant fraction of carriers of at least one allele of this mutation should be expected. To predict their responsiveness to anticholinesterase treatment, we examined the susceptibility of AChE, as compared to that of BuChE and the "atypical" BuChE variant, towards several anticholinesterases in use for AD treatment. IC_{50} values and rate constants reflecting inhibitor susceptibilities were calculated for various recombinant human cholinesterases produced in Xenopus oocytes and immobilized on microtiter plates through selective monoclonal antibodies. The reversible amino acridinium compound Tacrine, currently in use for AD therapy, displayed a 300-fold higher IC50 value for the "atypical" enzyme than for BuChE (1mM BtCh as substrate). Pseudo first order rate constants for inhibition of BuChEs by the carbamates heptyl-physostigmine (0.139 min^{-1}, 10nM inhibitor), physostigmine (0.3 min^{-1}, 1 μM inhib.) and SNZ-ENA713 (0.139 min^{-1}, 10 μM inhib.) were found to be higher than or equal to those of AChE, suggesting that BuChE serves as a second primary target for these drugs. Moreover, the"atypical" variant of BuChE displayed considerably slower inactivation rates to these drugs (0.01 min^{-1}, 0.025 min^{-1}, and 0.01 min^{-1}, respectively) as compared with the wild type BuChE. These findings predict that carriers of the D70G BuChE mutation would vary from other patients in their susceptibility to the above drugs, which potentially contributes to the wide variability of responses observed in clinical trials. Screening patients for D70G carriers should therefore precede anticholinesterase treatment.

REFERENCES

Loewenstein, Y., Gnatt, A., Neville, L.F., & Soreq, H. (1993) J. Mol. Biol. 234:289-296.
Neville, L.F., Gnatt, A., Loewenstein, Y., Seidman, S., Ehrlich, G., & Soreq, H. (1992) EMBO J. 11:1641-1649.
Soreq, H., Gnatt, A., Loewenstein, Y., & Neville, L.F. (1992) Trends in Biochem. Sci. 17:353-358.

HORSE SERUM BUTYRYLCHOLINESTERASE DOES NOT DISRUPT PASSIVE AVOIDANCE LEARNING OR SPONTANEOUS MOTOR ACTIVITY IN RATS

Raymond F. Genovese, Averi R. Roberts, William E. Fantegrossi, Roberta Larrison, and Bhupendra P. Doctor

Divisions of Neuropsychiatry and Biochemistry
Walter Reed Army Institute of Research
Washington, D.C. 20307-5100

Recent advances in the treatment of organophosphorus (OP) toxicity have focussed on the administration of an exogenous cholinesterase (ChE) to act as a scavenger for the OP agent (Doctor, et al., 1991). As compared with treatment therapies that involve counteracting the effects of increased cholinergic activity following OP exposure (e.g., atropine), the method offers the advantage of neutralizing the OP before target organs are affected. Since maximum efficacy is achieved when the exogenous ChE is administered prior to OP exposure, possible deficits in performance produced by ChE administration alone is an important consideration. We have previously demonstrated that, in contrast to physostigmine or atropine (Genovese, et al., 1990), administration of horse serum butyrlcholinesterase (HS-BChE) fails to disrupt performance of rats on an operant conditioning task (Genovese, et al., 1993). The present study further investigated the possibility that HS-BChE might produce deficits in performance. Specifically, we assessed the effects of HS-BChE, in rats, on spontaneous motor activity and on a passive avoidance memory task. Motor activity was monitered continuously by using photo-emitter / detector pairs mounted in the home cages of the rats. A 12h:12h light:dark cycle was maintained and activity counts (i.e., consecutive photobeam breaks) were collected in 5-min intervals. Under baseline conditions, approximately 80%, or more, of the total daily activity occurred during the dark period, demonstrating a typical circadian pattern of activity. Two groups of rats were injected (IP) with either HS-BChE (7500U, n = 4) or vehicle (n = 4), and motor activity was monitered for 10 consecutive days. No statistically significant differences were observed between groups for measures of total daily activity or the circadian pattern of activity. Comparisons between pre- and post-injection measures of activity were also not significantly different for either group. In order to assess the potential effects of HS-BChE on short-term memory, rats were trained on a passive avoidance procedure where traversal from a lighted compartment into a dark compartment was followed by mild foot shock. Following a one-trial learning procedure, rats were tested for traversal (without the administration of foot shock) 24-, 72-, and 168-h following training. In one group of rats (n = 5), HS-BChE (5000U, IP) was admistered 4-h prior to training, while two additional groups received atropine sulfate (10 mg/kg, IP, n = 10) or vehicle (IP, n = 7), respectively, immediately following training. HS-BChE was administered prior to training so that near maximal blood-BChE activity would be present following the training procedure (as established by previous pharmacokinetic experiments). Retention tests revealed no significant differences between treatment groups for the 24- and 72-h test. Rats treated with atropine, but not rats treated with HS-BChE, however, showed significantly less retention at 168-h, as compared with vehicle-treated control rats. These results confirm and extend earlier reports demonstrating that, in rats, HS-BChE may be devoid of detrimental cognitive and motor effects. Since the doses of HS-BChE used in the present studies have been shown to confer protection against OP toxicity (e.g., Brandeis, et al., 1993), these results also support the viability of prophylactic administration of exogenous BChE as a therapy against OP exposure.

REFERENCES

Brandeis, R., Raveh, L., Grunwald, J., Cohen, E., and Ashani, Y. , 1993, *Pharmacol Biochem Behav* 46:889-896.
Doctor, B.P., Raveh, L., Wolf, A.L., Maxwell, D.M., and Ashani, Y., 1991, *Neurosci Biobehav Rev* 15: 123-128
Genovese, R.F., Elsmore, T.F., and Witkin, J.M., 1990, *Pharmacol Biochem Behav* 37:117-122.
Genovese, R.F., Lu, X-C., Gentry, M.K., Larrison, R. and Doctor, B.P., 1993, *Proc Med Defense Biosci Rev*:1035-1042.

ACETYLCHOLINESTERASE INHIBITORS (AChE-I) AS A POTENTIAL USE FOR ALZHEIMER'S DISEASE (AD) THERAPY

Differences in Mechanisms of Enzyme Inhibition

Albert Enz

Preclinical
Sandoz Pharma Ltd.
Basel, Switzerland

The rationale underlying utilization of AChE inhibitors in treatment of Alzheimer's disease (AD) is based on the assumption that inhibition of AChE results in reduced rate of removal of ACh from the synaptic cleft, thereby increasing the probability for effective encounter between neurotransmitter and the reduced number of muscarinic receptors in this disease. In the history of experimental clinical trials, compounds exerting different mechanisms of cholinesterase inhibition were investigated. From the experience with clinically used AChE inhibitors in AD several problems can be encountered, possibly related to the intrinsic individual properties of these drugs:

i. Non-selective inhibitors have a low therapeutic index, and the inhibition of peripheral cholinesterases (mainly BChE) in heart, muscle and plasma contribute to adverse peripheral effects.

ii. Fast and multiple (complicated) drug metabolism has the potential to lead to organ toxic effects and can influence an efficient long-lasting AChE inhibition.

iii. The desired enzyme inhibition both in terms of duration and selectivity is dependent on the mechanism of inhibition.

Regarding the different inhibition mechanisms, the compounds can be divided into three main classes: reversible inhibitors (like tacrine), pseudo-irreversible inhibitors (carbamates) and irreversible inhibitors (organophosphate compounds).

SDZ ENA 713, an AChE inhibitor currently under clinical investigation, is a carbamate and inhibits preferentially brain AChE. The drug inhibits the enzyme in cortex and hippocampus to a greater extent than in striatum and pons/medulla. In doses showing clear signs of central cholinergic effects in animals, this drug has no significant effects on the cardiovascular system in rats or monkeys (Enz et al., 1993). The brain selectivity observed in animals is confirmed in ongoing human studies by sustained AChE inhibition in cerebrospinal fluid with no effects on plasma BChE or AChE in red blood cells. No relevant liver toxicity or effects on cardiovascular paramaters have been seen. The results obtained with SDZ ENA 713 suggest that the disadvantages of AChE inhibitors might be overcome by improving CNS selectivity and thereby decreasing the peripheral cholinergic effects and toxicity.

REFERENCES

Enz, A., Amstutz, R., Boddeke, H., Gmelin, G. & Malanowsky, J. (1993) in Cholinergic Function and Dysfunction, Progress in Brain Research (Cuello, A.C., ed.) Vol 98, pp 431-438, Elsevier Science Publisher, Amsterdam.

COMPARATIVE INHIBITION OF ACETYLCHOLINESTERASE BY TACRINE, PHYSOSTIGMINE AND HUPERZINE IN THE ADULT RAT BRAIN

H. S. Ved, J. M. Best, J. R. Dave, and B. P. Doctor

Divisions of Biochemistry and Neuropsychiatry
Walter Reed Army Institute of Research
Washington, D.C. 20307

Alzheimer's Disease (AD) is the most common type of adult-onset dementia. It is associated with a progressive and irreversible loss of memory and cognitive functions proceeding for years. The general malfunction of the cholinergic regions of the brain invariably leads to death. There seems to be a correlation between the impairment of cognitive function with an increase in the number of neuritic plaques and neurofibrillary tangles in the hippocampus, amygdala and cerebral cortex (Coyle et al., 1983; Mullan and Crawford, 1993). Moreover, the severity of the disease parallels the reduction in levels of AChE and ChAT in the frontal and temporal cortices (Perry et al.,1978). A diminished number of cholinergic neurons in basal forebrain nuclei and decreased ACh production in the brain of AD patients are thought to cause some of the characteristic cognitive impairments. Three broad strategies have been considered in dementia therapy: (1) replacement of lost neurotransmitter, (2) trophic support to slow the degeneration of the nervous system, and (3) intervention, i.e., prevention of the local inflammatory response thought to be caused by complement activation in the amyloid plaques (Davis et al., 1993). It has been suggested that anti-cholinergic drugs impair the memory of healthy individuals in a manner parallel to that observed early in the development of AD. Therefore, the principal current AD therapeutic approach, and the most promising one in the short term, is the stimulation of the cholinergic system. The first anti-ChE drug to be approved for use in AD therapy in the USA is Cognex® (1-2-3-4-tetrahydro-9 aminoacridine, Tac, tacrine). Recently, there has been a report of a multi-center, double-blind, placebo-controlled trial of Tac therapy, which included 663 patients suffering from mild to moderate AD. Huperzine A (Hup A) was recently shown to be a potent, reversible inhibitor of AChE (Ashani et al; 1992). The fairly long half-life ($T_{0.5}$ = 35 min) for AChE-Hup A complex is in marked contrast to the rapid on/off rates that characterize other reversible inhibitors of AChE with similar potency (Taylor and Radić; 1994). In the present studies, we have observed the *in vitro* differences in the ability to inhibit AChE by Hup-A, tacrine and physostigmine in the various anatomical regions of the adult rat brain homogenates. The findings show that while all these compounds inhibited AChE in a dose-dependent manner, the degree of inhibition varied from inhibitor to inhibitor. The rank order of potency for the inhibition of whole rat brain AChE was physostigmine > Hup-A > tacrine. The AChE inhibition varied in the different brain regions. Hup-A was most effective in the cortex > hypothalamus > cerebellum > hippocampus. In contrast, tacrine was most effective in the cerebellum > hypothalamus > cortex > hippocampus, whereas physostigmine's rank order of inhibition was cerebellum > cortex > hippocampus > hypothalamus. Further enzyme kinetic studies show that these compounds are uncompetitive inhibitors at the effective inhibitory concentrations, whereas they act as competitive inhibitors at lower concentrations. Moreover, the enzyme kinetic parameters (K_m, K_i, K_{on} and K_{off}, etc.) for each of the compounds are different in the different regions of the rat brain. These findings suggest that different AChE inhibitors may have varying degree of effectiveness, which could partially be explained by their differential interaction with AChE, in the distinct anatomical regions of rat brain.

REFERENCES

Ashani, Y., Peggins, J.O., and Doctor, B.P., 1992, *Biochem Biophy Res Com* 184: 719-726
Coyle, J.T., Price, D.L., and DeLong, M.R., 1983, *Science* 219:1184-1190.
Davis, R,E,, Emrnerling, M.R., Jaen, J.C., Moos, W.H., and Spiegel, K., 1993, *Rev. Neurobiol* 7:4183.
Mullan, M., and Crawford, F., 1993, *Proc. Natl. Acad. Sci. USA.* 88:11315-319.
Perry, E.K., Tomilinson, B.E., Blessed, G., Bergrnann, K., Gibson, P.H., and Perry, R.H., 1978, *Br. Med. J.* 2: 1437-39.
Taylor, P., and Radič, Z., 1994, *Annu. Rev. Pharmacol. Toxicol.* 87:281-320.

ACETYLCHOLINESTERASE GENE SEQUENCE AND COPY NUMBER ARE NORMAL IN ALZHEIMER'S DISEASE PATIENTS TREATED WITH THE ORGANOPHOSPHATE METRIFONATE

C. F. Bartels,[1] P. L. Moriearty,[2] R. E. Becker,[2] C. P. Mountjoy,[1] and O. Lockridge[1]

[1] Eppley Institute
University of Nebraska Medical Center
Omaha, Nebraska 68198-6805 USA;
[2] Department of Psychiatry
Southern Illinois University School of Medicine

ABSTRACT

Metrifonate, an organophosphate drug, has been used in clinical studies to inhibit AChE in patients with Alzheimer's disease (AD) in the hope that it would stop or inhibit progression of the disease.

Gene amplification can occur with drug application. Gene amplification has been suggested for acetylcholinesterase and butyrylcholinesterase following accidental exposure to organophosphorous insecticides. We looked for ACHE and BCHE gene amplification in 18 patients with AD who received metrifonate for 3 to 24 months. No gene amplification was detected in Southern blots hybridized with human ACHE and BCHE probes.

Several differences between AChE from normal brains and AChE from AD brains have been published: differences in optimal pH for activity, differences in inhibition by BW284C51, tacrine, physostigmine, indoleamines and proteases, and a difference in banding patterns on isoelectric focusing gels.

To see if these differences were due to mutations within the acetylcholinesterase molecule itself, we sequenced white blood cell DNA from two patients with Alzheimer's disease and compared the sequences to a control sample. There were no mutations that caused amino acid changes to distinguish patients with Alzheimer's disease from the control.

ACKNOWLEDGMENTS

Supported by U.S. Army Medical Research and Development Command Grant DAMD17-94-J-4005

COMPILATION OF EVALUATED MUTANTS OF CHOLINESTERASES

A. Shafferman, C. Kronman, and A. Ordentlich

Israel Institute for Biological Research
Ness-Ziona, Israel

The understanding of the correspondence between the molecular structure and the catalytic function or inhibition characteristics of cholinesterases has been tremendously enhanced by molecular biology techniques and in particular by site-directed mutagenesis studies. Such studies have also led to a better understanding of folding, oligomerization, and other post-translation processes of ChEs.

Here we have compiled information on over 190 mutations either generated by random mutagenesis (rand.) site-directed mutagenesis (s.d.m) or isolated as natural mutations (natr.). The compilation includes mutations in Human AChE (A), Human BChE (B), Drosophila ChE (D), Mouse AChE (M), and Torpedo AChE (T). Amino acid numbering is based on the recommended nomenclature (14). This list provides a basis for establishment of a data bank as has been agreed upon during the cholinesterase meeting (Madras, India; September 1994). Access to the data bank is through Dr. A. Chattonet-INRA as follows:
Client gopher - *gopher.montpellier.inra.fr(port 70)*
     Client WWW - *gopher: /gopher.montpellier.inra.fr: 70/1/cholinesterase*
WWW- *http: //gopher.montpellier.inra.fr: 70/1/cholinesterase*
E. mail - *chatonne@montpellier.inra.fr*
cousin@montpellier.inra.fr

Torpedo#	MUTATION	COMMENT	REFERENCE
39	P75L(D/s.d.m.)	folding (heat-sensitive)	42
39	P37S (B/natr.)	silent	18
44	R44E (T/s.d.m)	folding	38
44	R44K (T/s.d.m)	folding	38
70	Y72A (A/s.d.m.)	PAS	37,59
70	Y72N (M/s.d.m.)	PAS	32,44
72	D74N (A/s.d.m.)	PAS/signal transduction/allosteric modulation/substrate inhibition	20,21,22,33,59

Enzymes of the Cholinesterase Family, Edited by Daniel M. Quinn et al.
Plenum Press, New York, 1995

72	D74K (A/s.d.m.)	PAS/signal transduction/allosteric modulation/substrate inhibition	21,22
72	D74G (A/s.d.m.)	PAS/signal transduction/allosteric modulation/substrate inhibition	20,21,22
72	D74E (A/s.d.m.)	PAS/signal transduction/allosteric modulation	21,22
72	D74N (M/s.d.m.)	PAS	32,44
72	D70G (B/natr.;s.d.m.)	PAS	1,5,17,53
72	Y109D (D/s.d.m.)	PAS	16
72	Y109G (D/s.d.m.)	PAS	16
72	Y109K (D/s.d.m.)	PAS	16
78	F115S (D/natr.)	OP-resistance	43
82	E84Q (A/s.d.m.)	carboxyl/attraction studies	20,22, 40,48
84	W86A (A/s.d.m.)	anionic subsite/signal transduction	21,22,30,33,37, 40,54,55,59
84	W86F (A/s.d.m.)	anionic subsite/signal transduction	55,59
84	W86E (A/s.d.m.)	anionic subsite/signal transduction	21,55,59
84	W86A (M/s.d.m.)	anionic subsite	57
84	W86F (M/s.d.m.)	anionic subsite	57
84	W86Y (M/s.d.m.)	anionic subsite	57
89	N126D (D/s.d.m.)	N-glycosylation	12,15
92	E92Q (T/s.d.m.)	folding, salt-bridge with R44	9,38
92	E92L (T/s.d.m.)	folding, salt-bridge with R44	9,38
93	D95N (A/s.d.m.)	reduced biosynthesis	20,22
93	D93N (T/s.d.m.)	folding	9,38
93	D93V (T/s.d.m.)	folding	38
116	Y114A (B/s.d.m.)	restores function to D70 mutants	17
119	G117H (B/s.d.m.)	OP-resistance	51
121	Q119Y (B/s.d.m.)	PAS	53
121	Y124A (A/s.d.m.)	PAS	37,59
121	Y124Q (M/s.d.m.)	PAS	32,44
128	D131N (A/s.d.m.)	external charge	22,40
129	V132A (A/s.d.m.)	"backdoor" hypothesis	40
129	V132K (A/s.d.m.)	"backdoor" hypothesis	40
130	Y133A (A/s.d.m.)	maintenance of anionic subsite	55
130	Y133F (A/s.d.m.)	maintenance of anionic subsite	55
-	N174S (D/s.d.m.)	N-glycosylation	15
129	I199V (D/natr.)	OP-resistance	43
172	D175N (A/s.d.m.)	folding/salt bridge/non-producer	20,22
172	D248N (D/rand.)	folding	42
199	E202A (A/s.d.m.)	H-bond network/acylation, phosphylation,aging,substrate inhibition	21,22,33,59
199	E202Q (A/s.d.m.)	H-bond network,substrate inhibition	21,22,31,33, 36,54,55,59
199	E202D (A/s.d.m.)	H-bond network/active center modulation substrate inhibition	21,22,33,59
199	E199D (T/s.d.m.)	acylation,phosphylation	2,19
199	E199Q(T/s.d.m.)	acylation,phosphylation	2,19,45,58
199	E199H (T/s.d.m.)	acylation, phosphylation	2

200	S200V (T/s.d.m.)	catalytic triad	2
200	S200C (T/s.d.m.)	catalytic triad	2
200	S203A (A/s.d.m.)	catalytic triad	20,22
200	S203C (A/s.d.m.)	catalytic triad	20
200	S203T (A/s.d.m.)	catalytic triad	20
200	S198C (B/s.d.m.)	catalytic triad	39
200	S198T (B/s.d.m.)	catalytic triad	39
200	S198Q (B/s.d.m.)	catalytic triad	39
200	S198H (B/s.d.m.)	catalytic triad	39
200	S198D (B/s.d.m.)	catalytic triad	39
227	G303A (D/natr.)	OP-resistance	43
255	N331D (D/s.d.m.)	N-glycosylation	15
258	N265Q (A/s.d.m.)	N-glycosylation	24,34
278	E285A (A/s.d.m.)	PAS	37,48,59
279	A277H (B/s.d.m.)	PAS	53
279	A277W (B/s.d.m.)	PAS	53
279	W286A (A/s.d.m.)	PAS/substrate inhibition	21,22,30,33,36,37,59
279	W286A (M/s.d.m.)	PAS	32
279	W286R (M/s.d.m.)	PAS	32,44
279	W279A (T/s.d.m.)	PAS	13
285	G283D (B/s.d.m.)	PAS	53
285	E292A (A/s.d.m.)	"electrostatic attraction" hypothesis	48
288	F295A (A/s.d.m.)	acyl pocket,OP specificity	30,33,36,37,54,59
288	F295L (A/s.d.m.)	acyl specificity	30,33
288	F295L (M/s.d.m.)	acyl specificity	35,44,56
288	F295Y (M/s.d.m.)	acyl specificity	32,56
288	L286K (B/s.d.m.)	acyl specificity	39
288	L286Q (B/s.d.m.)	acyl specificity	39
288	L286R (B/s.d.m.)	acyl specificity	39
288	L286D (B/s.d.m.)	acyl specificity	39
289	R296S (M/s.d.m.)	acyl specificity	35
290	F297A (A/s.d.m.)	acyl specificity/no OP-specificity	30,33,37,54
290	F297V (A/s.d.m.)	acyl specificity	30,33
290	F297I (M/s.d.m.)	acyl specificity/substrate inhibition	35,44,56
290	F297Y (M/s.d.m.)	acyl specificity/substrate inhibition	32,44,56
290	F368Y (D/rand.)	OP-resistance	11,12,43
293	V300G (M/s.d.m.)	acyl specificity	35
314	S374F (D/rand.)	folding (cold-sensitive)	42
315	H322N (A/natr.)	YT blood group antigen	25,41
326	D333N (A/s.d.m.)	triad definition	20,22

327	E334A (A/s.d.m.)	catalytic triad	20
327	E334D (A/s.d.m.)	catalytic triad	20
327	E334N (A/s.d.m.)	catalytic triad	20,22
327	E327D (T/s.d.m.)	catalytic triad	10
327	E327Q (T/s.d.m.)	catalytic triad	10
330	Y337A (A/s.d.m.)	signal relay/ substrate inhibition	21,22,30,31,33, 36,37,54,55,59
330	Y337F (A/s.d.m.)	signal relay/ substrate inhibition	21,22,30, 33,36,37,55
330	Y337A (M/s.d.m.)	signal relay	32,44,46
330	Y337F (M/s.d.m.)	signal relay	32,46
331	F338A (A/s.d.m.)	hydrophobic subsite	21,22,30,33,37,59
331	F338G (M/s.d.m.)	substrate inhibition	32,44
331	F329C (B/s.d.m.)	OP-resistance	39
331	F329L(B/s.d.m.)	OP-resistance	39
331	F329Q (B/s.d.m.)	OP-resistance	39
331	F329D (B/s.d.m.)	OP-resistance	39
334	Y341A (A/s.d.m.)	PAS	21,22,37,59
342	D349N (A/s.d.m.)	conserved/attraction studies	20,22,48
343	N350Q (A/s.d.m.)	N-glycosylation	24,34
351	E358Q (A/s.d.m.)	"electrostatic attraction" hypothesis	48
382	E389Q (A/s.d.m.)	"electrostatic attraction" hypothesis	48
383	D390N (A/s.d.m.)	"electrostatic attraction" hypothesis	48
392	G390V (B/natr.)	fluoride-2 variant/ succinylcholine resistance	29
397	D397N (T/s.d.m.)	salt-bridge/non-producer	6
397	D404N (A/s.d.m.)	salt-bridge/non-producer	20,22
425	H425Q (T/s.d.m.)	triad definition and production	2
425	H432A (A/s.d.m.)	triad definition and production	20,22
439	M437D (B/s.d.m.)	abolishment of activity	39
440	H440Q(T/s.d.m.)	catalytic triad	2
440	H447A (A/s.d.m.)	catalytic triad	20,22
442	Y440D (B/s.d.m.)	OP-resistance	39
443	E450A (A/s.d.m.)	H-bond network/active center modulation	31,54,59
453	N531D (D/s.d.m.)	N-glycosylation	12,15
457	N464Q (A/s.d.m.)	N- glycosylation	24,34
473	W471R (B/natr.)	non-producer	18
502	Y500 termination frameshift (B/natr.)	non-producer	18
520	Q518L (B/natr.)	non-producer	18
537	C615R (D/s.d.m.)	subunit dimerization	12,15
537	C537S (T/s.d.m.)	folding	9, 38
537	C537D (T/s.d.m.)	folding	9
537	C537G (T/s.d.m.)	folding	9
538	stop (T/s.d.m.)	truncated/glycophospholipid anchorage	9

541	A539T (B/natr.)	K-variant/low activity	8
542	541 stop (B/s.d.m.)	subunit assembly	50
544	stop (T/s.d.m.)	truncated/glycophospholipid anchorage	9
550	stop (T/s.d.m.)	truncated/glycophospholipid anchorage	9
559	stop (T/s.d.m.)	truncated/glycophospholipid anchorage	9
563	F561Y (B/natr.)	silent	4,17,23
563	F561Y (B/s.d.m.)	restores function to D70 mutants	17
572	C541A (B/s.d.m.)	subunit assembly	50
572	C580A (A/s.d.m.)	subunit dimerization	7,27,52,49,60
572	C580K (A/s.d.m.)	intracellular retention	49,52
572	C580S (A/s.d.m./E. coli)	subunit dimerization	26
70,121	Y72A/Y124A (A/s.d.m.)	PAS	37,59
70,121	Y72N/Y124Q (M/s.d.m.)	PAS	32,44
121,279	Y124Q/W286R(M/s.d.m.)	PAS	32,44
70,278	Y72A/E285A (A/s.d.m.)	PAS	37,59
70,279	Y72A/W286A (A/s.d.m.)	PAS	37,59
70,279	Y72N/W286R (M/s.d.m.)	PAS	32,44
72,279	D74N/W286A (A/s.d.m.)	PAS	59
72,334	D74N/Y334A (A/s.d.m.)	PAS	59
72,427	D70G/S425P (B/natr.)	succinylcholine/dibucaine resistance	4,5,17,47
72,563	D70G/F561Y (B/natr.)	succinylcholine resistance	4,17
82,285	E84Q/E292A (A/s.d.m.)	"electrostatic attraction" hypothesis	48
258,343	N265Q/N350Q (A/s.d.m.)	N-glycosylation	24,34
258,457	N265Q/N464Q (A/s.d.m.)	N-glycosylation	24,34
278,279	E285A/W286A (A/s.d.m.)	PAS	37,59
279,285	A277W/G283D(B/s.d.m.)	PAS	53
279,334	W286A/Y341A (A/s.d.m.)	PAS	59
288,290	F295L/F297V (A/s.d.m.)	acyl specificity	30,33,36
288,290	F295L/F297I (M/s.d.m.)	acyl specificity	35
288,290	F288L/F290V (T/s.d.m.)	acyl specificity	13
342,351	D349N/E358Q (A/s.d.m.)	"electrostatic attraction" hypothesis	48
343,457	N350Q/N464Q (A/s.d.m.)	N-glycosylation	24,34
382,383	E389Q/D390N (A/s.d.m.)	"electrostatic attraction" hypothesis	48
425,440	H425Q/H440Q (T/s.d.m.)	no activity (catalytic triad)	2
443,445	E441G/E443G (B/s.d.m.)	catalysis/dibucaine resistance	17,23
499,541	E497V/A539T (B/natr.)	J-variant/low level in serum	8
574,575	D582A/L583V (A/s.d.m.)	intracellular retention	49,52

70,121,279	Y72N/Y124Q/ W286A (M/s.d.m.)	PAS	32
70,121,279	Y72N/Y124Q/ W286R (M/s.d.m.)	PAS	32,44
70,278,279	Y72A/E285A/ W286A (A/s.d.m.)	PAS	37,59
72,116,427	D70G/Y114H S425P (B/natr.)	succinylcholine resistance	17,47
72,116,563	D70G/Y114H F561Y (B/natr.)	succinylcholine resistance	17
72,279,334	D74N/W286A Y341A (A/s.d.m.)	PAS	59
82,342,351	E84Q/D349N/ E358Q (A/s.d.m.)	"electrostatic attraction" hypothesis	48
121,278,279	Y124A/E285A/ W286A (A/s.d.m.)	PAS	37,59
258,343,457	N265Q/N350Q/ N464Q (A/s.d.m.)	N-glycosylation	24,34
285,342,351	E292A/D349N/ E358Q (A/s.d.m.)	"electrostatic attraction" hypothesis	48
285,382,383	E292A/E389Q/ D390N (A/s.d.m.)	"electrostatic attraction" hypothesis	48
288,289,290	F295L/F296S/ F297I (M/s.d.m.)	acyl specificity	35
537,538,539	C537T/D538T/ G539T (T/s.d.m.)	glycophospholipid anchorage	9
542,543,544	S542T/S543T S544T (T/s.d.m.)	glycophospholipid anchorage	9
70,72,121,279	Y72N//D74N/ Y124Q/W286A(M/s.d.m.)	PAS	32
70,72,121,279	Y72N/D74N/ Y124Q/W286R(M/s.d.m.)	PAS	32
82,285,342,351	E84Q/E292A/ D349N/E358Q (A/s.d.m.)	"electrostatic attraction" hypothesis	48
342,351,382,383	D349N/E358Q/ E389Q/D390N (A/s.d.m.)	"electrostatic attraction" hypothesis	48
82,278,285,342, 351	E84Q/E285A/ E292A/D349N/ E358Q (A/s.d.m.)	"electrostatic attraction" hypothesis	48
82,342,351,382, 383	E84Q/D349N/ E358Q/E389Q/ D390N (A/s.d.m.)	"electrostatic attraction" hypothesis	48
285,342,351,382, 383	E292A/D349N/ E358Q/E389Q/ D390N (A/s.d.m.)	"electrostatic attraction" hypothesis	48

82,285,342,351, 382,383	E84Q/E292A/ D349N/E358Q/ E389Q/D390N (A/s.d.m.)	"electrostatic attraction" hypothesis	48
82,278,285,342, 351,382,383	E84Q/E285A/ E292A/D349N/ E358Q/E389Q/ D390N (A/s.d.m.)	"electrostatic attraction" hypothesis	48
533,534,537,539 542,543,544,545	N533T/A534T/C537T G539T/S542T/S543T S544T/G545R	glycophospholipid anchorage	9
CSKDEL-C terminus (A/s.d.m.)		intracellular retention	49,60
ASKDEL-C terminus (A/s.d.m.)		intracellular retention	49,52,60

List of deletions mutants and chimeras of ChEs

	MUTATION	COMMENT	REFERENCE
Deletions	148-166 deletion(D)	protein cleavage	15
	167-180 deletion(D)	protein cleavage	15
	148-180 deletion(D)	protein cleavage	15
	exon 4 deletion , exon 3-5 linkage(T)	glycophospholipid anchored inactive enzyme	3
Chimeras	[B:5-174] into A	PAS mapping	35,44
	[B:5-174] and [B:488-575] into A	PAS mapping	35
	[A:62-138] into B	A:B intermediate	28

Abbreviations: OP - organophosphate compound; PAS - peripheral anionic site.

REFERENCES

1. Mcguire *et al.* (1989) Proc. Natl. Acad. Sci. USA 86, 953-957.
2. Gibney *et al.* (1990) Proc. Natl. Acad. Sci. USA 87, 7546-7550.
3. Gibney, G. and Taylor, P. (1990) J. Biol. Chem. 265, 12576-12583.
4. Gnatt *et al* (1990) Cancer Res. Vol. 50, 1983-1987.
5. Neville *et al* (1990) J. Biol. Chem. 265, 20735-20738.
6. Krejci *et al.* (1991) Proc Natl. Acad. Sci. USA 88, 66437-6651.
7. Velan *et al.* (1991) J. Biol. Chem. 266, 23977-23984.
8. Bartels *et al* (1992) Am. J. hum. Genet. 50, 1104-1114.
9. Bucht *et al* (1992) In Multidisciplinary Approaches to Cholinesterase Functions (Shafferman, A. and Velan, B. eds, Plenum Press New-York) pp 185-188.
10. Duval *et al.* (1992) FEBS Lett. 309, 421-423.
11. Fournier *et al* (1992) J. Biol. Chem. 267, 14270-14274.
12. Fournier *et al* (1992) In Multidisciplinary Approaches to Cholinesterase Functions (Shafferman, A. and Velan, B. eds, Plenum Press New-York) pp 75-82.
13. Harel *et al.* (1992) Proc. Natl. Acad. Sci. USA 89, 10827-10831.
14. Massoulie *et al* (1992) In Multidisciplinary Approaches to Cholinesterase Functions (Shafferman A. and Velan, B., eds, Plenum Press, New-York) pp 285-288.
15. Mutero A. and Fournier, D. (1992) J. Biol. Chem. 267, 1695-1700
16. Mutero *et al.* (1992) Neuroreport 3, 39-42.

17. Neville *et al.* (1992) EMBO J. 11, 1641-1649.
18. Primo-Parmo *et al* . (1992) In Multidisciplinary Approaches to Cholinesterase Functions (Shafferman, A. and Velan, B. eds, Plenum Press New-York) pp 61-64.
19. Radic *et al.* (1992) Biochemistry 31, 9760-9767.
20. Shafferman *et. al.* (1992a) J. Biol. Chem. 267, 17640-17648.
21. Shafferman *et. al.* (1992b) EMBO J. 11, 3561-3568.
22. Shafferman *et al.* (1992) In Multidisciplinary Approaches to Cholinesterase Functions (Shafferman, A. and Velan, B. eds, Plenum Press New-York) pp 165-175.
23. Soreq *et al* (1992) Trans Biochem. Sci. 17, 353-358.
24. Velan *et al.* (1992) In Multidisciplinary Approaches to Cholinesterase Functions (Shafferman, A. and Velan, B. eds, (Plenum Press New-York) pp 39-47.
25. Bartels *et al.* (1993) Am. J. Hum. Genet. 52, 928-936.
26. Fischer *et al.* (1993) Cell. Mol. Neurobiol. 13, 25-38.
27. Kerem *et al.* (1993) J. Biol. Chem. 268, 180-184.
28. Loewenstein *et al* (1993) J. Mol Biol. 234, 289-296.
29. Masson *et al* (1993) J. Biol. Chem. 268. 14329-14341.
30. Ordentlich *et. al.* (1993) J. Biol. Chem. 268, 17083-17095.
31. Ordentlich *et. al.* (1993) FEBS Lett. 334, 215-220.
32. Radic *et al.* (1993) Biochemistry 32, 12074-12084.
33. Shafferman *et al.* (1993) Proceedings of Medical Defence Bioscience Review 3, 1097- 1108.
34. Velan *et al.* (1993) Biochem. J. 296, 649-656.
35. Vellom *et al.* (1993) Biochemistry 32, 12-17.
36. Ashani *et al.* (1994) Mol. Pharmacol. 45, 555-560.
37. Barak *et al.* (1994) J. Biol. Chem. 264, 6296-6305.
38. Bucht *et al* (1994) Biochim. Biophys. Act. in press.
39. Gnatt *et al.* (1994) J. Neurochem.62, 749-755.
40. Kronman *et al* (1994) J. Biol. Chem. in press.
41. Masson *et al.* (1994) Blood 83, 3003-3005.
42. Mutero *et al* (1994) Mol Gen Genet. 243, 699-705.
43. Mutero *et al* (1994) Proc. Natl. Acad. Sci. USA. 91, 5922-5926.
44. Radic *et al.* (1994) J. Biol. Chem.269, 11233-11239.
45. Saxena *et al.* (1994) Biochem. Biophys. Res. Com. 197, 343-349.
46. Saxena *et al.* (1994) Protein Sci. n press.
47. Schwarz *et al* (1994) J. Neurochem. 63, (Suppl. 1): S80D.
48. Shafferman *et. al.* (1994) EMBO J. 13, 3448-345.
49. Velan *et al..* (1994) J. Biol. Chem. 269, 22719-22725.
50. Blong and Lockridge (1995) In Enzymes of the Cholinesterase Family (Balasubaramanian, A.S., Doctor, B.P., Taylor, P., Quinn, D.M., Eds.) Plenum Publishing Corp. in press.
51. Broomfield *et al* (1995) In Enzymes of the Cholinesterase Family (Balasubaramanian, A.S., Doctor, B.P., Taylor, P., Quinn, D.M., Eds.) Plenum Publishing Corp. in press.
52. Kronman *et al.* (1995) In Enzymes of the Cholinesterase Family (Balasubaramanian, A.S., Doctor, B.P., Taylor, P., Quinn, D.M., Eds.) Plenum Publishing Corp. in press.
53. Masson *et al* (1995) In Enzymes of the Cholinesterase Family (Balasubaramanian, A.S., Doctor, B.P., Taylor, P., Quinn, D.M., Eds.) Plenum Publishing Corp. in press.
54. Ordentlich *et al.* (1995) In Enzymes of the Cholinesterase Family (Balasubaramanian, A.S., Doctor, B.P., Taylor, P., Quinn, D.M., Eds.) Plenum Publishing Corp. in press.
55. Ordentlich *et al.* (1995) J. Biol. Chem. in press.
56. Pickering *et al.* (1995) In Enzymes of the Cholinesterase Family (Balasubaramanian, A.S., Doctor, B.P., Taylor, P., Quinn, D.M., Eds.) Plenum Publishing Corp. in press.
57. Quinn *et al.* (1995a) In Enzymes of the Cholinesterase Family (Balasubaramanian, A.S., Doctor, B.P., Taylor, P., Quinn, D.M., Eds.) Plenum Publishing Corp. in press.
58. Quinn *et al.* (1995b) In Enzymes of the Cholinesterase Family (Balasubaramanian, A.S., Doctor, B.P., Taylor, P., Quinn, D.M., Eds.) Plenum Publishing Corp. in press.
59. Shafferman *et al* (1995) In Enzymes of the Cholinesterase Family (Balasubaramanian, A.S., Doctor, B.P., Taylor, P., Quinn, D.M., Eds.) Plenum Publishing Corp. in press.
60. Velan *et al.* (1995) In Enzymes of the Cholinesterase Family (Balasubaramanian, A.S., Doctor, B.P., Taylor, P., Quinn, D.M., Eds.) Plenum Publishing Corp. in press.

A DATABASE OF SEQUENCES RELATED TO ACETYLCHOLINESTERASE/LIPASE/α:β HYDROLASE SUPERFAMILY WITH PUBLIC ACCESS ON INTERNET

Xavier Cousin,[1,3] Thierry Hotelier,[2] Catherine Mazzoni,[2] Martine Arpagaus,[3] Jean-Pierre Toutant,[3] and Arnaud Chatonnet[3]

[1] Unité des venins
Institut Pasteur
Paris
France
[2] Centre de calcul
[3] Différenciation cellulaire et Croissance
INRA
Montpellier
France

INTRODUCTION

Comparison of sequences of distantly related proteins is useful to understand evolution of gene families. The more distant are the organisms, the higher is the functional significance of conserved regions between homologous proteins.

Alignment of sequences (Cousin et al. 1991; Krejci et al. 1991; Langin and Holm 1993; Haemilä et al. 1994) as well as comparison of tridimensional structures of acetylcholinesterase (Sussman et al. 1991), *Geotrichum* lipase (Shrag et al. 1991) and pancreatic lipase (Winkler et al. 1990) led to the definition of a superfamily of related sequences: the α:β hydrolase family (Ollis et al. 1992; Cygler et al. 1993). The structure of this superfamily is certainly a primordial structure, and it belongs to the 730 Ancient evolutionary Conserved Regions (ACRs, Green et al. 1993).

Recovery of sequences from generalist databases such as GENBANK, EMBL, PIR or SWISS-PROT is becoming tedious as the sizes of these databases increase. Relevant homologies are difficult to detect as random matching occurs more often. Duplicated sequences due to cloning of alternatively-spliced mRNA, or genetic variants, increase the number of entries which are recovered by homology searching using programs such as FASTA or BLAST. This tends to mask homologies between more distant sequences of relevant signification.

Enzymes of the Cholinesterase Family, Edited by Daniel M. Quinn et al.
Plenum Press, New York, 1995

CHEDB, A CHOLINESTERASE DATA BASE

To overcome some of these difficulties, we developed a database of sequences homologous to the cholinesterases and belonging to the acetylcholinesterase/lipase/α:β hydrolase family. We decided to develop access to this database (CHEDB) through Internet. The server works both as gopher and World Wide Web server. These softwares give now a popular access to lots of biological data (Parker 1993).

Connection

To use a Gopher or Mosaic client software you must have a computer connected to Internet via TCP/IP. Current versions of gopher software can be obtained by anonymous ftp from boombox.micro.umn.edu in the /pub/gopher directory. The CHEDB gopher server IP address is gopher.montpellier.inra.fr on port 70 selector: 1/cholinesterase. The address of the CHEDB WWW server is http://www.montpellier.inra.fr . We highly recommand using the latter type of connection as future developments will use extensively these friendly users possibilities. Information through the www server is presented in hypertext and highlighted items can be selected to browse easily in the database. Figure 1 shows the home page that appears on the screen upon connection.

The Menu

At present you can choose to open:

- a readme text file to help navigation in the directory.
- a table of correspondences between accession numbers and names of sequences in databases.

The aim of this table is to present an exhaustive compilation of all entries in databases for proteins showing significant homologies to cholinesterases. This table contains also incomplete sequences, gene fragments and genetic variants. Sequences were added when PROSITE blocks PS00120 and PS00122 (Bairoch 1992) were recognized by the BLOCK program of Henikoff and Henikoff (1991, 1994).

- a reduced table which contains only one entry per protein (the most complete and defined from the previous table). We use a nomenclature very close to the SWISS-PROT nomenclature. The names have the same number of characters in order to help in alignment processing. Five letters define the species (for example: ratno for *Rattus norvegicus* or human for *Homo sapiens*. Five letters define the enzyme: acche for acetylcholinesterase. Thus rat acetylcholinesterase is ratno-acche.
- an output of the PS00120 and PS00122 blocks from the BLOCK and PROSITE databases.
- a directory of outputs of individual sequences from the different databases. These files can be searched by keywords or strings of sequence.
- a FASTA format text file of all the sequences usable for different comparison and alignment programs.
- a tentative alignment of 62 different sequences from the reduced table constructed by the PILEUP program (Gap weight: 2.0; gap length: 0.0) of the GCG package (Devereux et al. 1984).

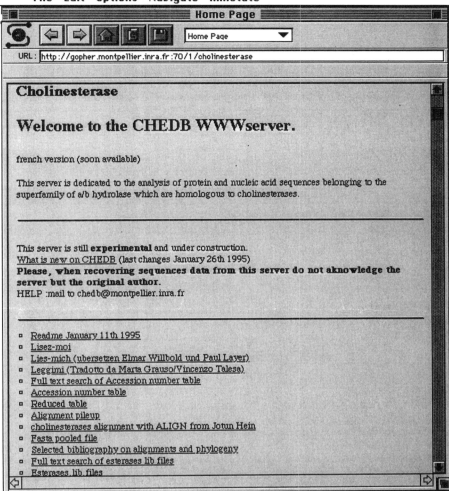

Figure 1. Home page of CHEDB. Underlined items can be opened and corresponding pages appear on the screen.

- a directory contains the structural data of the proteins in the family which have been crystallized. Images of molecular models in gif format can also be recovered here.
- the compilation of evaluated mutants of cholinesterases by A. Shafferman, C. Kronman and A. Ordentlich (this volume) is also included in the server as a table. An alignment of sequences of human AChE, human BChE, mouse, *Torpedo* and *Drosophila* AChEs, which have been used in mutation experiments, is available here. Amino acid positions for which a mutant has been studied are highlighted. Selecting one of those positions gives a file with the relevant information (this alignment is available on www only).
- a file gives the latest changes that we made to the server.

Help

Help is available through e-mail at : chedb@montpellier.inra.fr

PERSPECTIVES

We are working on different new features. In particular, it will be possible for a distant user to compare directly its personal sequence to the CHEDB using the BLAST programm of Altschul et al. (1990). The result will be sent directly to the distant client. A form would allow users to register their address for a directory of the cholinesterase community.

Although we tried to be exhaustive, data may have escaped our search and some errors may have resisted our proofreading. However we hope that the gopher server will improve the accuracy of the information and permit an easy access to it.

REFERENCES

Altschul S.F., Miller W., Myers E.W. and Lipman D.J. (1990) J. Mol. Biol. 215, 403-410.

Bairoch A. (1992) Nucleic Acids Res. 20, 2013-2018.

Cousin X., Toutant J.-P., Jbilo O., Arpagaus M., and Chatonnet A. (1991) In Cholinesterases: Structure, Function Genetics and Cell Biology, (Massoulié, J. et al. eds) p. 195, American Chemical Society, Washington, DC.

Cygler M., Schrag J.D., Sussman J.L., Harel M., Silman I., Gentry M.K. and Doctor B.P. (1993) Protein Sci. 2, 366-382.

Devereux J., Haeberli P. and Smithies O. (1984) Nucleic Acids Res. 12, 387-395.

Green P., Lipman D., Hillier L., Waterston R., States D., and Claverie J.M. (1993) Science 259, 1711-1716.

Hemilä H., Koivula T.T. and Palva I. (1994) Biochim. Biophys. Acta 1210, 249-253.

Henikoff S. and Henikoff J.G. (1991) Nucleic Acids Res. 19, 6565-6566.

Henikoff S. and Henikoff J.G. (1994) Genomics 19, 97-107.

Krejci E., Duval N., Chatonnet A., Vincens P. and Massoulié J. (1991) Proc. Natl. Acad. Sci. U.S.A. 88, 6647-6651.

Langin D. and Holm C. (1993) TIBS 18, 466-467.

Ollis D.L., Cheah E., Cygler M., Dijkstra B., Frolow F., Franken S.M., Harel M., Remington S.J., Silman I., Shrag J., Sussman J.L., Verschueren K.H.G. and Goldman A.(1992) Protein Engineering 5, 197-211

Parker M. (1993) TIBS 18, 485-486.

Shrag J.D., Li Y., Wu S., and Cygler M. (1991) Nature 351, 761-765.

Sussman J.L., Harel M., Frolow F., Oefner C., Goldman A., Toker L., and Silman I. (1991) Science 253, 872-879.

Winkler F.K., D'Arcy A., and Hunziker W. (1990) Nature 343, 771-774.

AMINO ACID ALIGNMENT OF CHOLINESTERASES, ESTERASES, LIPASES, AND RELATED PROTEINS

Mary K. Gentry and B. P. Doctor

Division of Biochemistry
Walter Reed Army Institute of Research
Washington, D.C.

The alignments previously published (Gentry and Doctor, 1991; Cygler et al., 1993), nine and 32 sequences respectively, have been further expanded by the addition of 22 newly-found sequences (Table 1). References and protein sequences were found by searching on the term "acetylcholinesterase" using the software package Entrez, an integrated citation and sequence retrieval system (National Center for Biotechnology Information, NLM, Bethesda, MD). Entrez uses the "neighboring" concept to find algorithmically precomputed similarities within a given database and has established hard links to other databases.

After retrieval of the sequences, they were aligned using the PileUp program in the GCG package (Genetics Computer Group, Madison, WI). The order of the alignment shown in the appendix (Figure 1) was determined by the program. PileUp calculates pairwise similarity scores in the process of aligning multiple sequences. On the basis of this calculation, the sequences shown here for human cholesterol esterase (Balipase), human carboxyl ester lipase (Humancel), and human pancreas carboxyl ester lipase (Celipase) are identical. The sequences for rat pancreas cholesterol esterase (gene, Ratchest), rat pancreas cholesterol esterase (Ratcholest), and rat liver cholesterol esterase (Ratsterest) are 99% similar; as are pI 6.1 rat liver carboxylesterase (Rat61) and rat carboxylesterase (Ratcoxys); and rat and mouse acetylcholinesterases (Ratache and Mouseache). Human carboxylesterase (Humance), human liver carboxylesterase (Humlivce), human liver carboxylesterase form A (Humcoxysa), human liver carboxylesterase form B (Humancoxysb), and human mono-

---

Figure 1. Sequence alignment of 54 homologous proteins. Amino acids of the catalytic triad are marked with asterisks. Lengths of 26 and 38 amino acids have been removed from *A. stephensi* and *Drosophila* acetylcholinesterases at position 133 for ease in alignment. Likewise, two lengths, 47 and 114 amino acids, have been removed from the sequence for chicken acetylcholinesterase starting at positions 380 and 390, respectively. (Next eight pages.)

Enzymes of the Cholinesterase Family, Edited by Daniel M. Quinn et al.
Plenum Press, New York, 1995

Multiple sequence alignment (residue positions 1–100):

```
                 1                                                                                      100
Geolip1       ....QAPRPS LNGNEVISGV LEGK...... .........VDTFKGIP FADPLNDLR FKHPQPFTG. .SYOG.LKA NDFSPACMQL DPGNSITILD KALGLAKVIP   Geolip1
Geolip2       ....QAPTAV LNGNEVISGV LEGK...... .........VDTFKGIP FADPPVGDLR FKHPQPFTG. .SYOG.LKA NDFSSACMQL DPGNAISLLD KVVGLGKIIP   Geolip2
Cclip1        ....APTAT LANGDTITGL NAII...... .........NEAFLGIP FAEPPVGNLR FKDVPVYSG. .SLDG.QKF TSYGPSCMQQ NPEGTYBENL PKAALDIVMQ    Cclip1
Crlip4        VSVAAAPTAT LANGDTITGL NAII...... .........NEAFLGIP FAQPPVGNLR FKPVPYSA. .SLNG.QKF TSYGPSCMQM NPLGNWDSSL PKAAINSIMQ    Crlip4
Cclip2        ....APTAT LANGDTITGL NAIV...... .........NEKFLGIP FAEPPVGTLR FKPPVPYSA. .SLNG.QQF TLYGPLCMQM NPMGSFEDTL PKNARHLVLQ    Cclip2
D2            ....RKRSYI KNTDTSIVVT QPGAIKGIVE DTHRVFYGVP FAQPPVGNLR WENPID.LK. .PWENVRET LTQKSQCAQR CKLGPGVCS.                     D2
Dictyocrys    AAKKFGRKGI RTLGDNEVLL SDGAIRGTVT DTHRVFYGIP FARPPIDELR YEDPQP.FK. .PWSIYRDG TKQRDQCIQD CKLGKGSCS.                      Dictyocrys
Balipase      ....AKLGAV YTEGGFVEGV NKKLGLL.G. DSVDIFKGIP FAAPTKA... LENPQPHP.. .GWQGTLKA KNFKKRCLQA TITQDSTYG.                      Balipase
Celipase      VASAAKLGAV YTEGGFVEGV NKKJGLL.G. DSVDIFKGIP FAAPTKA... LENPQPHP.. .GWQGTLKA KNFKKRCLQA TITQDSTYG.                      Celipase
Humancel      VASAAKLGAV YTEGGFVEGV NKKLGLL.G. DSVDIFKGIP FAAPTKA... LENPQRHP.. .GWQGTLKA TDFKKRCLQA TITQDDTYG.                      Humancel
Ratchest      AACAAKLGAV YTEGGFVEGV NKKLSLLGG. DSVDIFKGIP FATA.KT... LENPQRHP.. .GWQGTLKA TDFKKRCLQA TITQDDTYG.                      Ratchest
Ratsterest    ....AKLGAV YTEGGFVEGV NKKLSLLGG. DSVDIFKGIP FATA.KT... LENPQRHP.. .GWQGTLKA TDFKKRCLQA TITQDDTYG.                      Ratsterest
Ratcholest    ....AKLGAV YTEGGFVEGV NKKLSLF.G. DSVDIFKGIP FAAAPKA... LEKPERHP.. .GWQGTLKA KSFFKRCLQA TITQDDTYG.                      Ratcholest
Bovchole      ....AKLGAV YTEGGFVEGE NKKLSLLGD. DSVDIFKGIP FATPA.I... LENPQRHP.. .GWQGTLKA KDFKKRCLQA TLTQDSTFG.                       Bovchole
Rabbitchole   SGACGDLGPV YISLEGF.V.                                                                                             Rabbitchole
Mousece       .HSLLPPVV DTTGKVLGK YISLEGF.T. QPVAVFLGVP FAKPPLGSLR FAPPQP.AE. .PWSFVKNA TSYPPMCSQD AGWAKILSDM FSTEKELL.              Mousece
Ratce         .HPSSPPVV DTTKGKVLGK YVSLEGF.T. QPVAVFLGVP FAKPPLGSLR FAPPQP.AE. .PWSSVKNA TSYPPMCFQD PVTGQIVNDL LTNRKEKI.             Ratce
Rat61         .YPSSPPVV NTVKGKVLGK YVNLEGF.A. QPVAVFLGVP FAKPPLGSLR FAPPQP.AE. .PWINFVKNT TSYPPMCSQD AVGGQVLSEL FTNRKENI.            Rat61
Ratcoxys      .GYPSSPPVV NTVKGKVLGK YVNLEGF.A. QPVAVFLGIP FAKPPLGSLR FSEPQP.PE. .PWINFVKNT TSYPPMCSQD AVGGQVLSEL FTNRKENI.            Ratcoxys
Humance       .GHPSSPPVV DTVHGKVLGK FVSLEGF.A. QPVAIFLGIP FAKPPLGPLR FTPPQP.AE. .PWSVKNT TSYPPMCTQD PKAGQLLSEL FTNRKENI.             Humance
Humlivce      .GHPSSPPVV DTVHGKVLGK FVSLEGF.A. QPVAIFLGIP FAKPPLGPLR FTPPQP.AE. .PWSFVKNA TSYPPMCTQD PKAGQLLSEL FTNRKENI.            Humlivce
Humcoxysa     .AGHPSSPPVV DTVHGKVLGK FVSLEGF.A. QPVAIFLGIP FGKPPLGPLR FTPPQP.AE. .PWSFVKNA TSYPPMCTQD PKAGQLLSEL FTNRKENI.           Humcoxysa
Humcoxysb     .GHPSSPPVV DTVHGKVLGK FVSLEGF.A. QPVAIFLGIP FGKPPLGPLR .....G FTPPQP.AE. .PWSFVKNA TSYPPMCTQD PKAGQLLSEL FTNRKENI.     Humcoxysb
Hummonmacss   .                                                                                                                Hummonmacss
Humcoxys      AGQPASPPVV DTAQGRVLGK YVSLEGL.A. QPVAVFLGVP FAKPPLGSLR FAPPQP.AE. .PWSFVKNT TSYPPMCCQD PVVEQMTSDL FTNGKERL.            Humcoxys
Pigcoxys      AGHPSSPPMV DTVQGKVLGK YISLEGF.T. QPVAVFLGVP FAKPPLGSLR FAPPQP.AE. .PWSSVKNA TSYPPMCFQD PVTGQIVNDL LTNRKEKI.            Pigcoxys
Est-22        .HPSAPPVV VTNYGSVRGV QVKVNAA.E. RSVNVFLGIP FAKPPLGSLR FSEPQP.PE. .PWSHVKNT TSYPPMCSSD AVSGHMLSEL FTNRKENI.            Est-22
Anasthios     .QDSASPIR NTHTGQVRGS LVHVEGT.D. AGVHTFLGIP FAPPEP.AE. .AWSGVRDG TSLPAMCLQN LAIMDQDVLL LHFTPSI.                        Anasthios
Rablivest2    IGPKVTQPEV DTPLGRVRGR QVGVKVT.D. RMVNVFLGIP FAQPLGPLR FKEPVP.AE. .PWEGVRDA SINPPMCLQD .VERMSNSR LHFTPSI.                Rablivest2
Mouselivcoxy  .IIDRLVV QTSSGPIRGR STMVQ...G. REVHVNGVP FAKPPVDSLR TRLPPSCIQE .RYEYFPGF FTLNERMK.                                  Mouselivcoxy
Anopha        .IDRLVV QTSSGPVRGR SVTVQ...G. REVHVTGIP YAKPPVEDLR TGLSATCVQE .RYEYFPGF AGEMWNP.                                    Anopha
Drosache      .EDDII ATKNGKVRGM NLTV..F.G. GTVTAFLGIP YAQPPLGRLR FKKPQSLTK. .PWHGVLDA TKYANSCCQN IDQSFPGF SGEIWNP.                  Drosache
Humanbche     .EDVII TTKNGRIRGI NLFV..F.G. GTVTAFLGIP YAQPPLGRLR FKKPQSLTK. .WSDIWNA TKYANSCCQN .IDQSFPGF HGSEMWNP.                 Humanbche
Rabbitbche    .EEDFII TKTGRVRGL SMFV..L.G. GTVTAFLGIP YAQPPLGPLR FKKPQPLNK. .WSDIWNA TQYANSCYQN .IDQAFPGF HGSEMWNP.                 Rabbitbche
Mousebche     .DDHSELLV NTKSGKVMGT RVPV..L.S. SHISAFLGIP FAEPPVGNWR FRPEP.KK. .PWSGVWNA STYPNNCQQY QGSEMWNP.                       Mousebche
Tcalifache    .DDDSELLV RVRGGQLRGI RLKA..P.G. GPVSAFLGIP FAEPPVGSRR FMPPEP.KR. .PWSGVLDA TTFQNVCYQY .VDTIYPGF PGSEMWNP.             Tcalifache
Tmarmorache   .EGREDPQLLV MVRGELRGL RLMA..P.R. GPVSAFLGIP FAEPPVGNNR FRPEP.KR. .PWHGILDA TTFQNVCYQY .VDTIYPGF EGTEMWNP.             Tmarmorache
Mouseache     .EGREDAELLV TVRGRLRGI RLKT..P.G. GPVSAFLGIP YAQPPLGPLR FKKPVT.VD. .PWSGVVDA DTQNVCYQY YKYGPACVQ. EQIAGPRT.          Mouseache
Ratache       .SAPNRPE..V RTTTGSVRGL LIPAGPS.G. .VDGVLGIP YAKPPVDNLR FKKPVT.VD. .PWSGVVDA DTQNVCYQY YKYGPACVQ. EQIAGPRT.           Ratache
Fbsache       .LTPEGNVEAL KASCGPVWGV EYGAAEV... .FLAIP FAKPPVDNLR FEKPEA.PE. .PWEDVYQA TQFRNDC... .TPHY RLVAQFSS.                  Fbsache
Humanache     .FGACWAGPVV NTNYGKVMGT EYGAAEV... .FLAIP FAKPPVDNLR FEKPEA.PE. .PWEDVYQA TQFRNDC... .TPHY RLVAQFSS.                  Humanache
Chickache     .SDTDDPLIV QLPQGKLRG. .R....DN SOAIQVGTSW KPVDQFLGVY YAAPIGREX YAEPTGDIR FRAPEHLN. .WTGSWEA TKPRARCWQP GIRTFPPG.     Chickache
Nematest      .SEADPLIV QVHSGEIAGL .K....DN GLYYSYESIP YAPGIGREX YAEPTGDIR WILGVWNA TKTPVACLQW D..QFTPG.  .AN.                     Nematest
Nematest2     .TCSASNTPKV QVHSGEIAG. .GFEYTYNG RKIYSFLGIP YAQPPVQNNR FKEPQP.VQ. .HWTDVFNA TQSPVBCMQW N..QFINE.  .NN.               Nematest2
Bovthyro      .TNSRSVVA HLDSG.IIR. .GFEYTYNG IKPASFLGVP YASPPVONNR FKEPQP.VQ. .PWLGVWNA TVPGSACLGI E...FGS.  .GS.                  Bovthyro
Drosestp      .MSLESLTV QTKYGPVRG. .KRNVSLLG QEYVSFQGIP YARAPEGELR FKELEP.LE. .PWDNILNA TNEGPICFQT DVLYGRLMA.                      Drosestp
Drosestp6     .YAQPEAVVQ APEVGQIIG. .ISGHKTIAN RPVNAFLGIR YGTVGGGLAR FQAAQPI... .KWTETLDC TQCCEPCFHF DRRLQ....                      Drosestp6
Aphidestfe4   PLRREGRYIMA VTGCGPVRGV K.........                                                                                  Aphidestfe4
```

```
101                                                                                          200
Geolip1       EEFRGPLYDM AKGTVSMNED CLYLNVFRPA GT.....KPD AKLPVMWIY  GGAFVYGSS. ..AAYPGNSY VKESINMGQP VVFVSINYRT GPFGFLGGDA   Geolip1
Geolip2       DNLRGPLYDM AQGSVSMNED CLYLNVFRPA GT.....KPD AKLPVMWIY  GGAFVFGSGT .ASYPGNGY  VKESLEMGQP VVFVSINYRT GPYGFLGGDA   Geolip2
Cclip1        SKVFEAV... ....SPSSED CLTINVVRPP GT.....KAG ANLPMLWIF  GGGFEVGGT. .STFPPAQM  ITKSIAMGKP IIHVSVNYRV SSWGFLAGDE   Cclip1
Crlip4        SKLFQAV... .....LPNGED CLTINVVRPS GT.....KPG ANLPVMWIF GGGFEVGGS. .SLFPPAQM  ITASVLMGKP IIHVSMNYRV ASWGFLAGPD   Crlip4
Cclip2        SKIFQVV... .....LPNDED CLTINVIRPP GT.....RAS AGLPVMLWF GGGFLGGS.. ....FIYIF  MMLLNLLHSS VIHVSMNYRV ASWGFLAGPD   Cclip2
D2            .......... ....PMGTSED SLYQDIFTPK DA.....RPN SKYPVIVIP  GGAFSVGSG. ....FLFLF  VAKSVLMGKP VIVVNLHSS  GVLGLMGTD.  D2
Dictyocrys    .......... ..EVGTSED CLYLDVFIPR TV.....NPG SKVPVMVFIP  GGAFTQGTG. ...S.CPL   YDGLKFANSS VIVVNVNYRL GVLGFLCTG.  Dictyocrys
Balipase      .......... .......DED CLYLNIWPQ  G.....RKQVS RDLPVMNYLY GGAFLMGSGH GANFLNNYLY DGEEIATRGN VIVTFNYRV  GPLGFLSTGD  Balipase
Celipase      .......... .......DED CLYLNIWPQ  G.....RKQVS RDLPVMIWIY GGAFLMGSGH GANFLNNYLY DGEEIATRGN VIVTFNYRV  GPLGFLSTGD  Celipase
Humancel      .......... ......DED  CLYLNIWPQ  G.....RKQVS RDLPMIWIY  GGAFLMGSGH GANFLNNYLY DGEEIATRGN VIVTFNYRV  GPLGFLSTGD  Humancel
Ratchest      .......... ......QED  CLYLNIWPQ  G.....RKQVS RDLPVMIWIY GGAFLMGSGQ GANFLKNYLY DGEEIATRGN VIVTFNYRV  GPLGFLSTGD  Ratchest
Ratesterest   .......... ......QED  CLYLNIWPQ  G.....RKQVS HDLPVIWIY  GGAFLMGSGQ GANFLKNYLY DGEEIATRGN VIVTFNYRV  GPLGFLSTGD  Ratesterest
Ratcholest    .......... ......QED  CLYLNIWPQ  G.....RKQVS HDLPVMIWIY GGAFLMGSGQ GANFLKNYLY DGEEIATRGN VIVTFNYRV  GPLGFLSTGD  Ratcholest
Bovchole      .......... ......NED  CLYLNIWPQ  G.....RKEVS HDLPVMIWIY GGAFLMGASQ GANFLSNYLY DGEEIATRGN VIVTFNYRV  GPLGFLSTGD  Bovchole
Rabbitchole   .......... ......DQD  CLYLNIWPQ  G.....RKEVS HNLPVMIWIY GGAFLMGSSQ GANFLSNYLY DGEEIATRGN VIVTFNYRV  GPLGFLSTGD  Rabbitchole
Mousece       .......... ...PLKISED CLYLNIYSPA DL.....TKN  SRLPVMVWIH GGGLIIGGA. ....S..PY  SGLALSAHEN VVVTIQYRL  GPGLFSTGD   Mousece
Ratce         .......... ...PLEFSED CLYLNIYSPA DL.....TKN  SRLPVMVWIH GGGLVVGGA. ....S..PY  DGQVLSAHEN VVVTIQYRL  GPFGFFSTGD  Ratce
Rat61         .......... ...PLQFSED CLYLNVTPA  DL.....TKN  SRLPVMVWIH GGGLVVGA.. ....S..TY  DGQVLSAHEN VVVTIQRL   GIWGFFSTGD  Rat61
Ratcoxys      .......... ...PLQFSED CLYLNITTPA DL.....TKK  NRLPVMWVIH GGGLVVGAA. ....S..TY  DGLALAAHEN VVVTIQRL   GIWGFFSTGD  Ratcoxys
Humance       .......... ...PLKLSED CLYLNITTPA DL.....TKK  NRLPVMWVIH GGGLMVGAA. ....S..TY  DGLALAAHEN VVVTIQRL   GIWGFFSTGD  Humance
Humlivce      .......... ...PLKLSED CLYLNITTPA DL.....TKK  NRLPVMWVIH GGGLMVGAA. ....S..TY  DGLALAAHEN VVVTIQRL   GIWGFFSTGD  Humlivce
Humcoxysa     .......... ...PLKLSED CLYLNITTPA DL.....TKK  NRLPVMWVIH GGGLMVGAA. ....S..TY  DGLALAAHEN VVVTIQRL   GIWGFFSTGD  Humcoxysa
Humcoxysb     .......... ...PLKLSED CLYLNITTPA DL.....TKK  NRLPVMWVIH GGGLMVGAA. ....S..TY  DGLALAAHEN VVVTIQRL   GIWGFFSTGD  Humcoxysb
Hummonmacss   .......... ...PLKLSED CLYLNITTPA DL.....TKK  NRLPVMWVIH GGGLMVGAA. ....S..TY  DGLALAAHEN VVVTIQRL   GIWGFFSTGD  Hummonmacss
Humcoxys      .......... ......EH   CLYLNITTPA DL.....TKK  NRLPVMWVIH GGGLMVGAA. ....S..TY  DGLALAAHEN VVVTIQRL   GIWGFFSTGD  Humcoxys
Pigcoxys      .......... ...TLEFSED CLYLNITTPA DL.....TKR  GRLPVMWVIH GGGLVLGA.. ....P..MY  DGVVLAAHEN VVVAIQRL   GIWGFFSTGD  Pigcoxys
Est-22        .......... ...PLQFSED CLYLNITTPA DL.....TKS  DRLPVMWVIH GGGLVLGA.. ....S..TY  DGLVLSTHEN VVVVIQRL   GIWGFFSTGD  Est-22
Rabbitce      .......... ...PLKFSED CLYLNVTPV  DL.....TKR  EKLPVFVWIH GGGLVSGAA. ....S..TY  DGLALSAHEN VVVTIQRL   GIGGFGFNID  Rabbitce
Anasthios     .......... ...RLQISED CLYLNVTTPV ST.....EEQ  SDLPVMWVIH GGGLTMGWA. ....S..SY  DGSALAAFDN VIVASFQYRV GIAGYFSTGD  Anasthios
Rablivest2    .......... ...PM..SED CLYLNVWAPA HA.....REG  GSSLRVGSS. ...........  ....S..MY  DGSALAAFED VVVTIQRL   GVLGFFSTGD  Rablivest2
Mouselivcoxy  .......... ...IFPISED CLTLNIYSPT EI.....TAG  DKRPVMWVIH GGGFMSGTS. ....T..SH  DGSALAAYGD VIVASMQYRV GIFGFLSTGD  Mouselivcoxy
Anopha        .......... ...NTNVSED CLYLNIWPT  KT(-26)QSK GGLAMLWIY  GGGFMTGSA. ...TLDIY   NAEILAAVGN VIVASMQYRV GAFGFLYLAP  Anopha
Drosache      .......... ...NTNVSED CLYLNVWAPA KA(-38)RLR HGI..LIWIY GGGFMTGSA. ...TLDIY   NADIMAAVGN VIVASFQYRV GALGFLHLAP  Drosache
Humanbche     .......... ...NTDLSED CLYLNIWIPA P......KPK NAT.VLIWIY GGGFQTGTS. ...SLHVY   DGKFLARVER VTVVSMNYRV GALGFLALPG  Humanbche
Rabbitbche    .......... ...NTDLSED CLYLNIWIFT P......KPK NAT.MIWIY  GGGFQTGTS. ...SLQVY   DGKFLTRVER VIVVSMNYRV GALGFLALPG  Rabbitbche
Mousebche     .......... ...NTNLSED CLYLNIWIPA P......KPK NAT.VMWIY  GGGFQTGTS. ...SLPVY   DGKFLARVER VTVVSMNYRV GALGFLAFPG  Mousebche
Tcalifache    .......... ...NREMSED CLYLNIWIPS P......RPK STT.VMVWIY GGGFYSGSS. ...TLDVY   NGKYLATTEE VVLVSLSYRV GAFGFLALHG  Tcalifache
Tnarmorache   .......... ...NREMSED CLYLNIWIPS P......RPK SAT.VHLWIY GGGFYSGSS. ...TLDVY   NGKYLATTEE VVLVSLSYRV GAFGFLALHG  Tnarmorache
Mouseache     .......... ...NRELSED CLYLNIWTPY P......RPA SPTPVLIWIY GGGFYSGAA. ...SLDVY   DGRFLAQVEG AVLVSMNYRV GTFGFLALPG  Mouseache
Ratache       .......... ...NRELSED CLYLNIWTPY P......RPT SPTPVLIWIY GGGFYSGAS. ...SLDVY   DGRFLAQVEG TVLVSMNYRV GTFGFLALPG  Ratache
Fbsache       .......... ...NRELSED CLYLNIWTPY P......RPS SPTPVLVWIY GGGFYSGAS. ...SLDVY   DGRFLVQAEG TVLVSMNYRV GAFGFLALPG  Fbsache
Humanache     .......... ...NRELSED CLYLNIWTPY P......RPT SPTPVLVWIY GGGFYSGAS. ...SLDVY   DGRFLVQAER TVLVSMNYRV GAFGFLALPG  Humanache
Chickache     .......... ...NRELNMSED CLYLNVFTPK NA.....SSEFK NGRPVMVYIH GGGYELCASS DFCAYSL..  DGRYLAAAAE AVVVSMNYRV GSLGFLALAG  Chickache
Nematest      .......... ...PTPEEAG CLTLNVFTPR NA.....SSEFK ..PVLWFH   GGGYEIGSGS DFCAYSL..  ..SGTLPLKD VVVSINYRL  GVFGFLTTGD  Nematest
Nematest2     .......... ...YSGED.. CITLNVIKPK TI.....EKKL ..PVLWFH   NAAEGKGSGD QHGYEFF..  ..ADRYTSQG VIVTIQYRL  GFMGFFSEGT  Nematest2
Bovthyro      .......... ....VSED  CLTVSVYKP. NM......A  PNASVLVFFH GGAFMFGSG. RPAV....W. DGSFLAAVGN LIVVTASYRT GIFGFLSSGS  Bovthyro
Drosest6      .......... ...KLVGEBD CLTVSVXKP. .....KNSKR NSFPVVAHIH GGAFMFGSG. ....W..QN  GHENNMREGK FILVKISYRL GPLGFVS..T  Drosest6
Drosestp      .......... ...KLMGDED CLFLNVTP.  .....KKPNR SSFPVVLIH  GGAFMFGSG. ....S..IY  GHDSIMREGT LLIVKISYRL GPLGFAS..T  Drosestp
Aphideste4    .......... ...KIIGQBD CLFLNVTP.  KLFQENSAG  DLMNIVHHIH GGGYFPEG.  ....ILY    GPHYLLDNND FVYYSINYRL GVLGFAS..T  Aphideste4
Aphidestfe4   .......... ...KIIGQBD CIFANIYHW  QSLPRVRGTI DLMNIVHHIH GGGYFGBG.  ....ILY    GPHYLLDNND FVYYSINYRL GVLGFAS..T  Aphidestfe4
Heliojh       .......... ...NREMSEA CIFANLVTP. .....PIRPI LVFIH      GGGFAFGSL. ...HEDLH   GPEYLVTKNV IV.ITFNYRL NVFGFLS..M  Heliojh
Culexestbl    .......... ...KIVGCED SLKINV.... .....FAKEINPS TPLPVMLYI GGGFTEGTS. ...GTELY   GPDFIVQK.D IVLVSFNYRI GALGFLCCQS  Culexestbl
Glutactin     .......... ......VDD  CLYLDIYAPE G.........A NQLPVLVFVH GEMLFDGVGP E.........  AQPDYVLEKD VLLVSINYRL APFGFLSA..  Glutactin
Neurotactin   .......... ........ED CLYLDVTPH  V......RYN NPLPVVLIG  AESLAGPSPG ....ILR    PSARYSRSHD VIFVRPNFRL GVFGFLAL.D  Neurotactin
```

```
       201                                                                                                                    300
Geolip1      ...ITAEG NTNAGLHDQR KGLEWVSDNI ANFGGDPDKV MIFGESAGAM SVAHQLIAYG GDNTYNGKKL FHSAILQSGG PLPYHDSSSV GPDISYNRFA  Geolip1
Geolip2      ...ITAEG NTNAGLHDQR KGLEWVSDNI ANFGGDPDKV MIFGESAGAM SVAHQIVAYG GDNTYNGKQL FHSAILQSGG PLPYFDSTSV GPESAYSRFA  Geolip2
Cclip1       ...IKAEG SANAGLKDQR LGMQWVADNI AAFGGDPTKV TIFGESAGSM SVMCHLLWND GDNTYKGRPL FRAGIMQSGA MVPSDAVDGI YGNEIFDLLA  Cclip1
Cclip4       ...IKAEG SGNAGLHDQR LGLQWVADNI AGFGGDPSKV TIYGESAGAF STFVHLVWND GDNTYNGKPL FRAAIMQLGC MVPLDPVDGT YGTEINQVV   Cclip4
Cclip2       ...IQNEG SGNAGLHDQR LAMQWVADNI AGFGGDPSKV TIWGESAGAF SVAAHLTSTY SRQ......Y FNAAISSSSP LTVGLKDKTT ARGNANRFAT  D2
                                                            *
Dictyocrys   .....LL SGNFGFLDQI KALEWVYNNI GFLGGNKEMI TIWGESAGAF SVAAHLSEK SEG......K FHRALLSSTP YTVGLKSQTV ARGFAGRFSS  Dictyocrys
Balipase     .....ANL PGNYGLRDQH MALDWVQENI EVFGGDKNQV TIYGESAGAF SVSLQTLSPY NKG......L IRRAISQSGV ALSPWVIQKN PLFWAKKVAE  Balipase
Celipase     .....ANL PGNYGLRDQH MAIAWVKRNI AAFGGDPNNI TLFGESAGGA SVSLQTLSPY NKG......L IRRAISQSGV ALSPWVIQKN PLFWAKKVAE  Celipase
Humancel     .....ANL PGNYGLRDQH MAIAWVKRNI AAFGGDPNNI TLFGESAGGA SVSLQTLSPY NKG......L IRRAISQSGV ALSPWVIQEN PLFWAKTIAK  Humancel
Ratchest     .....ANL PGNFGLRDQH MAIAWVKRNI AAFGGDPNNI TLFGESAGGA SVSLQTLSPY NKG......L IRRAISQSGV ALSPWAIQEN PLFWATIAK   Ratchest
Ratsterest   .....ANL PGNFGLRDQH MAIAWVKRNI AAFGGDPDNI TLFGESAGGA SVSLQTLSPY NKG......L IRRAISQSGV ALSPWAIQEN PLFWARTIAK  Ratsterest
Ratcholest   .....ANL PGNFGLRDQH MAIAWVKRNI EAFGGDPDNI TLFGESAGGA SVSLQTLSPY NKG......L IKRAISQSGV GLCPWAIQQD PLFWAKRIAE  Ratcholest
Bovchole     .....SNL PGNYGLWDQH MAIAWVKRNI AAFGGDPDNI TLFGESAGGA SVSLQTLSPY NKG......L IRRAISQSGV ALSPWDIQKN PLFWAKKIAE  Bovchole
Rabbitchole  .....ANL PGNYGLRDFH MAIAWVKANI AAFGGDPDNI TIFGESAGGI SVSLVLVLSPL GKD......L FHRAISESGV VINTNVGKKN IQAVNEIIAT  Rabbitchole
Mousece      .....EHS RGNWAHLDQL ANFGGNPDSV TIFGESAGGV SVSALVLSPL AKN......L FHRAISESGV VLITNLDKKN TQAVAQMIAT  Mousece
Ratce        .....EHS RGNWAHLDQV AALRHWVQDNI ANFGGNPGSV TIFGESAGGF SVSALVLSPL AKN......L FHRAISESGV VLITSALITTD SKFIAKNIAT  Ratce
Rat61        .....EHS QGNWGHLDQV AALHWVQDNI ANFGGNPGSV TIFGESAGGV SVSALVLSPL AKN......L FHRAISESGV VLITSALITTD SKFIANLIAT  Rat61
Ratcoxys     .....EHS RGNWGHLDQV AALRWVQDNI ASFGGNPGSV TIFGESAGGE SVSVIVLSPL AKN......L FHRAISESGV ALTSVLVKKG DVKPLAEQIA  Ratcoxys
Humlivce     .....EHS RGNWGHLDQV AALRWVQDNI ASFGGNPGSV TIFGESAGGE SVSVIVLSPL AKN......L FHRAISESGV ALTSVLVKKG DVKPLAEQIA  Humlivce
Humcoxysa    .....EHS RGNWGHLDQV AALRWVQDNI ASFGGNPGSV TIFGESAGGE SVSVIVLSPL AKN......L FHRAISESGV ALTSVLVKKG DVKPLAEQIA  Humcoxysa
Humcoxysb    .....EHS GGNWGHLDQV AALRWVQDNI ASFGGNPGSV TIFGESAGGE SVSVIVLSPL AKN......L FHRAISESGV ALTSVLVKKG DVKPLAEQIA  Humcoxysb
Hummonmacss  .....EHS RGNWGHLDQV AALRWVQDNI ASFGGNPGSV TIFGESAGGE SVSVIVLSPL AKN......L FHRAISESGV ALTSVLVKKG DVKPLAEQIA  Hummonmacss
Humcoxys     .....EHS RGNWGHLDQV AALRHWVQDNI AKFGGDPGSV TIFGESAGGV SVSIILLSPL AKN......L FHRAISESGV ALTVALVRK. DMKAAAKQIA  Humcoxys
Pigcoxys     .....EHS RGNWGHLDQV AALQWIQENI IHFRGDPGSV TIFGESAGGQ SVSVIVLSPL AKN......L FHRAISESGV ALTVALVRK. NTRPLAEKIA  Pigcoxys
Est-22       .....E.. ....LFIV AVNRWVQDNI ANFGGDPGSV TIFGESAGGV SVSILLLSPL TKN......L FHRAISESGT ALLSSLFRK. NTKSLAEKIA  Est-22
Rabbitce     .....KHA RGNWGYLDQV AALQWIQENI TIFGESAGGV SVSALVLSPL AKG......L FHKAISESGT AVRILFTEQP EEQA..QRIA  Rabbitce
Anasthios    .....QHA TGNHGYLDQV AALRHWVQRNI AHFGGNPGRV TIFGNSAGGT SVSSHVLSPM SQG......L FHGAIMESLV ALLPGLITSS SEVV..STVA  Anasthios
Rablivest2   .....NVC PGNFGLWDQT LALKWVQDNI SSFGGDPNCV TVFGQSAGGS STDLLSLSPH SRD......L FHRAISESGS AWSE.AQNFA  Rablivest2
Mouselivcoxy .....SDA PGNYGLFPQA LALRFVKENI GNFGGDPDDI TIWGYSAGAA SVSQLTMSPY THD......L YSKAIIMSAS SFVGWA..TG PNVIDTSKQL  Mouselivcoxy
Anopha     YI.NGYEEDA PGNMGWWDQA LAIRWLKENA HAFGGNPEWM TLFGESAGSS SVNAQLMSFV TRG......L VKRGMMQSQT MNAPWSHMTS EKAVEIGRAL  Anopha
Drosache   EMPSEFAEEA PGNVGLWDQA LAIRWLKDNA HAFGGDPKTV TLFGESAGSS TLFGESAGSS SVNAQLMSFV TRG......L VKRGMMQSQT MNAPWSHMTS EKAVEIGRAL  Drosache
Humanbche  .....NPEA PGNMGLFDQQ LALQWVQKNI AAFGGNPKSV TLFGESAGAA SVSLHLLSPG SHS......L FTRAILQSGS FNAPWAVTSL YEARNRTLNL  Humanbche
Rabbitbche .....NPEA PGNMGLFDQQ LALQWVQRNI AAFGGNPKSV TLFGESAGAA SVSLHLLSPH SHP......L FTRAILQSGS SNAPWEVMSL HEARNRTLTL  Rabbitbche
Mousebche  .....NPDA PGNMGLFPQQ LALQWVQKNI AAFGGNPKSI TIFGESAGAA SVSLHLLCPQ SYP......L FTRAILESGS SNAPWAVKHP EEARNRTLTL  Mousebche
Tcalifache .....SQEA PGNVGLLDQR MALQWVHDNI QFFGGDPKTV TIFGESAGAA SVGRHLLSPG SRD......L FHRAVLQSGT PNCFWASVSV AEGRRAVEL  Tcalifache
Tmarmorache .....SREA PGNVGLLDQR MALQWVHDNI QFFGGDPKTV TIFGESAGAA SVGRHLLSPG SRD......L FHRAVLQSGT PNCFWASVSV AEGRRRAVEL  Tmarmorache
Mouseache  .....SREA PGNVGLLDQR LALQWVQDNI AAFGGDPMSV TLFGESAGAA SVGRHLLSPG SRS......L FHRAVLQSGT PNGPWATVGV GEARRRATL  Mouseache
Ratache    .....SREA PGNVGLLDQR LALQSVQRNI AAFGGDPTSV TLFGESAGAA SVGRHLLSPP SRG......L FHRAVLQSGA PNGPWATVGM GEARRRATL  Ratache
Fbsache    .....SREA PGNVGLLDQR LALQWVRDNA EAFGGDPLI SVGFHLLSPH SKD......L FHRAVLQSGS PNGPWATIGA AEGRRRAAAL  Fbsache
Humanache  .....HRDA PGNVGLWDQR LALQWVQDNI AAFGGDPTSV TLFGESAGAA SVGFHLLSPP SRG......L FHRAVLQSGA PNGPWATVGM GEARRRAAAL  Humanache
Chickache  .....HRDA PGNVGLWDQR LALQWVRDNA SSFGGDPNCV TVFGQSAGGS STDLLSLSPH SRD......L FORFIPISGT AHCDFAIRAS ENQAKIFREF  Chickache
Nematest   .....NVC PGNFGLWDQT LALKWVQDNI VAFGGDPNSV TIFGMSAGAS SVHNHLISPM SKG......L FNRAIIQSGS AFCHWS.TAE NV.AQKTYI   Nematest
Nematest2  .....SEL PGNYGLFPQA VALTVQTDNI LALKWIKQNI VLIGHSAGGA SVHLQWLRED FGQ......L ARAAFSFSGN RKGAISVSGN ALDPWVIQQG GRRRAFE..L  Nematest2
Bovthyro   G....DRDL PGNYGLKDQR LALKWIKONI AHFGGNPDNV VLIGHSAGGA SAHLQLLHED FKH..ARKGA FNRAIIQSGS AFCHWS.TAE NV.AQKTYI   Bovthyro
Drosest6   G....DGVL PGNVGLLDQR LALKWIQQNI VAFGGDPNSV TTGMSAGAS SVQYHLISDA SKD......L FKRAILMSGT GMSYFF.TTS PLPAAYISKQ  Drosestp
Drosestp   E....TTKI PGNAGLKDQN LALRWVLENI VAFGGDPKRV TLAGHSAGAA LAHLLSKA AGN......L FQRAVMSGS TYSSWSIIRQ RNWVEKLARA  Aphideste4
Aphideste4 E....QDGV PGNALSDLQ LAEWLQRNV VHFGRAGAT TLLGHRAGAT SVQYHLISDA AGN......L FQLILQSGT ALNPLIIDNQ PLDTLSTFAR  Aphidestfe4
Heliojh  ...LIDEL PGNVALSDLQ VHFGGDPGSV TLLGHRAGAT LVTLIVNSQK VKG......L YTRAWASGS AILPGKPLSE SGKQNEQLMA  Heliojh
Culexestbl  Culexestbl
Glutactin  Glutactin
Neurotactin  Neurotactin
```

```
                                                                                    400
Geolip1      QYA.GCDTSA ......SAND TLECLRSKSS SVLHDAQNSY DLKDLFGLLP Q...FlGFG P...RPDGNI IPDAAYELFR SGRYAKVY. ......        Geolip1
Geolip2      QYA.GCDASA .......GDNE TLACLRSKSS DVLHSAQNSY DLKDLFGLLP Q...FlGFG P...RPDGNI IPDAAYELYR SGRYAKVP. ......        Geolip2
Cclip1       SNA.GCGSAS .........D KLACLRGVSS DTLEDATN.. NFPGFLAYSS L...RLSYL P...RPDGVN ITDDMYALVR EGKYANIP  ......        Cclip1
Crlip4       ASA.GCGSAS .........D KLACLRSISN DKLFQATS.. DTPGALAYPS L...RLSFL P...RPDGTF ITDDMFKLVR DGKCANVP  ......        Crlip4
Cclip2       ASA.GCGSAS .........D KLACLRGLLQ DTLYQATS.. DTPGVLAYPS L...RLLYL P...RPDGTF ITDDMYALVR DGKYAHVP. ......        Cclip2
D2           NV..GCNIED ....... LTCLRGKSM DEIIDGPRKI ....GLTFGY KILDAFTIWS P...VIDGDI IPMQTLTDSK GRSKHMHF  ......        D2
Dictyocrys   KI..GCDLED I...... .DCHRSKSP EEIIAIQKEL ...GLAIGD KILDAFTIWS P...VVDGIN VNEQPLTMIK QGTTHDVP. ......        Dictyocrys
Balipase     KV..GCPVGD A........AR MAQCLKVTDP RAL......TL AYKVPLAGLE YPMLHYVGFV P...VIDGDF IPADPINLYA NAA...... .D       Balipase
Celipase     KV..GCPVGD A........AR MAQCLKVTDP RAL......TL AYKVPLAGLE YPMLHYVGFV P...VIDGDF IPADPINLYA NAA...... .D       Celipase
Humancel     KV..GCPVGD A........AR MAQCLKVTDF RAL......TL AYKVPLAGLE YPMLHYVGFV P...VIDGDF IPADPINLYA NAA...... .D       Humancel
Ratchest     KV..GCPTED T........AR MAGCLKITDP RAL......TL AYRLPLKSQE YPIVHYLAFI P...VVDGDF IPDDPINLYD NAA...... .D       Ratchest
Ratsterest   KV..GCPTED T........AK MAGCLKITDP RAL......TL AYRLPLKSQE YPIVHYLAFI P...VVDGDF IPDDPINLYD NAA...... .D       Ratsterest
Ratcholest   KV..GCPTED T........AK MAGCLKITDP RAL......TL AYRLPLKSQE YPIVHYLAFI P...VVDGDF IPDDPINLYD NAA...... .D       Ratcholest
Bovchole     KV..GCPVDD T........SK MAGCLKITDP RAL......TL AYKLPLGSTE YPKLHYLSFV P...VIDGDF IPDDPVNLYA NAA...... .D       Bovchole
Rabbitchole  KV..GCPLDY T........AT MAQCVKITDP HSL......TL AYNFPLAGLA YPMVHYLGFI P...VVDGDF LPEDPIILYG NAA...... .D       Rabbitchole
Mousece      LS..GCNDTS S........AA MVQCLRQKTE SELLEISGKL VQY....... ...NISLS T...MIDGVV LPKAPEILA EKSFNTVPY. ......        Mousece
Ratce        LS..GCNNTS S........AA MVQCLRQKTE AELLELTVKL D......... ..NTSMS T...VIDGVV LPKTPEEILT EKSFNTVPY. ......        Ratce
Rat61        LS..GCKTTT S........AV MVHCLRQKTE DELLETSLKL .NLFKLDLLG NPKESYPFLP T...VIDGVV LPKTPEEILA EKSFNTVPY. ......        Rat61
Ratcoxys     LS..GCKTTT S........AV MVHCLRQKTE DELLETSIKL .NLFKLDLLG NPKESYPFLP T...VIDGVV LPKTPEEILA EKSFNTVPY. ......        Ratcoxys
Humance      ITA.GCKTTT S........AV MVHCLRQKTE EELLETTLKM .KFLSLDLQG DPRESQPLLG T...VIDGML LLKTPEELQA ERNFHTVPY. ......        Humance
Humlivce     ITA.GCKTTT S........AV MVHCLRQKTE EELLETTLKM .KFLSLDLQG DPRESQPLLG T...VIDGML LLKTPEELQA ERNFHTVPY. ......        Humlivce
Humcoxysa    ITA.GCKTTT S........AV MVHCLRQKTE EELLETTLKM .KFLSLDLQG DPRESQPLLG T...VIDGML LLKTPEELQA ERNFHTVPY. ......        Humcoxysa
Humcoxysb    ITA.GCKTTT S........AV MVHCLRQKTE EELLETTLKM .KFLSLDLQG DPRESQPLLG T...VIDGML LLKTPEELQA ERNFHTVPY. ......        Humcoxysb
Hummonmacss  ITA.GCKTTT S........AA MVHCLRQKTE EELLETTLKI GNSYLWTYRE TQRES.TLLG T...VIDGML LLKTPEELQR ERNFHTVPY. ......        Hummonmacss
Humcoxys     ITA.GCKTTT S........AA FVHCLRQKSE DELLDILTKM .KFLTLDFHG DQRESHPFLP T...VVDGVL LPKMPEEILA EKDFNTVPY. ......        Humcoxys
Pigcoxys     VLA.GCKTTT S........AA MVHCLRQKTE DELLGTTLKL .NLFKLDLHG DSRQSHPFVP T...VLDGVL LPKMPEEILA EKNFNTVPY. ......        Pigcoxys
Est-22       VIS.GCKNTT S........AA MVHCLRQKTE EELMEVTLKM .KFMALDIVG DPKENTAFLT T...VIDGVL LPKAPAEIYE EKKYNMLPY. ......        Est-22
Rabbitce     IEA.GCEKSS S........AA LVECLREKTE AEMKQITLKM PPM....... .....FIS A...SLDGVF FPKSPRQLLS EKVINAVPY. ......        Rabbitce
Anasthios    AAA.GCEKSS S........ET LVRCLRAKSE EEMLAITQVF M......... ...LIP Y...VVDGVF LPRHPEEILA LADFQPVPS  ......        Anasthios
Rablivest2   NLS.RCGQVD S........AE LVQCLLQKEG KDLIT..... .SYSGILSFP SA..P...TIDGVF MTADPMTML. .REANL. EG                     Rablivest2
Mouselivcoxy NSV.ACGSAS P........AE LVQCLLQKEG QSPET..... .AVETCGWT TGTIDILRWS P...FPQRPQKLLA NKQFPTVPY.                  Mouselivcoxy
Anopha       IDDCNCNLIM L...KESPST VMQCMRNVDA KTISVQQWN. .SYSGILGFP SA..P...TIDGVF MTADPMTML. .REANL. EG                     Anopha
Drosache     INDCNCNASM L...KTNPAH VMSCMRSVDA KTISVQQWN. .SYSGILSFP .SA..P...TIDGAF LPADPMTIM. .KTADL. KD                    Drosache
Humanbche    AKLFGCSRE. ......NETE IIKCLRNKDA QBILLNEAFV VPYGTFLSVN FG..P...TVDGDF LTDMPDILL. .ELGGF. KK                     Humanbche
Rabbitbche   AKFVGCSTE. ......NETE IIKCLRNKDA QBILLNEVFV VPFDSLLSVN FG..P...TVDGDF LTDMPDTLL. .QLGQL. KK                     Rabbitbche
Mousebche    AKFTGCSKE. ......NEME MIKCLRSKDP QBILRNERFV LPSDSLSIN FG..P...TVDGDF LTDMPHTLL. .QLGKV. KK                      Mousebche
Tcalifache   GRNLNCNLNS .......DEE LIHCLREKKP QELIDVEWNV LPFDSIFRFS FV..P...VIDGEF FPTSLESML. .NSGNF. KK                    Tcalifache
Tmarmorache  GRNLNCNLNS .......DEE LIQCLREKKP QELIDVEWHV LPFDSIFRFS FV..P...VIDGEF FPTSLESML. .NAGNF. KK                   Tmarmorache
Mouseache    ARLVGCPPGG .AGGNDTE LIACLRTRPA QDLVDHEWHV LPQESIFRFS FV..P...VVDGDF LSDTPEALI. .NTGDF. QD                       Mouseache
Ratache      ARLVGCPPGG .AGGNDTE LISCLRTRPA QDLVDHEWHV LPQESIFRFS FV..P...VVDGDF LSDTPDALI. .NTGDF. QD                       Ratache
Fbsache      ARLVGCPPGG .AGGNDTE LVACLRARPA QDLVDHEWRV LPQEHVFRFS FV..P...VVDGDF LSDTPEALI. .NAGDF. VG                        Fbsache
Humanache    AHLVGCPPGG .TGGNDTE LVACLRTRPA QVLVNHEWHV MPPGSVFRFS FV..P...VVDGDF LSDTPEALI. .NAGDF. HG                       Humanache
Chickache    GRAVGCPYGG .NETE FLGCLRGKRA ADVLEGEGVV KSISGLTFI P....NLDGDF FPKFLDELRK EAPKK.                                   Chickache
Nematest     AEBIGCPWPG SSA...... .LFKWYQE QSPET..... .AVETCGWT TGTIDILRWS P...VIDGQY LPKNPENLIN DAPIK....                 Nematest
Nematest2    AKEVGCPSSS AKE...... .CMKKKTL HEIFD..... .PFHYWG Y..VVDGQY LRETPARVLQ RAP......                                 Nematest2
Bovthyro     GRNVGCE... .SAEDSTS LKKCLKSKPA SELVTAVRKF LIFSYVPFAP FSPVLEPSDA P....DAI ITDDPRIVIK SGKFGQVPW RVK               Bovthyro
Drosest6     GRIVGCG... .HTNVSAE LKDCLKSKPA SDLVSAVRGF LVFSYVPFSA FGLIVEPSDA .DAF LTEDPRAVIK SGKFAQVPW                       Drosest6
Drosestp     ANLMGCP... .TNNSVE IVECLRSRPA KAIAKSYLNF MFWNRPFPTP FGPTVE..VA G....Y..EKF LPDIPEKLVP H...DIPV.                Drosestp
Aphidestfe4  ANLLGCP... .TNNSVE IVECLRSRPA KAIAKSYLNF MFWRNFPFTP FGPTVE..VA G....Y..EKF LPDIPEKLVP H...DIPV.                Aphidestfe4
Aphidestfe4  LLQILGNQ.. ...RDGSEE IHRQLLDPA EKLNENANVL IE..QIGLTT FLPIVESPLP G....V..TTI IDDDPELLIA EGRGKNVPL.              Aphidestfe4
Heliojh      IGWDGQGG.. .ESGALR FLRAAKPEDI VAHQEKLLTD QDMQDDIFTP FGPFVEPYLH D....QCI IPKAPFEMAR TAWGDKIDI.                  Heliojh
Culexestbl   LAR..CPPFS INPSAQGLKPL YDCLARLFT SQLVAAFEQL LLQNEHLGLI QLGGFKLVVG D...PLG..F LPSHPASLAT NSSLA..... QD         Culexestbl
Glutactin    ........ .QCLREASS ERLWAATFDT WLHFVVDLPQ PQEANASGSR HEWLVLDGDV VFEHPSDTWK REQANDKP. VG                           Glutactin
Neurotactin  ..TLECADI. ......                                                                                             Neurotactin
301
```

```
401                                                                                                                                                          500
Geolip1      ..TISGNQED EGTAFAPVAL ..........N ATTTPHVKKW LQYIFYDASE ASIDRVLSLY PQTLSVGSPF RT....GILN ALTPQFKRVA AILSDMLFQS  Geolip1
Geolip2      ..YITGNQED EGTILAPVAI ..........N ATTTPHVKKW LKYICSEASD ASLDRVLSLY PGSWSEGAPF RT....GILN ALTPQFKRIA AIFTDLLFQS  Geolip2
Cclip4       ..VIIGDQND EGTFFGTSSL ..........N VTTDAQAREY FKQSFVHASD AEIDTLMTAY PGDITQGSPF DT....GILN ALTPQFKRIS AVLGDLGFTL  Cclip1
Crlip1       ..VIIGDQND EGTLFGLLLL ..........N VTTDAQARQY FKESFIHASD AEIDTLMAAY PSDITQGSPF DT....GIFN ALTPQFKRIA AVLGDLAFTL  Crlip4
Cclip2       ..VIIGDQND EGTLFGLLLL ..........N VTTDAQARAY FKQLFIHASD AEIDTLMAAY TSDITQGLPF DT....GIFN ALTPQFKRL  ALLGDLAFTL  Cclip2
D2           ..QHYWKCKH EAIPFI..... ..........    ..YSFSKIV VGIDYYRVLV AIVFPL.NBM KILPLYPRAP RGQDSRPILS ELITDYLFRC  D2
Dictyocrys   ..TIIGDNQD EALLFV..... ..........    ..IMTYKNV VIFSSYRTMV HVLFGIANGN KVLEHYPLPG FLKDSRPILS KLITDYLFRC  Dictyocrys
Balipase     IDYIAGTNNM DGHIFASIDM PA..INKGNK KVTEEDFYKL VSEFTITKGL RGAKTTFDVY TESWAQDPSQ ENKK.....  ..KTVV DFETDVLFLV  Balipase
Celipase     IDYIAGTNNM DGHIFASIDM PA..INKGNK KVTEEDFYKL VSEFTITKGL RGAKTTFDVY TESWAQDPSQ ENKK.....  ..KTVV DFETDVLFIV  Celipase
Humancel     IDYIAGTNNM DGHIFASIDM PA..INKGNK KVTEEDFYKL VSEFTITKGL RGAKTTFDVY TESWAQDPSQ ENKK.....  ..KTVV DFETDVLFIV  Humancel
Ratchest     IDYLAGINDM DGHLFATVDV PA..IDKAKQ DVTEEDFYRL VSGHTVAKGL KGTQAIFDIY TESWAQDPSQ ENMK.....  ..KTVV AFETDILFLI  Ratchest
Ratsterest   IDYLAGINDM DGHLFATVDV PA..IDKAKO DVTEEDFYRL VSGHTVAKGL KGTQAIFDIY TCAWAQDPSQ ENMK.....  ..KTVV AFETDILFLI  Ratsterest
Ratcholest   IDYLAGINDM DGHLFATVDV PA..IDKAKO DVTEEDFYRL VSGHTVAKGL KGTQAIFDIY TESWAQDPSQ ENMK.....  ..KTMV DLETDILFLI  Ratcholest
Bovchole     VDYIAGTNDM DGHLFVGMDV PA..INSNKQ DVTEEDFYKL VSGLTVTKGL RGANATYEVY TEPWAQDPSQ ETRK.....  ..KTMV DLETDILFLI  Bovchole
Rabbitchole  IDYLAGTNDM DGHLFATVDM PA..IDKSYK DISDQDFYKL VSGMTVTKGS EGAQATYSIY TESWAQDSSQ QNKK.....  ..KT.V DLETDILFLI  Rabbitchole
Mousece      ..IVGFNKQ EFGWIIPM.M LQNLLPEGRM NEETASLLLR RFHSEL...N ISESMIPAVI ISESMIPAVI EQYLRGVDDP AKKS.....  .ELLL DMFGDIFFGI  Mousece
Ratce        ..IVGINKQ EFGWIIPT.M MGNLLSEGRM NERMASSFLK RFSFNL...N ISESVIPAI  ISESVIPAI  EKYLRGTDDP AKRK.....  .ELLL DMFSDVFFGI  Ratce
Rat61        ..IVGINKQ EFGWIIPT.L MGYPLSEGKL DQKTAKSLLW KSYPTL...K ISEKMIPVVA ISEKMIPVVA EKYFGGTDDP AKRK.....  .DLFQ DLVADVIFGV  Rat61
Ratcoxys     ..IVGINKQ EFGWIIPT.L MGYPLSEGKL DQKTAKSLLW KSYPTL...K ISEKMIPVVA ISEKMIPVVA EKYFGGTDDP AKRK.....  .DLFQ DLVADVMFGV  Ratcoxys
Humance      ..MVGINKQ EFGWLIPMQL MSYPLSEGQL DQKTAMSLLW KSYPLV...C IAKELIPEAT IAKELIPEAT EKYLGGTDDT VKKK.....  .DLFL DLIADVMFGV  Humance
Humlivce     ..MVGINKQ EFGWLIPMQL MSYPLSEGQL DQKTAMSLLW KSYPLV...C IAKELIPEAT IAKELIPEAT EKYLGGTDDT VKKK.....  .DLFL DLIADVMFGV  Humlivce
Humcoxysa    ..MVGINKQ EFGWLIPM.L MSYPLSEGQL DQKTAMSLLW KSYPLV...C IAKELIPEAT IAKELIPEAT EKYLGGTDDT VKKK.....  .DLFL DLIADVMFGV  Humcoxysa
Humcoxysb    ..MVGINKQ EFGWLIPM.L MSYPLSEGQL DQKTAMSLLW KSYPLV...C IAKELIPEAT IAKELIPEAT EKYLGGTDDT VKKK.....  .DLFL DLIADVMFGV  Humcoxysb
Hummonmacss  ..MVGINKQ EFGWLIPM.L MSYPLSEGQL DQKTAMSLLW KSYPLV...C IAKELIPEAT IAKELIPEAT EKYLGGTDDT VKKK.....  .DLIL DLIADVMFGV  Hummonmacss
Humcoxys     ..MVGINKQ EFGWLIPMQL MSYPLSEGQL DQKTAMSLLG SPIPLF..A  IAKELIPEAT IAKELIPEAT EKYLGGTDDT VKKK.....  .DLIL DLMGDVVFGV  Humcoxys
Pigcoxys     ..IVGINKQ EFGWLIPT.M MGFPLSEGKL DQKTATSLLW KSSFLI...N IPEELIPVAT IPEELIPVAT DKYLGGTDDP VKKK.....  .DQLL ELIGDVVFGV  Pigcoxys
Est-22       ..IVGINKQ EFGWLIPT.M MNYPFSDVKL DQKTAMSLIK KSSFLL...N IPEDIAVAI  IPEDIAVAI  EKYLRDKDYT GRNK.....  .DLFL DLIADLLFGV  Est-22
Rabbitce     ..MVGINQQ EFGWIIPMQM LGYPLSEGKL DQKTADELIW KSYPIV...N VSKELTPVAT NEYIGVAENR AQVR.....  .DGLL DMLADLLFGV  Rabbitce
Anasthios    ..IIGINND EYGWIIPKLL LAIDPQEER. DRQAMREIMH QATKQL.M..  LPPALGDLLM DEYMGSNEDP KHLM.....  .AQFQ EMMADAMFVM  Anasthios
Rablivest2   ..LLGVTNN EFGWL....L LKFWNILDKM EHLSQEDLLE NSRPLLAHMQ LPPELIMPTVI DEYLDNGSDE SATR.....  .YALQ ELIGDTTIVI  Rablivest2
Mouselivcoxy IDILVGSNRD EGTYFLLYDF IDYFEKDAAT SLPRDKFLEI MNTIFNKASE PEREALIFQY TGWESG.NPG YQNQ.....  .HQVG RAVGDHFFIC  Mouselivcoxy
Anopha       IDILMGNVRD EGTYFLLYDF IDYFDKDDAT ALPRDKYLBI MNNIFGKATQ AEREALIFQY TSWE.G.NPG YQNQ.....  .QQIG RAVGDHFFTC  Anopha
Drosache     YQILVGVNKD EGTAFLVVGA PG.FSKDNNS IITRKEFQEG LKIFFPGVSE FGKESILFHY TDWVDD.QRP ENVR.....  .EALG DVVGDYNFIC  Drosache
Humanbche    TQILVGVNKD EGTAFLVVGA PG.FSKDNTS IITRKEFQEG LKIFFPGVSE FGKESILFHY TDWVDE.QRP ENYR.....  .EALD DVVGDYNFIC  Humanbche
Rabbitbche   AQILVGVNKD EGSFFLLYGA PG.FSKDNDS LITRKEPQEG LNNYFPGVSR LGREAVLFIY VDWLGE.QSP EVYR.....  .DALD DVIGDYNIIC  Rabbitbche
Mousebche    TQILLGVNKD EGSFFILYGA PG.FSKDSES KISREDFMSG VKLSVPHAND LGLDAVTLQY TDWMDD.HNG IKNR.....  .DGLD DIVGDHNVIC  Mousebche
Tcalifache   LQVLVGVVKD EGSYFIVYGV PG.FSKDNES LISRAQFLAG VRIGVPQASD LAAEAVVLHY TDWLHP.EDP THLR.....  .DAMS AVVGDHNVVC  Tcalifache
Tmarmorache  LQVLVGVVKD EGSYFIVYGV PG.FSKDNES LISRAQFLAG VRIGVPQASD LAAEAVVLHY TDWLHP.EDP AHLR.....  .DAMS AVVGDHNVVC  Tmarmorache
Mouseache    LQVLVGVVKD EGSYFIVYGA PG.FSKDNES LISRAQFLAG VRVGVPQVSD LAAEAVVLHY TDWLHP.EDP ARWR.....  .EALS DVVGDHNVVC  Mouseache
Rataache     LQVLVGVVKD EGSYFLVYGV PG.FSKDNES LISRAFFLAG VRVGVPQVSD LAAEAVVLHY TDWLHP.EDP ARLR.....  .EALS DVVGDHNVVC  Rataache
Fbsache      VEVLLGAVRV EGSYFLVVGA PG.FSKDNES LISREEFLGG VRMGVPQATR LAAEAVVLHY TDWLDA.DNP VKNR.....  .EALD DIVGDHNVVC  Fbsache
Humanache    .QMMTGVTEY EGIMLLASNNP AFSPADVGLT LMPQGIYGKD VVSNPDEIQK IFYEKVBGV  DKSDELAMR.                       .KKLC EALGDEFFNV  Humanache
Chickache    .PTLIGMSNK EGSYFATMNM GRVIADFGLS PEEIPKVDED FISEIIDRKL LYNNRYGENR QKVWDQILDY YTKQGKPERD LNGFYVDRYA ELLSDITFNV  Chickache
Nematest     VDLLIGSSQD KQFEESQGRT SSKTAFYQAL QNSLGEAAD  AGVQAAATWY YSLEHDSDY  ASFS.....  .RALE QATRDYFIIC  Nematest
Nematest2    ...AVSVTE DGGYNAALLL KERK..SGIV IDLLNDEWLD LAPYLFLFYD AKKT.......                                 .RLFTDILFKN  Nematest2
Bovthyro     ...AVTVTTE DGGYNAAQLL ERNKLTGESW FSTFLGLENG FNELNNNWNE HLPHILDYNY TISN.......                       .RMFTDVLFKN  Bovthyro
Drosestp     ..LISIAQD EG.....LI FSTFLGLENG FNELNNNWNE HLPHILDYNY TISN.......                                   .KMISDRSFGY  Drosestp
Drosest6     ..LIGFTSS EC.....LI FTFNRLLNFDL FNELNNNWNE HLPHILDYNY TISN.......                                   .KMISDRSFGY  Drosest6
Aphideste4   ..MIGGTSE EG.....LL LLQKIKLHPE LLSHPHLDYN IIPPKLLF..  MTP.......                                   .ISID...NFV KSCSDGFYEY  Aphideste4
Aphideste4   LPMIIGATKD DQLARLQSRN TAHEAHTEKL RELHANWTRE EVRAYLENSQ IGALGLTDEV IEKYNASSYA SLVSIISDIR SVCPLLTNAR QQPSVFYVV   Aphideste4
Heliojh      ..LISIAQD EG.....LI FSTFIGLENG FNELNNNWNE HLPHILDYNY TISN.......                                   .KMISDRSFGY  Heliojh
Culexestbl   ..MIGGTSE EG.....LL LLQKIKLHPE LLSHPHLDYN IIPPKLLF..                                               .NFV KSCSDGFYEY  Culexestbl
Glutactin    LPMIIGATKD ASAFIVSRIY DQLARLQSRN TAHEAHTEKL RELHANWTRE EVRAYLENSQ IGALGLTDEV                       PGLL ELSNYILYRA  Glutactin
Neurotactin  ..VLVMGA                                                                                                            Neurotactin
```

```
        501                                        *                                                                          600
Geolip1      PRRVMLSATK DVNRWTYLST HLH...... ......NIV PFLGTFHGNE LIFQFHVNIG P......... .........A NSYLRYFISF ANHHDPNVGT   Geolip1
Geolip2      PRRVMLNATK DVNRWTYLAT QLH...... ......NIV PFLGTFHGSD LLFQYYAGFW S......... .........S SAYRRYFISF ANHHDPNVGT   Geolip2
Cclip1       ARRYFLNHYT GGTKYSFLSK QLS...... .......GL. PVLGTFHSND IVPQDYLLGS S......... .........S LIYNNAFIAF ATDLDPNTAG   Cclip1
Cclip4       PRRYFLNHFQ GGTKYSFLSK QLS...... .......GL. PVLGTHHAND IVWQDFLVSH S......... .........S AVYNNAFIAF ANDLDPNKAG   Cclip4
Cclip2       ARRYFLNYYQ GGTKYSFLLK QLIL..... .......GL. PVLGTFHGND IIWQDYIVGS ......... .........S VIYNNAFIAF ANDLDPNKAG   Cclip2
D2           PDRYHTVTNA KKLSSPTYHY HYVHVKSTGH SLDACDDK.. ...VCHGTE LSLFFNS... ...YE.IMGE IDINNYIVNL QLLINFN..T                D2
Dictyocrys   PGRYHVSKSA QANESPIYHY QYKQVLSGGH SFEACEGL.. ...VCHGTE LPMVFNT... .YESALDL DLEEEEEFA EQLANNYVNF IKYSNPSHPN   Dictyocrys
Balipase     PTEIALAQHR ANA.KSAKTY AYLFSHPSRM PVY.....P KWVGADHADD IQVVFGKPFA TPTG...... .YRPQDRTVS KAMIAYWTNF AKTGDPNMGD   Balipase
Celipase     PTEIALAQHR ANA.KSAKTY AYLFSHPSRM PVY.....P KWVGADHADD IQVVFGKPFA TPTG...... .YRPQDRTVS KAMIAYWTNF AKTGDPNMGD   Celipase
Humancel     PTEIALAQHR ANA.KSAKTY AYLFSHPSRM PIY.....P KWMGADHADD IQVVFGKPFA TPLG...... .YRAQDRTVS KAMIAYWTNF AKSGDPNMGN   Humancel
Ratchest     PTETALAQHR AHA.KSAKTY SYLFSHPSRM PIY.....P KWMGADHADD IQVVFGKPFA TPLG...... .YRAQDRTVS KAMIAYWTNF AKSGDPNMGN   Ratchest
Ratsterest   PTEMALAQHR AHA.KSAKTY SYLFSHPSRM PIY.....P KWMGADHADD LQVVFGKPFA TPLG...... .YRAQDRTVS KAMIAYWTNF AKSGDPNMGN   Ratsterest
Ratcholest   PTEMALAQHR AHA.KSANTY SYLFSHPSRM PIY.....P KWMGADHADD LQVVFGKPFA TPLG...... .YRAQDRTVS KAMIAYWTNF AKSGDPNMGN   Ratcholest
Bovchole     PTKIAVAQHR SHA.KSANTY TYLFSQPSRM PIY.....P KWMGADHADD LHDIFGKPFA TPTG...... .YRAQDRTVS KAMIAYWTNF ARTGDPNTGH   Bovchole
Rabbitchole  PTEIALAQHR ANS.STAKTY AYLFSHPSRM PIY.....P SWMGADHADD IFFVFGAPLL .KEG...... .ASEEETNLS KTLIAYRTNF ARTGDPNTGF   Rabbitchole
Mousece      PAVLLSRSLR DAG.VSTYMY EFRYRPSFVS DKR.....P QTVEGDHGDE IFFVFGAPLL .KEG...... .ASEEETNLS KMVMKFWANF ARNGNPNGE.   Mousece
Ratce        PAVLMSRSLR DAG.APTYMY EFQYRPSFVS DQR.....P QTVQGDHGDE IFSVFGTPFL .KEG...... .ASEEETNLS KLVMKFWANF ARNGSPNGG.   Ratce
Rat61        PSVMVSRSHR DAG.APTFMY EFEYRPSFVS AMR.....F KTVIGDHGDE LFSVFGSPFL .KDG...... .ASEEETNLS KMVMKFWANF ARNGNPNGG.   Rat61
Ratcoxys     PSVMVSRSHR DAG.APTFMY EFEYRPSFVS DMK.....P KTVIGDHGDE LFSVFGSPFL .KDG...... .ASEEEIRLS KMVMKFWANF ARNGNPNGE.   Ratcoxys
Humance      PSVIVARNHR DAG.APTYMY EFQYRPSFSS DMK.....P KTVIGDHGDE LFSVFGAPEL .KEG...... .ASEEEIRLS KMVMKFWANF ARNGNPNGE.   Humance
Humlivce     PSVIVARNHR DAG.APTYMY EFQYRPSFSS DMK.....P KTVIGDHGDE LFSVFGAPEL .KEG...... .ASEEEIRLS KMVMKFWANF ARNGNPNGE.   Humlivce
Humcoxysa    PSVIVARNHR DAG.APTYMY EFQYRPSFSS DMK.....P KTVIGDHGDE LFSVFGAPEL .KEG...... .ASEEEIRLS KMVMKFWANF ARNGNPNGE.   Humcoxysa
Humcoxysb    PSVIVARNHR DAG.APTYMY EFQYRPSFSS DMK.....P KTVIGDHGDE LFSVFGAPEL .KEG...... .ASEEEIRLS KMVMKFWANF ARNGNPNGE.   Humcoxysb
Hummonmacss  PSVIVARNHR DAG.APTYMY EFQYRPSFSS DMK.....P KTVIGDHGDE LFSVFGAPEL .KEG...... .ASEEEIRLS KMVMKFWANF ARNGNPNGE.   Hummonmacss
Humcoxys     PSVIVARNHR DAG.APTYMY EFQYRPSFSS DMK.....P KTVIGDHGDE IFSVFGFPLL .KEG...... .ASEEEIRLS KMVMKFWANF ARNGNPNGK.   Humcoxys
Pigcoxys     PSVTVAROHR DAG.APTYMY EFQYRPSFSS DKK.....P KTVIGDHGDE IFSVFGFPLL .KGD...... .APEEVSLS KTVMKFWANF ARSGNPNGK.   Pigcoxys
Est-22       PSVIVSRGHR DAG.APTYMY EFQYSPSFSS EMK.....P DTVVGDHGDE IYSVFGAPLL .RGG...... .TSEEEINLS KMMMKFWANF ARNGNPNGQ.   Est-22
Rabbitce     PSVTVARNHR DAG.APTYMY EYRYRPSFVS DMR.....P KTVIGDHGDE IFSVLGAPFL .KEG...... .ATEEEIKLS KHVMKWWANF ARNGNPNGE.   Rabbitce
Anasthios    SAVEVARHHR DAG.NEVYFY EFQHRPSSAA GVV.....P EFVKADHADE IAFVFGKPFL .AGN...... .ATEEEAKLS RTVMKWTNF ARNGNPNGE.   Anasthios
Rablivest2   PALRVAHLQR .SH.APTYFY EFQHRPSFTK DLR.....P PHVRADHGDE VVFVFRSHLF GSKVP..... .LTEEEELLS RVMKYWANF ARNRNPNGK.   Rablivest2
Mouselivcoxy PTLIFSKYLQ DAG.CFVFLY EFOHTPSSFA KFK.....P AWVKADHSSE VEYIFGQPMN TDESSLLAFP EATEEEKQLS LTMVAAQWSQ ARTGNPNGK.   Mouselivcoxy
Anopha       PTNEFALGLT ERG.ASVHYY YFTHRTS..T SLW.....G EWMGVLHGDE VEYIFGQPMN NSLQ...... .YRPVERELG KRMLSAVIEF AKTGNP....   Anopha
Drosache     PTNEYAQALA EWG.ASVHYY YFEHRS...K LPW.....P EWMGVMHGYE IEFVFGLPLE RRDN...... .YTKAEEILS RSIVKRWANF AKYGNPNETQ   Drosache
Humanbche    PALEFTKKFS EWG.NNAFFY YFEHRS...K LPW.....P EWMGVMHGYE IEFVFGLPLG RRVN...... .YTKAEEILS RSIMKRWANF AKYGHPNGTQ   Humanbche
Rabbitbche   PALEFTKKFS EWG.NNAFFY FFEHRSS..K LVW.....P EWMGVMHGYE IEFVFGLPLG RRVN...... .YTRAEEIFS RSIMKTWANF AKYGHPNGTQ   Rabbitbche
Mousebche    PALEFTKKFA KFG.NGTYLY FFNHRAS..N LVW.....P EWMGVIHGYE IEFVFGLPLV KELN...... .YTAEEEALS RRIMHWATF AKTGNPNEPH   Mousebche
Tcalifache   PLMHFVNKYT KFG.NGTYLY FFNHRAS..N LVW.....P EWMGVIHGYE IEFVFGLPLV KELN...... .YTAEEEALS RRIMHWATF AKTGNPNEPH   Tcalifache
Tmarmorache  PLMHFVNKYT KFG.NGTYLY IFEHRAS..T LTW.....P EWMGVIHGYE IEFIFGLPLD PSLN...... .YTTEERIFA QRLMKWTNF ARTGDPNDPR   Tmarmorache
Mouseache    PVAQLAGRIA AQG.ARVYAY IFEHRAS..T LTW.....P LWMGVPHGYE IEFIFGLPLD PSLN...... .YTTEERIFA QRLAKQWTNF ARTGDPNDPR   Mouseache
Ratache      PVAQLAGRIA AQG.ARVYAY IFEHRAS..T LTW.....P LWMGVPHGYE IEFIFGLPLD PSLN...... .YTVEERIFA QRLMQWTNF ARTGDPNDPR   Ratache
Fbsache      PVAQLAGRIA AQG.ARVYAY IFEHRAS..T LSW.....P LWMGVPHGYE IEFIFGLPLE PSLN...... .YTIEERTFA QRLMRYWANF ARTGDPNDPR   Fbsache
Humanache    PVAQLAGRIA AQG.GRVYAY VFEHRAS..T LSW.....P LWMGVPHGYE IEFIFGLPLD PSRN...... .YTAEEKIFA QRLMRYWANF ARTGDPNEPR   Humanache
Chickache    PLMAFAQRWA QRG.GRVYAY LFDHRRS..T LLW.....P SWMGVPHGYE IEFVFGGPEL PRNN...... .YTRMREVELS RRIMRYWGNF ARTGDPNGGV   Chickache
Nematest     GVIQAAKNAA KHG.NEVYFY TFFEVVNFDSF GWWDGM...M PFKAAVHCTE LRYLLGEGVY SKFP.P.... .TEEDRKVM ETTTTLFSNF AKYGNPNGKG   Nematest
Nematest2    PILREITARV ERK.TPVWTY RFDHYNEQIW KKYIPE...Q A.KGSPHANE VYLFNMPVM AQIDFK.... .KEPESWLQ RDLIDMVVSF AKTGVPHIQD   Nematest2
Bovthyro     PVIDMA.SHW ART.VRGNVF MYHAPESYSH S........ .SLELLTD VLYAFGLPFY PAYEG..... .QFTLEEKSLS LKIMQYFSNF IRSGNPNYPH   Bovthyro
Drosest6     STQESLDLHR KYGKSPVYSF VYDNPTDSGV GRLLSNRTDV HF.GTVHGDD YFLIFENFVR DV...EMRP DEEVISKKFI GMLED.FALN DKGTLTFGEC   Drosest6
Aphideste4   GTSKAAQHIA AKNTAPVYFY EFGYSGNYSY VAFFDPKSYS RGSSPTHGDE TNYVLKV... .D....GFTV YDNEEDRKMI KTMVNIWATF IKSGVPDTEN   Aphideste4
Aphidestfe4  GTSKAAQHIA AKNADPALFY EFGYSGNYSY VAFFDEKSYS RGSSPTHGDE LTYVFKVNSM SE....ALHA SPSENDVKMK NIMTGYFLNF IKCSQPTCED   Aphidestfe4
Heliojh      PALKLAQKHA ARSRARTFVY RICLDSEF.Y NHYRIMMIDP KLRGTAHADS .EEGVHHGDE LSYLFS..NF TQ...QVPG KETFEYRGL. VINGDPNCGM   Heliojh
Culexestbl   GLHRTILARA SYRSVPAYLY TFDYRGEHHR GRYEPHTVEQ FGVDASLSDD SVTLFPYPPE ASR....... .LNPLDRSLS QTLVDVFSAF ATTGVRNPSS   Culexestbl
Glutactin    PYINSISQ..                                                                              RALVTMWVNF ATTGVNFPSS   Glutactin
Neurotactin  TQGEGPDQIA TVDADVQAIL GRYEPHTVEQ RRFVSAMQQL FYYYVSHGTV QS                                                      Neurotactin
```

601
```
Geolip1      NLLQWDQY.. T....DEGKEM LEIHMTDNVM RTDDYRIEGI SNFETDVNLY G.....
Geolip2      NLKQWDMY.. T..DSGKEM LQIHMIGNSM RTDDFRIEGI SNFESDVTLF G.....
Cclip1       LLVKWPEY.. TSSSQSGNNL MMINALGLYT GKDNFRTAGY DALFSNPPSF FV....
Crlip4       LLVNWPKY.. TSSSQSGNNL LQINALGLYT GKDNFRTAGY DALFTNPSF  FV....
Cclip2       LWTNWPFY.. TSSLQIGNNL MQINGLGLYT GKDNFRPDAY SALFSNPPLF FV....LT  YYRNQVRPU.
D2           GLDVPVQWRQ ..VTCTQNS  TLILETTIET KVSFTNDPKC NALD....LT  YYRNQVRPU.
Dictyocrys   GLPFTPKVWNP ..TTKTNT  SLVMKLGFEV KDLITNDPKC DLFDSLSYNG  YTKDQNRMRK  SKK.....
Balipase     SAVP..TH.. WEPYTTEN  SGYLEITKKM GSSSMKRSLR TNFIR....Y  WTLTYLALPT VTDQEATPVP PTGDSEATPV VPPTGDSGAP
Celipase     SAVP..TH.. WEPYTTEN  SGYLEITKKM GSSSMKRSLR TNFLR....Y  WTLTYLALPT VTDQEATPVP PTGDSEATPV VPPTGDSGAP
Humancel     SAVP..TH.. WEPYTTEN  SGYLEITKKM GSSSMKRSLR TNFLR....Y  WTLTYLALPT VTDQEATPVP PTGDSEATPV VPPTGDSGAP
Ratchest     SPVP..TH.. WYPYTMEN  GNYLDINKKI TSTSMKEHLR EKFLK....F  WAVTFEMLPT VVGDHT...P PEDDSEAAPV PPFTDDSQTT VPPTDDSQTT
Ratsterest   SPVP..TH.. WYPYTMEN  GNYLDINKKI TSTSMKEHLR EKFLK....F  WAVTFEMLPT VVGDHT...P PEDDSEAAPV PPFTDDSQVP VPPTDDSQTT
Ratcholest   SPVP..TH.. WYPYTMEN  GNYLDINKKI TSTSMKEHLR EKFLK....F  WAVTFEMLPT VVGDHT...P PEDDSEAAPV PPFTDDSQTT VPPTDDSQTT
Bovchole     STVP..AN.. WDPYTLED  GNYLEINKQM DSNSMKLHLR TNYLQ....F  WTQTYQALPT VTSAGASLLP PEDNSQASPV PPADNSGAPT EPSAGDSEVA
Rabbitchole  RKCP..PL.. .GALHPGER  Q..LPGDQQE DESGLHEVPP EKQLPAASCS SGPDLQALPV VLED..PETDP  PTDDAEPSPG ARDDS....  TPPADDSVAA
Mousece      ..GL..PH.. WPEYDEQE  G.YLQIGATT .........QQA QKLKAEEVAF WTELLAKN...  ....PPQTEHT  EHTEHK.
Ratce        ..GL..PH.. WPKYDQKE  G.YLQIGATT .........QQA QKLKGEEVAF WTELLAKN...  ....PPQTEHT  EHT.
Rat61        ..GL..PH.. WPEYDQRE  G.YLKIGAST .........QAA QRLKDKEVAF WSELRAKEAA E.EPSHWKHV  EL.
Ratcoxys     ..GL..PH.. WPEYDQKE  G.YLKIGAST .........QAA QRLKDKEVAF WSELRAKEAA E.EPSHWKHV  EL.
Humance      ..GL..PH.. WPEYNQKE  G.YLQIGANT .........QAA QKLKDKEVAF WTNLFAKKAV E.KPPQTEHI  EL.
Humlivce     ..GL..PH.. WPEYNOKE  G.YLQIGANT .........QAG QKLKDKEVAF WTNLFAKKAV E.KPPQTEHI  EL.
Humcoxysa    ..GL..PH.. WPEYNQKE  G.YLQIGANT .........QAG QKLKDKEVAF WTNLFAKKAV E.KPPQTEHI  EL.
Humcoxysb    ..GL..PH.. WPEYNQKE  G.YLQIGANT .........QAG QKLKDKEVAF WTNLFAKKAV E.KPPQTEHI  EL.
Hummonmacss  ..GL..PH.. WPEYNOKE  G.YLQIGANT .........QAA QKLKDKEVAF WTNLFAKKAV E.KPPQTEHI  EL.
Humcoxys     ..GL..PH.. WPEYNQKE  G.YLQIGANT .........QAA KRLKGEEVAF WNDLLSKEAA K.KPPKIKHA  EL.
Pigcoxys     ..GL..PH.. WPMYDOEE  G.YLQIGVNT .........QQA QKLKEEKEVAF WTELLAKKQL  .....PTEHT  EL.
Est-22       ..GL..PH.. WPEYDQKE  G.YLQIGATT .........QQA QKLKEKEVAF WTELWAKEAA ..RPRETEHI
Rabbitce     ..GL..PH.. WPAIDVKE  G.YLQIGVNT .........QAA KKLKERKMEF WMGL..TEQI M.SDRRRKHT  DI.
Anasthios    ..GL..VH.. WPQIDMDE  R.YLEIDLTQ .........KAA QALKARRLQF WHTILPQRVQ ELRGTEQKHT  EL.
Rablivest2   ..GL..AH.. WPLFDLDQ  R.YLQLNMQP .........AVG VKLKGRLQF WTETLPRKIQ EWHREQRSRK  VPEEL.
Mouselivcoxy ..GL..PP.. Q.YLEIGLEP .........RTG VKLKKGRLQF WTETLPRKIQ EWHREQRSRK  VPEEL.
Anopha       ALEG..EH.. WPLYTREN  PIFFIFNAEG EDDLRGEKYG RGPMATSCAF WNDFLPRLRA WSVPSKSPCN LLEQMSIASV  SSTMPIVOMV  VLVLIPLCAW
Drosache     AQDG..EE.. WFNFSKED  PVYYIFSTDD .........KIEKIA RGPLAARCSF WNDYLPKVRS WAGTCDGDS. ...GSASI  SPRLQLLGIA ALIYI..CAA
Humanbche    N.NS..TS.. WFVFKSTE  QKYILTNTES T......RIM TKLRAQOCRF WTSFFPKVLE MTGNIDEAEW EW.....  KAGFHRWNNY  MMDWKNQFND
Rabbitbche   N.NS..TR.. WPVFKSTE  QKYILTNTES P......RIY TKLRAQOCRF WTLFFPKVLE MTGDIDETEQ EW.....  KAGFHRWNNY  MMAWKNHFND
Mousebche    G.NS..TM.. WPVFTSTE  QKYILNTEK  S......KIY SKLRAPQCQF WRLFFPKVLE WNQFLPKLLN ATETIDEAER QW.....  MDWQNQFND
Tcalifache   .SQE..SK.. WPLFTTKE  QKFIDLNTEP M......KVH QRLRVQMCVF WNQFLPKLLN ATETIDEAER QW.....  KTEFHRWSSY  MMHWKNQFDH
Tmarmorache  .SQE..SK.. WPLFTTKE  QKFIDLNTEP I......KVH QRLRVQMCVF WNQFLPKLLN ATETIDEAER QW.....  KTEFHRWSSY  MMHWKNQFDQ
Mouseache    DSKS..PQ.. WPPYTTAA  QQYVSLNLKP L......EVR RGLRAQTCAF WNRFLPKLLS ATDTLDEAER QW.....  KAEFHRWSSY  MVHWKNQFDH
Ratache      DSKS..PR.. WPPYTTAA  QQYVSLNLKP L......EVR RGLRAQTCAF WNRFLPKLLS ATDTLDEAER QW.....  KAEFHRWSSY  MVHWKNQFDH
Fbsache      APKA..PQ.. WPPYTAGA  QQYVSLNLRP L......GVP QASRAQACAF WNRFLPKLLN ATDTLDEAER QW.....  KAEFHRWSSY  MVHWKNQFDH
Humanache    DPKA..PQ.. WPPYTAGA  QQYVSLDIRP L......EVR RGLRAQACAF WNRFLPKLLS ATDTLDEAER ATGPPEDAER EW.....  KAEFHRWSSY  MVHWKNQFDH
Chickache    G..G..PR.. WPPYTTPSG  QRYAHLNARP L......SVG HGLRTQICAF WTRFLPKLLN ATGPPEDAER EW.....  RLEFHRWSSY  MGRWRTQFEH
Nematest     ATPAEIWEX.. YSLNRPER  HYRISYPKCE .........MR DVYHEGRIQF LEKIDGDSD. ..KYONWSME DIVDPAYSKT TSNSEKDEL
Nematest2    V....EW.. RPVSDPDD  VNFLNFQSSG VS.....VK HGLFOEPLDF WNNLREREGF DIVDPAYSKT TSNSEKDEL
Bovthyro     EFSRRAPEFA AFWPDFVPKI .........AESY KELSVLLPNR QGLKKADCSF WSKYIQSLKA SADETKDGPS ADSEEDQPA  GGGLTEDLLG LPELASKTYS
Drosestp     DFKDSV.... ..GSEKFQ  LLAIYIDAAR IGSMNWFRKL HE.....
Drosest6     NFQNNV.... ..NSKEYQ  VARISRNACK NEEYARFP..
Aphidestp    S...EI.... ..WLPVSK  NLADPFRFTK ITQQQTFEAR EQSTTGIMNF GVAYH.
Aphidestfe4  S...EI.... ..WLPVSK  MOYEDIVSPT IIRSKEFASR QQ...DIIEF FDSFTSNSPL
Heliojh      NNSLEV.... ..WPANNG  QTKPTFKCLN IA.NDGVAFV DYPDADRLDM WDAMYVNDEL
Culexestbl   TAKGGV.... ..VFEPNA  QUTKPTFKCLN FGEGYLPNY  RVIYKPTNF  SPPITTTTT  F..... YAYNPYANWQ NRPSCQHPNW HPADPEVVRA
Glutactin    GWWPQATSEY GPFLRFTNNQ QSPLELDFH.     SPFITTTTT  T..... YAYNPYANWQ NRPSCQHPNW HPADPEVVRA
Neurotactin
```

700
```
Geolip1
Geolip2
Cclip1
Crlip4
Cclip2
D2
Dictyocrys
Balipase
Celipase
Humancel
Ratchest
Ratsterest
Ratcholest
Bovchole
Rabbitchole
Mousece
Ratce
Rat61
Ratcoxys
Humance
Humlivce
Humcoxysa
Humcoxysb
Hummonmacss
Humcoxys
Pigcoxys
Est-22
Rabbitce
Anasthios
Rablivest2
Mouselivcoxy
Anopha
Drosache
Humanbche
Rabbitbche
Mousebche
Tcalifache
Tmarmorache
Mouseache
Ratache
Fbsache
Humanache
Chickache
Nematest
Nematest2
Bovthyro
Drosestp
Drosest6
Aphidestp
Aphidestfe4
Heliojh
Culexestbl
Glutactin
Neurotactin
```

	701
Geolip1	....
Geolip2	....
Cclip1	....
Crlip4	....
Cclip2	....
D2	....
Dictyocrys	....
Balipase	PVPPTGDSG.
Celipase	PVPPTGDSGA P.
Humancel	PVPPTGDSGA P.
Ratchest	PVPPTDNSQA G.
Ratsterest	PVPPTDNSQA G.
Ratcholest	PVPPTDNSQA G.
Bovchole	QMPVVIGF.
Rabbitchole	QMPMAIGF.
Mousece	....
Ratce	....
Rat61	....
Ratcoxys	....
Humance	....
Humlivce	....
Humcoxysa	....
Humcoxysb	....
Hummonmacss	....
Humcoxys	....
Pigcoxys	....
Est-22	....
Rabbitce	....
Anasthios	....
Rablivest2	....
Mouselivcoxy	....
Anopha	WWAIKKNKTP P.
Drosache	LRTKRVF....
Humanbche	YTSKKESCVG L.
Rabbitbche	YTSKKERCAG F.
Mousebche	YTSKKESCTA L.
Tcalifache	Y.SRHESCAE L.
Tmarmorache	Y.SRHENCAE L.
Mouseache	Y.SKQERCSD L.
Ratache	Y.SKQBRCSD L.
Fbsache	Y.SKQDRCSD L.
Humanache	Y.SKQDRCSD L.
Chickache	Y.SRQQPCAT L.
Nematest	....
Nematest2	....
Bovthyro	K.........
Drosest6	....
Drosestp	....
Aphideste4	....
Aphidestfee4	....
Heliojh	....
Culexestb1	....
Glutactin	QEARQQEFI.
Neurotactin	....

Table 1. Protein sequences appearing in the alignment

Code	Family	Source	Reference
Geolip1	Lipase	*Geotrichum candidum* gene 1	Shimada et al., 1990
Geolip2		*Geotrichum candidum* gene 2	Shimada et al.,1989
Cclip1		*Candida cylindracea* gene 1	Kawaguchi et al., 1989
Crlip4		*Candida rugosa*	Lotti et al., 1993.
Cclip2		*Candida cylindracea* gene 2	Longhi et al., 1992
D2	Esterase	*Dictyostelium* D2	Rubino et al., 1989
Dictyocrys		*Dictyostelium* crystal protein	Bomblies et al., 1990
Ratchest	Cholesterol esterase	Rat pancreas gene	Fontaine et al., 1991
Ratsterest		Rat liver gene	Chen et al., unpublished
Ratcholest		Rat pancreas	Han et al., 1987;
			Kissel et al., 1989
Balipase		Human	Hui et al., 1990;
			Nilsson et al., 1990;
			Baba et al., 1991
Celipase	Carboxyl ester lipase	Human pancreas gene	Reue et al.,1991
Humancel		Human CEL gene	Lidberg et al., 1992
Bovchole	Cholesterol esterase	Cow	Kyger et al., 1989
Rabbitchole		Rabbit pancreas	Colwell et al., 1993
Mousece	Carboxylesterase	Mouse isoenzyme	Ovnic et al., 1991
Ratce		Rat liver esterase E1	Long et al., 1988;
			Takagi et al., 1988
Rat61		Rat liver, pI 6.1	Robbi et al., 1990
Ratcoxys		Rat	Medda et al., 1992
Humcoxysb		Human liver	Kroetz et al., unpublished
Hummonmacss	Serine esterase	Human monocyte/macrophage	Zschunke et al., 1991
Humance	Carboxylesterase	Human	Munger et al., 1991; Long et al., 1991
Humlivce		Human liver gene	Shibata et al., 1993
Humcoxysa		Human liver gene	Kroetz et al., 1993
Humcoxys		Human liver gene	Riddles et al., 1991
Pigcoxys		Pig liver gene	Matsushima et al., 1991
Rabbitce		Rabbit liver esterase 1	Korza et al., 1988
Est-22	Esterase	Mouse gene	Ovnic et al., 1991
Anasthios		*Anas platyrhynchos* gene	Hwang et al.,1993
Rablivest2	Carboxylesterase	Rabbit liver esterase 2	Ozols, 1989
Mouselivcoxy		Mouse liver gene	Aida et al., 1993
Anopha	Acetylcholinesterase	*Anopheles stephensi*	Hall et al., 1991
Drosache		*Drosophila*	Hall et al., 1986
Humanbche	Butyrylcholinesterase	Human	Lockridge et al., 1987
Rabbitbche		Rabbit	Jbilo et al., 1990
Mousebche		Mouse	Rachinsky et al., 1990
Tcalifache	Acetylcholinesterase	*Torpedo californica*	Schumacher et al., 1986
Tmarmorache		*Torpedo marmorata*	Sikorav et al., 1987
Mouseache		Mouse	Rachinsky et al., 1990
Ratache		Rat	Legay et al., 1993
Fbsache		Fetal bovine	Doctor et al., 1990
Humanache		Human	Soreq et al., 1990
Chickache		Chicken brain and muscle	Randall et al., unpublished
Nematest	Esterase	*C. elegans* gut gene	Arpagaus et al., 1994
Nematest2		*C. elegans, C. briggsae* gut	Kennedy et al., 1993
Bovthyro	Nonhydrolytic protein	Bovine thyroglobulin	Mercken et al., 1985
Drosest6	Carboxylesterase	*Drosophila* (esterase 6)	Oakeshott et al., 1987
Drosestp		*Drosophila* (esterase P)	Collet et al., 1990
Aphideste4	Esterase	*Myzus persicae* (esterase E4)	Field et al., 1993
Aphidestfe4		*M. persicae* (esterase FE4)	Field et al., 1993
Heliojh	Carboxylesterase	*Heliothis* juvenile hormone	Hanzlik et al., 1989
Culexestb1		*Culex* (esterase B1)	Mouches et al., 1990
Glutactin	Nonhydrolytic protein	*Drosophila*	Olson et al., 1990
Neurotactin		*Drosophila*	de la Escalera et al., 1990

	80	84	86,87	92,93	96,97	151-156	245-247	278,279	285	335-338	340,341	343	346-349	414,415	417-419	533	547-549
Geolip1	L	N	LT	AL	AK	GGAFVY	ESA	SG	H	DAQN	YD	K	FGLL	AF	PVA	.	HGN
Geolip2	L	N	IS	VV	GK	GGAFVY	ESA	SG	F	SAQN	YD	K	FGLL	IL	PVA	.	HGS
Cclip1	Q	G	YE	KA	DL	GGGFEV	ESA	SG	.	DATN	.N	P	LAYS	FF	TSS	.	HSN
Crlip4	M	G	WD	KA	NS	GGGFEV	ESA	SG	.	QATS	.D	P	LAYP	VF	LSS	.	HAN
Cclip2	M	G	FE	KN	HL	GGGFEL	ESA	SG	.	QATS	.D	P	LAYP	LF	LLL	.	HGN
D2	K	G	GV	..	..	GGAFSV	ESA	SS	.	DGPR	I.	.	LTFG	PF	...	D	HGT
Dictyocrys	D	G	GS	..	..	GGAFTQ	ESA	ST	.	AIQK	L.	.	LAIG	LF	...	E	HGT
Balipase	A	Q	ST	..	..	GGAFLM	ESA	SG	W	....	LA	K	LAGL	IF	SID	Y	HAD
Celipase	A	Q	ST	..	..	GGAFLM	ESA	SG	W	....	LA	K	LAGL	IF	SID	Y	HAD
Humancel	A	Q	ST	..	..	GGAFLM	ESA	SG	W	....	LA	K	LAGL	IF	SID	Y	HAD
Ratchest	A	Q	DT	..	..	GGAFLM	ESA	SG	W	....	LA	R	LKSQ	LF	TVD	Y	HAD
Ratsterest	A	Q	DT	..	..	GGAFLM	ESA	SG	W	....	LA	R	LKSQ	LF	TVD	Y	HAD
Ratcholest	A	Q	DT	..	..	GGAFLM	ESA	SG	W	....	LA	R	LKSQ	LF	TVD	Y	HAD
Bovchole	A	Q	ST	..	..	GGAFLM	ESA	SG	W	....	LA	K	LGST	LF	GMD	Y	HAD
Rabbitchole	A	Q	ST	..	..	GGAFLM	ESA	SG	W	....	LA	N	LAGL	LF	TVD	Y	HAD
Mousece	D	A	IL	ST	EI	GGGLVI	ESS	SG	.	EISG	LV	Y	....	WI	PM.	R	HGD
Ratce	D	G	LL	ST	ES	GGGLII	ESA	SG	.	ELTV	LD	.	....	WI	PT.	R	HGD
Rat61	D	G	VL	TN	EN	GGGLVV	ESA	SG	.	ETSL	L.	L	LDLL	WI	PT.	R	HGD
Ratcoxys	D	G	VL	TN	EN	GGGLVV	ESA	SG	.	ETSL	L.	L	LDLL	WI	PT.	R	HGD
Humance	D	G	LL	TN	EN	GGGLMV	ESA	SG	.	ETTL	M.	F	LDLQ	WL	PMQ	K	HGD
Humlivce	D	G	LL	TN	EN	GGGLMV	ESA	SG	.	ETTL	M.	F	LDLQ	WL	PMQ	K	HGD
Humcoxysa	D	G	LL	TN	EN	GGGLMV	ESA	SG	.	ETTL	M.	F	LDLQ	WL	PMQ	K	HGD
Humcoxysb	D	G	LL	TN	EN	GGGLMV	ESA	SG	.	ETTL	M.	F	LDLQ	WL	PM.	K	HGD
Hummonmacss	D	G	LL	TN	EN	GGGLMV	ESA	SG	.	ETTL	M.	F	LDLQ	WL	PM.	K	HGD
Humcoxys	.	.	..	..	..	GGGLMV	ESA	SG	.	ETTL	IG	S	WTYR	WL	PMQ	K	HGD
Pigcoxys	D	E	MT	TN	ER	GGGLVL	ESA	SG	.	DLTL	M.	F	LDFH	WI	PT.	K	HGD
Est-22	D	G	IV	TN	EK	GGGLVL	ESA	SG	.	GTTL	L.	L	LDLH	WI	PT.	K	HGD
Rabbitce	D	G	ML	TN	EN	GGGLMV	ESA	SG	.	EVTL	M.	F	LDLV	WI	PMQ	R	HGD
Anasthios	D	G	YL	TN	EK	GGGLVS	ESA	SG	.	QITL	MP	M	....	WI	P.R	V	HAD
Rablivest2	N	M	QD	HF	PS	GGGLTM	ESA	SL	.	AITQ	FM	.	....	WI	PKL	R	HGD
Mouselivcoxy	D	E	MS	TL	KM	GGSLRV	NSA	SG	.	T...	..	.	....	WL	...	K	HSS
Anopha	E	Y	YF	GE	WN	GGGFMS	ESA	SG	W	VQQW	..	Y	ILGF	YF	LYD	W	HGD
Drosache	E	Y	YF	GE	WN	GGGFMT	ESA	SG	W	VQQW	..	Y	ILSF	YF	LYD	W	HGD
Humanbche	N	D	SF	GS	WN	GGGFQT	ESA	SG	W	LNEA	VV	Y	PLSV	AF	VYG	W	HGY
Rabbitbche	N	D	SF	GS	WN	GGGFQT	ESA	SG	W	LNEV	VV	F	LLSV	AF	VYG	W	HGY
Mousebche	N	D	AF	GS	WN	GGGFQT	ESA	SG	W	RNER	VL	S	ILSI	AF	VYG	W	HGY
Tcalifache	Y	D	QF	GS	WN	GGGFYS	ESA	SG	W	DVEW	VL	F	IFRF	FF	LYG	W	HGY
Tmarmorache	Y	D	QF	GS	WN	GGGFYS	ESA	SG	W	DVEW	VL	F	IFRF	FF	LYG	W	HGY
Mouseache	Y	D	LY	GT	WN	GGGFYS	ESA	SG	W	DHEW	VL	Q	IFRF	YF	VYG	W	HGY
Ratache	Y	D	LY	GT	WN	GGGFYS	ESA	SG	W	DHEW	VL	Q	IFRF	YF	VYG	W	HGY
Fbsache	Y	D	LY	GT	WN	GGGFYS	ESA	SG	W	DHEW	VL	Q	VFRF	YF	VYG	W	HGY
Humanache	Y	D	LY	GT	WN	GGGFYS	ESA	SG	W	NHEW	VL	Q	VFRF	YF	VYG	W	HGY
Chickache	M	D	TF	GS	WN	GGGFTG	ESA	SG	W	EGEG	VM	P	VFRF	YF	VYG	W	HGY
Nematest	.	.	.T	QI	PR	GGGYEL	QSA	SG	F	T....	..	L	VKGY	ML	SMN	W	HCT
Nematest2	.	.	.T	LV	FS	GGGYEI	YSA	SA	W	D....	..	A	TQGW	YF	TMN	Y	HAN
Bovthyro	P	T	TP	..	..	NAAEGK	DRG	GG	A	DAQT	LL	V	....	IN	AKA	.	LLT
Drosest6	W	.	FT	..	.A	GGAFMF	HSA	SG	W	TAVR	FL	F	VPFA	YN	ALL	V	HGD
Drosestp	W	.	FI	..	.N	GGAFMF	HSA	SG	W	SAVR	FL	F	VPFS	YN	AQL	L	HGD
Aphideste4	I	.	.F	..	.G	GGGYYF	MSA	SG	W	KSYL	FM	W	FPFT	..	..L	F	HGD
Aphidestfe4	I	.	.F	..	.G	GGGYYF	MSA	SG	W	KSYL	FM	W	FPFT	..	..L	F	HGD
Heliojh	T	Y	RL	..	.A	GGGFAF	QSA	SG	F	EANA	LI	.	IGLT	..	..E	K	HIE
Culexestb1	F	L	..	..	..	GGGFTE	HSA	SG	W	EKLL	DQ	M	DIFT	...	..L	Y	HAD
Glutactin	F	D	LR	RG	..	GEMLFD	QAG	SG	Y	AAFE	LL	Q	HLGL	FI	SRI	H	LSD
Neurotactin	R	G	TV	..	..	AESLAG	HRA	SG	.	AATP	TW	H	VDLP	.V	VMG	F	HGT

Figure 2. Comparison of amino acid residues lining the active-site gorge of *T. californica* AChE with aligned residues of other proteins. Numbering is based on the PileUp alignment of proteins.

cyte/macrophage serine esterase (Hummonmacss) form a group in which the similarity is between 99 and 100%.

The identification of amino acids lining the active-site gorge of *T. californica* acetylcholinesterase was made (N. Qian and B.P. Doctor, unpublished observations) using the three-dimensional model of this enzyme, as determined by Sussman et al. (1991). The comparison shown here (Figure 2) was made by lifting the aligned residues from the PileUp alignment.

REFERENCES

Aida, I., Moore, R., and Negishi, M., 1993, *Biochim. Biophys. Acta* 1174:72-74.

Arpagaus, M., Fedon, Y., Cousin, X., Chatonnet, A., Berge, J.B., Fournier, D., and Toutant, J.P., 1994, *J. Biol. Chem.*, 269:9957-9965.

Baba, T., Downs, D., Jackson, K.W., Tang, J., and Wang, C.-S., 1991, *Biochemistry* 30:500-510.

Bomblies, L., Biegelmann, E., Döring, V., Gerisch, F., Krafft-Czepa, H., Noefel, A.A., Schleicher, M., and Humbel, B.M., 1990, *J. Cell Biol.* 110:669-679.

Chen, X., Harrison, E.H., and Fisher, E.A., unpublished.

Collet, C., Nielsen, K.M., Russell, R.J., Karl, M., Oakeshott, J.G., and Richmond, R.C., 1990, *Mol. Biol. Evol.* 7:9-28.

Colwell, N.S., Aleman-Gomez, J.A., and Kumar, B.V., 1993, *Biochim. Biophys. Acta* 1172:175-180.

Cygler, M., Schrag, J.D., Sussman, J.L., Harel, M., Silman, I., Gentry, M.K. and Doctor, B.P., 1993, *Protein Sci.* 2,366-382.

de la Escalera, S., Bockamp, E.-O., Moya, F., Piovant, M., and Jiménez, F., 1990, *EMBO J.* 9:3593-3601.

Doctor, B.P., Chapman, T.C., Christner, C.E., Deal, C.D., de la Hoz, D.M., Gentry, M.K., Ogert, R.A., Rush, R.S., Smyth, K.K., and Wolfe, A.D., 1990, *FEBS Lett.* 266:123-127.

Field, L.M., Williamson, M.S., Moores, G.D., and Devonshire, A.L., 1993, *Biochem. J.* 294:569-574.

Fontaine, R.N., Carter, C.P., and Hui, D.Y., 1991, *Biochemistry* 30:7008-7014.

Gentry, M.K. and Doctor, B.P., 1991, In *Cholinesterases: Structure, Function, Mechanism, Genetics, and Cell Biology* (Massoulié, J., Bacou, F., Barnard, E., Chatonnet, A., Doctor, B.P. and Quinn, D.M., eds), pp. 394-398, American Chemical Society, Washington, DC.

Hall, L.M.C., and Malcolm, C.A., 1991, *Cell. Mol. Neurobiol.* 11:131-141.

Hall, L.M.C., and Spierer, P., 1986, *EMBO J.* 5:2949-2954.

Han, J.H., Stratowa, C., and Rutter, W.J., 1987, *Biochemistry* 26:1617-1625.

Hanzlik, T.N., Abdel-Aal, Y.A.I., Harshman, L.G., and Hammock, B.D., 1989, *J. Biol. Chem.*, 264:12419-12425.

Hui, D.Y., and Kissel, J.A., 1990, *FEBS Lett.* 26:131-134.

Hwang, C.-S., and Kolattukudy, P.E., 1993, *J. Biol. Chem.* 268:14278-14284.

Jbilo, O., and Chatonnet, A., 1990, *Nucleic Acids Res.* 18:3990.

Kawaguchi, K., Honda, H., Taniguchi-Morimura, J., and Iwasaki, S., 1989, *Nature* 341:164-166.

Kennedy, B.P., Aamodt, E.J., Allen, F.L., Chung, M.A., Heschl, M.F., and McGhee, J.D., 1993, *J. Mol. Biol.* 229:890-908.

Kissel, J.A., Fontaine, R.N., Turck, C.W., Brockman, H.L., and Huik, D.Y., 1989, *Biochim. Biophys. Acta* 1006:227-236.

Korza, G., and Ozols, J., 1988, *J. Biol. Chem.* 263:3486-3495.

Kroetz, D., McBride, O., and Gonzalez, F., 1993, *Biochemistry* 32:11606-11617.

Kroetz, D.L., McBride, O.W., and Gonzalez, F.J., 1993, unpublished.

Kyger, E.M., Wiegand, R., and Lange, L.G., 1989, *Biochem. Biophys. Res. Commun.* 164:1302-1309.

Legay, C., Bon, S., Vernier, P., Coussen, F., and Massoulié, J., 1993, *J. Neurochem.* 60:337-346.

Lidberg, U., Nilson, J., Stroemberg, K., Stenman, G., Sahlin, P., Enerbaeck, S., and Bjursell, G., 1992, *Genomics* 13:630-640.

Lockridge, O., Bartels, C.F., Vaughan, T.A., Wong, C.K., Norton, S.E., and Johnson, L.L., 1987, *J. Biol. Chem.* 262:549-557.

Long, R.M., Satoh, H., Martin, M., Kimura, S., Gonzalez, F.J., and Pohl, L.R., 1988, *Biochem. Biophys. Res. Commun.* 156:866-873.

Long, R.M., Calabrese, M.R., Martin, B.M., and Pohl, L.R., 1991, *Life Sci.* 48:43-49.

Longhi, S., Fusetti, F., Grandori, R., Lotti, M., Vanoni, M., and Alberghina, L., 1992, *Biochim. Biophys. Acta* 1131:227-231.

Lotti, M., Grandori, R., Fusetti, F., Longhi, S., Brocca, S., Tramontano, A., and Alberghina, A., 1993, *Gene* 124:45-55.

Matsushima, M., Inoue, H., Ichinose, M., Tsukada, S., Miki, K., Kurokawa, K., Takahashi, T., and Takahashi, K., 1991, *FEBS Lett.* 293:37-41.

Medda, S., and Proia, R.L., 1992, *Eur. J. Biochem.* 206:801-806.

Mercken, L. Simmond, M.-J., Swillens, S., Massaer, M., and Vassart, G., 1985, *Nature* 316:647-651.

Mouches, C., Pauplin, Y., Agarwal, M., Lemieux, L., Herzog, M., Abadon, M., Beyssat-Arnaouty, V., Hyrien, O., de St. Vincent, B.R., Georghiou, G.P., and Pasteur, N., 1990, *Proc. Natl. Acad. Sci. USA* 87:2574-2578.

Munger, J.S., Shi, G.-P., Mark, E.A., Chin, D.T., Gerard, C., and Chapman, H.A., 1991, *J. Biol. Chem.* 266:18832-18838.

Nilsson, J., Bläckberg, L. Carlsson, P., Enerbäck, S., Hernell, O., and Bjursell, G., 1990, *Eur. J. Biochem.*, 192:543-550.

Oakeshott, J.G., Collet, C., Phillis, R.W., Nielsen, K.M., Russell, R.J., Chambers, G.K., Ross, V., and Richmond, R.C., 1987, *Proc. Natl. Acad. Sci. USA* 84:3359-3363.

Olson, P.F., Fessler, L.I., Nelson, R.E., Sterne, R.E., Campbell, A.G., and Fessler, J.H., 1990, *EMBO J.* 9:1219-1227.

Ovnic, M., Swank, R.T., Fletcher, C., Zhen, L., Novak, E.K., Baumann, H., Heintz, N., and Ganschow, R.E., 1991, *Genomics* 11:956-67.

Ovnic, M., Tepperman, K., Medda, S., Elliott, R.W., Stephenson, D.A., Grant, S.G., and Ganschow, R.E., 1991, *Genomics* 9:344-354.

Ozols, J., 1989, *J. Biol. Chem.* 264:12533-12545.

Rachinsky, T.L., Camp, S., Li, Y., Elström, T.J., Newton, M., and Taylor, P., 1990, *Neuron* 5:317-327.

Randall, W.R., Rimer, M., and Gough, N.R., unpublished.

Reue, K., Zambaux, J., Wong, H., Lee, G., Leete, T.H., Ronk, M., Shively, J.E., Sternby, B., Borgstrom, B., Ameis, D., Schotz, M.C., 1991, *J. Lipid Res.* 32:267-276.

Riddles, P.W., Richards, L.J., Bowles, M.R., and Pond, S.M., 1991, *Gene* 108:289-292.

Robbi, M., Beaufay, H., and Octabe, J.-N., 1990, *Biochem. J.* 269:451-458.

Rubino, S., Mann, S.K.O., Hori, R.T., Pinko, C., and Firtel, R.A., 1989, *Dev. Biol.* 131:27-36.

Schumacher, M., Camp, S., Maulet, Y., Newton, M., MacPhee-Quigley, K., Taylor, S.S., Friedmann, T., and Taylor, P., 1986, *Nature* 319:407-409.

Shibata, F., Takagi, Y., Kitajima, M., Kuroda, T., and Omura, T., 1993, *Genomics* 17:76-82.

Shimada, Y., Sugihara, A., Tominaga, Y., Iizumi, T., and Tsunasawa, S., 1989, *J. Biochem.* 106:383-388.

Shimada, Y., Sugihara, A., Iizumi, T., and Tominaga, Y., 1990, *J. Biochem.* 107:703-707.

Sikorav, J.-L., Krejci, E., and Massoulié, J., 1987, *EMBO J.* 6:1865-1873.

Soreq, H., Ben-Aziz, B., Prody, C.A., Seidman, S., Gnatt, A., Neville, L., Lieman-Hurwitz, J., Lev-Lehman, E., Ginzberg, D., Lapidot-Lifson, Y., and Zakut, H., 1990, *Proc. Natl. Acad. Sci. USA* 87:9688-9692.

Sussman, J.L., Harel, M., Frolow, F., Oefner, C., Goldman, A., Toker, L., Silman, I., 1991, *Science* 253:872-879.

Takagi, Y., Morohashi, K., Kawabata, S., Go, M., and Omura, T., 1988, *J. Biochem.*, 104:801-806.

Zschunke, F., Salmassi, A., Kreipe, H., Buck, F., Parwaresch, M.R., and Radzun, H.J., 1991, *Blood* 78:506-512.

PARTICIPANTS

Jaya Adhikari
ETL, Dept. of Zoology
Visva Bharati University
Santiniketan 731235, West Bengal, India
Tel: 03463-52751, x88
Fax: 91-03463-52672

Michael Adler
Neurotoxicology Branch, ATTN:
  SGRD-UV-YN
U.S. Army Research Institute of Chemical
  Defense
Aberdeen Proving Ground, MD
  21010-5425, USA
Tel: 410-671-1913
Fax: 410-671-1960

Gabi Amitai
Div. of Chemistry
Israel Institute for Biological Research
P. O. Box 19, Ness-Ziona 70450, Israel
Tel: 972-8-381506
Fax: 972-8-401404

Lili Anglister
Dept. of Anatomy & Embryology
Hebrew University, Hadassah Medical
  School
Jerusalem 91120, Israel
Tel: 972-2-758450
Fax: 972-2-757451

Alain Anselmet
Laboratoire de Neurobiologie Moléculaire
  et Cellulaire
CNRS URA 1857, Ecole Normale
  Supérieure
46 rue d'Ulm, 75005 Paris, France
Tel: 33144323749
Fax: 33144323887

Yacov Ashani
Israel Institute for Biological Research
P. O. Box 19
Ness-Ziona 70450, Israel
Tel: 972-8-381455
Fax: 972-8-401404

A. S. Balasubramanian
Dept. of Neurological Sciences
Christian Medical College and Hospital
Vellore 632 004, Tamil Nadu, India
Tel: 91-416-22102
Fax: 91-416-22103

K. Balasubramanian
Dept. of Neurological Sciences
Christian Medical College and Hospital
Vellore 632 004, Tamil Nadu, India
Tel: 91-416-22102
Fax: 91-416-22103

Cynthia Bartels
University of Nebraska Medical Center,
  Eppley Institute
600 S. 42nd St., Box 986805
Omaha, NE 68198-6805, USA
Tel: 402-559-6014
Fax: 402-559-4651

Hendrick P. Benschop
TNO Prins Maurits Laboratory
PO Box 45
2280 AA Rijswijk, The Netherlands
Tel: 31-15-843529
Fax: 31-15-843991

Harvey A. Berman
Dept. of Biochemical Pharmacology,
  School of Pharmacy
SUNY at Buffalo, 335 Hochestetter Hall,
Buffalo, NY 14260, USA
Tel: 716-645-2859
Fax: 716-645-3850

Dhanasekaran Bhanumathy
Dept. of Neurological Sciences
Christian Medical College and Hospital,
Vellore - 632 004, Tamil Nadu, India
Tel: 91-416-22102
Fax: 91-416-22103

B.K. Bhattacharya
Defence Research & Development
  Establishment
Tansen Road
Gwalior-474002, India
Tel: 91-751-340354
Fax: 91-751-341148

Shelley Bhattacharya
ETL, Dept. of Zoology, Visva Bharati
  University
Santiniketan-731 235
West Bengal, India
Tel: 091-03463-52751, x88
Fax: 091-03463-52672

Renee Blong
University of Nebraska Medical Center,
  Eppley Institute
600 S. 42nd St., Box 986805
Omaha, NE 68198-6805, USA
Tel: 402-559-6014
Fax: 402-559-4651

Rathanam Boopathy
DST Scheme, Department of
  Bio-Technology
Bharathiar University
Coimbatore - 641 046, Tamil Nadu, India
Tel: 91-422-42222-258
Fax: 91-422-42387

Cassian Bon
Unité des Venins, Institut Pasteur
25 rue du Dr. Roux
75724 Paris Cedex 15, France
Tel: (33-1) 45688685
Fax: (33-1) 40613057

Shambhunath Bose
ETL, Dept. of Zoology
Visva Bharati University
Santiniketan-731 235, West Bengal, India
Tel: 091-03463-52751, x88
Fax: 091-03463-52672

William S. Brimijoin
Department of Pharmacology, Mayo Clinic
200 First Street S.W.
Rochester, Minnesota 55905, USA
Tel: 507-284-8165
Fax: 507-284-9111

Robert S. Brodie
Dept. of Anatomy and Neurobiology
College of Medicine, University of
  California
Irvine, CA 92717, USA
Tel: 714-856-6050
Fax: 714-856-8549

Clarence A. Broomfield
Biochemical Pharmacology Branch
U.S. Army Research Institute of Chemical
  Defense
Aberdeen Proving Ground, MD
  21010-5425, USA
Tel: 410-671-2626
Fax: 410-671-1960

Göran Bucht
National Defence Research Establishment
Dept. of NBC Defense
S-901 82 Umeå, Sweden
Tel: 46-90106672
Fax: 46-90106800

Timothy Budd
Dept. of Pharmacology, Oxford University
Mansfield Road
Oxford OX1 3QT, UK
Tel: 0865-271627
Fax: 0865-271853

Shelley Camp
Dept. of Pharmacology, 0636, UCSD
9500 Gilman Rd.
La Jolla, CA 92093-0636, USA
Tel: 619-534-1367
Fax: 619-534-8248

Arnaud Chatonnet
Différenciation Cellulaire et Croissance,
  INRA
Place Viala 34060
Montpellier Cedex 1, France
Tel: 33-67 41 60 99
Fax: 33-67 54 56 94

Didier Clarençon
CRSSA Émile Pardé, 3P 87
38702, La Tronche Cedex, France
Tel: 33-76-36-69-00
Fax: 33-76-63-69-01

Cecile Clery
CRSSA, Biochemistry Unit, BP 87
38702 La Tronche Cedex, France
Tel: 33-76-63-59-59
Fax: 33-76-63-69-61

Ephraim Cohen
Dept. of Entomology, Faculty of
  Agriculture
The Hebrew University
Rehovot 76100, Israel
Tel: 972-8-481118
Fax: 972-8-466768

Barbara Coleman
Dept. of Pharmacology, 0636, UCSD
9500 Gilman Rd.
La Jolla, CA 92093-0636, USA
Tel: 619-534-1367
Fax: 619-534-8248

Austin Cooper
Janeway Child Health Centre
St. Johns, Newfoundland, A1AIR8
Canada
Tel: 709-778-4530
Fax: 709-778-4333

Francoise Coussen
Laboratoire de Neurobiologie Moléculaire
  et Cellulaire
CNRS URA 1857, Ecole Normale
  Supérieure
46 rue d'Ulm, 75005 Paris, France
Tel: 33144323748
Fax: 33144323887

Miroslaw Cygler
Biotechnology Research Institute, NRCC
6100 Royalmount Ave.
Montreal, Quebec H4P 2R2, Canada
Tel: 514-496-6321
Fax: 514-496-5143

Jitendra Dave
Dept. of Medical Neurosciences
Walter Reed Army Institute of Research
Washington, DC 20307-5100, USA
Tel: 202-782-0534
Fax: 202-782-6910

Bhupendra P. Doctor
Div. of Biochemistry
Walter Reed Army Institute of Research
Washington, DC 20307-5100
Tel: 202-782-3001
Fax: 202-782-6304

S. N. Dube
Pharmacology & Toxicology Division
Defence Research & Development
  Establishment
Tansen Road, Gwalior-474002, India
Tel: 91-751-340354
Fax: 91-75-1341148

Heinrich Egghart
USARDSC-UK European Research Office
223 Old Marylebone Rd.
London NW1 5TH, UK
Tel: 44-71-514-4423
Fax: 44-71-724-1433

Albert Enz
Preclinical Research, Sandoz Pharma. Ltd.
Bldg. 386 762
CH-4002 Basel, Switzerland
Tel: 61-324-4705
Fax: 61-324-4787

Didier Fournier
Lab. d'Entomologie, Université Paul
  Sabatier
118 route de Narbonne
310162 Toulouse, France
Fax: 33-61-55-60-00

John George
Dept. of Zoology, University of Guelph
Guelph, Ontario
Canada N1G 2W1
Tel: 519-824-4120, x347
Fax: 519-767-1656

Damon Getman
Dept. of Pharmacology, 0636, UCSD
9500 Gilman Rd.
La Jolla, CA 92093-0636, USA
Tel: 619-534-1367
Fax: 619-534-6833

Ezio Giacobini
Dept. of Pharmacolology
Southern Illinois University School of
  Medicine
P. O. Box 19230, Springfield, IL
  62794-9230
Tel: 217-785-2185
Fax: 217-524-0145

Kurt Giles
Dept. of Pharmacology, Oxford University
Mansfield Road
Oxford OX1 3QT, UK
Tel: 44-865-271857
Fax: 44-865-271853

Victor Gisiger
Département d'Anatomie, Université de
  Montréal, C.P. 6128,
Succursale Centre-Ville
Montréal, Québec H3C 3J7, Canada
Tel: 514-343-5765
Fax: 514-343-2459

Maurice Goeldner
Laboratoire de Chimie Bio-Organique,
  URA 1386 CNRS
Faculté de Pharmacie, ULP Strasbourg
BP 24-67401 Illkirch cedex, France
Tel: 3388676991
Fax: 3388678891

N. Gopalan
Defence Research & Development
  Establishment
Tansen Road
Gwalior-474002, India
Tel: 91-751-340354
Fax: 91-751-341148

Patrick Gourmelon
Institut de Protection et de Surete
  Nucleaire DPMD
92260 Fontenay aux Roses Cedex, France
Tel: 33 1 46 54 7401
Fax: 33 1 46 54 4610

Jacques Grassi
Service de Pharmacologie et
  d'Immunologie, DRIPP
C.E Saclay
91191 Gif sur Yvette cedex, France
Tel: 33-1-69-08-28-71
Fax: 33-1-69-08-59-07

Susan A. Greenfield
University Department of Pharmacology
Mansfield Road
Oxford OX 1 3QT, England
Tel: (0)865-271628
Fax: (0)865-271853

Zoran Grubič
Institute of Pathophysiology, School of
  Medicine
Zaloška 4
61105 Ljubljana, Slovenia
Tel: (386) 61 310 841
Fax: (386) 61 302 272

Rajinder Gupta
Delhi University
Delhi, India

S. Das Gupta
Indian Science Congress Asssociation
14, Dr Biresh Guha St.
Calcutta, 700017, India
Tel: 91-33-247-4530
Fax: 91-33-402551

Brennie E. Hackley, Jr.
U.S. Army Research Institute of Chemical
  Defense
Aberdeen Proving Ground, MD
  21010-5425, USA
Tel: 410-671-3276
Fax: 410-671-1960

Michal Harel
Department of Structural Biology
Weizmann Institute of Science
Rehovot 76100 , Israel
Tel: 972-8-342647
Fax: 972-8-344159

Ferdinand Hucho
Frei Universitat Berlin, Institut fur
  Biochemie
Thielallee 63, 14195
Berlin, Germany
Tel: 49-30-838-5545
Fax: 49-30-838-3753

Lakshmanan Jaganathan
Dept. of Biotechnology
Bharathiar University
Coimbatore, India Pin Code: 641046
Tel: 91-422-42222-258
Fax: 91-422-42387

Bernard J. Jasmin
Dept. Physiology, University of Ottawa
451 Smyth Rd.
Ottawa, Ontario K1H 8M5, Canada
Tel: 613-787-6544
Fax: 613-787-6718

K. Jeevratnam
Defense Research & Development
  Establishment
Tansen Road
Gwalior 474002, India
Tel: 91 751-340354
Fax: 91-751-341148

Subramanya Karanth
Toxicology Unit IC & P Dept.
CFTRI
Mysore - 570 013, India
Tel: 091 0821 37150
Fax: 091 821 27697 2447

Andis Klegeris
University Dept. of Pharmacology
Mansfield Rd.
Oxford OX1 3QT, UK
Tel: (0865) 271 627
Fax: (0865) 271 853

Ildiko Kovach
Dept. of Chemistry
The Catholic University of America
Washington, DC 20064, USA
Tel: 202-319-6550
Fax: 202-319-5381

Alan Kozikowski
89 Headquarters Plaza
North Tower, 14th Floor
Morristown NJ 07960, USA
Tel: 201-984-1777
Fax: 201-993-1685

Eric Krejci
Laboratoire de Neurobiologie, Ecole
  Normale Supérieure
46 rue d'Ulm
75005 Paris, France
Tel: 33-1-44323748
Fax: 33-1-44323887

David Kreimer
Dept. of Neurobiology
Weismann Institute of Science
Rehovot 76100, Israel
Tel: 972-8-342128
Fax: 972-8-344131

Chanoch Kronman
Israel Institute for Biological Research
P.O. Box 19
Ness-Ziona 70450, Israel
Tel: 972-8-381522
Fax: 972-8-401404

Madhavan Kutty
Janeway Child Health Centre, Janeway
  Place
St. Johns, Newfoundland
Canada A1A1R8
Tel: 709-778-4512
Fax: 709-778-4333

Paul G. Layer
Technical University Darmstadt, Faculty of
  Biology
Schnittspahnstrasse 3
Darmstadt, Germany
Tel: 0049-6151-163800
Fax: 0049-6151-166548

Claire Legay
Laboratoire de Neurobiologie, Ecole
  Normale Supérieure
46 rue d'Ulm
75005 Paris, France
Tel: 33 44323748
Fax: 33 44323887

Bogmuil Lezczunski
Agricultural & Pedagogic University,
  Dept. of Biochemistry
ul. Prusa 12
PL-08110 Siedlce, Poland
Tel: 48 25 445959
Fax: 48 25 44 5959

Yael Loewenstein
Dept. of Biological Chemistry
Institute of Life Sciences, The Hebrew
  University
Jerusalem 91904, Israel
Tel: 972-2-585450
Fax: 972-2-666804

Oksana Lockridge
Department of Biochemistry
University of Nebraska-Lincoln
The Beadle Center, Rm. N200
P.O. Box 880664
Lincoln, NE 68588-0664
Tel: 402-559-6032
Fax: 402-559-4651

Zhigang Luo
Dept. of Pharmacology, 0636, UCSD
9500 Gilman Rd.
La Jolla, CA 92093-0636, USA
Tel: 619-534-1367
Fax: 619-534-6833

Patrick Masson
CRSSA Biochemistry Unit
BP87 - 38702 la Tronche cedex, France
Tel: 33-76-63-69-59
Fax: 33-76-63-69-01

Pascale Marchot
Dept. of Pharmacology, 0636, UCSD
9500 Gilman Rd.
La Jolla, CA 92093-0636, USA
Tel: 619-534-1367
Fax: 619-534-6833

Jean Massoulié
Laboratoire de Neurobiologie, Ecole
  Normale Supérieure
CRNC-URA 295, 46 rue d'Ulm
75005 Paris, France
Tel: (33)144323891
Fax: (33)144323887

Donald M. Maxwell
Biochemical Pharmacology Branch
U.S. Army Medical Research Institute of
  Chemical Defense
Aberdeen Proving Ground, MD 21010-
  5425
Tel: 410-671-1315
Fax: 410-671-1960

Marcel Mesulam
Dept. of Neurology, Northwestern Univ.
  Medical Center
233 E. Erie Street, Suite 614
Chicago, IL 60611
Tel: 312-908-8266
Fax: 312-908-8789

Shawli Mondal
ETL, Dept. of Zoology
Visva Bharati University
Santiniketan-731 235, West Bengal, India
Tel: 03463-52751, x88
Fax: 91-03463-52672

David H. Moore
U.S. Army Medical Research &
  Development Command
Fort Detrick, Frederick, MD 21703-5012
Tel: 301-619-7201
Fax: 301-619-2416

Banibrata Mukhopadhyay
ETL, Dept. of Zoology
Visva Bharati University
Santiniketan-731 235, West Bengal, India
Tel: 03463-52751, x88
Fax: 91-03463-52672

Annick Mutero
Dept. of Pharmacology, 0636, UCSD
9500 Gilman Rd.
La Jolla, CA 92093-0636, USA
Tel: 619-534-1367
Fax: 619-534-6833

Haridarsan Nair
Department of Chemistry
University of Iowa
Iowa City, Iowa 52242, USA
Tel: 319-335-1335
Fax: 319-335-1270

William Neidermyer
Riverbend, Rte 5, Box 148B8
Alexander City, AL 35010, USA

Anna Oomen
Dept. of Neurological Sciences
Christian Medical College and Hospital
Vellore 632 004, Tamil Nadu, India
Tel: 91-416-22102
Fax: 91-416-22103

Arie Ordentlich
Israel Institute for Biological Research
P.O. Box 19
Ness-Ziona 70450, Israel
Tel: 972-8-381442
Fax: 972-8-401404

Stephanie Padilla
Cellular and Molecular Toxicology Branch
Neurotoxicology Division (MD-74B), US
  EPA
Research Triangle Park, NC 27711, USA
Tel: 919-541-3956
Fax: 919-541-4849

Ranganathan Parthasarathy
3732 Beaufort Lane
Louisville, KY 40207
Tel: 502-894-6187
Fax: 502-894-6155

Natalie Pickering
Dept. of Pharmacology, 0636, UCSD
9500 Gilman Rd.
La Jolla, CA 92093-0636, USA
Tel: 619-534-4026
Fax: 619-534-6833

Cary Pope
School of Pharmacy
Northeast Louisiana University
Monroe, LA 71209-0470
Tel: 318-342-1723
Fax: 318-342-1686

Daniel M. Quinn
Department of Chemistry
University of Iowa
Iowa City, Iowa 52242, USA
Tel: 319-335-1335
Fax: 319-335-1270

I. Rabinovitz
Dept. of Pharmacology
Israel Institute of Biological Research
P. O. Box 19, Ness-Ziona, Israel
Tel: 972-8-381506
Fax: 972-8-401404

Zoran Radić
Dept. of Pharmacology, 0636, UCSD
9500 Gilman Rd.
La Jolla, CA 92093-0636, USA
Tel: 619-534-4021
Fax: 619-534-6833

Satish Rao
Dept. of Neurological Sciences
Christian Medical College and Hospital
Vellore 632 004, Tamil Nadu, India
Tel: 91-416-22102
Fax: 91-416-22103

Terrone L. Rosenberry
Dept. of Pharmacology
Case Western Reserve University
Cleveland, OH 44106-4965, USA
Tel: 216-368-3494
Fax: 216-368-3395

Susana Rossi
Dept. of Cell Biology and Anatomy
(R-124)
Univ. of Miami School of Medicine, 1600
N.W. 10th Ave.
Miami, FL 33136-1015, USA
Tel: 305-547-6940
Fax: 305-545-7166

Richard L. Rotundo
Dept. of Cell Biology and Anatomy
(R-124)
Univ. of Miami School of Medicine, 1600
N.W. 10th Ave.
Miami, FL 33136-1015, USA
Tel: 305-547-6940
Fax: 305-545-7166

P. Santhoshkumar
Toxicology Unit, IC & P Dept.
CFTRI
Mysore-570 013, India
Tel: 091-0821-37150
Fax: 091-821-27697, 2447

Debapriya Sarkar
ETL, Dept. of Zoology
Visva Bharati University
Santiniketan-731 235, West Bengal, India
Tel: 03463-52751, x88
Fax: 91-03463-52672

Ashima Saxena
Div. of Biochemistry
Walter Reed Army Institute of Research
Washington, DC 20307-5100, USA
Tel: 202-782-0087
Fax: 202-782-6304

Sutapa Sen
ETL, Dept. of Zoology
Visva Bharati University
Santiniketan-731 235, West Bengal, India
Tel: 03463-52751, x88
Fax: 91-03463-52672

Avigdor Shafferman
Israel Institute for Biological Research
P.O. Box 19
Ness-Ziona 70450, Israel
Tel: 972-8-381518
Fax: 972-8-401404

T. Shivanandappa
Toxicology Unit, IC&P Dept.
CFTRI
Mysore - 570 013, India
Tel: 091-0821-37150
Fax: 091-821-27697, 2447

A.K. Sikder
Synthetic Chemistry Division
Defence Research and Development
  Establishment
Gwalior-474002, India
Tel: 91-751-340354
Fax: 91-75-1341148

Israel Silman
Department of Neurobiology
Weizmann Institute of Science
76100 Rehovot, Israel
Tel: 972-8-343649
Fax: 972-8-344161

Janez Sketelj
Institute of Pathophysiology
Zaloska 4
61105 Ljubljana, Slovenia
Tel: 386-61-310-841
Fax: 386-61-302-272

David H. Small
Department of Pathology
University of Melbourne
Parkville, Victoria 3052, Australia
Tel: 61-3-344-4205
Fax: 61-3-344-4004

Satu M. Somani
Southern Illinois Univ. School of Medicine
P.O. Box 19230
Springfield, IL 62794-9230
Tel: 217-785-2196
Fax: 217-524-0145

Hermona Soreq
Dept. of Chemistry, The Life Sciences
  Institute
The Hebrew University
91904 Jerusalem, Israel
Tel: 972 2 585109
Fax: 972 2 520258

Madhavan Soundararajan
Dept. of Biochemistry
University of Nebraska
East Campus, NE 68583-0718
Tel: 402-472-2939
Fax: 402-472-7842

Malathi Srivatsan
MS 510, Dept. of Physiology
Univ. of Kentucky Medical Ctr.
Lexington, KY 40536-0084
Tel: 606-323-5447
Fax: 606-258-1070

Kevin Stopps
School of Biol. Sciences
Queen Mary &Westifield College, Mile
  End Rd.
London, UK E1 4NS
Tel: 071-775-3013
Fax: 081-983-0973

Man-Ji Sun
Institute of Pharmacology and Toxicology
Tai-Ping Road No. 27
Beijing 100850, China
Fax: 861-821-1656

Joel Sussman
Dept. of Structural Biology
Weizmann Institute of Science
Rehovot 76100 Israel
Tel: 972-8-342647
Fax: 972-8-344159

Vincenzo Talesa
Unite Différenciation Cellulaire et
  Croissance, INRA
Place Viala
34060 Montpellier, cedex 1, France
Tel: 67 61 26 87
Fax: 67 54 56 94

Palmer Taylor
Dept. of Pharmacology, 0636, UCSD
9500 Gilman Rd.
La Jolla, CA 92093-0636, USA
Tel: 619-534-1366
Fax: 619-534-6833

Jean-Pierre Toutant
Physiologie Animale, INRA
Place Viala
34060 Montpellier, France
Tel: (33)67 61 26 87
Fax: (33)67 54 56 94

Latha Unni
Southern Illinois Univ. School of Medicine
Neurochemistry Lab
801 N. Rutledge, Rm. 4326
Springfield, IL 62702
Tel: 217-782-8103
Fax: 217-782-0988

Lalit Vaya
Torrent Pharmaceuticals Ltd
Torrent House, Off Ashram Rd.
Ahmedabad-380 009, India
Tel: 079-405090, 44323
Fax: 079-460048

Haresh Ved
Div. of Biochemistry
Walter Reed Army Institute of Research
Washington, DC 20307-5100, USA
Tel: 202-782-0956
Fax: 202-782-6304

Baruch Velan
Israel Institute for Biological Research
P. O. Box 19
70450 Ness-Ziona, Israel
Tel: 972-8-381518
Fax: 972-8-401404

R. Vijayaraghavan
Defence Research and Development
  Establishment
Gwalior - 474 002, India
Tel: 91-751-340354
Fax: 91-75-1341148

Elmar Willbold
Technische Hochschule Darmstadt
Institut für Zoologie Schnittspahnstrasse 3
64287 Darmstadt, Germany
Tel: 49-6151-163800
Fax: 49-6151-166548

Stanislaw Wlodek
Chemistry Dept., University of Houston
4800 Calhoun Rd.
Houston, TX 772204-5614
Tel: 713-743-3320
Fax: 713-743-2709

Barry W. Wilson
Dept. of Avian Science
University of California at Davis
Davis, CA 95616
Tel: 916-752-3519
Fax: 916-752-8960

Sean M. Wilson
Dept. of Avian Science
University of California at Davis
Davis, CA 95616
Tel: 916-752-3519
Fax: 916-752-8960

AUTHOR INDEX

SUBJECT INDEX

DATE DUE

DEMCO, INC. 38-2971